T0342266

Charged Aerosol Detection for Liquid Chromatography and Related Separation Techniques

Charged Aerosol Detection for Liquid Chromatography and Related Separation Techniques

Edited by Paul H. Gamache

Thermo Fisher Scientific, Chelmsford, MA, USA

This edition first published 2017

Registered Office
John Wiley & Sons, Inc., 111 River Street, Hoboken, NJ 07030, USA

Editorial Office
111 River Street, Hoboken, NJ 07030, USA

For details of our global editorial offices, customer services, and more information about Wiley products visit us at www.wiley.com.

Wiley also publishes its books in a variety of electronic formats and by print-on-demand. Some content that appears in standard print versions of this book may not be available in other formats.

Library of Congress Cataloging-in-Publication Data

Names: Gamache, Paul H., 1957-
Title: Charged aerosol detection for liquid chromatography and related separation techniques / edited by Paul H. Gamache, Thermo Fisher Scientific, USA.
Description: First edition. | Hoboken, NJ : John Wiley & Sons, Inc., 2017. | Includes bibliographical references and index.
Identifiers: LCCN 2017007285 (print) | LCCN 2017008536 (ebook) | ISBN 9780470937785 (cloth) | ISBN 9781119390695 (pdf) | ISBN 9781119390701 (epub)
Subjects: LCSH: Liquid chromatography. | Atmospheric aerosols. | Aerosols. | Electrostatic precipitation. | Separation (Technology)
Classification: LCC QD79.C454 C43 2017 (print) | LCC QD79.C454 (ebook) | DDC 543/.84–dc23
LC record available at https://lccn.loc.gov/2017007285

Cover design by Wiley

Set in 10/12pt Warnock by SPi Global, Pondicherry, India

Printed in the United States of America

10 9 8 7 6 5 4 3 2 1

I would like to dedicate this book to my family. To Dad whose courage, integrity, and selflessness makes him a perfect example of America's greatest generation. To Mom for her undying devotion to her family and her great cooking. To my brother, Dan, and his family for their love and support. My deepest gratitude goes to my wife, Anne, and children, Chelsea and Michael. I am so proud to see the accomplishments and growth of both Chelsea and Michael. Finally, to Anne, the love of my life and one of the most genuine and unselfish people that I have ever met.

Contents

List of Contributors

Ian N. Acworth
Thermo Fisher Scientific
Chelmsford, MA
USA

Stefan Almeling
European Directorate for the
Quality of Medicines & HealthCare
(EDQM)
Strasbourg
France

Amber Awad
Dominion Diagnostics
North Kingstown, RI
USA

Bruce Bailey
Thermo Fisher Scientific
Chelmsford, MA
USA

Agata Blazewicz
Pharmaceutical Chemistry
Department
National Medicines Institute
Warsaw
Poland

Ton Brooijmans
DSM Coating Resins
Waalwijk
The Netherlands

Sophie Brossard
European Directorate for the
Quality of Medicines & HealthCare
(EDQM)
Strasbourg
France

Pierre Chaminade
Lip(Sys)^2 Lipids, Analytical and
Biological Systems, Chimie
Analytique Pharmaceutique
Université Paris-Sud, Université
Paris-Saclay
Châtenay-Malabry
France

Hitesh P. Chokshi
Roche Innovation Center
New York, NY
USA

Paul Cools
DSM Coating Resins
Waalwijk
The Netherlands
Current address: The Dow Chemical
Company
Freeport, TX
USA

Chris Crafts
Thermo Fisher Scientific
Chelmsford, MA
USA

Greg W. Dicinoski
Australian Centre for Research on Separation Science (ACROSS), School of Chemistry, Faculty of Science, Engineering and Technology
University of Tasmania
Hobart
Tasmania, Australia

Mark Emanuele
Westford, MA
USA

Zbigniew Fijalek
Pharmaceutical Chemistry Department
National Medicines Institute
and
Warsaw Medical University
Warsaw
Poland

Paul H. Gamache
Thermo Fisher Scientific
Chelmsford, MA
USA

Paul R. Haddad
Australian Centre for Research on Separation Science (ACROSS), School of Chemistry, Faculty of Science, Engineering and Technology
University of Tasmania
Hobart
Tasmania, Australia

Sylvie Héron
Lip(Sys)^2 Lipids, Analytical and Biological Systems, LETIAM
Université Paris-Sud, Université Paris-Saclay, IUT d'Orsay
Orsay
France

Ulrike Holzgrabe
Institute of Pharmacy and Food Chemistry
University of Wuerzburg
Wuerzburg
Germany

Joseph P. Hutchinson
Australian Centre for Research on Separation Science (ACROSS), School of Chemistry, Faculty of Science, Engineering and Technology
University of Tasmania
Hobart
Tasmania, Australia

David Ilko
Institute of Pharmacy and Food Chemistry
University of Wuerzburg
Wuerzburg
Germany

Yong Jiang
State Key Laboratory of Natural and Biomimetic Drugs
Peking University
Beijing
China

Arul Joseph
Gilead Sciences, Inc.
Foster City, CA
USA

Stanley L. Kaufman
Retired from TSI Inc.
Shoreview, MN
USA

Shinichi Kitamura
Graduate School of Life and Environmental Sciences
Osaka Prefecture University
Osaka
Japan

William Kopaciewicz
Thermo Fisher Scientific
Chelmsford, MA
USA

Dawen Kou
Genentech Inc.
South San Francisco, CA
USA

Lijuan Liang
Beijing Friendship Hospital
Capital Medical University
and
State Key Laboratory of Natural and Biomimetic Drugs, School of Pharmaceutical Sciences
Peking University Health Science Center
Beijing
China

Danielle Libong
Lip(Sys)^2 Lipids, Analytical and Biological Systems, Chimie Analytique Pharmaceutique
Université Paris-Sud, Université Paris-Saclay
Châtenay-Malabry
France

Xiaodong Liu
Thermo Fisher Scientific
Sunnyvale, CA
USA

Gerald Manius
Retired from Hoffmann-La Roche Inc.
Nutley, NJ
USA

Robert C. Neugebauer
European Directorate for the Quality of Medicines & HealthCare (EDQM)
Strasbourg
France

Marc Plante
Thermo Fisher Scientific
Chelmsford, MA
USA

Christopher A. Pohl
Thermo Fisher Scientific
Sunnyvale, CA
USA

Magdalena Poplawska
Warsaw Medical University
Warsaw
Poland

Jeffrey S. Rohrer
Thermo Fisher Scientific
Sunnyvale, CA
USA

Abu Rustum
Merck & Co Inc.
Summit, NJ
USA

Katarzyna Sarna
Pharmaceutical Chemistry
Department
National Medicines Institute
Warsaw
Poland

Michael Swartz
Analytical Development
Validation Science
Uxbridge, MA
USA

Alain Tchapla
Lip(Sys)^2 Lipids, Analytical and
Biological Systems, LETIAM
Université Paris-Sud, Université
Paris-Saclay, IUT d'Orsay
Orsay
France

David Thomas
Thermo Fisher Scientific
Chelmsford, MA
USA

Alan K. Thompson
Nalco Champion, an Ecolab
Company
Aberdeen
UK

Hung Tian
Novartis
East Hanover, NJ
USA

Pengfei Tu
State Key Laboratory of Natural and
Biomimetic Drugs
Peking University
Beijing
China

Michael Türck
Merck KGaA
Darmstadt
Germany

Malgorzata Warowna-Grzeskiewicz
Pharmaceutical Chemistry
Department
National Medicines Institute
and
Warsaw Medical University
Warsaw
Poland

Ke Zhang
Genentech Inc.
South San Francisco, CA
USA

Preface

It has been approximately 12 years since charged aerosol detection (CAD) first became widely available for use with liquid-phase separations. CAD is perhaps best described as a universal detection technique for quantitative analysis. In addition to characteristics such as dynamic range, ease of use, reproducibility, and sensitivity, such techniques are often chosen for the broadness of their detection scope and their ability to provide analyte-independent sensitivity (i.e., signal/amount). Since its introduction, CAD has been widely adopted as evidenced by over 250 peer-reviewed articles that encompass several fields of application. These descriptions and the pioneering work of Dr. Stan Kaufman, Dr. Roy Dixon, and others have led to an increased understanding of the theory and practice of CAD. The capabilities of CAD technology have also continued to evolve along with those of relevant separation technologies. In many respects, CAD is still a new technique and its fundamental performance characteristics are not fully understood. The primary objectives of this book are therefore to further elucidate CAD theory, to objectively describe its advantages and limitations, and to provide detailed recommendations for its practical use.

This is the first book devoted to the topic of CAD and is intended to be a primary resource for analytical chemists in a variety of disciplines. The book was developed in collaboration with many scientists with expertise in research and/or application of separation science, detection, and aerosol measurement. CAD can be coupled with a range of separation techniques in numerous configurations where many variables can influence its performance. The practical value of CAD also depends on its intended use and the alternative tools available for a given application. The key aspects of this book are therefore to relate fundamental properties and operational variables of CAD to its performance in a variety of conditions and to provide expert insight and different perspectives on use of CAD to address specific analytical problems.

This book is arranged into three main sections each having several chapters. Each chapter was written by a different contributor or group and begins with a summary and/or table of contents to provide a quick view of the information provided.

Section 1 includes a theoretical description (Chapter 1), a review of current literature (Chapter 2), a guidance for practical use (Chapter 3), and an overview of universal detection in the context of comprehensive multicomponent analysis. Chapter 1 provides significant new insight to CAD theory making use of aerosol size distribution measurements to develop a detailed semiempirical model that describes its response. This chapter mainly focuses on newer CAD instrument designs, the evolution of which is further described in Chapter 2. Chapter 3 is complementary to Chapter 1 and provides detailed practical recommendations, which include method transfer approaches from older to newer instrument models. This chapter also provides basic requirements for configuring multi-detector systems often used to extend both the quantitative and qualitative information obtained from each sample. The role of universal detectors in multi-detector configurations together with emerging trends in separation science is then further elucidated in Chapter 4. This includes discussion of capillary and ultrahigh pressure liquid chromatography, supercritical and subcritical fluid chromatography, and multidimensional separations to address more complex analyses and to minimize the environmental impact of laboratory testing. Many of these techniques are further exemplified throughout the later sections of the book.

Section 2 consists of separate chapters each focusing on specific analyte classes: lipids (Chapter 5), ions (Chapter 6), carbohydrates (Chapter 7), polymers and surfactants (Chapter 8), and diverse, putatively bioactive species in herbal medicines (Chapter 9). For each analyte class, an informative review is provided on theory and experimental approaches to both separate and quantify species present in different sample matrices and in different areas of application (e.g., biosciences, chemical, food and beverage, natural products, and pharmaceutical). Examples are used to illustrate the advantages and limitations of CAD compared with alternative techniques. Taken together these chapters include descriptions of the use of CAD with many different separation techniques including those discussed in Chapter 4 along with normal-phase liquid chromatography (NPLC), size exclusion chromatography (SEC), hydrophilic interaction chromatography (HILIC), and mixed-mode techniques that, for example, combine ion exchange with either reversed-phase or HILIC retention mechanisms.

Section 3 describes the use of CAD to address specific analytical problems. Since CAD is widely used within pharmaceutical and biopharmaceutical industries, this section includes an overview of its use during various stages of drug development and informative guidance on method development, validation, and transfer in regulated environments (Chapter 10). This is followed by examples describing development and validation of pharmaceutical methods for analyses involving aminoglycoside antibiotics (Chapter 12), quaternary ammonium muscle relaxants (Chapter 13), and the anticonvulsant topiramate (Chapter 11). Each of these applications pose significant challenges in terms of both separation and detection, and their descriptions provide valuable insight to routine use of CAD in a regulated environment. The final two chapters discuss very interesting fields of application outside of the pharmaceutical industry. Chapter 14 describes analysis of scale-inhibiting polymers in oil-field chemistry applications and includes thorough validation of a gel permeation chromatography (GPC) method with CAD for routine analysis of residual levels in oilfield brine samples. Lastly, Chapter 15 describes various approaches to characterize industrial synthetic polymers ranging from smaller oligomers to larger and more complex formulations. In addition to SEC, this chapter includes discussion of liquid chromatography at critical conditions (LCCC) and gradient polymer elution chromatography (GPEC) to analyze complex polymer compositions. This includes examples of the combined use of CAD with complementary devices to address the very difficult task of characterizing raw material, intermediate compositions, and end products. Chapters 14 and 15 along with Chapter 8, which describes analysis of polymers and surfactants primarily as pharmaceutical excipients and reagents, are particularly recommended to readers interested in use of CAD in polymer analysis.

Throughout development of this book, significant advances have continued to be made in both CAD and relevant separation technologies. In addition to the chapters within Sections 2 and 3, it is recommended that the reader also refer to Chapters 1–3 for updated information on a given topic. Chapter 2 provides a comprehensive literature review, which includes more recent examples of the applications of CAD, while Chapters 1 and 3 describe specific technology advancements and their implications for practical use. It is sincerely hoped that this book will provide a valuable resource to all workers in the field of liquid-phase separations and will help to stimulate further research in all aspects of universal detection.

Acknowledgment

This book would not have been possible without the significant efforts and expertise of more than 45 contributors. I am very grateful to Stan Kaufman for the many fruitful discussions and for his brilliant input and time devoted to Chapter 1. The early and continued success of CAD is largely due to the devotion and talents of my colleagues from ESA Inc., several of whom are contributors to the book and many others who have greatly supported this effort. I would especially like to acknowledge and thank Ian Acworth, Bob Kwiatkowski, Ryan McCarthy, Nick Santiago, and John Waraska for their essential contributions.

Section 1

Fundamentals of Charged Aerosol Detection

1

Principles of Charged Aerosol Detection

Paul H. Gamache[1] and Stanley L. Kaufman[2]

[1] *Thermo Fisher Scientific, Chelmsford, MA, USA*
[2] *Retired from TSI Inc., Shoreview, MN, USA*

CHAPTER MENU

1.1 Summary

This chapter provides a brief history and detailed overview of charged aerosol detection (CAD) and a semiempirical model describing its response and expected performance under various analytical conditions. CAD and other evaporative aerosol detectors involve the same successive steps of primary spray droplet formation from an eluent stream, conditioning by inertial impaction to remove droplets too large to evaporate during passage through the instrument, and evaporation of remaining droplets to form residue particles each comprised of nonvolatile background impurities and any nonvolatile analyte present. Detection of the residue particles produces the detector signal. In CAD, the aerosol is given a charge dependent on the particle size, and the total charge carried by the aerosol is measured as a current; in ELSD, the aerosol is detected by its light scattering properties. Both detection methods produce a response that is approximately mass-flow dependent. The analyte dry bulk density affects the residue particle size for a given eluting mass, which has a minor effect on the mass sensitivity of both detector types.

Charged Aerosol Detection for Liquid Chromatography and Related Separation Techniques, First Edition. Edited by Paul H. Gamache.

Other analyte properties in particular optical properties (e.g., refractive index (RI)) for the ELSD likewise affect the sensitivity. Detection selectivity for evaporative aerosol detectors is based on differences in vaporization of components within an eluent. Accordingly, these techniques are expected to have very similar detection scope, eluent requirements, and solvent dependency of response. The unique characteristics of CAD are due to the aerosol measurement technique, which includes diffusion charging of residue particles and detection of the current due to deposition of particles with their charge in an aerosol-electrometer filter. Aerosol charging by diffusion mechanisms is well known to have only a minor dependence on particle material (i.e., analyte properties), which is the basis for uniform response capabilities of CAD. Like ELSD, CAD response (e.g., peak area vs. mass injected (m_{inj})) can be described by a power law function with a variable exponent b. Linear response, never perfectly achieved by either methods, would correspond to $b = 1$. For both techniques, the exponent b is at its maximum at the lowest m_{inj} and decreases with increasing m_{inj}. This is attributed to smaller residue particles that have a higher power law exponent (β_1) of response and are more prevalent with low m_{inj} and the low concentration that occurs near the edges of any peak. For ELSD this corresponds to Rayleigh light scattering for particle diameters (d) typically < 50 nm where $\beta_1 = 6$, while for CAD corresponds to aerosol charging of $d < \sim 9$ nm where $\beta_1 \sim 2.25$. For CAD, the lower d transition and smaller β_1 (closer to 1) underlies the widely observed lower detection limits, wider dynamic range, and less complex response curve than ELSD. Newer CAD designs produce an even smaller relative proportion of residue particles of $d < \sim 9$ nm, thus further simplifying the response curve, enabling lower sensitivity limits and a wider quasi-linear response range.

1.2 History and Introduction to the Technology

The technique that is now most commonly called charged aerosol detection (CAD) was first described in 2001 by Kaufman at TSI Inc. in a provisional patent application that ultimately led to US patent 6,568,24 [1]. This device was termed an evaporative electrical detector (EED) and was based on coupling liquid chromatography (LC) and other separation techniques with TSI's well-established electrical aerosol measurement (EAM) technology [2]. Around the same time, Dixon and Peterson at California State University were pursuing a similar avenue of innovation with a laboratory-built device that coupled LC with an earlier generation of TSI's EAM instruments. Dixon and Peterson described their device, termed aerosol charge detector (ACD), in the *Journal of Analytical Chemistry* in 2002 [3]. In both instances, the primary objective was to exploit the advantages, well described in aerosol science literature [4], of EAM over direct light scattering for measuring the very small (i.e., low nm diameter range) particles typically produced by LC detectors. The approach was therefore mainly geared toward addressing some of the limitations of evaporative light scattering detection (ELSD), which at the time had been used for LC detection for about 20 years. Subsequent collaboration between TSI and ESA Biosciences, Inc. led to the introduction of the first commercial instrument, the Corona® CAD®, in 2005 [5]. While there are some differences among these early EAM-based LC devices and with newer commercial instruments, the basic detection process remains the same. Therefore, Kaufman's patent disclosure and Dixon's article are acknowledged as the primary theoretical descriptions of CAD.

Since its commercial introduction in 2005, CAD has been widely adopted for a broad range of chromatographic applications. CAD and other aerosol techniques, including ELSD and condensation nucleation light scattering detection (CNLSD) [6], are described as "universal" since response depends primarily on aerosol particle size and number concentration (e.g., number of particles/cubic centimeter of gas) rather than individual analyte properties. These "common property" measurement characteristics provide significant advantages over other devices whose *detection scope* (*viz.*, range of chemicals for which a useful response can be obtained) and *sensitivity* (*viz.*, signal output per unit mass or per unit concentration) are highly dependent on analyte nature such as optical properties (e.g., ultraviolet (UV) absorption, fluorescence (Fl)) or propensity to form gas-phase ions (e.g., electrospray ionization with mass spectrometry (ESI-MS)). While UV detection remains a primary technique for many LC analyses, its detection scope is limited to compounds with a sufficient UV chromophore, and its sensitivity varies widely among analytes. Likewise, the detection scope and sensitivity of MS strongly depends on ion source and operational conditions, analyte (e.g., basicity, surface activity), and eluent conditions (e.g., pH). In many respects, CAD is still considered to be relatively new, and its performance for a given application is not completely understood or fully explained by current theory. The objective of this chapter is to provide an updated description of CAD theory that draws upon the primary references, aerosol science literature, insights presented throughout this book, and more than 200 literature references and review articles that describe its applications in chromatographic analysis [7–11]. Relevant theory and background of aerosol particle measurement technology in the context of non-LC applications will be briefly discussed. Because these non-LC applications typically involve measurement of steady-state or slowly changing aerosols as opposed to the rapidly changing aerosols encountered with LC, simplified models that address transient versus steady-state measurements will be introduced to help describe and predict the expected performance of CAD.

Understanding the properties of aerosols is of great significance in many fields, including those within environmental, industrial, health, and medical disciplines. To a practicing chromatographer, the properties and behaviors of aerosols may be unfamiliar or seem counterintuitive. Fortunately, significant knowledge exists within the field of aerosol science. For readers interested in additional background on aerosol technology, the most recent editions of *Aerosol Technology: Properties, Behavior, and Measurement of Airborne Particles* [4] and *Aerosol Measurement: Principles, Techniques, and Applications* [12] are highly recommended. To facilitate more detailed descriptions of theory and practice of CAD, basic information about aerosols along with some definitions and conventions used for other topics within this chapter are described in the succeeding text.

An aerosol is defined as a suspension of solid particles and/or liquid droplets in a gas and is most commonly described as a two-phase system (gas and condensed phase). For LC detection, the gas includes the flow that is supplied to the detector and any component from the liquid eluent that is vaporized. The evaporation process with LC detection involves a transition from liquid droplets to, typically, solid particles, but, in some cases, the "stable particles" that remain after evaporation are still in liquid form (e.g., oils). In describing LC detection, we will assume that evaporation is complete before the aerosol exits the evaporation tube and, unless otherwise specified, will use the terms *droplet* for species before evaporation and *residue particle* for those after evaporation, irrespective of the state of matter. Aerosol size distribution is an important and experimentally accessible parameter of an aerosol, contributing strongly to all of its properties. Except for materials such as extended fibers or agglomerates, the properties and behavior of an aerosol particle are usually assumed to be equivalent to that of a sphere. Unless otherwise indicated, we will assume that droplets and particles are spherical. It is common to distinguish between the two phases of an aerosol based solely on size where the condensed phase spans a diameter range of ~1 nm to 100 μm. Table 1.1 provides some perspective on the number of

Table 1.1 Estimated number of molecules for a given particle diameter.

	Molar mass (g/mol)	ρ Density (g/cm^3)	Number of molecules			
		d (nm)	2	5	10	50
Albumin	*66,500*	*1.37*	0	1	6	812
Insulin	*5,808*	*1.37*	1	9	74	9.30×10^3
Polysorbate 80	*1,310*	*1.07*	2	32	257	3.22×10^4
Azithromycin	*749*	*1.18*	4	62	497	6.21×10^4
Cholesterol	*387*	*1.05*	7	107	856	1.07×10^5
Glucose	*180*	*1.54*	22	337	2.7×10^3	3.37×10^5
NaCl	*58*	*2.17*	94	1.46×10^3	1.17×10^4	1.46×10^6

molecules comprising a particle of a given diameter calculated by Equations 1.1 and 1.2 and assuming that all particles are spherical:

$$V = \left(\frac{1}{6}\right)\pi d^3 \tag{1.1}$$

$$\text{Number of molecules} = \frac{V\rho N_A}{M} \tag{1.2}$$

where V = volume (cm^3); d = particle diameter (cm); ρ = density (g/cm^3); M = molar mass (g/mol); and N_A = Avogadro's constant (6.022×10^{23} mol^{-1}).

For a high molecular weight protein such as albumin (66.5 kDa), a $d = 10$ nm may consist of only a few molecules, while the same size particle may consist of >2,500 glucose or >10,000 NaCl molecules.

Within the majority of the aerosol size range, many of the properties and behaviors of a condensed-phase component are more determined by physical size than chemical composition. This includes stability, motion in gravitational and electrical fields, drag forces in gases, adhesion, and, for CAD and other EAM techniques, interaction with gas-phase ions through Brownian diffusion. Within an aerosol, interaction of a particle or droplet with the surrounding gas molecules is a very significant determinant of its behavior. This is not surprising if one considers that a 1 or 100 nm diameter particle will typically experience about 10^{11} or 10^{14} collisions per second, respectively, with gas molecules. A key parameter in that regard is the mean free path (λ—defined as the average distance traveled between collisions) of the gas molecules, roughly 65 nm in air at atmospheric pressure. Approaching the lower end of the aerosol size range, the properties and behavior of a particle, as one might envision from Table 1.1, will begin to reflect more of its chemical makeup. For much of the previously stated aerosol size range, Stokes' law describes the primarily viscous forces that are fundamental to motion of particles in a fluid (a gas). For particle sizes of order λ and smaller, the motion begins to deviate from Stokes' law where that of $d < 20$ nm is typically said to be in the molecular kinetic region. Likewise, charging of particles is closely related to the relationship between size and λ of the ions (λ_{ion}—typically 14.5 nm for air ions at atmospheric pressure). This relationship is fundamental to the process of diffusion charging, which is central to the performance of CAD and other widely used EAM devices. The topic of diffusion charging will be discussed in more detail in Section 1.3, especially in terms of charge level as a function of d, its dependence on particle (analyte + background) material, and changes in these relationships as d approaches that of individual molecules. Especially toward this extreme, the process of converting individual solute molecules into gaseous ions for mass spectrometry (e.g., by ESI or atmospheric pressure chemical ionization (APCI)) becomes more relevant. APCI, in particular, is sometimes viewed within the same context as CAD since both use nebulization and corona discharge. Basic differences between diffusion charging of aerosol particles and gaseous ion formation as they relate to detection scope and response uniformity will be discussed at the end of this chapter.

Most aerosols are "polydisperse," that is, they have a broad and often nonsymmetrical size distribution. "Monodisperse" aerosols have a relatively narrow size range and are typically generated in the laboratory by specialized techniques. The original development of CAD is partly based on the understanding obtained by diffusion charging of electrical-mobility-classified, and therefore monodisperse, aerosols. Polydisperse aerosols on the other hand may include a range of sizes that typically span more than an order of magnitude. This is typical of the aerosols produced by pneumatic nebulizers such as those used in CAD. For LC detection this means that a "snap shot" of the aerosol at any point location within the detection flow path will always present a rather broad size distribution, whose peak or modal size depends on the time-dependent concentration of nonvolatile material in the effluent. As a result, the time-dependent signal of a detector such as CAD or ELSD will represent signal contribution from a range of particle sizes. This is an important aspect to consider in descriptions of LC-aerosol detection theory.

A *universal detector* for chromatography, as defined by the International Union of Pure and Applied Chemistry [13], is "a detector which responds to every component in the column effluent except the mobile phase." This definition seems to address chemical detection scope while not specifically addressing relative sensitivity (e.g., peak area/ng) among analytes, which is also a very important consideration for many analytical methods. As widely described in literature, a significant need exists for LC detectors that can be used with a range of chromatographic conditions and can provide *both* broad detection scope *and* sensitivity that is independent of analyte properties (defined here as *uniform response*). Uniform response provides the possibility to quantify multiple analytes with a single calibration standard or calibration model [14–16]. This is desired for many analyses where individual standards are unavailable (e.g., pharmaceutical impurities and natural products) or for applications (e.g., lipid class and some polymer separations) where single chromatographic peaks typically consist of several species. Considering the entire chemical scope of potential analytes, the wide diversity of solvents and additives used with LC, and the common presence of impurities (e.g., from solvents, additives, column bleed, sample matrix), it is easy to recognize that an LC detector whose sensitivity is completely independent of chemical and physical properties will not be very useful since signal will always remain constant. That is, there must be some level of detection *selectivity* or *specificity* in order to obtain a differential response between a component of interest (analyte) and other components of the column eluent. For evaporative aerosol detectors, selectivity is based on relative volatilization of components in an eluent. Chapter 4 of this book provides an excellent overview of the advantages and limitations of aerosol-based detectors and other techniques such as RI, short wavelength UV absorbance, and chemiluminescent nitrogen detection in terms of detection scope, uniform response, and overall quantitative performance. This chapter will describe the theory that underlies the performance of CAD mostly within this same context with particular focus on solvent dependency of response, shape of the response curve, chromatographic peak shape, sensitivity limits, response uniformity, and the dependence of these characteristics on analytical conditions and analyte properties.

1.3 Charged Aerosol Detection Process

This section will provide an overview of CAD and a rationale for assumptions made to model its response (Section 1.4). Key assumptions are described toward the end of each subtopic and all are listed at the end of this section. Unless otherwise stated, this discussion will be primarily based on the design of the Corona™ Veo™ and Vanquish™ series CADs (Thermo Fisher Scientific Inc.), here referred to collectively as "VCAD." In most cases, the basic principles are the same as earlier models (e.g., Corona® Ultra® RS), here collectively referred to as "UCAD," and relevant differences will be described as appropriate. Further details of the various models are described in Chapter 2. The general detection scheme of nebulization, aerosol conditioning, evaporation, aerosol charging, particle selection, and aerosol charge measurement is depicted in Figure 1.1.

1.3.1 Nebulization

CAD and other aerosol detectors typically use pneumatic nebulization in which a high velocity gas flow provides the energy to break up a liquid stream into droplets. First generation commercial UCAD instruments used a cross-flow nebulizer design (Figure 1.2a) in which the gas flow is directed at a right angle to the tip of a liquid nozzle. More recent models use a concentric design (Figure 1.2b) in which the liquid flows through a central capillary that is surrounded by a gas flow conduit. In this configuration, parallel flows merge at the nebulizer tip where high gas velocities provide the shear force to break up the liquid. The cross-flow nebulizer spray is directed toward a convex-shaped impactor surface located in close proximity to the nebulizer nozzle, while that of the concentric nebulizer is directed toward a more distant impactor surface. In both designs, the surface is assumed to impose a size cut on the generated droplet distribution.

Pneumatic nebulization is widely characterized as a physical process (i.e., independent of solute composition or concentration) that produces polydisperse aerosols exhibiting a nearly lognormal size distribution (*viz.*, logarithm of size is approximately normally distributed). While average diameter is usually sufficient to describe a monodisperse aerosol, geometric mean

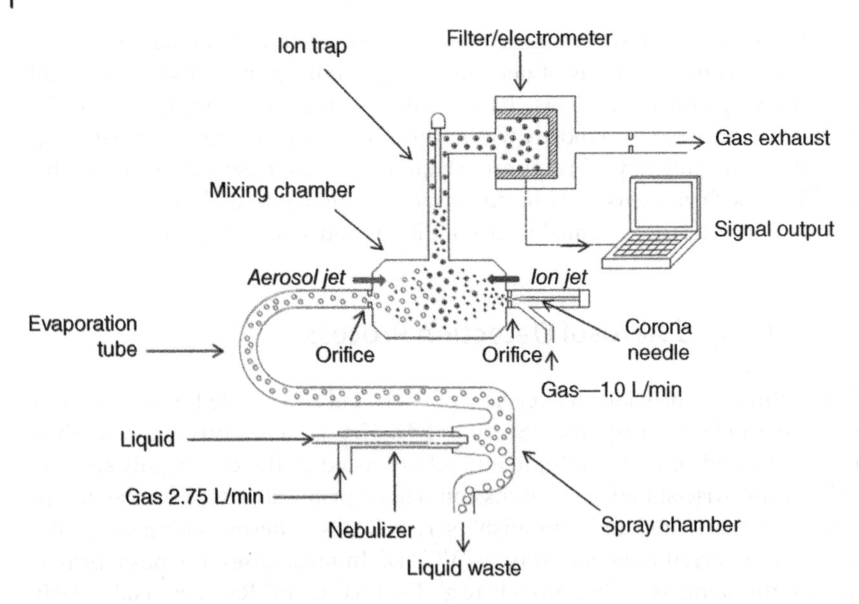

Figure 1.1 CAD involves pneumatic nebulization of liquid column eluent, aerosol conditioning within a spray chamber, solvent evaporation, diffusion charging of resultant aerosol residue within a mixing chamber using an opposing ion jet formed via corona discharge, removal of excess ions and high mobility charged particles in an ion trap, and measurement of aggregate charge of aerosol particles with a filter/electrometer.

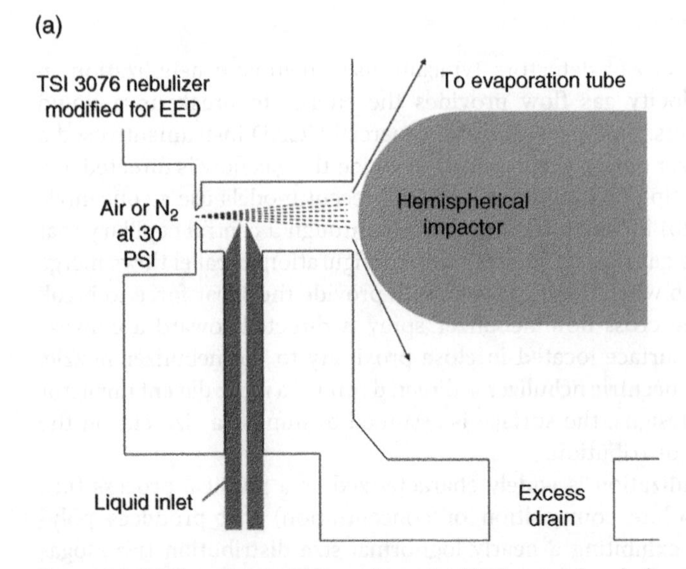

Figure 1.2 (a) Cross-flow nebulizer + impactor. (b) Concentric nebulizer and spray chamber.

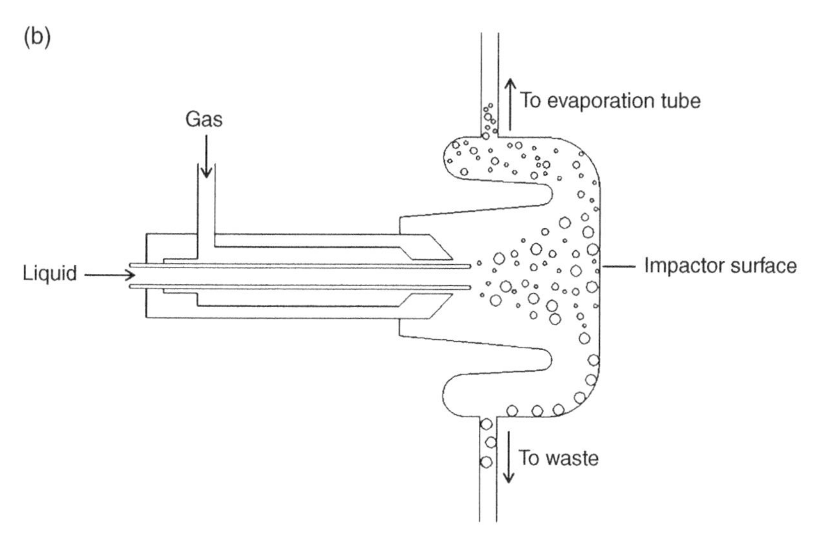

Figure 1.2 (Continued)

diameter (GMD) and geometric standard deviation (GSD) are more appropriate for lognormal and nearly lognormal distributions. GMD (i.e., nth root of the product of n diameters) typically provides a better estimate of central tendency than arithmetic mean. A true lognormal distribution is symmetrical on a logarithmic scale, and therefore count median diameter (CMD) and GMD are equivalent. GSD (i.e., inverse log of the standard deviation of the log-transformed values) is likewise often preferred to estimate width or shape of the distribution.

The characteristics (e.g., GMD, GSD, etc.) and stability of the *primary aerosol* (*viz.*, formed at the nebulizer tip) are critical to the performance of CAD and other aerosol-based detectors. Several diagnostic techniques including cascade impaction, EAM, and laser Fraunhofer diffraction have been used to characterize pneumatically generated aerosols and to develop semiempirical models to predict aerosol production and transport [17, 18]. This chapter will include data obtained from an EAM-based device (scanning mobility particle sizer (SMPS) spectrometer TSI Model 3938) to help describe CAD response. Several semiempirical models have been developed to predict aerosol characteristics generated by different nebulizer designs and as a function of solvent and conditions [19–21]. The widely described Nukiyama–Tanasawa (N-T) model is shown in Equation 1.3. As with other models, there are reported limitations to its predictive accuracy for a given nebulizer design. Even so, it is useful to at least qualitatively describe the main control parameters of the primary aerosol, which can be classified according to LC conditions (i.e., eluent volumetric flow rate, surface tension, and viscosity), nebulizer dimensions

Table 1.2 Parameters and property values for N-T model.

Parameter	Water	Methanol	Acetonitrile
σ—liquid surface tension (dyn/cm)	72.8	22.6	29
ρ—liquid density (g/cm^3)	1.00	0.79	0.78
η—dynamic viscosity (P, poise)	0.010	0.006	0.003
Q_l—volumetric flow rate of liquid (L/min)	LC flow rate		
Q_g—volumetric flow rate of gas (L/min)	Typically 2.75		
V_l = liquid axial velocity (m/s)	Defined by Q_l and nebulizer capillary i.d.		
V_g = gas axial velocity (m/s)	Defined by Q_g and nebulizer annular area		

(i.e., axial velocity of gas and liquid in the nebulizer), and nebulizer conditions (i.e., gas volumetric flow rate). Table 1.2 lists the parameters and property values of common LC solvents:

$$\mathrm{SMD} = \frac{585\sigma^{0.5}}{(v_g - v_l)\rho^{0.5}} + 597\left[\frac{\eta}{(\sigma\rho)^{0.5}}\right]^{0.45} \times \left(1000 \times \frac{Q_l}{Q_g}\right)^{1.5} \tag{1.3}$$

The N-T model predicts the Sauter mean diameter (SMD), which is the diameter of a spherical droplet whose ratio of volume to surface area is equal to that of the complete spray sample. For a liquid flow rate of 0.6 mL/min and nominal dimensions of a VCAD concentric nebulizer, SMD (μm) predictions are 9.5 for water and 7.0 for acetonitrile (CH_3CN). This illustrates that, for a given condition, solvents with relatively low surface tension and low viscosity tend to produce smaller droplets. To model CAD response (Section 1.4), we have found that N-T predicted SMDs along with empirical data from an SMPS were suitable to predict the primary aerosol characteristics for both water and CH_3CN. SMDs for water and CH_3CN were converted to GMDs (μm) of 3.6 and 2.7, respectively, using equations described in Hinds book [4] and assuming a GSD, for both solvents, of 1.85.

From the previous discussion, the following assumptions are made regarding nebulization:

- Assumption 1—In accordance with a solute-independent nebulization process, all droplets in a primary aerosol *at a given instant* have the same solute mass concentration and differ only in size.
- Assumption 2—Primary aerosol droplet distributions remain constant throughout isocratic elution.

1.3.2 Aerosol Conditioning

In this context, aerosol conditioning refers to modifications to a primary aerosol that produces a *secondary aerosol*, defined here as that entering the evaporation tube. This discussion will mainly focus on *aerosol transport*, which is the fraction of the primary aerosol that is transported to the evaporation tube and thus remains to form an aerosol residue. Especially for solvent gradients, aerosol transport is a key factor contributing to baseline drift and changes in analyte sensitivity (e.g., peak area/ng).

1.3.2.1 Solvent Load Reduction

A main consideration with aerosol detectors for LC is the solvent load, especially when using higher liquid flow rates. Water is the most common LC solvent and also has relatively low volatility. From the saturation vapor pressure (P_s) at reference conditions (e.g., 23.8 torr at 25°C), it can be estimated that a ~25 μL/min flow of water should saturate a 1.0 L/min flow of gas. As a consequence, the flow path of an aerosol detector can potentially be saturated (100% relative humidity (%RH)), which would inhibit evaporation leading to inferior results. The VCAD spray chamber (Figure 1.2b) is designed so that the aerosol spray is directed toward a surface, which is arranged to guide the flow in an upward and reversed direction. This serves as an inertial impactor, such as those widely used in aerosol applications. In this arrangement, larger droplets with higher inertia are unable to follow the sharply curved streamlines of the flow and therefore collide with the chamber wall and coalesce and are directed to waste. SMPS measurements indicate that, for a 0.6 mL/min flow of water, a maximum droplet diameter (D) of ~5 μm is transported to the VCAD evaporation tube. Under these conditions ~90% of the liquid volume is typically directed to waste.

1.3.2.2 Secondary Processes

Discussed here are additional processes, which have been described [3] as potentially influencing transport and characteristics of pneumatically generated aerosols.

Droplet coagulation can result from collisions due to relative motion and/or electrostatic interaction, the latter being a consequence of natural spray electrification. The overall effect of coagulation is to increase aerosol droplet size and decrease number concentration. Droplets that grow sufficiently via coagulation prior to reaching the impactor may be removed by the impactor. Spray electrification is expected to occur in most LC conditions since ions are distributed more or less randomly within a liquid. When a droplet is formed by pneumatic nebulization, it samples the liquid and its ions. Statistically, some droplets are therefore formed with an imbalance of charge. Coagulation rate should be highest within the spray chamber where droplet number concentration is highest. Our studies using a bipolar ion source within a VCAD spray chamber to neutralize the aerosol (Section 1.3) suggest that natural spray electrification does not significantly influence CAD response. Also, from estimates of maximum droplet concentration produced with the VCAD nebulizer (i.e., >10^8 droplets/cm^3), it is expected that coagulation due to differences in relative velocity may have only a minor influence on aerosol size distribution.

Evaporation can start at the point of nebulization and is considered here in the context of aerosol transport. For example, a reduction in D before impaction would lead to higher transport than predicted by impactor cut size (D_{cut}). A main factor influencing evaporation rate (R_e) is the saturation ratio (SR) (i.e., relative humidity (RH)), which is the ratio of the partial pressure of the vapor to P_S. Porstendörfer [22] and Cresser [23] examined initial evaporation of nebulized pure water and aqueous salt solutions. They argued that equilibrium conditions of 100% RH are established within typical spray chambers. As shown in Figure 1.4 (next section), lifetime of a pure water droplet near a D_{cut} of 5 μm is on the order of 10 s, which is quite long compared with estimated transport time before impaction (i.e., <0.2 s). The presence of solutes should further slow R_e relative to pure water by reducing P_s [22, 23]. Also, both the rapid adiabatic expansion of gas at the nebulizer nozzle and the endothermic process of evaporation are expected under most conditions to have an overall cooling effect. This should also reduce P_s, thereby even further slowing R_e. Taken together, this suggests that pre-impactor evaporation may not significantly influence aerosol transport when nebulizing highly aqueous solutions. However, our measurements with an SMPS indicate that, for more volatile solvents, D_{cut} and therefore transport is effectively increased due to pre-impactor evaporation. For example, D_{cut} determined for CH_3CN was ~11 μm as compared with ~5 μm for water. For a flow rate of 0.6 mL/min, ~30% of pure CH_3CN is directed to waste (70% transport) as compared with 10% transported for water, as described in Section 1.3.2.1. The influence of solvent properties on aerosol transport, at least with the VCAD design, is therefore a function of both nebulization (Equation 1.3) and pre-impactor evaporation. In general, aerosol transport should be highest for solvents with low viscosity and low surface tension since these properties are typically associated with relatively high P_s and also, by Equation 1.3, with production of smaller droplets.

Figure 1.3 compares particle size distributions measured by an SMPS at the outlet of Corona Veo and Corona ultra RS evaporation tubes for a constant 1.0 mL/min flow of 1.0 μg/mL theophylline (THEO) in 20%/80% v/v CH_3OH/water. This illustrates that, for the same solution concentration and flow rate, the VCAD design produces a dried aerosol distribution with a significantly larger particle size and number concentration (GMD 18.8 nm; 1.3×10^6 particles/cm^3) than the UCAD design (GMD 8.8 nm; 0.39×10^6 particles/cm^3). While the cross-flow nebulizer is expected to produce a course (i.e., large characteristic droplet size) primary aerosol, the impactor design and location

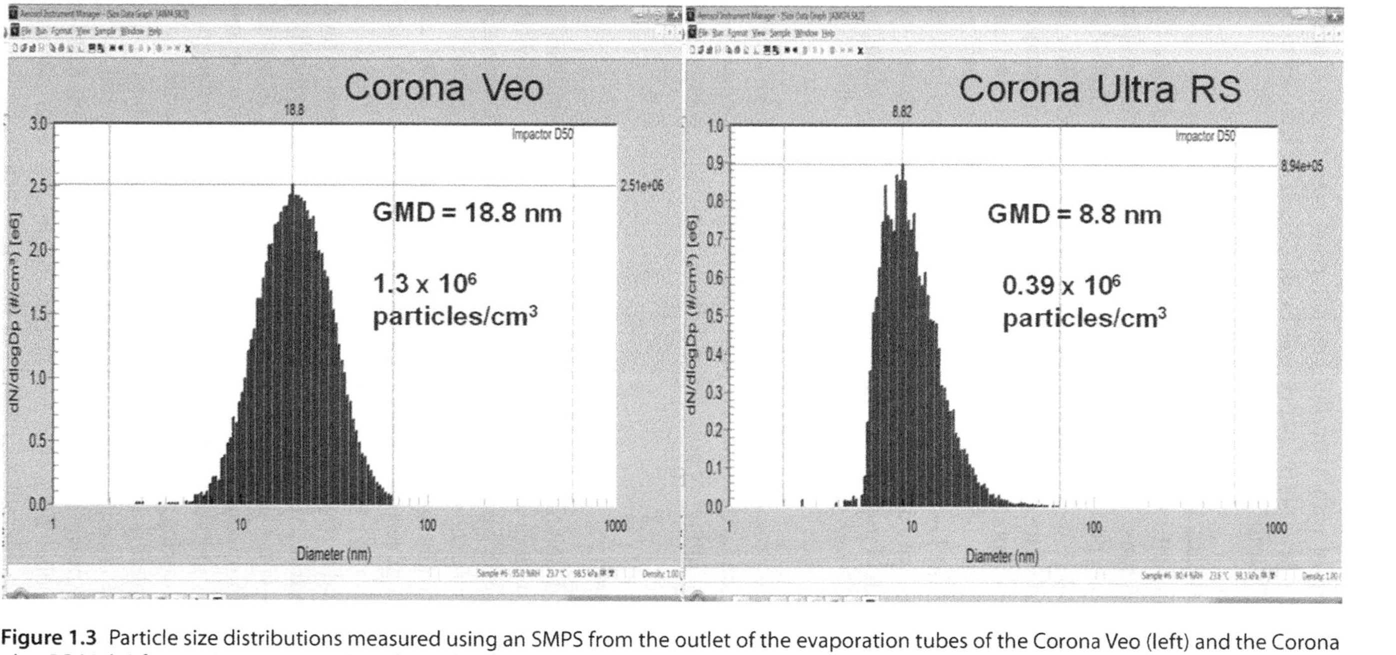

Figure 1.3 Particle size distributions measured using an SMPS from the outlet of the evaporation tubes of the Corona Veo (left) and the Corona ultra RS (right) for continuous 1.0 mL/min flow of 1.0 µg/mL theophylline (THEO) in 20% v/v aqueous CH_3OH.

imposes a smaller D_{cut} than the VCAD. This results in lower aerosol transport and a dried aerosol with a smaller characteristic size. Implications of these differences to relative performance will be discussed in later sections of this chapter and in Chapter 3.

1.3.2.3 Summary: Aerosol Transport

The main function of a spray chamber is to remove droplets above a D_{cut}, which is required under most LC conditions to ensure complete solvent evaporation. It follows from Assumption 1 that the fraction of analyte removed from the primary aerosol is equivalent to the volume fraction of solvent removed. The volume of liquid flowing to waste can therefore be used to quantitatively estimate analyte transport to the evaporation tube. A consequence of solvent reduction is lower detection sensitivity (signal/amount) for eluents with higher surface tension, higher viscosity and lower volatility (e.g., aqueous), and changes in baseline level and analyte sensitivity during solvent gradients. Differences between solvents in the concentration of nonvolatile impurities can also give rise to changes in baseline level and noise during solvent gradients. This latter factor will be discussed in more detail in Chapter 3. A key assumption regarding aerosol transport is:

- Assumption 3—Removal of primary aerosol droplets above a certain D_{cut} to form a secondary aerosol is solute independent but depends on eluent properties that influence both nebulization (Equation 1.3) and pre-impactor evaporation.

1.3.3 Evaporation

CAD, ELSD, and CNLSD share the same basic principle of nebulizing an eluent and drying it to leave an aerosol residue. Since detection selectivity is, in all cases, based on relative vaporization of components within an eluent (e.g., analyte vs. solvent), the observed detection scope and eluent requirements are, as expected, practically the same for these detectors [24, 25]. Understanding the specific process of aerosol evaporation and factors influencing relative solute partitioning between gas and condensed phases can be useful in method development and in predicting response—for example, to develop more robust calibration models. Also, since baseline noise and drift typically increase monotonically with background signal, the best sensitivity limits are usually obtained by minimizing the concentration of nonvolatile impurity solutes and by optimizing evaporation conditions for a given analysis. Challenges with predicting and optimizing analytical performance can however arise due to factors such as incomplete knowledge of relevant analyte properties; the diversity, complexity, and changing composition of eluents encountered with LC; the presence of unknown impurities; and aspects of evaporation that are rather specific to aerosols. This section will discuss the process of aerosol evaporation with a focus on the solute properties and analytical conditions that may influence size and stability of aerosol particles.

1.3.3.1 Aerosol Evaporation Process

An aerosol droplet produced by nebulizing an LC eluent is most often a binary (e.g., aqueous-organic) solvent mixture containing a variety of solutes (i.e., analyte, additives, and impurities). Evaporation decreases the droplet size, which concentrates the less volatile components and dynamically changes the solvent composition of the droplet. The process enhances solute–solute interactions, precipitation, and other effects, producing a condensed phase—solid in many cases. Predicting whether or not a given component will contribute to, and remain in, a condensed-phase particle requires consideration of the component's solubility, ionization state, and enthalpy of fusion in addition to commonly used parameters such as P_s and boiling point.

Aerosol evaporation has other special considerations including kinetic processes and the Kelvin effect, which both become more significant for small droplets and particles (e.g., $\leq 1.0\,\mu m$). The Kelvin effect describes an increase in P_s due to a decrease in the attractive forces (e.g., surface tension) between molecules on sharply curved surfaces. A resulting increase in evaporation rate (R_e) explains interesting phenomena such as the tendency for water to evaporate even in conditions above 100% RH. Kinetic processes, which become more important as particle size approaches the mean free path of vapor molecules, have the effect of decreasing R_e. This effect can be quantified by a Fuchs' correction factor (ϕ), which can be calculated as described by Davies [26]. A further consideration is the importance of condensation, which always accompanies aerosol evaporation. For example, heterogeneous condensation under conditions of vapor supersaturation and with aerosol particles that can serve as nuclei is the basis for the exponential particle growth techniques employed in condensation particle counting (CPC) with light scattering detection. CPC, for example, is a component of both CNLSD and the SMPS device used in studies described in this chapter. Aerosol condensation can also take place inadvertently with LC detection, leading to baseline noise and spiking (e.g., when evaporation temperature (T_e) is set low enough to create a decreasing T gradient within the detection flow path).

1.3.3.2 Evaporation Rate (R_e)

For aerosol detection with LC, a primary goal is to quickly and completely evaporate all components of the mobile phase while preserving analytes as stable particles whose size increases monotonically with injected mass. For that reason, the evaporation rate (R_e) of mobile phase components is of primary interest. Figure 1.4 shows theoretical water droplet lifetimes, and by inference evaporation rate, as a function of initial diameter calculated from Equation 1.4 and using additional correction factors (e.g., Kelvin effect, kinetic processes) that are described in more detail in Hinds [4]:

$$t = \frac{R\rho_s D^2}{8D_v M\left(P_D/T_D - P_v/T_v\right)} \tag{1.4}$$

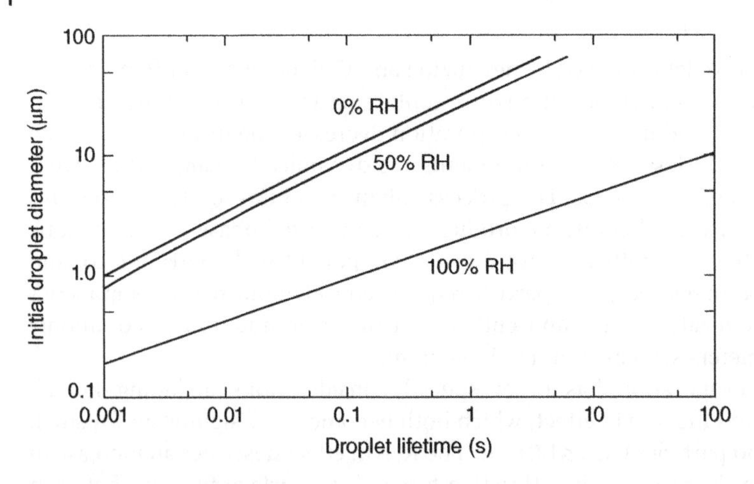

Figure 1.4 Water droplet lifetimes as a function of droplet size for 0, 50, and 100% relative humidity at 293°K (20°C) [4].

where t = evaporation time, R = universal gas constant, ρ_S = solvent density, D_V = diffusion coefficient of a component's vapor molecules, M = molar mass, D = droplet diameter, P_D = partial pressure of the vapor near the droplet surface, $T_D = T$ at the droplet surface, and P_v and T_v = partial pressure and T, respectively, of the vapor well away from the surface.

Figure 1.4 indicates that water droplet lifetime depends strongly on both droplet size and SR. For example, at a given %RH, the lifetime of a 1.0 μm droplet is ~100-fold less than that of a 10 μm droplet. This follows from Equation 1.4 where droplet lifetime is proportional to D^2 (i.e., surface area). Also, the lifetime of a given droplet size is ~1000-fold lower at 50% RH than in saturated conditions. Similar dependence of droplet lifetime on size and SR is observed for other solvents. This highlights the importance of the spray chamber both to remove larger droplets and to decrease SR. In general, spray chambers and impactors are designed to ensure that droplet size and SR of a secondary aerosol are sufficiently low to allow complete solvent evaporation within a detector's specified working range of solvents, temperatures, and flow rates. This instrumental working range specification is usually based on a mobile phase of 100% water at a given maximum flow rate (e.g., 2.0 mL/min for VCAD). Temperature-controlled evaporation then provides flexibility to accommodate a range of eluents and analytes and a means of method optimization. For the discussion in the following text, we will assume that evaporation of the bulk solvent is essentially complete within the evaporation tube and will discuss factors that determine the size, composition, and (size-based) stability of dried aerosol particles.

1.3.3.3 Dried Particle Size

We assume that the volume of a dried particle remaining after droplet evaporation is proportional to the starting volume of the droplet and its (v/v) concentration of nonvolatile solutes, including analytes and impurities. Equation 1.5 is widely used to describe the relationship between dried particle diameter (d), primary aerosol droplet diameter D, nonvolatile solute mass concentration C_S, and solute "bulk" density (ρ_S). This indicates that the volume of each dried particle is proportional to the nonvolatile solute mass concentration within its corresponding primary droplet, corrected for density. The particle's diameter is then proportional to the cube root of its volume. It is assumed that all droplets include at least some level of nonvolatile impurity, and the subscript (s) refers to solute, which is taken to mean "nonvolatile impurity + analyte, if present":

$$d = D\left(\frac{C_S}{\rho_S}\right)^{1/3} \tag{1.5}$$

Key aspects of Equation 1.5 include:

1) The exponent (1/3) indicates a nonlinear relationship between d and C_S. Table 1.3 provides an example of this relationship for a given D and range of solute concentrations. This is a main factor that determines both the shape of a response curve and chromatographic peak shape as will be discussed in Section 1.4.
2) Solute density (ρ_S) influences dried particle diameter (d) weakly because of the cube root (e.g., Table 1.3). Since ρ_S differs among analytes and response depends on d, it then follows that ρ_S is a factor that is expected to contribute to differences in sensitivity (e.g., peak area/ng) among analytes. A recent study by Matsuyama *et al.* [27] clearly showed the dependence of CAD response on analyte density. This topic will be further discussed in Section 1.5.5.

Table 1.3 Dried particle size dependency on solute concentration and density.

	Analyte A	Analyte B
$D = 5.0\ \mu m$	$\rho = 1.0$	$\rho = 2.0$
(μg/mL)	d (nm)	d (nm)
0.1	23	19
1.0	50	40
10	108	86
100	233	185
1000	501	398

1.3.3.4 Volatility and Detector Response

The ability to predict the response of a given solute as a function of experimental (e.g., mobile phase composition) conditions is of great interest for many laboratories. Several studies have examined LC-aerosol detection scope and relative sensitivity for a diverse range of chemicals in relation to common indicators of volatility. This includes M, P_s, boiling point (B_p), and enthalpy of vaporization (ΔH_v) obtained from literature and/or predicted from chemical structure using quantitative structure–property relationship (QSPR) algorithms (e.g., SPARC [28]). Most studies have shown an approximate value above or below which solutes are reliably detected, and these are typically described as *nonvolatile* (e.g., $B_p > {\sim}400°C$, [7, 16] or $M > 350\,g/mol$ *and* $\Delta H_v > 65\,kJ/mol$ [15]). Near these limits (e.g., B_p from 300 to 400°C, [7]), weaker correlation (e.g., lower sensitivity or no response) has been observed. Solutes within this range are typically considered *semivolatile* and those beyond considered *volatile*. The lines between these categories, however, are certainly blurred, and there are unexpected outliers and differences in sensitivity among solutes. Additional insight is provided from these and other studies [29–34] that have examined the influence of variables related to solvents, additives (e.g., pH modifiers, buffers, ion-pairing agents), conditions (e.g., liquid flow rate, evaporation T), and solute (e.g., log D, pK_a, charge state) on LC-aerosol detection scope, sensitivity, and uniformity of response. Some of these aspects are further discussed in the succeeding text.

1.3.3.5 Particle Size Dependency

Since evaporation time is proportional to surface area (D^2 in Equation 1.4), a given component is expected to evaporate more readily from a smaller droplet or particle. Accordingly, a "volatility range" associated with a given parameter (e.g., P_s) is mostly relevant to a specific detector design, analytical method, and mass range of study. For example, under the same LC conditions, response curves obtained (Figure 1.5) for caffeine (CAF) and THEO while similar with a VCAD are quite different with a UCAD. The latter is highly characteristic of differences in the shape of response curves between nonvolatile (THEO) and semivolatile (CAF) analytes. Since the Veo is designed to produce overall larger particles (e.g., Figure 1.3), it follows that a greater proportion of CAF should remain in the condensed phase. In this example, better semivolatile detection was obtained with the VCAD design at a higher evaporation T (T_e), which suggests that the term "sub-ambient" T evaporation [35] is mainly relevant to specific detector designs perhaps those that produce aerosols with a relatively small size distribution. An additional consequence related to particle size dependency follows from Equation 1.4 where differences in sensitivity among analytes and the influence of parameters such as T_e should be more pronounced for lower analyte levels [7]. An example of the latter is shown in Chapter 3 where an increase in T_e resulted in reduced response that was only significant for lower levels of a given analyte.

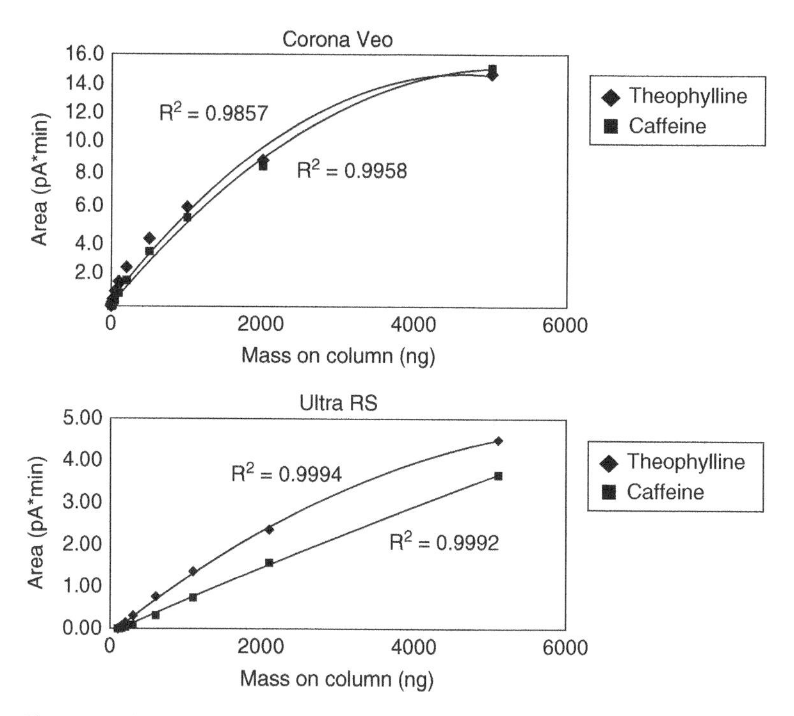

Figure 1.5 Response curves obtained with different CAD instrument designs. Under the same chromatographic conditions (isocratic 20% CH_3OH, 1.0 mL/min), the Corona ultra RS response for caffeine (more volatile) is lower and exhibits a higher power law exponent (Sections 1.4 and 1.5) than the less volatile theophylline. VCAD response curves are similar for both analytes even at higher Te (35°C) than UCAD (presumed < ~20°C due to cooling from nebulization). This is attributed to the larger overall particles measured with the VCAD design.

1.3.3.6 Ionizable Solutes

Evaporation of aerosol droplets produced by nebulizing an LC eluent leads to a rapid increase in the concentration of less volatile components thereby favoring solute–solute interactions. Several studies have shown that volatile acidic and/or basic mobile phase additives can be used to increase detection scope and influence sensitivity for ionizable analytes. This can be attributed to salt formation and therefore greater "combined" mass partitioning within the condensed phase [30–34]. This can be advantageous by enabling better detection of volatile analytes (e.g., volatile basic analyte + volatile acidic modifier) or disadvantageous by leading to higher background signal and noise (e.g., volatile basic impurity + volatile acidic modifier) or decreased response uniformity among analytes (e.g., differential interaction between analytes and a given volatile modifier). While the choice of mobile phase additive is usually driven by requirements of the chromatographic separation technique (i.e., pH modification, pH buffering, ion pairing), possible influences on aerosol detection should also be considered. The previous references, which include more recent studies with CAD [33, 34], provide useful insight to the main factors influencing response for ionizable solutes. For example, Cohen *et al.* [33] showed that when using relatively weak acidic modifiers (e.g., acetic acid (AA)—pK_a 4.79; formic acid (FA)—pK_a 3.4), little to no response was obtained for volatile basic analytes (pK_a for conjugate acid > ~8). However, high sensitivity was obtained for most analytes of interest when using stronger acidic modifiers (e.g., HCl, trichloroacetic acid (TCA), trifluoroacetic acid (TFA), and heptafluorobutyric acid (HFBA)). These authors described the main physicochemical properties influencing detectability and sensitivity as (i) pK_a's of both the acidic modifier and conjugate acid of basic analyte, (ii) number of ionizable functional groups on a given molecule, and (iii) steric effects (i.e., accessibility of ionizable site). While this study addressed basic analytes, the same principles apply to analysis of volatile acidic analytes where volatile basic additives can be used to increase detection scope. Furthermore, this principle can apply simultaneously to acidic and basic analytes with the use of volatile buffers. Clear examples of this are the use of hydrophilic interaction chromatography (HILIC) with CAD for simultaneous analysis of anions and cations as further described in Chapters 3 and 6. Within this context, Russell *et al.* [34] studied a diverse set of acidic, basic, and neutral analytes spanning a range of properties related to volatility (e.g., MW, B_P, M_P). For nonvolatile analytes, somewhat higher response was obtained for basic compared with neutral analytes when using volatile buffers such as ammonium formate and ammonium acetate or acidic modifiers FA, TFA, and HFBA—also attributed to salt formation with volatile acidic mobile phase additives. As previously reported [32, 36], response for basic analytes increased monotonically with molar mass of the acidic modifier (i.e., FA < TFA < HFBA), but the increase, at least in Russell's study [34], was not stoichiometric.

1.3.3.7 Background Solutes: Impurities

The possible influence of impurities on detector response is often difficult to assess since the exact nature and concentration is usually unknown. Some insight is provided from LC-MS applications, which share a requirement for volatile eluents. Commonly observed impurities include Na, Cl, K, Fe, phthalates, polyethylene glycol, polysiloxanes, and fluoropolymers arising from various sources including solvents, additives, system components, and general labware [37]. Molecular clusters and adducts of these contaminants that are frequently observed with LC-MS may also be relevant to aerosol detection [38]. Recommended practices to achieve a low and consistent background level are specifically discussed in Chapter 3. With good laboratory practices, impurities can usually be kept at concentrations low enough to produce particles, in the absence of analyte, in the low nm size range. Examples of this are shown in Figure 1.6. In addition to the previously described solvent property-related effects (Equations 1.3 and 1.4), differences between solvents in the

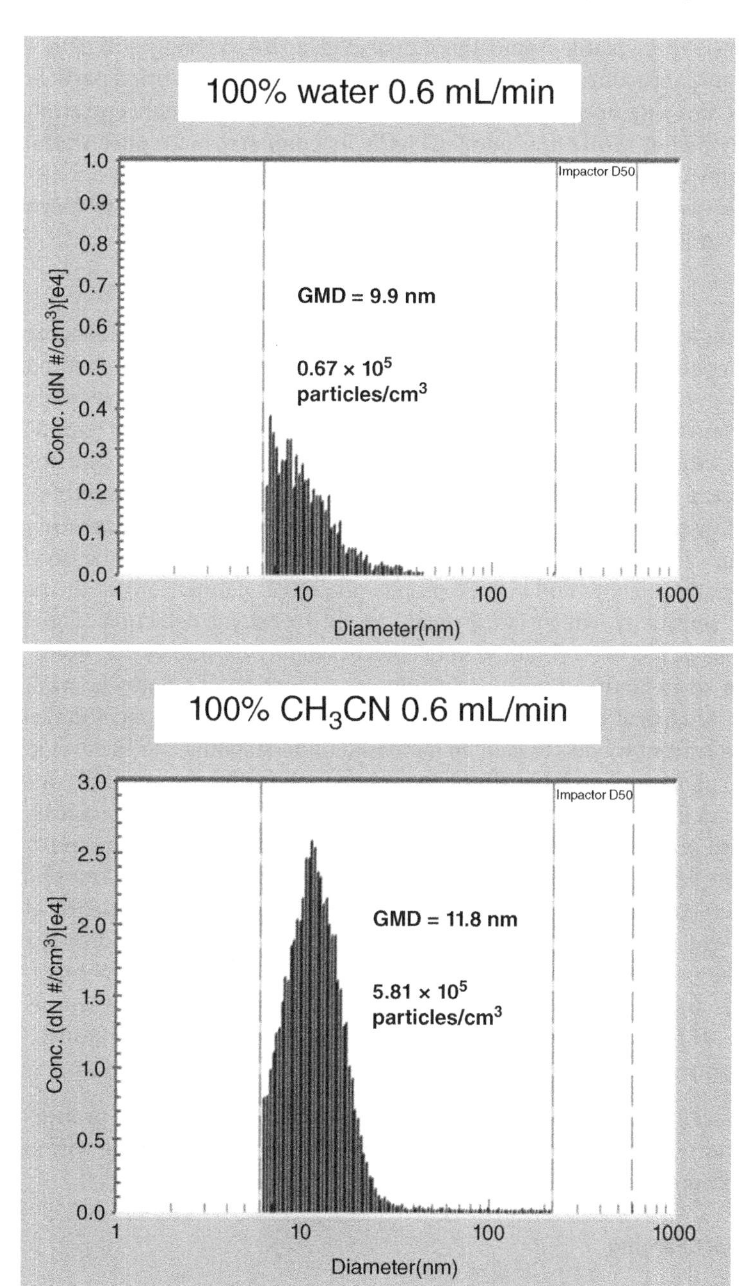

Figure 1.6 Particle size distributions measured using an SMPS from the outlet of the evaporation tubes of the Corona Veo for constant 0.6 mL/min flow of water (left) and CH_3CN (right).

concentration of nonvolatile impurities can also give rise to changes in detector baseline and noise during solvent gradients. Figure 1.6 shows dried particle distributions showing approximately ninefold higher number concentration and a slightly higher GMD for "pure" CH_3CN as compared with pure water. While this appears to be mainly attributed to aerosol transport (sevenfold difference as previously noted), it suggests that, in this experiment, CH_3CN also had a higher concentration of nonvolatile impurities.

1.3.3.8 Summary

Understanding aerosol-specific aspects of evaporation and factors influencing relative gas-/condensed-phase partitioning of solutes can be useful for method development and for predicting response. Some factors influencing sensitivity are mainly related to aerosol particle size. Assessments based on volatility parameters (e.g., P_s) should therefore be made in the context of a specific detector design, analytical method, and mass range of study. For a given analytical method, typical volatility parameters (e.g., P_s) are often adequate to predict an approximate boundary beyond which solutes are considered nonvolatile. Interactions between ionizable analytes and volatile mobile phase (i.e., counterion) additives can increase sensitivity, which is especially useful to extend detection scope. Significant insight has been obtained in LC-aerosol detection studies that examine detection scope and response uniformity as a function of solute chemical diversity and analytical conditions. However, it is apparent that opportunities exist to refine current models to gain an increased understanding and to develop more robust calibration models. Efforts, for example, to further understand control parameters in spray drying applications for pharmaceutical manufacturing [39, 40] and to predict gas to particle partitioning of atmospheric organic aerosols seem quite relevant to this topic [41–43]. Aspects of these efforts that may be worth further consideration include entropic and enthalpic components of solid–liquid–gas-phase changes (i.e., ΔH_{fus}, ΔH_{vap}, crystalline vs. amorphous forms), prediction of parameters (e.g., P_s) from subcooled liquid state rather than extrapolation from higher T measurements, and component interactions (ionic, dipole–dipole, induction, H-bonding) involving more complex mixtures.

From the previous discussion, the following is assumed for evaporation:

- Assumption 4—Each dried aerosol residue particle's diameter is proportional to the 1/3 power of the v/v concentration of nonvolatile solute within its primary aerosol droplet.

1.3.4 Aerosol Charging

1.3.4.1 Mechanisms

Charging of aerosols can occur by several mechanisms. Most relevant to CAD are static electrification, field charging, and diffusion charging. Static electrification of liquids is based on an electrostatic charge at the surface that is carried away as aerosol droplets separate from the bulk liquid. ESI and natural spray electrification are both examples of static electrification. The mechanisms of field and diffusion charging, by contrast, involve *collisions* between ions and particles within an aerosol. Field charging, which takes place in the presence of a strong electric field (i.e., potential gradient), can produce a high charge density on aerosol particles, but the level and distribution of charges significantly depends on the particle material (i.e., dielectric constant). Diffusion charging involves collision of ions and aerosol particles when there is no appreciable electric field and is also the dominant mechanism for $d < 100\,\mathrm{nm}$ even in the presence of an electric field [4]. This process leads to capture of ions, resulting in a net electrical charge on aerosol particles. An important advantage of diffusion charging is that the process has minimal dependence on the particle material [44], which provides the basis for uniform response capabilities of CAD. Also, while field charging (i.e., mean charge per particle) is largely proportional to particle surface area (d^2), unipolar diffusion charging was found empirically to be proportional to $d^{1.133}$ over a wide range of particles sizes. This relationship combined with that for evaporation (Equation 1.5) is one of the main factors that underlie the shape of a CAD response curve.

1.3.4.2 Diffusion Charging Overview

Diffusion charging has widespread application in the use, control, and study of aerosol particles, making it of great interest in many fields (e.g., nanotechnology, industrial hygiene, environmental studies, (bio) pharmaceutical, and semiconductor manufacturing). Diffusion charging with electrical mobility classification is widely accepted as an essential technique to study aerosols in the low nm ($<100\,\mathrm{nm}$) size range and is used, for example, to measure aerosol size distributions, to separate or classify particles of a given size, and to monitor and assess potential health effects of industrial and atmospheric aerosols [44–46]. A critical requirement to effectively implement these techniques is to have an in-depth understanding of the particle charging process, and a common objective is to produce a known charge distribution on aerosol particles. When the charge distribution is fully predictable, it becomes possible to obtain absolute concentrations of particles over the range of the instrument. Ions for diffusion charging include those produced by ionizing radiation (e.g., α radiation from ^{210}Po, β radiation from ^{85}Kr, soft X-rays, photo-ionization sources (e.g., UV)) and electrical discharges of the corona and dielectric barrier type. Bipolar charging, also termed charge neutralization, involves particle exposure to both positive and negative ions. Initially neutral particles may acquire charge, while particles with initially high charge levels may discharge through capture of oppositely charged ions. This typically leads to a steady-state distribution (the "Fuchs distribution") consisting of mainly neutral particles and singly charged particles of both signs. As previously mentioned, a ^{210}Po alpha emitter has been used within the VCAD spray chamber as a bipolar ion source to study the potential influence of natural spray electrification on CAD response. Also, aerosol size distribution measurements with TSI's SMPS described within this chapter were obtained using a soft X-ray bipolar ion source combined with electrical mobility classification and CPC.

1.3.4.3 Unipolar Diffusion Charging Theory

CAD is based on *unipolar* diffusion charging in which the aerosol is exposed to ions of only one polarity (i.e., positive for CAD). This technique involves a non-equilibrium steady-state charge distribution in which higher charging levels and higher charging efficiency can be attained more readily than with bipolar charging, which is especially important for ultrafine particles (i.e., $d < 20\,\text{nm}$ [47]). The magnitude of charge depends mainly on the ion concentration produced by the ion source, time of aerosol exposure to ions, and particle size (i.e., proportional to $d^{1.133}$). The theory describes a process of charging that results from random Brownian motion of ions in proximity to aerosol particles. It is assumed that (i) each ion–particle collision leads to charge transfer to the particle and that (ii) charge is not transferred back to adjacent gas molecules that, as discussed in Section 1.5.6, may help to explain some differences between diffusion charging and APCI. With diffusion charging, particles can acquire multiple charges, and, as charge accumulates, the rate slows due to increased repulsive forces on the ions. Ions are quickly thermalized on formation, acquiring a Boltzmann distribution of velocities; thus there are always some ions with high velocities, making the charging rate always greater than zero [48]. The process of acquiring multiple charges and the evolution of a charge distribution among particles of a given size is described by a theoretical model that utilizes an infinite set of differential equations [49] (the so-called birth-and-death model). These equations include combination coefficients of ions with particles carrying n elementary charges.

Diffusion charging depends strongly on the relationship between λ_{ion} and d and, based on this, is said to occur in three regimes. Pui [50] studied the nature and mobility of ions produced in air by various sources including corona discharge and, although a variety of ions may be produced, found 14.5 nm at atmospheric pressure to be a useful equivalent λ_{ion}. Based on this, the three regimes are approximately continuum where $d > 200\,\text{nm}$, transition where $20\,\text{nm} > d < 200\,\text{nm}$, and free molecular where $d < 20\,\text{nm}$. The basis for diffusion charging in the continuum regime is well described by a rigorously validated macroscopic diffusion-mobility theory. In the transition and free molecular regions, the process is more complex and several theories exist and continue to be studied and debated. Among these theories, the Fuchs' limiting-sphere theory [51] is the most widely used and validated. Briefly, Fuchs' theory in the transition regime is based on an imaginary sphere that surrounds a given particle at a distance of approximately λ_{ion} from the particle surface. Outside of this sphere, ion motion follows the same theory as in the continuum regime. Inside the sphere, motion depends only on initial (thermal) speed as ions enter the limiting sphere and their interaction potential with the particle.

The ion–particle interaction potential at a given distance from the center of the particle includes contributions from a Coulombic force and an image force. The image force is a weak function of the dielectric constant of the material, while the Coulombic force is material independent. The possible relevance of particle dielectric constant to material-dependent charging and therefore CAD response uniformity will be further discussed in the succeeding text.

1.3.4.4 CAD "Corona Jet" Charger Design

The specific design of a unipolar diffusion charger plays a significant role in the performance of a given EAM technique, especially with regard to ultrafine aerosols ($d < 20$ nm). The design used in all current commercial CAD instruments is based on that described by Medved *et al.* [2] and is also used in the TSI Model 3070A electrical aerosol detector (EAD) and other TSI devices (e.g., nanoparticle surface area monitor (NSAM)). The device uses a corona discharge, which is the most effective technique for continuously producing high ion concentrations.

Corona Discharge A positive corona discharge is produced by creating a nonuniform electrostatic field between a sharp needle tip and an orifice plate. In existing CAD designs, the needle is positive with respect to the orifice plate. At the needle tip a high enough field strength is established to break down the normally insulating surrounding gas such that it becomes conductive. In the region near the tip, electrons flow with high enough velocity to knock electrons from gas molecules and create positive ions and additional free electrons. These electrons are, in turn, accelerated toward the positive polarity needle tip, creating enough energy to knock electrons off other gas molecules. This leads to a self-sustaining chain reaction called the Townsend avalanche that creates high concentrations of electron-positive ion pairs. Some electrons recombine with positive ions releasing energy in the form of blue and UV photons, which serve to ionize other molecules and help sustain and stabilize the electron avalanche. This also creates the characteristic blue-white glow of a positive corona discharge. The CAD ion source is isolated from the aerosol (vapor and analyte) and supplied with particle-free gas, either N_2 or air. Several studies, including those using APCI-MS, have examined the nature of ions produced by corona discharge with these surrounding gases. The most consistently reported primary ions are $N_2^{+\bullet}$ and $N_4^{+\bullet}$, and, with even trace levels of water vapor (e.g., 3 ppm) [52], the most abundant secondary ions are hydrated protons $H_3O^+(H_2O)_n$ [53, 54]. Pui found that $H_3O^+(H_2O)_6$ is most probable for positive discharge with air under 10% RH [50] and calculated the λ_{ion} as 14.5 nm [46], which, as discussed in Section 1.3.4.3, is a generally accepted equivalent λ_{ion}. With organic solvent vapors, other types of cluster ions with different mobility are likely to occur [55], thus possibly influencing the charging process. Additional study is required to evaluate the significance of this relative to the large solvent dependency of CAD response associated with nebulization and aerosol transport.

1.3.4.5 Corona Ion Jet and Aerosol Particle Jet

Positive ions produced as described previously are repelled away from the needle tip and attracted toward the negative polarity orifice plate. In the corona jet design, positive ions are additionally entrained by a flow of gas (in the case of CAD 1.0 L/min air or N_2) through a subsonic orifice into a separate mixing chamber with conductive walls. In this manner, a steady turbulent flow of positive ions is established within a region with little to no electric field, thus facilitating aerosol charging by diffusion mechanisms. The aerosol produced by nebulizing and evaporating LC eluent is introduced to the mixing chamber through a second subsonic orifice to form a turbulent aerosol particle jet that opposes (collides with) the corona ion jet. In this process, convective mixing of the turbulent jets allows diffusion charging to proceed unhindered. This technique differs from the diffusion charger used in Dixon's description of ACD [3], which lacked the described convective mixing component. Briefly, the EAA (TSI, Inc.) device used by Dixon was based on corona discharge established along the length of a wire within a concentric tube. A weak electric field between that and an additional outer concentric tube was used to maintain the migration of ions into a region with aerosol flow between the two concentric tubes. In that design, the flow of ions was laminar rather than turbulent, and the ion flow converged with rather than opposed the aerosol flow.

Charger Performance Characteristics Medved *et al.* [2] studied transmission efficiency (fraction of particles leaving vs. entering the charger) and charging efficiency (fraction of particles charged) of the corona jet device as a function of d over the range of 3–50 nm. They showed high transmission efficiency and, by inference, minimal particle loss. Charging efficiency of this *unipolar* charger was higher than that predicted from Fuchs' theory for *bipolar* charging over the entire size range, as expected, and decreased rapidly for $d < 10$ nm. Several studies with this device [2, 56–59] have shown (e.g., Figure 1.9) that, in accordance with theory, mean charge per particle is found to be proportional to $d^{1.133}$ over the range from 10 nm to 1.0 μm and for $d < \sim 9$ nm, proportional to surface area (d^2) or a higher exponent. This well-established empirical relationship is a main factor that determines both the shape of a response curve and the chromatographic peak shape obtained with CAD as will be discussed in Sections 1.4 and 1.5.

As mentioned in Section 1.3.4.3, especially in the low nm diameter range, particle dielectric constant may contribute to the image force described in limiting sphere theory. This may therefore influence response uniformity of CAD. Shin *et al.* [60] used the corona jet charger to study compact, roughly spherical aerosol particles of sucrose, NaCl, and Ag, which have dielectric constants (ε) of 3.3, 6.1, and infinity (metallic), respectively. Over a range of 10–200 nm, they observed a difference in slope of mean charge per particle versus d of 10.5% between Ag and sucrose and a difference of 5.2% between Ag and NaCl. Li *et al.* [61] reported similar results of ~3% difference between NaCl and Ag particles over a mean d range of 20–200 nm. In both studies, material dependency was less than predicted from Fuchs' theory, and this may be attributed to the high efficiency of the specific corona jet charger for which it is theorized that once particles acquire a first charge, the resultant material-independent Coulombic force becomes the dominant factor of the interaction potential. As previously described, aerosol particles encountered with LC are rarely expected to consist of a single pure substance, a factor that may serve to dampen the influence of analyte dielectric constant on response uniformity. Considering this effect and the previously described dependency, it seems reasonable to assume that analyte dielectric constant is a relatively minor factor influencing the response uniformity of CAD.

1.3.5 Summary of Aerosol Charging

It is well established that for diffusion charging with CAD, mean charge per particle is proportional to $d^{1.133}$ over the range from 10 nm to 1.0 μm while for $d < \sim 9$ nm is proportional to a higher power of d. This has significant implications to the shape of a CAD response curve. Studies of particle material dependence on diffusion charging suggest that differences in analyte dielectric constant may have a minor influence on response uniformity (discussed in Section 1.5). The final steps of the CAD process, after aerosol charging, are removal of excess ions with an ion trap and measurement of aggregate charge with a filter/electrometer. These steps will be described in more detail in Section 1.4.

1.3.6 Summary of CAD Process

Here, we briefly summarize the CAD process by tracing discrete volumes of eluent to a column outlet and then by describing subsequent aerosol characteristics at different stages within the detection flow path. Figure 1.7 depicts a column with three solute bands at different positions along a column and a chromatogram that represents a completed separation. Isocratic conditions are assumed where the concentration of nonvolatile solute within a given volume is the sum of its time-dependent analyte concentration and a constant impurity concentration. The time-dependent analyte concentration results from the mass injected and the dilution and dispersion of the separation process (described in Section 1.4). The key assumptions described previously are:

1) The size distribution of a primary aerosol is solute independent. Accordingly, all droplets, *at a given instant,* have the same solute mass concentration and differ only in size.
2) Primary aerosol droplet distributions remain constant throughout isocratic elution but change during solvent gradients due to the changing solvent properties.
3) Removal of primary aerosol droplets by inertial impaction above a certain cut size to form a secondary aerosol is solute independent but depends on

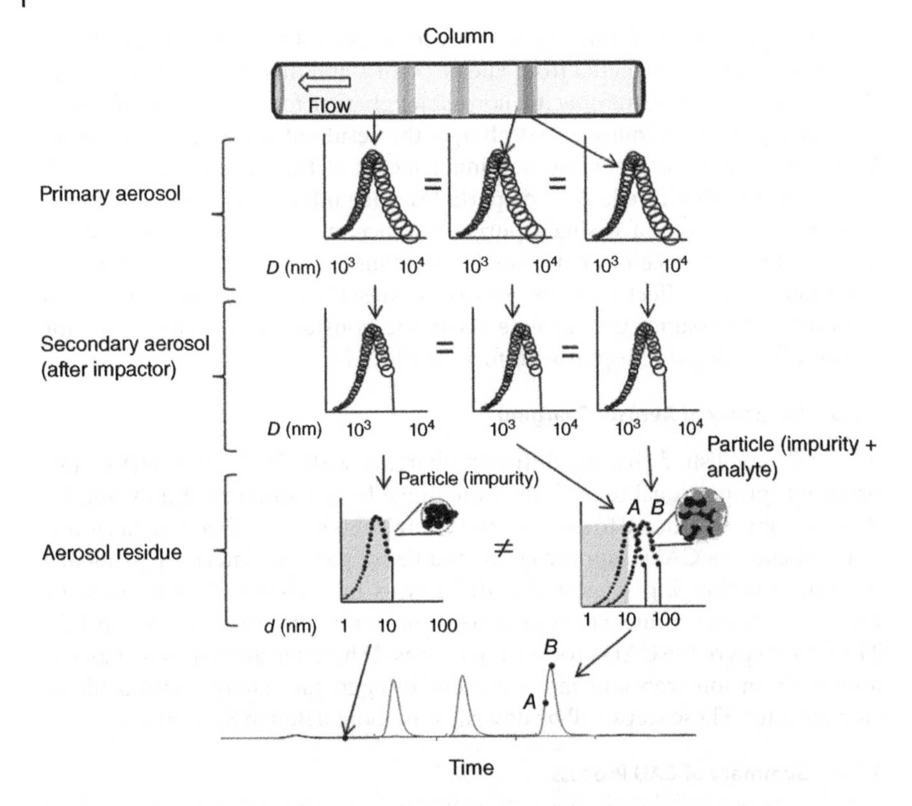

Figure 1.7 CAD is depicted for an isocratic separation where three discrete eluent volumes (baseline and two points within a single analyte solute band) are projected to the outlet of a column and traced through the main steps of the process. Particle cartoons are magnified to show impurity or analyte + impurity components. Shaded areas of each aerosol residue distribution reflect the relative proportion of $d < \sim 9$ nm, which have a higher aerosol charging power law exponent, thereby influencing the overall power law exponent of CAD response.

eluent properties that influence both nebulization (Equation 1.3) and pre-impaction evaporation. Empirically determined "effective" cut sizes are used in Section 1.4 for modeling CAD response.

4) The effective diameter of each dried aerosol residue particle is equal to the diameter of its corresponding primary droplet times the cube root of its v/v concentration of its nonvolatile constituents.

   Note: All of the previous assumptions are also generally valid for both ELSD and CNLSD.

5) Signal (current) is based on diffusion charging of dried aerosol particles and measurement of aggregate charge reaching the electrometer filter where mean charge per particle is proportional to $d^{1.133}$ over the range from 10 nm

to 1.0 μm while for $d < \sim 9$ nm is proportional to a higher exponent. Loss of particles in the ion trap also plays a role (further described in Section 1.4). Analyte dielectric constant is a relatively minor factor influencing the response uniformity of CAD.

Figure 1.7 shows that, for isocratic conditions, all three of the illustrative primary aerosols have the same droplet size distribution (Assumption 2) and all droplets within a given primary aerosol have the same solute concentration (Assumption 1). Since the "effective" impactor cut size remains constant in isocratic conditions (Assumption 3), all secondary aerosols also have identical size distributions. Number concentrations are the same for all dried aerosols and identical to that of the secondary aerosols. The size of each residue particle reflects the v/v concentration of nonvolatile solute in its corresponding primary droplet (Assumption 4). Representative residue particles are magnified to depict a small particle consisting of many background impurity molecules and a larger particle with the same number of background molecules whose size is increased by the presence of many analyte molecules. The residue distributions for the two volumes within a single solute band reflect a size shift proportional to a 1/3 power law exponent of the instantaneous nonvolatile solute v/v concentrations at peak maximum and at ~50% of peak maximum. The difference in response is mainly due to the difference in the charge accepted and carried by the particles. Differences between these distributions in the relative proportion of particles below $d \sim 9$ nm underlie changes in the overall power law exponent of CAD response over a range of analyte amounts as further described in Section 1.4.

1.4 CAD Response Model

To construct a model of CAD, we will consider the successive steps of primary spray droplet formation from an eluent stream, impaction to remove droplets larger than a specified cut size, evaporation of the remaining droplets to form residue particles of the analytes, charging of the residue particles, removal of excess ions, and detection of the current due to deposition of the particles with their charge in an aerosol-electrometer filter. Each step of this process depends on the droplet or residue particle size, and each of these dependencies is evaluated based on a theoretical model or on empirical data.

The model predicts primarily the time-dependent response of the detector to the concentration of analyte in the eluent. This response function is then used to predict the time-dependent current during passage of an eluting peak, and the current is integrated over the peak to predict the net charge delivered as a function of the injected analyte mass. For simplicity, we assume only isocratic conditions.

1.4.1 Primary Droplet Size Distribution

Droplets of the eluent are formed by a pneumatic nebulizer as discussed earlier. We assume that the analyte concentration is sufficiently low so as not to influence properties of the mobile phase such as surface tension, viscosity, and so on. This ensures that the behavior of the nebulizer, and thus the size distribution of these "primary droplets," is not affected by the presence or absence of an analyte solute band.

We assume further that this primary droplet size (D) distribution is lognormal. The fundamental parameters of this distribution are, besides overall amplitude, the CMD, and the GSD. The normalized form is

$$\text{Lognormal}\left(D, \text{CMD}, \text{GSD}\right) = \frac{e^{-1/2\cdot\left[\frac{\ln(D/\text{CMD})}{\ln(\text{GSD})}\right]^2}}{\sqrt{2\cdot\pi}\cdot D\cdot\ln\left(\text{GSD}\right)} \tag{1.6}$$

and the normalization property is

$$\int_0^\infty \text{Lognormal}\left(D, \text{CMD}, \text{GSD}\right) dD = 1 \tag{1.7}$$

1.4.2 Impactor

An impactor is integrated within the spray chamber to remove droplets that are too large to be fully evaporated in the downstream evaporation tube, thus forming a "secondary aerosol." In the inertial impactor, the primary droplets and carrier gas are made to flow around an obstacle, whereby the inertia of droplets above a specified diameter D_{cut} causes them to collide with the obstacle. The liquid from these droplets is led to a waste drain. We assume an ideal impactor with a sharp cutoff so that the secondary aerosol droplet size distribution is described by the same lognormal function below the cutoff and is zero above the cutoff. The actual value of the cutoff is dependent on details of the nebulizer geometry and on the mobile phase density, viscosity, and surface tension and is determined empirically.

The mean droplet volume in the primary distribution may be written as

$$V_{\text{dp}} = \int_0^\infty (x\cdot\text{CMD})^3\cdot\text{Lognormal}\left(x\cdot\text{CMD}, \text{CMD}, \text{GSD}\right)\cdot\text{CMD}\, dx \tag{1.8}$$

and the mean volume of the droplets remaining after impaction is

$$V_{\text{dc}} = \int_0^{\frac{D_{\text{cut}}}{\text{CMD}}} (x\cdot\text{CMD})^3\cdot\text{Lognormal}\left(x\cdot\text{CMD}, \text{CMD}, \text{GSD}\right)\cdot\text{CMD}\, dx \tag{1.9}$$

The fraction of eluent volume that is converted to aerosol is V_{dc}/V_{dp}, and the remainder goes to the waste drain.

1.4.3 Drying and Residue Formation

The secondary droplets from the nebulizer and impactor are led through a heated drying tube, where each droplet evaporates to leave a residue particle consisting of the nonvolatile dissolved material that was in the droplet. The simplest intuitive picture of this process requires that the volume of a droplet's residue particle is equal to the droplet volume times the volume/volume (v/v) concentration C of its nonvolatile constituents. In the approximation that the residue particle is spherical, its diameter will be $d_{res}(D, C) = D{\cdot}C^{\alpha}$ with $\alpha = 1/3$. Empirical studies of the shift in the measured residue size distribution versus (v/v) concentration persistently show a power law with an exponent $\alpha \sim 0.2$. There seems to be no known systematic bias in these measurements, but also no rationale for a value such as 1/5 is apparent. In view of this we will retain α as a parameter in the model formalism but will continue to assume $\alpha = 1/3$ for present purposes.

1.4.3.1 Residue Particle Parameters

The residue particles will have a size distribution identical to that of the secondary droplets, scaled according to $d_{res}(D, C) = D{\cdot}C^{\alpha}$. Thus the parameters of the distribution become

$$d_{cut} = D_{cut} \cdot C^{\alpha} \quad \text{cmd} = \text{CMD} \cdot C^{\alpha} \tag{1.10}$$

but the dimensionless GSD is identical to that of the droplets. Thus the residue size distribution is Lognormal(d, cmd, GSD) for $d < d_{cut}$ and zero otherwise. It should be noted that the terms "count mean diameter" and "geometric standard deviation" as used here apply only to the mathematical form of the distribution without the cutoff. Figure 1.8 shows the residue particle size distribution according to this model for isocratic elution with a fixed v/v concentration of nonvolatiles.

1.4.4 Charging of Residue Particles

The dried aerosol is injected as a jet into a mixing chamber along with an opposing jet of gas containing positive ions formed in a corona discharge in a separate chamber. The electric field in the mixing chamber is negligible, thus the ion–particle encounters are governed by diffusion. The mean charge per particle is dependent on particle diameter, and the dependence has been found empirically for the charger design used in all CAD models [2, 56–59] to follow

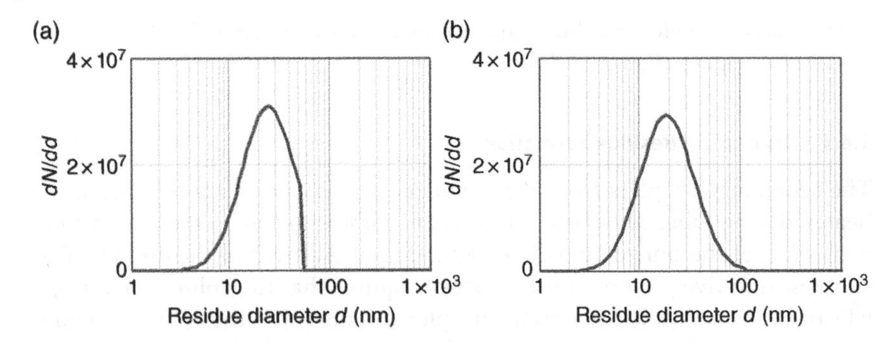

Figure 1.8 Distribution of dry residue particle diameters for v/v concentration = 10^{-6}. (a) Mobile phase = water; CMD = 3.6 µm; GSD = 1.85; D_{cut} = 5.0 µm, (b) mobile phase = acetonitrile; CMD = 2.7 µm; GSD = 1.85; D_{cut} = 11 µm.

a power law with an exponent $\beta_2 = 1.133$ for particles larger than about $d_0 = 9$ nm. For smaller particles, the exponent is larger, the limited data being consistent with $\beta_1 = 2.25$. For the model it will be useful to have an empirical function that represents this behavior. A reference point is that for particles of diameter $d_{ref} = 100$ nm, the mean charge per particle is $n_{ref} = 3.9$ electronic charge units. We use dimensionless units for the diameter, defining

$$x = \frac{d}{d_0} \quad x_{ref} = \frac{d_{ref}}{d_0} \tag{1.11}$$

Then the mean number of charge units per particle is represented by

$$n(x) = n_{ref} \cdot \frac{x^{\beta_1}}{x_{ref}^{\beta_2}} \quad \text{if } x < 1, \quad \text{and} \quad n(x) = n_{ref} \cdot \left(\frac{x}{x_{ref}}\right)^{\beta_2} \quad \text{otherwise.} \tag{1.12}$$

Figure 1.9 shows how $n(x)$ changes with residue particle diameter.

1.4.5 Ion Removal

Ions formed in the corona discharge of the unipolar charger are injected into the mixing chamber by a gas flow. These ions quickly accumulate small clusters of solvent vapor molecules, and it is these clusters that diffuse to the residue particles and impart their charge. The electrical mobilities of these species are known to be greater than about 0.5 cm^2/V·s. Excess ions can travel downstream with the charged aerosol particles, but it is important to keep them from reaching the electrometer detector where they could contribute a large and fluctuating background current. To this end, a simple coaxial electrostatic precipitator is included as an "ion trap" immediately downstream from the mixing chamber where the charging takes place. The trap has the parameters in Table 1.4.

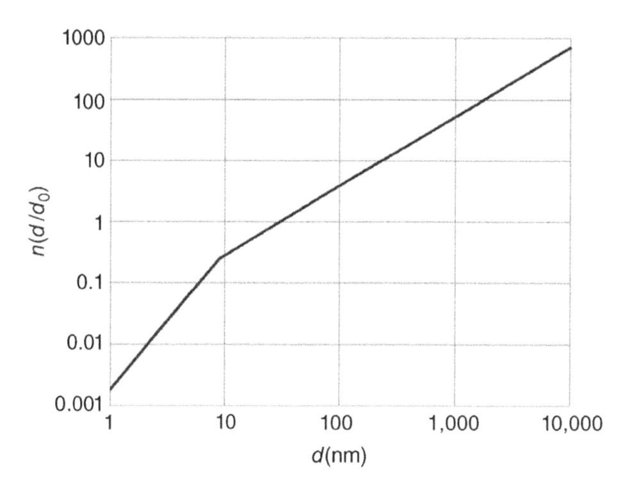

Figure 1.9 Mean number of electronic charge units per particle versus particle diameter in nanometers.

Besides its function in removing ions, the ion trap removes some of the residue particles. Its transmission for charged particles and ions is calculated in this section.

For simplicity we assume fully developed laminar flow in the annular space of the ion trap. The gas velocity is parallel to the axis denoted by z and is given by

$$v_z(r) = \frac{2 \cdot F}{\pi \cdot \left(r_2^2 - r_1^2\right)} \cdot \frac{r_2^2 - r^2 - 2 \cdot r_{\text{maxv}}^2 \cdot \ln(r_2 / r)}{r_2^2 + r_1^2 - 2 \cdot r_{\text{maxv}}^2} \tag{1.13}$$

where the radius of maximum v_z is

$$r_{\text{maxv}} = \sqrt{\frac{r_2^2 - r_1^2}{2 \cdot \ln(r_2 / r_1)}} \tag{1.14}$$

The electric field in the annular space is radial, outward directed, and given by

$$E_r(r) = \frac{V}{r \cdot \ln(r_2 / r_1)} \tag{1.15}$$

A particle or ion of electrical mobility Z moves downstream at the gas flow velocity, parallel to the z axis, while the electric field drives its migration across the flow at radial velocity $v_r(r, Z) = E_r(r) \cdot Z$. If the particle enters the

Table 1.4 Ion trap parameters.

Inner electrode radius (mm)	r_1	0.75
Outer electrode radius (mm)	r_2	2.7
Length (mm)	L	37
Applied voltage (V)	V	20
Total gas flow (L/min)	F_{gas}	3.75

precipitator at $z = 0$ and at a starting radial coordinate r_0, it follows a trajectory given by

$$z(r, r_0, Z) = \int_{r_0}^{r} \frac{V_z(r')}{V_r(r', Z)} dr' \quad r_1 < r_0 < r_2 \tag{1.16}$$

All ions or particles having electrical mobility greater than a critical value Z_{crit} will be precipitated onto the outer electrode of the ion trap. The value of Z_{crit} can be found by solving $z(r_2, r_1, Z) = L$ (L = trap length) for Z. With the parameters given by Table 1.4, $Z_{crit} = 0.172\,\text{cm}^2/\text{V·s}$. Thus the trap removes all of the ions, as desired. It was verified by observing the electrometer current while varying the trap voltage V that ion current observed at $V = 0$ vanished when $V = 10\,\text{V}$ was reached. The value 20 V was chosen to assure removal of all ions.

1.4.5.1 Attenuation of Particle Signal by Ion Trap

For particles with mobility $Z < Z_{crit}$, those entering outside a critical radial coordinate $r_{crit}(Z)$ will also be collected on the precipitator wall and lost; those entering at $r_0 < r_{crit}(Z)$ will pass through and can carry their charge into the electrometer to produce a signal. To obtain the value of r_{crit} given a particle diameter, we use the Millikan relationship giving the electrical mobility of a spherical particle of diameter d:

$$Z(d) = \frac{e \cdot C_c(\lambda, d)}{3 \cdot \pi \cdot \eta \cdot d} \quad \text{where} \quad C_c(\lambda, d) = 1 + \frac{2\lambda}{d} \cdot \left[1.165 + 0.483 \cdot e^{-0.997 \cdot \left(\frac{d}{2\lambda}\right)} \right] \tag{1.17}$$

C_c is the Cunningham slip correction [62], λ is the gas mean free path, η is the gas viscosity, and e is the electronic charge. Thus to find $r_{crit}(d)$, we solve $z(r_2, r, Z(d)) = L$ for r.

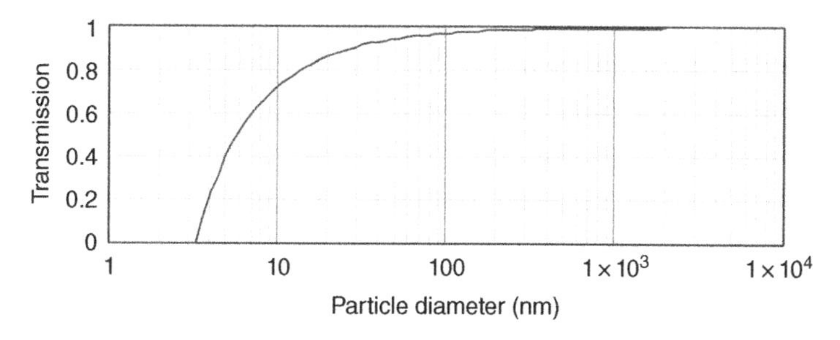

Figure 1.10 Ion trap transmission for singly charged particles.

Having found $r_{crit}(d)$, we note that the transmission for particles of diameter d is just the ratio of the cross-sectional area within $r_{crit}(d)$ to the total cross-sectional area of the trap's annular passage (Figure 1.10):

$$T(d) = \frac{r_{crit}(d)^2 - r_1^2}{r_2^2 - r_1^2} \tag{1.18}$$

1.4.6 Signal Current

Eluent flows into the nebulizer at a rate F_{liq}. For our model calculations we take $F_{liq} = 0.6\,\text{mL/min}$. Table 1.5 shows the mean droplet volume, the fraction of droplets remaining after impaction, and the resulting rate of residue particle formation for our example cases of water and CH_3CN.

Table 1.5 Parameters and results for the rate of residue particle formation for water and acetonitrile.

Mobile phase		Water	CH_3CN
Flow rate (mL/min)	F_{liq}	0.6	0.6
Mean droplet volume		Femtoliter	
Before impactor	V_{dp}	134.1	56.6
After impactor	V_{dc}	12.7	37.9
Fraction of droplet count after impactor			
	N_{cut}	0.703	0.989
Fraction of droplet volume after impactor			
	ϕ	0.095	0.669
Primary droplet rate (/s)	r_d	7.46E+07	1.77E+08
Residue particle rate (/s)	r_p	5.24E+07	1.75E+08

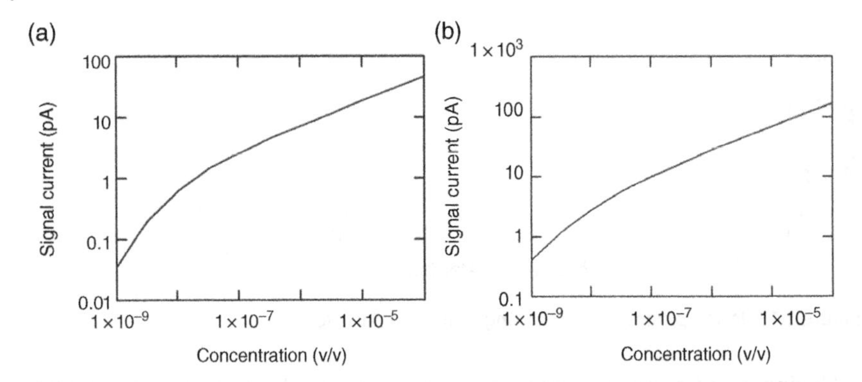

Figure 1.11 CAD detector signal as a function of concentration. Mobile phase: (a) water; (b) CH_3CN.

The particles so formed are charged, and their charge is carried to the electrometer's collection filter, where the arriving charge is measured as a current. The mean number of charge units carried to the electrometer by the entire distribution of residue particles produced at a given concentration C, including loss in the precipitator, is

$$n_{\text{mean}}(C) = \int_{\frac{d_{\text{crit}}}{d_0}}^{\frac{C^{\alpha} \cdot D_{\text{cut}}}{d_0}} n(x) \cdot T(d_0 \cdot x) \cdot \text{Lognormal}\left(d_0 \cdot x, \text{cmd}, \text{GSD}\right) \cdot d_0 \, dx \tag{1.19}$$

where the limits of integration are set by d_{crit}, where $T(d) = 0$, and d_{cut}, where the size distribution is reduced to zero because of the impactor. The detector current (Figure 1.11) is then given by

$$i(C) = e \cdot n_{\text{mean}}(C) \cdot r_{\text{p}} \quad \text{where} \quad r_{\text{p}} = \left(\frac{F_{\text{liq}}}{V_{\text{dp}}}\right) \cdot N_{\text{cut}} \cdot \tag{1.20}$$

1.4.7 Signal from an Eluting Peak: Peak Shape

The eluent consists of a mobile phase, dissolved mobile phase impurities (which we ignore for this model), and the analytes, which have been separated by the column. Following Equation 1.21, the instantaneous v/v concentration resulting from injected mass m_{inj} of an analyte whose bulk density is ρ may be represented by a Gaussian peak in concentration:

$$C(t, m_{\text{inj}}) = \left(\frac{m_{\text{inj}}}{\rho \cdot F_{\text{liq}} \cdot \sqrt{2 \cdot \pi} \cdot \sigma_{\text{t}}}\right) \cdot \exp\left[-\frac{1}{2}\left(\frac{t - t_{\text{r}}}{\sigma_{\text{t}}}\right)^2\right] \tag{1.21}$$

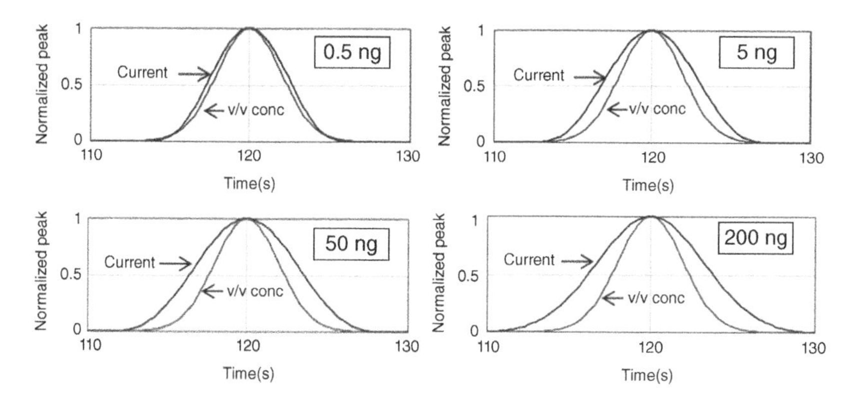

Figure 1.12 Broadening of the signal peak. The column retention time is 2 min and the peak standard deviation σ is 2 s. The mobile phase is acetonitrile and the analyte has an assumed density of 1.4 g/cm^3.

where t_r is the column retention time and σ_t is the peak's standard deviation in time. The time-dependent CAD signal current pulse is then given by $I(t) = i[C(t, m_{inj})]$. The nonlinear response $i(C)$ results in a broadening of the current-time profile compared with the concentration-time profile, and due to the change in slope of $i(C)$ as the peak concentration increases, the broadening becomes more significant with increasing injected mass. Figure 1.12 shows the peak in C and in I for several values of m_{inj}. The 200 µg case shows that broadening does not increase much at extremely high injected mass. It should be noted that the peak comparisons in Figure 1.12 are normalized with respect to C and I and do not accurately reflect differences in relative peak area.

1.4.8 Peak Area Versus Injected Mass

The total charge delivered to the electrometer during the elution of a peak is the area of the current versus time function $I(t)$. In the model this can be integrated for a range of injected mass values, leading to the following expression for peak area or total charge:

$$Q\left(m_{inj}\right) = \int i\left[C\left(t, m_{inj}\right)\right] dt \tag{1.22}$$

The result of this calculation is shown in Figure 1.13 for acetonitrile with $\rho = 1.4\,\text{g/cm}^3$.

1.4.9 Summary

We have presented a model of the salient characteristics of the CAD detector, tracing its behavior from the formation of spray droplets of eluent through impaction to remove the largest droplets, drying to form residue particles of

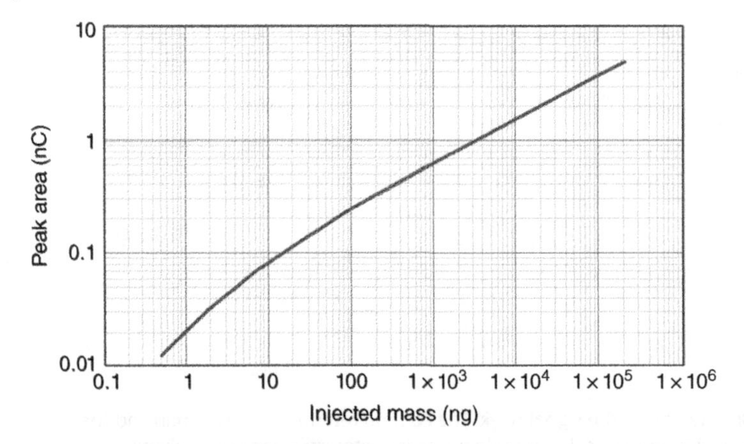

Figure 1.13 Peak area is shown in nanocoulombs for injected mass in the range of 0.5 ng to 200 μg. Mobile phase: acetonitrile; analyte bulk density: 1.4 g/cm³.

the analyte, charging the resulting aerosol in a unipolar charger, and removing the ions with attendant loss of some of the smallest residue particles. The resulting current shows a slight decrease in slope over the very wide dynamic range of the detector, which gives rise to a small degree of broadening of the time-dependent eluting peaks compared with the actual peak width of the analyte concentration. To make use of this model, some empirical knowledge of the behavior of the nebulizer and impactor is required, as well as the charging characteristics of the unipolar charger.

1.5 Performance Characteristics

This section will discuss performance expectations of CAD considering analyte properties, detector design, and experimental variables. Since CAD, ELSD, and CNLSD use the same basic principle to produce an aerosol residue, a key focus will be on characteristics of the downstream aerosol measurement technique that may explain differences in performance among these devices. This section will also describe basic differences between CAD and gaseous ion formation for MS especially in relation to detection scope and response uniformity.

1.5.1 Response Curve: Shape and Dynamic Range

We begin by discussing the shape of CAD response curves, that is, peak area response (A) versus analyte mass injected (m_{inj}), over the entire dynamic range of the detector. We will compare response curves from two different CAD designs and a representative ELSD (SEDEX 90LT, SEDERE, France).

Response can be described over segments of the range of each detector by a power law function:

$$A = a\left(m_{\text{inj}}\right)^{b}$$

The coefficient a indicates overall sensitivity and the exponent b describes the curve shape. Response is linear when $b = 1.0$ and a is then the slope of A/m_{inj}. Nonlinear response can be described as deviation from a straight line either by curving upward (*supralinear*) or by curving downward (*sublinear*). *Sublinear* response occurs when $b < 1$; *supralinear* response occurs when $b > 1$. The more b differs from 1.0 (lower or higher), the greater is the deviation from linearity or degree of curvature in either direction. Figure 1.14 shows response curves for the nonvolatile analyte THEO analyzed under the same conditions with either ELSD or CAD. Over its dynamic range, ELSD response (Figure 1.14a)

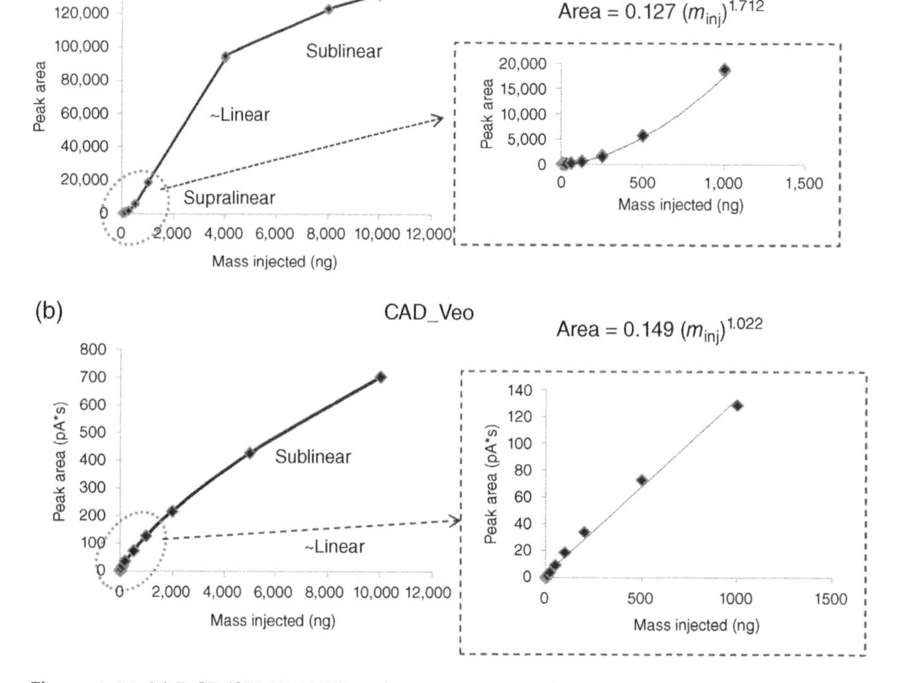

Figure 1.14 (a) ELSD (SEDEX 90LT) peak area versus m_{inj} theophylline—7.8–10,000 ng (left); 7.8–1,000 ng (right). Mobile phase 20% CH_3OH; 1.0 mL/min; 20 × 4.0 mm, 3 µm C18 column. Signal units depend on attenuation setting and thus not specified. (b) VCAD peak area versus m_{inj} theophylline—0.2–10,000 ng (left); 0.2–1,000 ng (right). Conditions as in (a). Note: VCAD signal output is $pA^{1.5}$ (see *power function* discussed in Section 1.5.1).

exhibited a characteristic sigmoidal shape. Over the lower range (m_{inj} 7.8–1000 ng), ELSD response was supralinear with a high degree of curvature ($b = 1.712$) and transitioned to quasi-linear and then sublinear ($b = 0.31$) toward higher m_{inj} (8,000–10,000 ng). Similar to ELSD, CAD response (Figure 1.14b) was sublinear at the higher range. For this analysis, however, CAD response (VCAD with signal output of $pA^{1.5}$; see *power function* (PF) discussed in the succeeding text) was approximately linear over nearly three orders of magnitude, from m_{inj} 0.2 to 1000 ng, as indicated by both the power law exponent ($b = 1.022$) and the coefficient of determination ($r^2 = 0.993$) from linear regression ($\text{Area} = 0.131 m_{inj} + 2.003$).

As described in Section 1.4, the power law exponent b for CAD response is derived from:

1) Droplet evaporation/residue formation power law exponent (α), which is the same for CAD and ELSD and here assumed to be 1/3
2) Aerosol charging power law exponents (β), which depend on particle diameter (d) where $\beta_2 = 1.133$ for $d > {\sim}9$ nm and $\beta_1 \sim 2.25$ for the smallest particles, that is, $d < {\sim}9$ nm

Similarly, the power law exponent b for ELSD response is derived from:

1) Droplet evaporation residue formation power law exponent (α), which, as previously mentioned, is assumed to be 1/3
2) Light scattering power law exponents, which likewise depend on d or more specifically involve three different light scattering mechanisms that depend on the relationship between d and incident light wavelength (λ) [29, 63]:
   a) Refraction and reflection: $\beta_3 = 2.0$ for $d/\lambda > 2.0$
   b) Mie: $\beta_2 = 4.0$ for $\sim 2.0 > d/\lambda > 0.1$
   c) Rayleigh: $\beta_1 = 6.0$ for $d/\lambda < 0.1$

Both CAD and ELSD involve continuous measurement of polydisperse aerosols whose distributions shift to larger size both with increasing m_{inj} and toward the apex of a given peak. Signal current (pA) results from an entire distribution of residue particles at each point, and peak area (pC) is the current integrated over a distribution of analyte concentrations across the width of a chromatographic peak. From this, it is expected that peak area response for ELSD always involves a mixture of light scattering mechanisms, while CAD likewise always involves measurement of charged particles above and below $d = 9$ nm. Theoretical limits for the overall power law exponents b can be approximated from the previously mentioned individual power law exponents α and β (see Section 1.4 for more explicit description) such that:

- VCAD "intrinsic" response: $0.38 < b < 0.75$
- ELSD: $0.67 < b < 2.0$

For both CAD and ELSD, smallest b is observed for highest m_{inj} where residue distributions contain predominantly larger particles (lower β). This explains the downward curving, or sublinear response curves, at the high end of the m_{inj} range (Figure 1.14) of both detectors. The response of both techniques then transitions toward higher b toward the lowest m_{inj} where the residue distributions now contain a greater proportion of small particles (higher β). The approximate particle sizes, d, where each technique transitions toward highest β are:

- CAD: $d < \sim 9\,\text{nm}$
- ELSD: $d < \sim 50\,\text{nm}$ (assuming $\lambda = 500\,\text{nm}$, the approximate center of visible light)

In comparison with CAD, the transition for ELSD is approximately fivefold higher in diameter and thus 125-fold higher in volume or mass. This together with the higher β ($\beta_1 = 6$) underlies the commonly observed signal "drop" that is steeper (greater curvature) and occurs at higher levels of m_{inj} than in CAD. This is associated with a higher limit of detection (LOD) : for this analysis estimated (see the following text) at ~7.8 ng for ELSD and at <1 ng for CAD. It is important to recognize that b and thus the degree of curvature also change with m_{inj}. When these same data are plotted as b versus m_{inj} (Figure 1.15), a change in b (slope) can be examined more closely, that is, between individual values of m_{inj}. This further shows the relative complexity of ELSD response

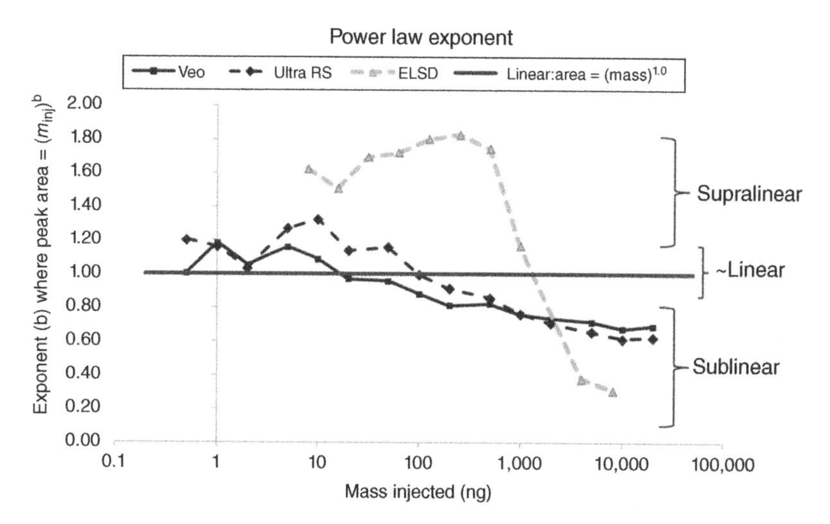

Figure 1.15 Power law exponent b versus m_{inj} obtained under the same conditions (see Figure 1.14) with two CAD designs and ELSD. Note: Veo signal output is $pA^{1.5}$ (see *power function* discussed in Section 1.5.1).

with an approximately sixfold (0.31–1.80) range of b over a relatively small dynamic range: for this analysis ~10^3 for ELSD versus ~10^5 for CAD. The maximum value (b_{max} = 1.8) for ELSD is similar to that commonly reported [7], while the minimum, 0.31, is lower than predicted and here attributed to photomultiplier saturation [63–65]. The higher (b) variance evident in Figure 1.15 at low m_{inj} for all three detectors is attributed to the higher signal variance expected near the lower sensitivity limits.

Older generation UCAD models produce a dried aerosol distribution with a smaller characteristic particle size and lower number concentration than that of newer designs (e.g., Figure 1.3). Under the same conditions, the older design therefore has a higher relative proportion of smaller particles (d < ~9 nm, β_1 ~ 2.25) and thus a higher "intrinsic" b. Figure 1.15 shows a 2.14-fold range of b (0.62–1.33) for UCAD, which indicates supralinear response for the low mass range. A signal drop can therefore be observed with this design but is less steep (b_{max} ~ 1.33) than that of ELSD (b_{max} ~ 1.8) and occurs at a lower m_{inj} (d transition ~9 nm vs. ~50 nm). For Ultra RS a signal drop is typically observed at m_{inj} < ~1 ng, but this more specifically depends on the conditions (e.g., chromatographic) that determine instantaneous solute (analyte + impurity) mass concentration. Newer CAD designs, under the same conditions, produce a lower relative proportion of small particles, thereby reducing the intrinsic b, simplifying the response curve, and shifting any signal drop to even lower mass levels. Since the intrinsic response of the VCAD design remains sublinear throughout its dynamic range, the instrument's digital signal processing scheme includes an internal PF that applies an exponent to the instantaneous signal (e.g., $pA^{1.5}$). The value of the PF exponent is chosen to maintain quasi-linear response over a similar ~10^2 mass range observed with older models while extending it to lower levels. This facilitates method transfer between instruments while expanding the useable quantitative range to lower levels. With this approach, VCAD consistently exhibits a smaller b range (here ~1.75-fold; 0.68–1.18; Figure 1.15) and therefore a simpler response curve than older models. For this analytical condition, VCAD also consistently exhibits improved sensitivity limits especially when eluents have low impurity concentrations (Figure 1.16). The topic of sensitivity limits is further discussed in Section 1.5.4 and in Chapter 3.

1.5.1.1 Semivolatile Analytes

Figure 1.16 shows that, for both CAD designs, CAF has lower sensitivity (A/m_{inj}) and, in Figure 1.5, a higher b than THEO. This is highly characteristic of a "semivolatile" analyte for which evaporation has a more pronounced influence on response. Since evaporation rate R_e is inversely proportional to particle surface area, this effectively increases b so that the response typically becomes supralinear at a much higher m_{inj} (e.g., below 100 ng instead of below 1 ng for nonvolatiles). The effect is most significant toward the lower end of the m_{inj} range, and

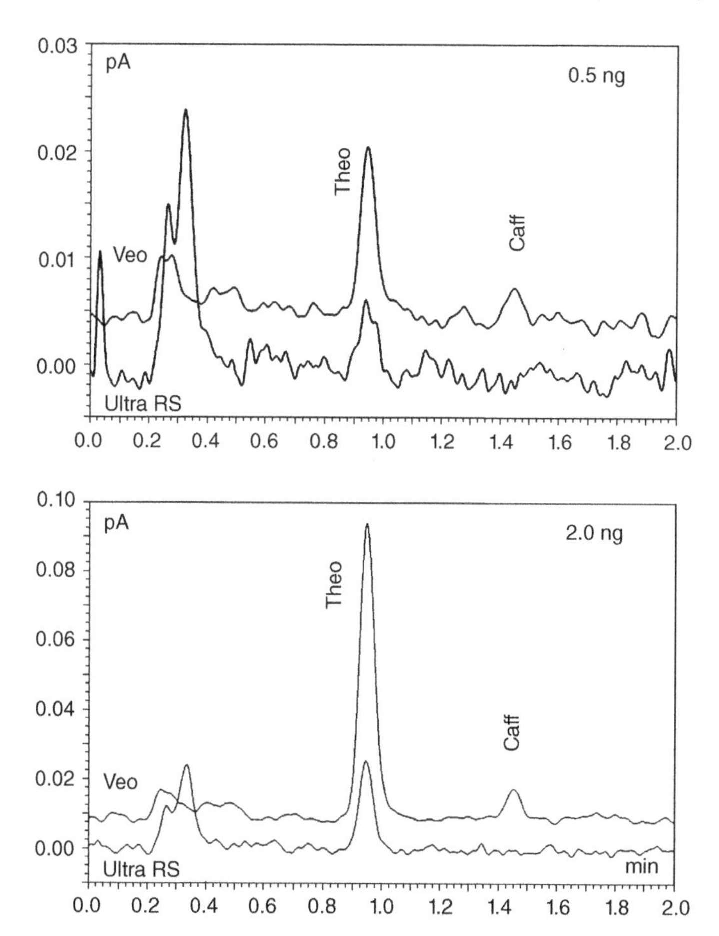

Figure 1.16 Comparison of Corona ultra RS and Corona Veo near the limit of detection for theophylline (Theo) and caffeine (Caff); $m_{inj}=0.5$ ng each (left); $m_{inj}=2.0$ ng (right).

thus response curves of semivolatile analytes are often observed as sigmoidal. As previously described, the magnitude of this effect depends not only on the volatility of the analyte but also on detector design and conditions.

1.5.1.2 Calibration

Calibration and quantitative analysis with CAD are discussed briefly in this section and in more detail within Chapter 3. As with any quantitative method, the choice of calibration approach depends on the response obtained for a given method, experimental design (e.g., level and spacing of calibrators), and requirements for quantitation (e.g., mass range, limit of quantitation (LOQ), precision, accuracy). Most chromatography data systems include

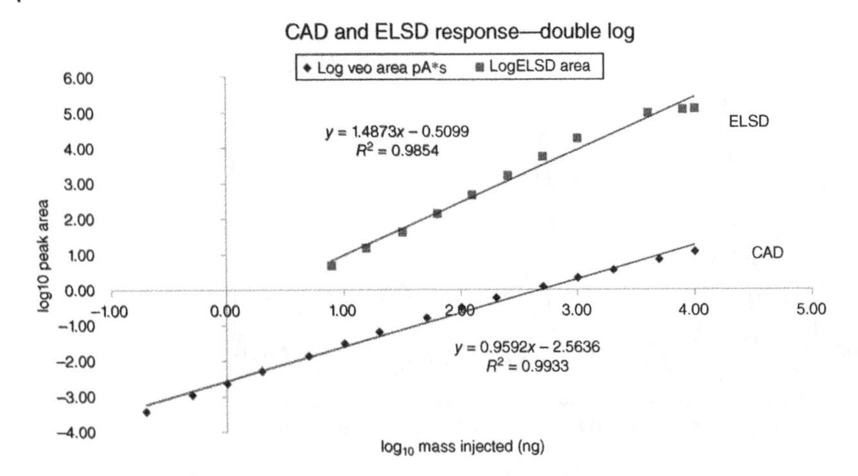

Figure 1.17 Log_{10} (area) versus $\log_{10}$ (m_{inj}) obtained under the same conditions (see Figure 1.14) with a Corona Veo ($pA^{1.5}$) and with ELSD.

linear, double logarithmic, quadratic (second-order polynomial), and other curve fitting options. A double logarithmic function is very often used to describe and calibrate ELSD response and is discussed here in relation to the power law function and data described in Figure 1.14. By taking the $\log_{10}$ of both sides of the power law equation, the exponent b, which describes the shape of the curve, now becomes the slope of the line (Equation 1.23). Likewise, $\log_{10}$ of the coefficient a now becomes the Y-intercept:

$$\log_{10}(\text{area}) = b\log_{10}(m_{\text{inj}}) + \log_{10}(a) \tag{1.23}$$

Figure 1.17 shows $\log_{10}$ area versus $\log_{10}$ m_{inj} where the slopes from linear regression can be viewed as an aggregate measure of the power law exponent b over the full experimental m_{inj} range (Veo – 2.5×10^5; ELSD 10^3). The obtained slopes reflect the previously described supralinear response of ELSD (slope ~1.49) and near to sublinear response of VCAD (slope ~ 0.96). Deviations from the line reflect changes in b. While these deviations are less visually apparent than in Figure 1.15, it should be recognized that the data are plotted on logarithmic scales, and therefore care must be taken with interpreting goodness of fit for accurate quantitative analysis (further discussed in Chapter 3).

A main consideration with CAD is the selection of a user-defined PF setting, an option that can be used to help "linearize" signal output for a given analysis. Figure 1.18 shows the effect of changing the setting for this parameter in terms of b versus m_{inj}. In this example, an instrumental setting of 1.33 has the effect of simply shifting the curve higher such that approximately linear response

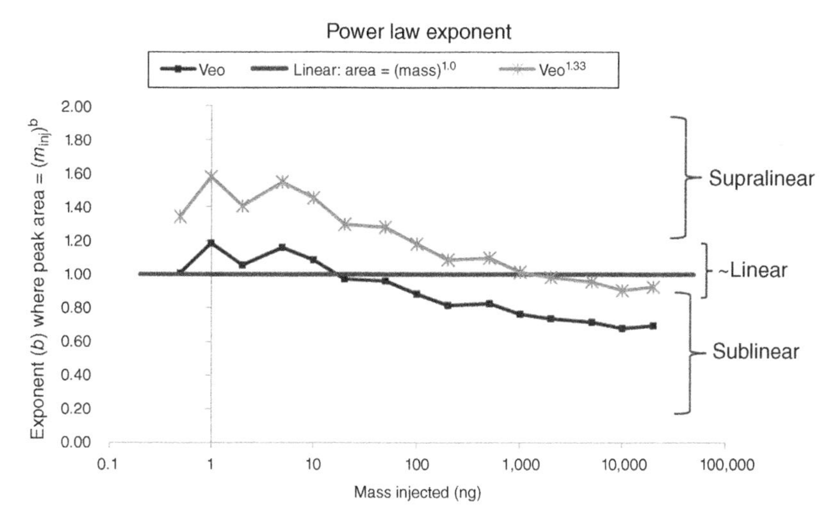

Figure 1.18 Power law exponent b versus m_{inj} for VCAD with different power function settings ($pA^{1.5}$ and $pA^{2.0}$; Exponent of $2.0 = 1.5 \times 1.33$).

would now be obtained over a higher m_{inj} range. A consequence of this is that response at the lower m_{inj} range would now be supralinear.

With supralinear response, sensitivity (e.g., A/m_{inj}) decreases (signal drop) with decreasing m_{inj}, which has significant practical consequences:

a) Quantitation by relative response (e.g., area% and other assumptions of linearity) based on sensitivity (A/m_{inj} or "response factor") from higher level calibrators will underestimate lower level components (e.g., drug impurities, minor components of a polymer distribution).
b) Achievable sensitivity limits (LODs, LOQs) are higher (worse) than estimated by the common practice of linear extrapolation of signal-to-noise ratio (SNR) for high level standards.
c) Precision and reproducibility are typically worse (e.g., higher % relative standard deviation, %RSD) along an exponentially decaying, that is, supralinear curve.
d) Chromatographic peaks are artificially sharpened (see Section 1.5.2), which brings into question common measures of chromatographic efficiency and resolution.

With sublinear response, the converse of the previously mentioned consequences (a) through (d) is true. This and other factors can lead to significant errors in quantitation and can also be a source of confusion when developing, evaluating, and optimizing methods. A strong recommendation is therefore to choose a PF setting *only to help linearize response within the mass range of interest.* It is not advisable to choose a setting when a main objective is to increase SNR or artificially sharpen chromatographic peaks since these parameters become less valid and often misleading when response is not linear. In the previous example, a PF setting of 1.33 would seem appropriate for analyses requiring minimum LOD of $m_{inj} \geq 100$ ng. In this case, a suitable fit may then be obtained by linear regression, or an improved fit may be obtained with other available curve fitting options (e.g., quadratic).

As described previously, semivolatile analytes exhibit sigmoidal response curves that are rather analyte specific. While a PF setting of <1.0 may help to linearize response over the low mass range of a single analyte, this would most likely have an adverse affect on its higher mass range and on the complexity of the response curves of other analytes. Also, since response obtained with newer CAD designs is sublinear over most of its range, the use of a PF of <1.0 is generally not advisable.

1.5.2 Peak Shape

As described in Section 1.4, sublinear response artificially distorts the chromatographic peak by broadening of the instantaneous signal versus time profile compared with the "true" mass concentration versus time profile. Also, because of the changing slope (described previously in terms of *b* vs. m_{inj}), this distortion becomes more significant with increasing m_{inj}. Figure 1.19 plots the ratio of peak full width at half maximum (FWHM) for "raw" current predicted by the CAD response model versus a "true" Gaussian concentration distribution

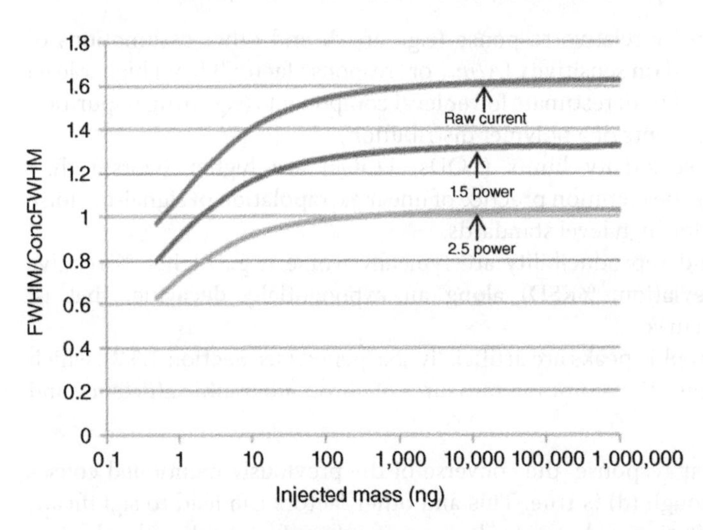

Figure 1.19 Ratio of peak full width at half maximum (FWHM) for "raw" current predicted by the CAD response model versus a "true" Gaussian concentration distribution over a range of m_{inj} and for different power functions.

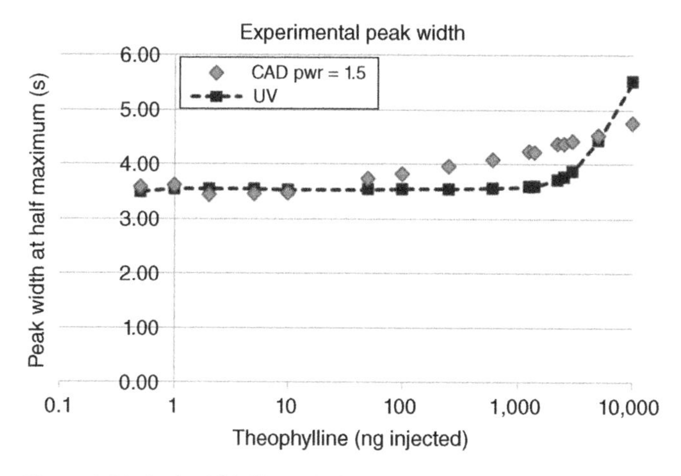

Figure 1.20 Peak width (FWHM) obtained with a UV detector in series with CAD (power function of $pA^{1.5}$) over a m_{inj} range of 0.5–10,000 ng theophylline showing gradual broadening of CAD peak area due to nonlinear response. Broadening of both signals at highest m_{inj} is attributed to column overload. Conditions as in Figure 1.14a.

over a range of m_{inj} and for different PFs. This shows the significant broadening from an inherent (sublinear) response and the effects of $current^{1.5}$ and $current^{2.5}$ PF exponents. As described, newer CAD designs coupled with application of an internal PF exponent of, for example, 1.5, can extend the quasi-linear response of CAD to lower analyte levels (e.g., $m_{inj} < 1$ ng). Experimental results (Figure 1.20) with a 1.5 internal exponent illustrate a slight peak broadening effect over most of the range. In this example, column overload at higher m_{inj} made an additional contribution to peak broadening as shown by comparison to an in-line UV detector.

1.5.3 Mass Versus Concentration Sensitivity

Chromatographic detectors are sometimes classified as either *mass-flow sensitive* (response is proportional to mass of analyte reaching the detector per unit time) or *concentration sensitive* (response is proportional to analyte concentration within the eluent at a particular time (e.g., UV, conductivity)). For flow injection analysis (i.e., no column) or for post-column changes in liquid flow, peak area for a true mass-flow-sensitive device is independent of liquid flow rate, and thus an increase in flow rate should result in temporal peak sharpening with increased height; for a true concentration-sensitive device, peak height is independent of flow rate, and thus an increase in flow rate should result in lower peak area. These behaviors should also be observed with changes in column flow rate provided that column efficiency (e.g., number of theoretical plates) is maintained. The dependency of CAD response on flow rate is of particular interest given the relatively common use of post-column flow splitting (e.g., to MS) or solvent addition to compensate for solvent gradient effects on response uniformity. Gorecki *et al.* [24] first described the use of a post-column "inverse gradient" approach with CAD where a second pump was used to deliver a solvent gradient that was the exact inverse of that used for the analytical separation. The two solvent streams were mixed post-column to deliver a constant solvent composition to the detector (i.e., the midpoint of the gradient, 50% CH_3CN) at twice the liquid flow. They showed that CAD behaved approximately as a mass-flow-sensitive device since response (average calibration curve for several analytes) obtained with 50% CH_3CN without compensation was very similar to that with compensation at twice the liquid flow. While the CAD response model (Section 1.4) is described in terms of the concentration-time profile of an eluting analyte, the time-dependent signal current results from the charge of an entire distribution of residue particles at a given time/concentration, and peak area results from the total charge delivered to the electrometer over the entire peak. Therefore, the response of CAD is related to the mass flow of analyte per unit time to the electrometer. However, deviation from true mass-flow-sensitive behavior is expected since only a fraction of the primary aerosol is typically transported to the electrometer, and, as described previously, nonlinear response distorts the chromatographic peak. For example, the described post-column addition would be expected to dilute the instantaneous analyte concentration and double the eluent volume per unit time delivered to the detector. This should result in a rather complex scenario involving a time-dependent solute concentration that depends on completeness of post-column mixing, an increase in the characteristic size of the primary aerosol distribution (e.g., SMD by Equation 1.3), and a related change in the volume removed by impaction. For the experimental conditions described by Gorecki *et al.* and also in several similar studies, these factors appear to combine such that CAD response more closely approximates that of a mass-flow-sensitive than a concentration-sensitive device. Most studies indicate, however, that CAD and other evaporative detectors do not produce a true mass-flow-dependent response and, as with many other techniques, the extent of deviation from this behavior depends somewhat on analytical conditions [66].

The extent to which CAD behaves as a mass-flow-dependent device is also of interest when scaling chromatographic methods to larger or smaller internal diameter (i.d.) columns. A common approach is to approximately scale both flow rate and injection volume in proportion to column volume. Figure 1.21 shows an example of this where the sample concentration was also adjusted so that m_{inj} was the same. Since, with a smaller i.d. column, analytes elute in a smaller eluent volume, their peak concentration is higher. The similar response obtained for equivalent m_{inj} of both analytes clearly shows that CAD behaves primarily as a mass-flow-dependent device. The slightly higher response for the lower flow condition, however, indicates some deviation from true

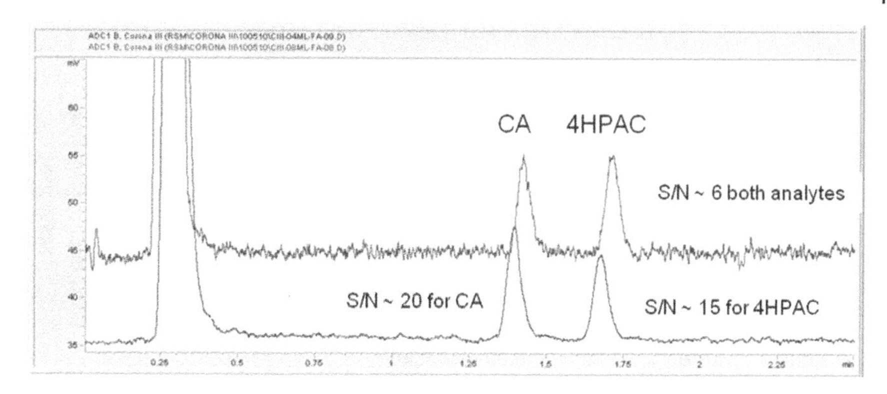

Figure 1.21 Chromatograms for the same m_{inj} (5 ng) of chlorogenic acid (CA) and 4-hydroxyphenylacetic acid (4HPAC) showing approximate mass flow versus time-dependent response of CAD with lower noise and thus higher signal-to-noise ratio at the lower flow rate. Top trace: flow rate—0.8 mL/min, 3.0 mm i.d. × 50 mm L, 2.2 μm C18 column; bottom trace: flow rate—0.4 mL/min, 2.1 mm i.d. × 50 mm L, 2.2 μm C18 column.

mass-flow response. Since nebulizer gas flow was the same, this may be due to production of a finer primary aerosol (e.g., lower SMD by Equation 1.3) and higher mass transport in the lower flow condition. It is significant to note that lower baseline current and noise and therefore higher SNR was achieved with the lower flow condition. This is attributed to a lower mass flow of nonvolatile impurity as further discussed in Section 1.5.4.

1.5.4 Sensitivity Limits

Sensitivity limits at the low m_{inj} range for a given LC-CAD method are typically expressed in terms of LOD and LOQ. These parameters can be determined by various procedures including those based on visual inspection, SNR, and relationship between the slope of a calibration curve and standard deviation of the response. Since response may become supralinear toward these limits, accurate estimation of LOD and LOQ usually requires actual analysis of appropriately low analyte levels rather than estimation from higher analyte levels. The latter can lead to misleading results especially when assuming linear response.

With any evaporative aerosol detector, the most important consideration is the concentration of nonvolatile and semivolatile impurities within an eluent. For these devices, highly sensitive detection can sometimes be viewed as a double-edged sword since; by definition for a universal detector, it should apply equally to impurities and analytes that are within a given method's volatility range. Figure 1.22 plots CAD baseline current and noise as a function of the v/v concentration of THEO, which was intentionally added to the mobile phase. Here, the noise, measured in pA peak to peak, is observed to be a consistent

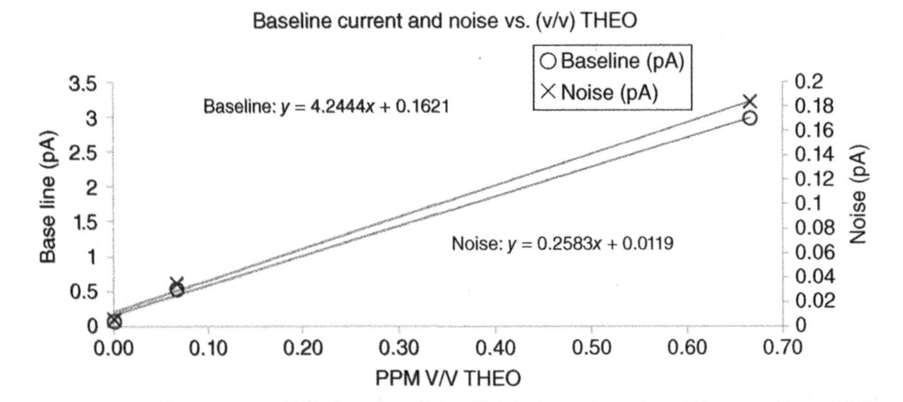

Figure 1.22 Baseline current and noise as a function of the v/v concentration of theophylline (THEO) intentionally added to the mobile phase. Conditions as in Figure 1.14a.

fraction (~6%) of the baseline signal over the measured concentration range. This clearly indicates that noise can be reduced by minimizing the concentration of nonvolatile impurities in the eluent. As previously noted, evaporation temperature (T_e) is an instrumental setting that can be used to optimize signal versus noise. For a given method, increasing T_e can often reduce the baseline current and noise as long as the analytes of interest are significantly less volatile than the impurities in the mobile phase. In practice, the nature of impurities is usually unknown, and therefore experimentation is required to optimize performance. The topics of optimization and recommended practices to minimize the concentration of impurities are further discussed in Chapter 3.

Especially with the older cross-flow nebulizer design, baseline noise with CAD can also arise due to either pressure fluctuations within the aerosol flow path or due to unstable nebulization. Intermittent noise "spikes" caused by pressure fluctuations within the flow path are often associated with irregular "plug" drainage of waste liquid from the spray chamber of older CAD models. Spray instability is most commonly associated with low mobile phase flow rates and especially when nebulizing eluents with low viscosity, low surface tension, and high volatility. This is typically identified and mitigated by increasing the liquid flow rate and/or decreasing the nebulizer gas flow rate.

1.5.5 Response Uniformity

Uniform response is defined here as sensitivity (e.g., peak area/m_{inj}) that is independent of analyte properties. The ability to obtain analyte-independent response provides the possibility to quantify multiple analytes with a single calibrator or calibration model, which is highly desirable for many analyses (e.g., drug impurities, natural products, lipids, polymers). The following variables are expected to significantly affect evaporative aerosol detector response uniformity: (i) solvent gradients, (ii) analyte volatility and salt formation, (iii) analyte density, and (iv) dependency of aerosol detector response on particle material. Since the first 3 of these factors should similarly influence all evaporative aerosol detectors, this discussion will mainly focus on characteristics of the downstream aerosol measurement technique that may explain differences in response uniformity between these devices.

1.5.5.1 Solvent Gradient Effects

As described in Section 1.3, gradient effects are due to changes in aerosol formation and transport, which are solvent dependent and analyte independent. All evaporative aerosol detectors share a necessity to reduce solvent load to ensure complete downstream solvent evaporation. While variations in design and options for nebulization and inertial impaction may lead to some differences in transport among devices, the overall magnitude of solvent effects on response uniformity is quite similar among all evaporative aerosol detectors. Several studies have shown up to a 10-fold and often nonlinear change in analyte sensitivity as a function of aqueous-organic composition [14, 67]. Inverse gradient post-column addition, as described previously, is one way to compensate for these changes. Another technique, which relies on the previously mentioned analyte independence, involves the injection of one or more "universal" calibrants at frequent (e.g., 1 min) intervals during a solvent gradient. The response obtained is then used to construct a three-dimensional calibration model (e.g., retention time – m_{inj} – peak area) for use in quantifying a range of analytes. Examples of this approach and some advantages and limitations of various techniques to improve response uniformity for solvent gradient methods are described in more detail in Chapters 3 and 4 of this book.

1.5.5.2 Analyte Volatility and Salt Formation

The relationships between analyte volatility and both detection scope and response uniformity are expected to be similar for all evaporative aerosol detectors. In all cases, reducing the evaporation temperature (T_e) is generally expected to produce more uniform response and a broader detection scope since, under these conditions, selectivity is effectively reduced. A consequence of this, however, may be higher baseline current and noise due to more sensitive detection of semivolatile impurities that are likely to be present in most experiments. Also, as described in Section 1.3, acidic and/or basic mobile phase additives can promote salt formation during droplet evaporation. For all evaporative aerosol detectors, this can lead to a broader detection scope and more uniform response for analytes through improved detection of volatile analytes that are ionized in solution. On the other hand, salt formation can also lead to higher sensitivity for ionized nonvolatile analytes compared with those that are neutral. The effect of this on response uniformity can generally be minimized by using lower MW mobile phase additives since they should contribute less to overall particle mass [32, 34, 36].

1.5.5.3 Analyte Density

Equation 1.5 and Table 1.3 indicate that solute density (ρ_s) influences dried particle diameter (d_p) and thus should have a similar, albeit minor, influence on the sensitivity of all evaporative aerosol detectors. Matsuyama *et al.* [27] showed an inverse relationship between analyte density and CAD response, which follows from Equation 1.5 since analytes with higher density produce smaller residue particles and thus lower response. A simple way to compensate for differences in analyte density is therefore to multiply the obtained response (peak area or peak height) for each analyte by the cube root of its density. For ionizable analytes, it may be preferable to use density of the predicted salt form (e.g., basic analyte + acidic mobile phase additive) rather than that of the free analyte to compensate for differences among analytes.

1.5.5.4 Dependence of Aerosol Measurement Technique on Residue Particle Material

As described previously, the influence of solvent gradients, analyte volatility, salt formation, and analyte density on response uniformity is expected to be similar among all evaporative aerosol detectors. When comparing the response uniformity of these devices, a main differentiating factor may therefore be the relative dependency of the downstream aerosol detector's response on residue particle material (i.e., primarily determined by analyte properties). Aerosol charging by diffusion mechanisms is well known to have only a minor dependence on particle material (Section 1.3). Accordingly, several studies have shown that, after accounting for other variables (e.g., solvent dependency, volatility), CAD response exhibits a minor dependence on analyte properties [5, 16, 24, 67] and consistently provides more uniform response than ELSD [7, 25, 68–73]. ELSD makes use of photometers to measure the combined angular light scattering of all aerosol particles within the detection flow path. Angular light scattering intensity is a very complex function of the light source intensity, scattering angle of measurement, aerosol particle concentration, RI of the particle material, particle shape, and ratio of particle diameter to the wavelength of incident light (d_p/λ) [63, 64]. The dependence of ELSD response on analyte RI has been previously described [63, 74]. Oppenheimer *et al.* [64] estimated that ELSD response may vary by a factor of 2 or more for analytes encompassing an RI range of 1.4–1.6. The RI of light-absorbing analytes also depends on their absorption coefficient [4]. Analyte-specific absorption and FI have been reported to influence ELSD response where variations are more pronounced when using monochromatic (e.g., laser diode) compared with polychromatic light sources [14, 75].

The technique of CNLSD is similar to ELSD in which the condensation of liquid vapor onto the dried aerosol particles is used to amplify particle size into a region of higher scattering efficiency [76]. The response of CNLSD, assuming use of serial diffusion screens to suppress background and to modulate the dynamic range, is a product of particle size distribution, particle number concentration, diffusion screen transmission efficiency, and condensation nucleation detection efficiency curve [11]. CNLSD commonly provides better sensitivity than ELSD, but studies have shown that the signal is significantly dependent on analyte nature [25, 77]. This was also observed in our own study (Figure 1.23) comparing CAD and CNLSD response for 24 chemically diverse analytes. The study included only nonvolatile analytes and was based on flow injection analysis with a constant solvent. In the absence of a chromatographic separation, peak area measurements from both detectors were corrected to account for the counterion content of a few analytes that were obtained as salts (amitriptyline, dibucaine, nortriptyline, and propranolol). While no correction was made for differences in analyte density, this is expected to be a minor factor that applies similarly to both detection techniques. Variation in response (%RSD) among analytes was found in this study to be ~8.0% for CAD and ~39.0% for CNLSD. The much larger variation observed with CNLSD is assumed to be mainly attributed to the downstream aerosol measurement technique. While the response of CNLSD should be less dependent on analyte RI, the dependence of this technique on analyte properties may be due to differences in the wettability and solubility of the particle material for a given condensing fluid (e.g., water, *n*-butanol). This is supported by several studies related to atmospheric aerosols where, for example, with water as the condensing fluid, significant differences in particle counting efficiency were found among organic aerosols, between water-soluble inorganic salts and water-insoluble aerosols, and between aerosols with and without trace impurities [78–80].

In summary, the more uniform response for nonvolatile analytes that is commonly observed with CAD compared with other detectors, including ELSD and CNLSD, is mainly attributed to the widely described minor dependence of diffusion charging on aerosol particle material. Primary factors that influence the response uniformity of all evaporative aerosol techniques include solvent dependency, analyte density, and volatility. Also, to obtain accurate quantification of multiple analytes with a single calibrator or calibration model, additional factors should be considered including detector linearity and precision over the mass range of interest; pre-analytical variables such as analyte purity, stability, and solubility; and analytical variables not specifically related to the detection technique such as column recovery, chromatographic resolution, and peak shape.

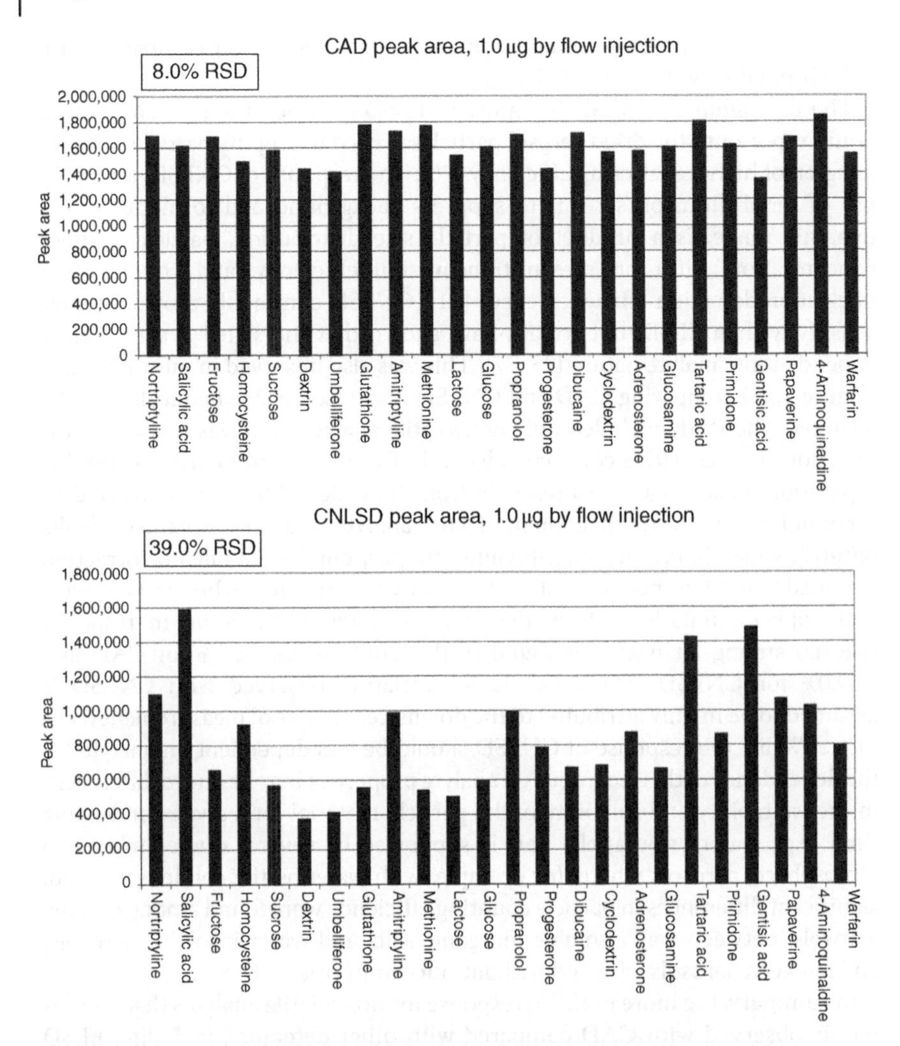

Figure 1.23 Peak area response obtained with CAD (top) and CNLSD (Quant, NQAD; bottom) from flow injection analysis of 1.0 µg of each analyte dissolved in mobile phase (50% aqueous CH_3OH, v/v). Flow rate = 1.0 mL/min. Evaporation temperature: CAD ambient; NQAD 35°C.

1.5.6 CAD Versus Formation of Gaseous Ions for MS

Here we will describe basic differences between diffusion charging of aerosol particles with CAD and gaseous ion formation for MS. Among the methods for interfacing LC and MS, perhaps most similar to CAD are APCI and, to a lesser extent, pneumatically assisted ESI. As with CAD, these techniques typically use a concentric nebulizer, produce a similar primary aerosol size distribution (e.g., D_0 in the low μm range), operate near atmospheric pressure, and, in positive ion mode, involve the same charging polarity as CAD.

1.5.6.1 Pneumatically Assisted ESI

With pneumatically assisted ESI, a metal liquid nebulizer capillary is commonly used and is held at a several kilovolts potential to induce charge on the liquid surface. Solvent evaporation from primary aerosol droplets intensifies droplet charge density until it reaches the Rayleigh limit where surface tension just balances Coulombic repulsion. Beyond this, repeated cycles of fission (Coulombic disintegration) and evaporation result in rapid formation of nanodroplets with $D < 20\,\text{nm}$ [81] with charge near the Rayleigh limit. The mechanism of subsequent gaseous ion formation is still debated, but it is generally accepted that low M solutes follow an ion evaporation model (IEM), while higher MW solutes follow either a charged residue model (CRM) or chain ejection model (CEM), each described in more detail in Refs. [81–83]. Briefly, the CRM applies mainly to globular proteins where Rayleigh-charged nanodroplets containing single molecules evaporate to dryness. The resulting analyte ion charge state is said to originate from the nanodroplet's Rayleigh surface charge and is minimally dependent on analyte structure. The other two models appear to be similar in terms of dependency of charge on analyte properties—only the IEM is discussed here. The IEM describes a process where the nanodroplet electric field is high enough to eject small solvated cluster ions from the droplet surface. The solvation shell is then lost in the interface to the MS analyzer. Reaction kinetics as described by the IEM model are strongly influenced by both experimental conditions and analyte chemical properties. For example, only positive droplets emit positive ions and only negative droplets emit negative ions; an appropriate solution pH is often required to ensure that analytes exist mainly as ionized solutes of the appropriate polarity, while neutral and ion-paired solutes either evaporate as neutral vapor molecules or precipitate, producing little or no ions for detection; ionic surfactants prefer a location near the droplet surface and more readily form gaseous ions, thereby responding more strongly than other analytes. Gas-phase reactions may also occur, especially when the order of gas-phase basicity among coexisting species is opposite that in solution.

1.5.6.2 APCI

APCI is described in more detail in the following articles [52, 84]. With most APCI sources, a primary aerosol is produced from a mobile phase with a concentric nebulizer. Solvent and solutes are rapidly evaporated at high T (e.g., >300°C probe T). A corona discharge located downstream within the vapor flow produces solvated cluster ions primarily from highly abundant solvent molecules. These collide with analyte molecules to form solvated analyte ions. The solvent shell is stripped away in the interface to the MS. With APCI, analytes are often ionized simultaneously by more than one mechanism, for example, proton transfer plus charge exchange. With proton transfer ionization, wide differences in analyte response may be observed because protonated analyte molecules (MH^+) are preferentially formed from species with the highest proton affinity. In charge exchange ionization, molecular ions ($M^{+\bullet}$) are formed possibly by direct ionization in the corona discharge or by charge exchange with radical ions. The ratio of charge exchange to proton transfer ionization is dependent on the solvent, gas, and other factors.

1.5.6.3 Main Differences between CAD and MS

LC-MS ion sources that involve pneumatic nebulization are expected to produce a solvent-dependent response that is generally similar to CAD. ESI and APCI are designed to produce single gas-phase analyte ions. For example, the mechanism of ESI involves the production nanodroplets $D \sim 20\,\text{nm}$ (i.e., $<10^{-18}\,\text{mL}$). For most of the encountered range of analyte MW and concentrations, it can be estimated by Equations 1.1 and 1.2 that such droplets should contain, at most, a single analyte ion or molecule; exceptions may be high concentrations of very small molecules. By contrast, CAD is designed to produce particles consisting of many molecules. This was seen in Tables 1.1 and 1.3, where for most of the encountered range of analyte MW and concentrations, nonvolatile analyte mass in approximately micrometre-sized primary droplets (e.g., d_0) should form particles comprised of many molecules; there may be exceptions with low concentrations of very large molecules.

APCI, like CAD, uses a corona discharge and charging is based on collisions with gas-phase ions. A main difference between these techniques is the relative size of the colliding "reagent" ion and analyte species. Premnath *et al.* [38] described an insightful study investigating the intersection between the particle diffusion charging that predominates in CAD and the chemical ionization mechanisms dominant in APCI. Briefly, cluster ions produced by ESI from different amino acids were collided with charge-neutral triethylamine (TEA) vapor molecules at the inlet of a MS. As described in Section 1.3.4.3, unipolar diffusion charging involves a steady-state, non-equilibrium charge distribution where it is assumed that each ion–particle collision leads to charge transfer to the particle and charge is not transferred back to the surrounding vapor molecules. By contrast, as described for chemical ionization, ions collide with vapor molecules, and the transfer of charge is highly dependent on the activation energy for charge transfer and steric effects during collision. Also, an equilibrium can be achieved because charge can rapidly transfer back to vapor molecules. TEA, by virtue of its high gas-phase basicity, was used to investigate an upper size limit for charge transfer back to the vapor molecules. Their studies showed that back reaction rates were highest for the smallest cluster ion size and decreased to negligible for clusters of $d_p > \sim 0.5\,\text{nm}$. This upper size limit is in general agreement with Fuch's limiting-sphere theory and also consistent with the performance of the corona jet charging technique as described previously.

In summary, both ESI and APCI involve conversion of solute to single gas-phase analyte ions through mechanisms that are very dependent on conditions (e.g., solution pH) and analyte properties (e.g., solvation energy, surface activity, solution basicity, gas-phase proton affinity) and often subject to matrix-related suppression or enhancement [85, 86]. The detection scope resulting from these techniques is therefore very dependent on conditions, and response can rarely be described as uniform. By contrast, CAD is designed to produce condensed-phase (typically solid) particles each consisting of many analyte molecules, and subsequent charging by diffusion mechanisms is known to have only minor dependence on particle material. The detection scope of CAD is limited to less volatile analytes and response is often described as uniform. However, several factors affecting response including analyte volatility and salt formation, analyte density, and solvent composition pose challenges to its use for true universal quantitation. The quantitative capabilities of CAD and qualitative capabilities of LC-MS are highly complementary, and use of these techniques in combination is facilitated by their similar eluent requirements.

References

1 Kaufman SL, Evaporative electrical detector. US 6,568,245. May 27, 2003.
2 Medved A, Dorman F, Kaufman S, and Pöcher A, A new corona-based charger for aerosol particles. Journal of Aerosol Science, 2000; 31; 616–617.
3 Dixon RW and Peterson DS, Development and testing of a detection method for liquid chromatography based on aerosol charging. Analytical Chemistry, 2002; 74; 2930–2937.
4 Hinds WC, Aerosol technology: properties, behavior, and measurement of airborne particles. John Wiley & Sons, Inc., Hoboken, 2012.
5 Gamache PH, McCarthy RS, Freeto SM, Asa DJ, Woodcock MJ, Laws K, and Cole RO, HPLC analysis of nonvolatile analytes using charged aerosol detection. LC GC North America, 2005; 23; 150–161.
6 Allen LB and Koropchak JA, Condensation nucleation light scattering: a new approach to development of high-sensitivity, universal detectors for separations. Analytical Chemistry, 1993; 65; 841–844.
7 Cohen RD and Liu Y, Advances in Aerosol-Based Detectors, in Grushka E and Grinberg N (eds.) Advances in chromatography. CRC Press, Boca Raton, 2014; p. 1.
8 Ligor M, Studzińska S, Horna A, and Buszewski B, Corona-charged aerosol detection: an analytical approach. Critical Reviews in Analytical Chemistry, 2013; 43; 64–78.

9 Vehovec T and Obreza A, Review of operating principle and applications of the charged aerosol detector. Journal of Chromatography A, 2010; 1217; 1549–1556.
10 Almeling S, Ilko D, and Holzgrabe U, Charged aerosol detection in pharmaceutical analysis. Journal of Pharmaceutical and Biomedical Analysis, 2012; 69; 50–63.
11 Magnusson L-E, Risley D, and Koropchak J, Aerosol-based detectors for liquid chromatography. Journal of Chromatography A, 2015; 1421; 68–81.
12 Kulkarni P, Baron PA, and Willeke K, Aerosol measurement: principles, techniques, and applications. John Wiley & Sons, Inc., Hoboken, 2011.
13 Ettre L, Nomenclature for chromatography (IUPAC Recommendations 1993). Pure and Applied Chemistry, 1993; 65; 819–872.
14 Mathews B, Higginson P, Lyons R, Mitchell J, Sach N, Snowden M, Taylor M, and Wright A, Improving quantitative measurements for the evaporative light scattering detector. Chromatographia, 2004; 60; 625–633.
15 Squibb AW, Taylor MR, Parnas BL, Williams G, Girdler R, Waghorn P, Wright AG, and Pullen FS, Application of parallel gradient high performance liquid chromatography with ultra-violet, evaporative light scattering and electrospray mass spectrometric detection for the quantitative quality control of the compound file to support pharmaceutical discovery. Journal of Chromatography A, 2008; 1189; 101–108.
16 Hutchinson JP, Li J, Farrell W, Groeber E, Szucs R, Dicinoski G, and Haddad PR, Universal response model for a corona charged aerosol detector. Journal of Chromatography A, 2010; 1217; 7418–7427.
17 Mclean J, Minnich M, and Iacone L, Nebulizer diagnostics: fundamental parameters, challenges, and techniques on the horizon. Journal of Analytical Atomic Spectrometry, 1998; 13; 829–842.
18 Zarrin F, Kaufman S, and Socha J, Droplet size measurements of various nebulizers using differential electrical mobility particle sizer. Journal of Aerosol Science, 1991; 22; S343–S346.
19 Mugele R and Evans H, Droplet size distribution in sprays. Journal of Industrial and Engineering Chemistry, 1951; 43; 1317–1324.
20 Rizk N and Lefebvre A, Spray characteristics of plain-jet airblast atomizers. Journal of Engineering for Gas Turbines and Power, 1984; 106; 634–638.
21 Kahen K, Acon BW, and Montaser A, Modified Nukiyama–Tanasawa and Rizk–Lefebvre models to predict droplet size for microconcentric nebulizers with aqueous and organic solvents. Journal of Analytical Atomic Spectrometry, 2005; 20; 631–637.
22 Porstendörfer J, Gebhart J, and Röbig G, Effect of evaporation on the size distribution of nebulized aerosols. Journal of Aerosol Science, 1977; 8; 371–380.
23 Cresser M and Browner R, A method for investigating size distributions of aqueous droplets in the range 0.5–10 μm produced by pneumatic nebulizers. Spectrochimica Acta Part B: Atomic Spectroscopy, 1980; 35; 73–79.

24 Gorecki T, Lynen F, Szucs R, and Sandra P, Universal response in liquid chromatography using charged aerosol detection. Analytical Chemistry, 2006; 78; 3186–3192.
25 Hutchinson JP, Li J, Farrell W, Groeber E, Szucs R, Dicinoski G, and Haddad PR, Comparison of the response of four aerosol detectors used with ultra high pressure liquid chromatography. Journal of Chromatography A, 2011; 1218; 1646–1655.
26 Davies C, Evaporation of Airborne Droplets, in Shaw DT (ed.) Fundamentals of aerosol science. John Wiley & Sons, Inc., New York, 1978; 135–164.
27 Matsuyama S, Orihara Y, Kinugasa S, and Ohtani H, Effects of densities of brominated flame retardants on the detection response for HPLC analysis with a Corona-charged aerosol detector. Analytical Sciences, 2015; 31; 61–65.
28 Hilal S, Karickhoff S, and Carreira L, Verification and validation of the SPARC model. US Environmental Protection Agency, Washington, DC, 2003.
29 Guiochon G, Moysan A, and Holley C, Influence of various parameters on the response factors of the evaporative light scattering detector for a number of non-volatile compounds. Journal of Liquid Chromatography, 1988; 11; 2547–2570.
30 Lantz MD, Risley DS, and Peterson JA, Simultaneous resolution and detection of a drug substance, impurities, and counter ion using a mixed-mode HPLC column with evaporative light scattering detection. Journal of Liquid Chromatography and Related Technologies, 1997; 20; 1409–1422.
31 Deschamps FS, Baillet A, and Chaminade P, Mechanism of response enhancement in evaporative light scattering detection with the addition of triethylamine and formic acid. Analyst, 2002; 127; 35–41.
32 Sinclair I and Gallagher R, Charged aerosol detection: factors for consideration in its use as a generic quantitative detector. Chromatography Today, 2008; 1; 5–9.
33 Cohen RD, Liu Y, and Gong X, Analysis of volatile bases by high performance liquid chromatography with aerosol-based detection. Journal of Chromatography A, 2012; 1229; 172–179.
34 Russell JJ, Heaton JC, Underwood T, Boughtflower R, and McCalley DV, Performance of charged aerosol detection with hydrophilic interaction chromatography. Journal of Chromatography A, 2015; 1405; 72–84.
35 McConville J, Bullock S, Warner F, and O'Donohue S, Sub-Ambient ELSD for Enhanced Detection of Semi-Volatile Compounds. www.agilent.com/cs/library/posters/Public/ACS%20Fall%202006%20ELSD.pdf (accessed April 6, 2017).
36 Megoulas NC and Koupparis MA, Enhancement of evaporative light scattering detection in high-performance liquid chromatographic determination of neomycin based on highly volatile mobile phase, high-molecular-mass ion-pairing reagents and controlled peak shape. Journal of Chromatography A, 2004; 1057; 125–131.

37 Keller BO, Sui J, Young AB, and Whittal RM, Interferences and contaminants encountered in modern mass spectrometry. Analytica Chimica Acta, 2008; 627; 71–81.

38 Premnath V, Oberreit D, and Hogan Jr CJ, Collision-based ionization: bridging the gap between chemical ionization and aerosol particle diffusion charging. Aerosol Science and Technology, 2011; 45; 712–726.

39 Vehring R, Foss WR, and Lechuga-Ballesteros D, Particle formation in spray drying. Journal of Aerosol Science, 2007; 38; 728–746.

40 Paudel A, Worku ZA, Meeus J, Guns S, and Van den Mooter G, Manufacturing of solid dispersions of poorly water soluble drugs by spray drying: formulation and process considerations. International Journal of Pharmaceutics, 2013; 453; 253–284.

41 Pankow JF, An absorption model of the gas/aerosol partitioning involved in the formation of secondary organic aerosol. Atmospheric Environment, 1994; 28; 189–193.

42 Booth A, Montague W, Barley M, Topping D, McFiggans G, Garforth A, and Percival C, Solid state and sub-cooled liquid vapour pressures of cyclic aliphatic dicarboxylic acids. Atmospheric Chemistry and Physics, 2011; 11; 655–665.

43 Schnitzler EG and McDonald KM, Characterization of low-temperature vapour pressure estimates for secondary organic aerosol applications. Atmospheric Environment, 2012; 56; 9–15.

44 Fissan H, Neumann S, Trampe A, Pui D, and Shin W, Rationale and principle of an instrument measuring lung deposited nanoparticle surface area. Journal of Nanoparticle Research, 2007; 9; 53–59.

45 Flagan RC, History of electrical aerosol measurements. Aerosol Science and Technology, 1998; 28; 301–380.

46 Pui D, Fruin S, and McMurry P, Unipolar diffusion charging of ultrafine aerosols. Aerosol Science and Technology, 1988; 8; 173–187.

47 Adachi M, Kousaka Y, and Okuyama K, Unipolar and bipolar diffusion charging of ultrafine aerosol particles. Journal of Aerosol Science, 1985; 16; 109–123.

48 Biskos G, Reavell K, and Collings N, Unipolar diffusion charging of aerosol particles in the transition regime. Journal of Aerosol Science, 2005; 36; 247–265.

49 Boisdron Y and Brock J, On the stochastic nature of the acquisition of electrical charge and radioactivity by aerosol particles. Atmospheric Environment (1967), 1970; 4; 35–50.

50 Pui DY-H, Experimental study of diffusion charging of aerosols. ERDA, Washington, DC, 1976.

51 Fuchs N, On the stationary charge distribution on aerosol particles in a bipolar ionic atmosphere. Geofisica Pura e Applicata, 1963; 56; 185–193.

52 Kolakowski BM, Grossert JS, and Ramaley L, Studies on the positive-ion mass spectra from atmospheric pressure chemical ionization of gases and solvents used in liquid chromatography and direct liquid injection. Journal of the American Society for Mass Spectrometry, 2004; 15; 311–324.

53 Sabo M and Matejčík S, Corona discharge ion mobility spectrometry with orthogonal acceleration time of flight mass spectrometry for monitoring of volatile organic compounds. Analytical Chemistry, 2012; 84; 5327–5334.
54 Sekimoto K and Takayama M, Fundamental processes of corona discharge. Journal of the Institute of Electrostatics Japan, 2009; 33; 38–42.
55 Maißer A, Thomas JM, Larriba-Andaluz C, He S, and Hogan CJ, The mass–mobility distributions of ions produced by a Po-210 source in air. Journal of Aerosol Science, 2015; 90; 36–50.
56 Jung H and Kittelson DB, Characterization of aerosol surface instruments in transition regime. Aerosol Science and Technology, 2005; 39; 902–911.
57 Woo K-S, Chen D-R, Pui DY, and Wilson WE, Use of continuous measurements of integral aerosol parameters to estimate particle surface area. Aerosol Science and Technology, 2001; 34; 57–65.
58 Li L, Chen D-R, and Tsai P-J, Use of an electrical aerosol detector (EAD) for nanoparticle size distribution measurement. Journal of Nanoparticle Research, 2009; 11; 111–120.
59 Kaufman S, Medved A, Pöcher A, Hill N, Caldow R, and Quant F. An electrical aerosol detector based on the corona-jet charger. in Poster PI2-07, Abstracts of the 21st Annual American Association for Aerosol Research (AAAR) Conference, Charlotte, NC, October 7–11, 2002.
60 Shin WG, Qi C, Wang J, Fissan H, and Pui DY, The effect of dielectric constant of materials on unipolar diffusion charging of nanoparticles. Journal of Aerosol Science, 2009; 40; 463–468.
61 Li L, Chen D-R, and Tsai P-J, Evaluation of an electrical aerosol detector (EAD) for the aerosol integral parameter measurement. Journal of Electrostatics, 2009; 67; 765–773.
62 Kim JH, Mulholland GW, Kukuck SR, and Pui DY, Slip correction measurements of certified PSL nanoparticles using a nanometer differential mobility analyzer (nano-DMA) for Knudsen number from 0.5 to 83. Journal of Research of the National Institute of Standards and Technology, 2005; 110; 31–54.
63 Mourey TH and Oppenheimer LE, Principles of operation of an evaporative light-scattering detector for liquid chromatography. Analytical Chemistry, 1984; 56; 2427–2434.
64 Oppenheimer LE and Mourey TH, Examination of the concentration response of evaporative light-scattering mass detectors. Journal of Chromatography A, 1985; 323; 297–304.
65 Van der Meeren P, Vanderdeelen J, and Baert L, Simulation of the mass response of the evaporative light scattering detector. Analytical Chemistry, 1992; 64; 1056–1062.
66 Hazotte A, Libong D, Matoga M, and Chaminade P, Comparison of universal detectors for high-temperature micro liquid chromatography. Journal of Chromatography A, 2007; 1170; 52–61.

67 Hutchinson JP, Remenyi T, Nesterenko P, Farrell W, Groeber E, Szucs R, Dicinoski G, and Haddad PR, Investigation of polar organic solvents compatible with Corona Charged Aerosol Detection and their use for the determination of sugars by hydrophilic interaction liquid chromatography. Analytica Chimica Acta, 2012; 750; 199–206.

68 Vervoort N, Daemen D, and Török G, Performance evaluation of evaporative light scattering detection and charged aerosol detection in reversed phase liquid chromatography. Journal of Chromatography A, 2008; 1189; 92–100.

69 Kou D, Manius G, Zhan S, and Chokshi HP, Size exclusion chromatography with Corona charged aerosol detector for the analysis of polyethylene glycol polymer. Journal of Chromatography A, 2009; 1216; 5424–5428.

70 Wipf P, Werner S, Twining LA, and Kendall C, HPLC determinations of enantiomeric ratios. Chirality, 2007; 19; 5–9.

71 Takahashi K, Kinugasa S, Senda M, Kimizuka K, Fukushima K, Matsumoto T, Shibata Y, and Christensen J, Quantitative comparison of a corona-charged aerosol detector and an evaporative light-scattering detector for the analysis of a synthetic polymer by supercritical fluid chromatography. Journal of Chromatography A, 2008; 1193; 151–155.

72 Shaodong J, Lee WJ, Ee JW, Park JH, Kwon SW, and Lee J, Comparison of ultraviolet detection, evaporative light scattering detection and charged aerosol detection methods for liquid-chromatographic determination of anti-diabetic drugs. Journal of Pharmaceutical and Biomedical Analysis, 2010; 51; 973–978.

73 Merle C, Laugel C, Chaminade P, and Baillet-Guffroy A, Quantitative study of the stratum corneum lipid classes by normal phase liquid chromatography: comparison between two universal detectors. Journal of Liquid Chromatography and Related Technologies, 2010; 33; 629–644.

74 Righezza M and Guiochon G, Effects of the nature of the solvent and solutes on the response of a light-scattering detector. Journal of Liquid Chromatography, 1988; 11; 1967–2004.

75 Righezza M and Guiochon G, Effect of the wavelength of the laser beam on the response of an evaporative light scattering detector. Journal of Liquid Chromatography, 1988; 11; 2709–2729.

76 Koropchak JA, Sadain S, Yang X, Magnusson L-E, Heybroek M, Anisimov M, and Kaufman SL, Peer reviewed: nanoparticle detection technology for chemical analysis. Analytical Chemistry, 1999; 71; 386A–394A.

77 Koropchak J, Heenan C, and Allen L, Direct comparison of evaporative light-scattering and condensation nucleation light-scattering detection for liquid chromatography. Journal of Chromatography A, 1996; 736; 11–19.

78 Hering SV, Stolzenburg MR, Quant FR, Oberreit DR, and Keady PB, A laminar-flow, water-based condensation particle counter (WCPC). Aerosol Science and Technology, 2005; 39; 659–672.

79 Kupc A, Bischof O, Tritscher T, Beeston M, Krinke T, and Wagner PE, Laboratory characterization of a new nano-water-based CPC 3788 and performance comparison to an ultrafine butanol-based CPC 3776. Aerosol Science and Technology, 2013; 47; 183–191.
80 Liu W, Kaufman SL, Osmondson BL, Sem GJ, Quant FR, and Oberreit DR, Water-based condensation particle counters for environmental monitoring of ultrafine particles. Journal of the Air and Waste Management Association, 2006; 56; 444–455.
81 Konermann L, Ahadi E, Rodriguez AD, and Vahidi S, Unraveling the mechanism of electrospray ionization. Analytical Chemistry, 2012; 85; 2–9.
82 Kebarle P and Verkerk UH, Electrospray: from ions in solution to ions in the gas phase, what we know now. Mass Spectrometry Reviews, 2009; 28; 898–917.
83 Kaufman SL, Electrospray diagnostics performed by using sucrose and proteins in the gas-phase electrophoretic mobility molecular analyzer (GEMMA). Analytica Chimica Acta, 2000; 406; 3–10.
84 Herrera LC, Grossert JS, and Ramaley L, Quantitative aspects of and ionization mechanisms in positive-ion atmospheric pressure chemical ionization mass spectrometry. Journal of the American Society for Mass Spectrometry, 2008; 19; 1926–1941.
85 King R, Bonfiglio R, Fernandez-Metzler C, Miller-Stein C, and Olah T, Mechanistic investigation of ionization suppression in electrospray ionization. Journal of the American Society for Mass Spectrometry, 2000; 11; 942–950.
86 Holčapek M, Jirásko R, and Lísa M, Recent developments in liquid chromatography–mass spectrometry and related techniques. Journal of Chromatography A, 2012; 1259; 3–15.

2

Charged Aerosol Detection

A Literature Review

Ian N. Acworth and William Kopaciewicz

Thermo Fisher Scientific, Chelmsford, MA, USA

CHAPTER MENU

2.1 Introduction

The most common detection technique for HPLC is UV/Vis spectrophotometry, due to its sensitivity, wide linear dynamic range, large range of analytes detected, and gradient method compatibility. However, this approach is most applicable when the analyte possesses a chromophore. For those that do not possess a chromophore, other detectors are used including refractive index (RI), mass spectrometry (MS), evaporative light scattering, and charged aerosol. All these approaches have some limitations. RI is easy to use but lacks sensitivity and is not gradient compatible and thus limited to less complex samples. MS, depending on the type being used, can be difficult to operate, requires that the compound be capable of forming gas-phase ions, and, without the availability of labeled standards, may not be quantitative. The evaporative light scattering detector (ELSD), one of the most widely used "universal" detectors, is nonlinear, relatively insensitive, has a fairly narrow linear dynamic range, and can show significant inter-analyte response variability [1, 2]. Charged aerosol detection (CAD) is also nonlinear but shows good sensitivity and offers advantages relative to ELSD and related HPLC detection technologies (e.g., condensation nucleation light scattering detector (CNLSD)) (see Table 2.1).

Charged Aerosol Detection for Liquid Chromatography and Related Separation Techniques, First Edition. Edited by Paul H. Gamache.

Table 2.1 Comparison of CAD performance to other detectors for different applications/markets.

Title	Application area	Overview	Conclusions	References
Comparison of two evaporative universal detectors for the determination of sugars in food samples by liquid chromatography	Food	Simple sugars. Fructose, glucose, maltose, lactose, and maltotriose in different sauces, confectionary, and dairy products were resolved on an NH_2-Kromasil® column (EKA Chemicals AB) using aqueous ACN (70%) and detected using either CAD or evaporative light scattering detection (ELSD) (Eurosep)	CAD was more sensitive than ELSD, but ELSD was slightly better for precision, repeatability, and reproducibility. ELSD acquisition and maintenance costs were suggested to be lower than with CAD	[3]
Charged aerosol detection and ELSD—fundamental differences affecting analytical performance	General methodology	Review. Comparison of performance between CAD and ELSD	CAD showed superior performance to ELSD including better sensitivity, wider linear dynamic range, improved reproducibility, and less inter-analyte response variability and was reported to be easier to use	[4]
Analysis of volatile bases by high-performance liquid chromatography (HPLC) with aerosol-based detection	General methodology	Extending range of analytes measured. Development of methods to detect low levels of volatile bases through the formation of nonvolatile salts. Twelve volatile bases that show weak ultraviolet (UV) absorbance response were resolved on a Sequant® ZIC®-pHILIC column (EMD Millipore Corporation) using a TFA/ACN/water (0.04 : 60 : 40 v/v/v) mobile phase and detection by either CAD or Nano Quantity Analyte Detector (NQAD™—Quant Technologies—a CNLSD)	CAD exhibited greater band broadening than NQAD, but CAD was more sensitive (LODs by 2.5 lower than NQAD)	[5]
HPLC determinations of enantiomeric ratios	General methodology	Chiral measurement. Accurate determination of enantiomeric ratios of non-UV active compounds. Racemic methyl carbamate rac-3 was analyzed using a Chiralcel® OD-H column (Daicel Corporation) using a hexane/isopropanol (95 : 5 v/v) mobile phase and detected by either CAD or ELSD	The nonlinear nature and narrow dynamic range of ELSD caused issues when measuring enantiomeric ratios (ER). A compound with an ER of 95 : 5 would appear to be pure by ELSD. No issues were found when using CAD	[6]

Comparison of the response of four aerosol detectors used with ultrahigh-pressure liquid chromatography	General methodology	Response characteristics of aerosol detectors. The analytical performance of CAD, ELSD (Varian), and NQAD was compared for a range of analytes using both flow-injection analysis and HPLC separation on an Acclaim® PolarAdvantage II column (Thermo Scientific) with an aqueous ACN gradient	ELSD and NQAD gave poorest precision. CAD and NQAD were less sensitive than UV under aqueous mobile phase conditions, but both were more sensitive than UV when using 80% ACN. The reproducibility of the detector response for 11 analytes over 10 consecutive separations was found to be ~5% for the charged aerosol detectors and ~11% for the light scattering detectors. The tested analytes included semivolatile species, which exhibited a more variable response on the aerosol detectors	[7]
Ion pair reversed-phase liquid chromatography with UV detection for analysis of UV transparent cations	General methodology	UV detection of non-chromophore analytes. The authors describe the use of an anionic ion-pair reagent (IPR) to improve the UV detection and hydrophobic retention of polar and UV transparent cations	Anionic IPR added to the mobile phase forms an ion pair with cations that upon formation of the ion pair causes a redshift in the absorption of wavelength, making it possible for direct UV detection cations that lack a chromophore. The LOQs of this approach, depending on the analyte, is either equivalent to or up to 10-fold better than measurement by CAD	[8]
Quantitative study of the stratum corneum lipid classes by normal-phase liquid chromatography: comparison between two universal detectors	Lipids	Fatty acids, ceramides, and cholesterol. Lipids extracted from the stratum corneum of human skin were measured using gradient HPLC with either ELSD (Eurosep) or CAD. Analytes were resolved on a YMC-Pack PVA-Sil column (YMC America, Inc.) using a heptane/chloroform—acetone gradient	External standardization was found to be suitable for a quantitative study of different lipid classes, using a representative standard for each class. CAD offered better sensitivity, repeatability, precision, and accuracy than ELSD	[9]
Comparison between charged aerosol detection and light scattering detection for the analysis of *Leishmania* membrane phospholipids	Lipids	Phospholipids. Phospholipid classes from *Leishmania* membranes were resolved on a YMC-Pack PVA-Sil column (YMC America, Inc.) using a *n*-heptane/isopropanol, chloroform/isopropanol, and aqueous methanol gradient and detected using either CAD or ELSD (Eurosep)	CAD was more sensitive and more precise than ELSD at the lower end of the calibration curve. LODs and LOQs were typically threefold lower with CAD than ELSD	[10]

(*Continued*)

Table 2.1 (Continued)

Title	Application area	Overview	Conclusions	References
Simple and efficient profiling of phospholipids in phospholipase D-modified soy lecithin by HPLC with charged aerosol detection	Lipids	Phospholipids. An HPLC-CAD method to measure six different classes of phospholipids in soy lecithin was developed. Analytes were separated on a Luna® silica column (Phenomenex® Inc.) using normal-phase conditions. Analytical performance was compared to HPLC-ELSD	The HPLC-CAD approach had a 10-fold broader linear range, lower LOD, and better precision and was able to detect low levels of lyso-phosphatidyl-inositol	[11]
Comparison between ELSD and charged aerosol detection for the analysis of saikosaponins	Natural products	Saikosaponins. Analysis of the botanical, *Bupleurum falcatum*. Ten saikosaponins were resolved on an Ascentis® Express C18 column (Supelco) using a gradient of eluent A (90% ACN) and eluent B (10% ACN) and detected by either CAD or Sedex ELSD (Sedere)	CAD had a wider linear dynamic range, higher sensitivity, and better reproducibility than ELSD. CAD was also found to be easier to use. The high temperature of the drift tube required by ELSD can be an issue for thermally labile analytes	[12]
Comparison of UV detection and charged aerosol detection methods for liquid chromatographic determination of protoescigenin	Natural products	Saponins. Analysis of escin, a complex mixture of pentacyclic triterpene saponins from the horse chestnut, *Aesculus hippocastanum*. Analytes were resolved on an Acquity UPLC® BEH C18 column (Waters) using ACN and 0.1% acetic acid (30:70 v/v) and detected using UV-CAD in series	The sensitivity of CAD was slightly greater than UV detection. CAD and UV showed linear response, both for narrow and a broad range of concentrations. CAD and UV were shown to be complementary with CAD being used to measure analytes with weak chromophores and DAD providing peak purity data	[13]
Performance evaluation of charged aerosol and ELSD for the determination of ginsenosides by LC	Natural products	Saponins. Analysis of the botanical, *Panax ginseng*. Seven triterpenoid saponins (ginsenosides) were resolved on a Zorbax Extend C18 column (Agilent Technologies), using a water–ACN gradient and detected by either CAD or ELSD (Sedere)	CAD was more sensitive than ELSD. The linearity of CAD was somewhat wider than ELSD. The intraday and interday precision by CAD was better than ELSD and closer to UV	[14]
Polyketide analysis using mass spectrometry (MS), ELSD, and charged aerosol detector systems	Natural products	Antibiotics. MS, ELSD (Alltech Associates), and CAD were evaluated for the measurement of 6-deoxyerythronolide B, a polyketide precursor to the antibiotic erythromycin	CAD showed the lowest LOD of all three detectors and a comparable dynamic range to MS. The authors noted that CAD was a viable alternative to MS for the analysis of polyketide production schemes with low titers	[15]

Comparison of two aerosol-based detectors for the analysis of gabapentin in pharmaceutical formulations by hydrophilic interaction chromatography	Pharmaceutical/ biopharmaceutical	Drug analysis. Four HILIC columns, coupled to either ELSD (Alltech Associates) or CAD, were evaluated for measurement of gabapentin in commercial tablets and capsules	Using HILIC conditions the ELSD was comparable with CAD in terms of linearity, sensitivity, precision, and accuracy. Using reversed-phase conditions CAD was much more sensitive than ELSD	[16]
Comparison of UV detection, ELSD, and charged aerosol detection methods for liquid chromatographic determination of antidiabetic drugs	Pharmaceutical/ biopharmaceutical	Drug analysis. Several chromophore-containing antidiabetic drugs were resolved on a GraceSmart™ RP-18 column using an aqueous ACN gradient and detected by UV, ELSD (Alltech), and CAD	CAD showed consistent inter-analyte response (UV and ELSD did not) CAD showed the best accuracy and LOD of all three detectors with similar precision to UV	[17]
Hydrophilic interaction chromatography with aerosol-based detectors (ELSD, CAD, NQAD) for polar compounds lacking a UV chromophore in an intravenous formulation. Journal of pharmaceutical and biomedical analysis	Pharmaceutical/ biopharmaceutical	Formulation. Characterization of an intravenous formulation. Three columns—a TSKgel Amide-80 (Tosoh Bioscience, LLC), a Sequant® ZIC®-HILIC (EMD Millipore Corporation), and a Trinity® Acclaim™ P1 (Thermo Scientific)—and three detectors, ELSD (Alltech Associates), CAD, and NQAD (Quant Technologies), were evaluated for reproducibility, linearity, and LOD	The HILIC-ELSD approach showed excellent linearity, accuracy, precision, specificity, robustness, and stability. CAD and NQAD were used for a subset of validation experiments. Due to their greater sensitivity, samples required dilution prior to analysis	[18]
Analysis of ionic surfactants by HPLC with ELSD and charged aerosol detection	Pharmaceutical/ biopharmaceutical	Formulation. Analysis of cationic and anionic surfactants. The resolution of five cationic and seven ionic surfactants was evaluated using different gradient mobile phase conditions on either an Acclaim® Surfactant (Thermo Scientific) or a Capcell Pak C18 (Shiseido) column. Analytes were detected using either CAD or ELSD (Alltech Associates)	CAD was ~1.4–2 times more sensitive than ELSD for both cationic and anionic surfactants. CAD showed better linearity than ELSD	[19]

(*Continued*)

Table 2.1 (Continued)

Title	Application area	Overview	Conclusions	References
Aerosol-based detectors for the investigation of phospholipid hydrolysis in a pharmaceutical suspension formulation	Pharmaceutical/ biopharmaceutical	Formulation. An HPLC method was developed to quantify free fatty acids in a pharmaceutical suspension formulated with phospholipids as stabilizing agents. Free fatty acids were resolved on a Zorbax SB-C18 column (Agilent Technologies) using an aqueous ACN (80%) mobile phase containing 0.1% acetic acid and detected by either ELSD (Alltech Associates) or CAD	CAD was superior to ELSD with respect to sensitivity, precision, recovery, and linearity	[20]
Quantitative comparison of a Corona-charged aerosol detector and an evaporative light scattering detector for the analysis of a synthetic polymer by supercritical fluid chromatography	Pharmaceutical/ biopharmaceutical	Formulation, excipient analysis. The performance of supercritical fluid chromatography with either ELSD (Alltech Associates) or CAD was evaluated using polyethylene glycol (PEG)-certified reference material and a well-defined equimass mixture of uniform PEG oligomers	The CAD approach was able to detect a 10-fold more dilute solution of uniform PEG oligomers than ELSD. Molecular mass data by ELSD was 4.6% lower than the certified value of PEG 1000. CAD results were virtually the same as the certified value	[21]
Size-exclusion chromatography with Corona-charged aerosol detector for the analysis of PEG polymer	Pharmaceutical/ biopharmaceutical	Formulation, excipient analysis. A size-exclusion chromatographic-CAD method was developed for measurement of PEG purity and polydispersity, and its performance was compared with methods using refractive index (RI) (Waters) or ELSD (Polymer Labs)	CAD was found to provide more accurate impurity and polydispersity profiles of PEG reagent lots that better differentiate their quality. RI lacked sensitivity and ELSD underestimated both impurity levels and polydispersity	[22]
Quantitative determination of nonionic surfactants with CAD	Pharmaceutical/ biopharmaceutical	Formulation, excipient analysis. Tween 80 and Span 80 profiles were evaluated using HPLC and UHPLC conditions with detection by either CAD or ELSD	CAD was 10 times more sensitive and showed a wider dynamic range	[23]

Control of impurities in L-aspartic acid and L-alanine by HPLC coupled with a Corona-charged aerosol detector	Pharmaceutical/ biopharmaceutical	Impurity testing. Fully validated method. Analytes were resolved on an Intersil® ODS 3 column (GL Sciences) using an aqueous methanol mobile phase containing perfluoroheptanoic acid. Analytes were detected using either CAD or ELSD (Polymer Labs)	The sensitivity of CAD was 3.6–42 times more sensitive than ELSD. The authors concluded that ELSD was not suitable for this method	[24]
Performance evaluation of ELSD and charged aerosol detection in reversed-phase liquid chromatography	Pharmaceutical/ biopharmaceutical	Impurity testing. Comparison of CAD, ELSD, and UV performance. Ten test analytes were resolved on an X-Bridge C18 column (Waters) using an ammonium acetate/ACN gradient. Analytes were detected using DAD, ELSD (Waters), or CAD	CAD was up to sixfold more sensitive than ELSD ELSD was somewhat less repeatable than CAD. CAD was reported to be more user friendly. Mass response factors were more consistent with CAD than UV	[25]
Evaluation of charged aerosol detection (CAD) as a complementary technique for high-throughput LC-MS-UV-ELSD analysis of drug discovery screening libraries	Pharmaceutical/ biopharmaceutical	Library screening. Single calibrant quantification. The authors used an LC-MS-UV-CAD platform to generate a linear calibration curve using three structurally diverse compounds with similar retention times. The calibration curve was then applied to a set of 20 chemically diverse samples	ELSD (Sedex) could not be used for this study due to sigmoidal calibration curves. CAD data from generalized calibration curves was sufficiently accurate to confirm that analytes were present in both the quantities and purities necessary to generate meaningful biological screening results	[26]

2.2 CAD History and Background

Charged aerosol detection involves the nebulization of the liquid stream coming from the HPLC column to form droplets, which are then desolvated to produce dried particles. Particles are charged, and the charge is measured using an electrometer (see Chapter 1 for greater detail of nebulization, particle formation, particle charging, and charge measurement). CAD uses an electrical charging method based on TSI Inc.'s (St. Paul, MN) electrical aerosol detector (EAD) [27]. The detector was codeveloped by ESA Biosciences Inc. (Chelmsford, MA) and TSI Inc. and was introduced to the scientific community at the Pittsburgh Conference in 2005. It was awarded a Silver Pittcon Editors' Award for best new product and later in 2005 was selected for a prestigious R&D 100 Award. A similar approach based on TSI's electrical aerosol analyzer, but called aerosol charge detection, was developed independently by Dixon and Peterson [28].

Continued research and engineering improvements in product design led to the evolution of a range of CADs with ever increasing capabilities (Table 2.2) (Figure 2.1).

Table 2.2 Charged aerosol detector development.

Model	Date range	Capabilities	Enhancement(s)
Corona® CAD®	2005–2013	First commercial charged aerosol detector for HPLC, Full control via front panel	Designed for use on any HPLC
Corona® *Plus*	2006–2010	Expanded solvent compatibility software drivers for many CDS systems	Heated nebulizer to handle water/THF gradients and external gas conditioning module
Corona® ultra	2009–2011	Stackable design, enhanced sensitivity, real-time display of chromatogram	Compatible with UHPLC and precision internal gas pressure regulation added
Corona® ultra® RS	2011–2013	Integrated into Dionex UltiMate 3000 LC platform, incorporation of onboard diagnostics and monitoring	Flow diversion system to eliminate waste overflow, power function algorithm for data linearization
Corona™ Veo™	2013–present	Expanded flow rate range; enhanced sensitivity	Total redesign including concentric nebulizer, heated evaporation, electronic gas regulation system
Vanquish™ Charged Aerosol detector	2015–present	Designed for direct integration into Vanquish UHPLC platform	New electronic control, slide-in module design, reduced flow path for optimum operation

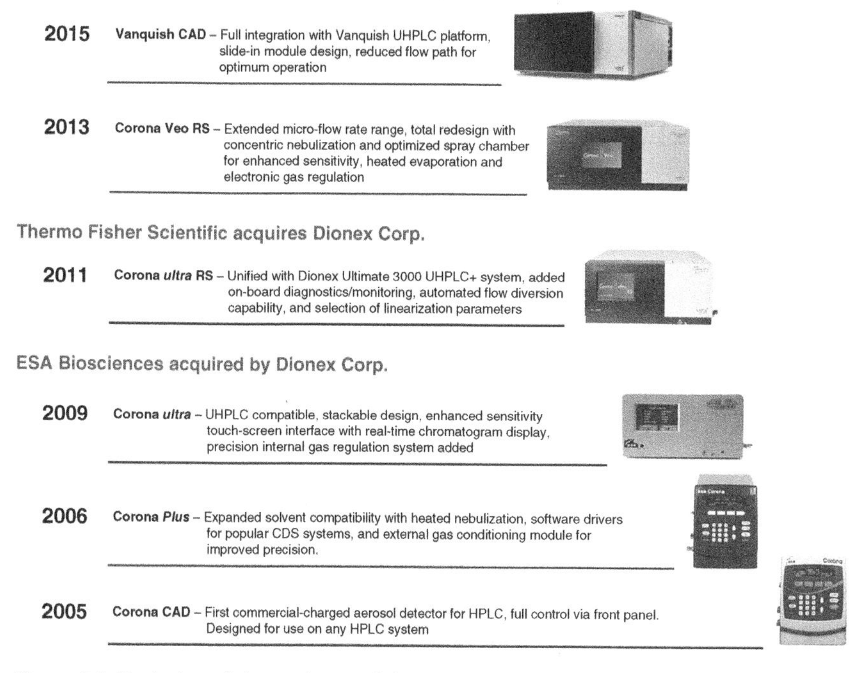

Figure 2.1 Evolution of charged aerosol detectors.

There are several publications that review the operating principles and performance of CAD and other aerosol-based detectors (Table 2.3). It should be noted that all of these references discuss the original design of CAD (Corona CAD, Corona *Plus*, Corona ultra, and Corona ultra RS) with its cross-flow pneumatic nebulizer and impactor. As presented in Table 2.2, with the introduction of the Corona Veo, CAD went through a major redesign of many of its key components. For example, it now uses a concentric (as opposed to cross-flow) nebulizer and a modified spray chamber. The operating principles of Corona Veo and Vanquish CADs and the performance impact of detector redesign are discussed further in Chapter 1.

Charged aerosol detection can be used with a variety of different chromatographic approaches including HPLC with reversed-phase (e.g., [34–38]), normal-phase [38–42], HILIC (e.g., [29, 43–47]), or mixed-mode techniques [48–53]; size-exclusion chromatography [22, 54, 55]; and supercritical fluid chromatography [21, 56, 57]. A prerequisite for CAD, ELSD, and MS is that the mobile phase *must be volatile.* A wide variety of solvents can be used including but not limited to water, acetonitrile, methanol, isopropanol, acetone, tetrahydrofuran, heptane, and chloroform. Solvent quality is essential, and any solvent (organic or aqueous) used must be free from particulates, and those with minimal residue after evaporation are preferred. Failure to address this impacts detector performance due to increased background noise [58, 59]. Volatile buffers are often used including ammonium formate and ammonium acetate. Volatile

Table 2.3 Review publications of the operating principles and performance of CAD and other nebulizer-based detectors.

Title	Overview	References
Advances in aerosol-based detectors	An in-depth review of different aerosol-based detectors including ELSD, CAD, the nucleation light scattering detector (CNLSD), and chemiluminscence aerosol detector (CLAD)—how they work, comparison of performance, limitations, response uniformity, and selected applications are discussed	[1]
Review of operating principle and applications of the charged aerosol detector	Background to the operating principles of CAD and a review of some selected applications	[29]
Aerosol-based detectors for liquid chromatography	Operating principles of ELSD, condensation (CNLSD), and CAD highlights their analytical performance and factors affecting response and discusses selected applications	[30]
Evaporative light scattering and charged aerosol detector	Theory of ELSD operation, parameters that influence ELSD response, some applications, and compare performance to CAD	[31]
Charged aerosol detection and ELSD—fundamental differences affecting analytical performance	Compares the analytical performance of CAD and ELSD including sensitivity, linear dynamic range, inter-analyte response variability, and precision	[2]
Universal response in liquid chromatography using charged aerosol detection	The authors discuss the effect of mobile phase composition on response of aerosol-based detectors and show that an inverse gradient applied post-column normalizes analyte response. Analysis of six sulfonamide drugs by gradient HPLC-CAD, with or without gradient compensation, was used as an example	[32]
Comparison of the sensitivity of evaporative universal detectors and LC/MS in the HILIC and the reversed-phase HPLC modes	The sensitivity of different aerosol-based detectors when used with reversed-phase (RP) or HILIC conditions were evaluated. As expected the higher organic composition of HILIC mobile phases led to improved sensitivity when compared with RP conditions, but the different detectors were affected to different degrees. ELSD was slightly more sensitive with HILIC than RP. CAD showed a 10-fold improvement and ESI-MS showed a 5–10-fold improvement when comparing HILIC with RP conditions	[33]

ion-pairing agents include trifluoroacetic acid and perfluoroheptanoic acid. In general, any mobile phase used with LC-MS can be used with CAD. Recently, the use of suppressors now expands the range of applications to include nonvolatile buffers for ion-exchange chromatography [60, 61]. Figure 2.2 presents an example application of anion-exchange CAD with sodium hydroxide eluent for simple analysis carbohydrates using online desalting. Figure 2.3 shows gradient elution with online desalting for the measurement of dahlia inulins.

The response of aerosol-based detectors is affected by the organic content of the mobile phase entering the detector during gradient operation, a result of changes in nebulization and aerosol transport efficiency due to reduced eluent surface tension and viscosity and higher volatility. The response for an analyte eluting during the aqueous portion of the gradient is typically suppressed relative to one eluting with higher amounts of organic solvent in the mobile phase. There are a couple of approaches to minimize this effect to normalize analyte response across the gradient. First, a constant amount of organic solvent can be added to a post-column, pre-detector using a "T"-connection. This approach can also increase the analyte signal during isocratic separations, but there is a risk of causing precipitation [62]. Second, a precise inverse gradient, sometimes called a compensatory gradient, flowing through the post-column "T"-connector ensures that the detector always experiences a consistent mobile phase composition. The second approach (Figure 2.4), although more complicated, largely mitigates the solvent effect and ensures similar response factors across the gradient [32, 37, 63, 64].

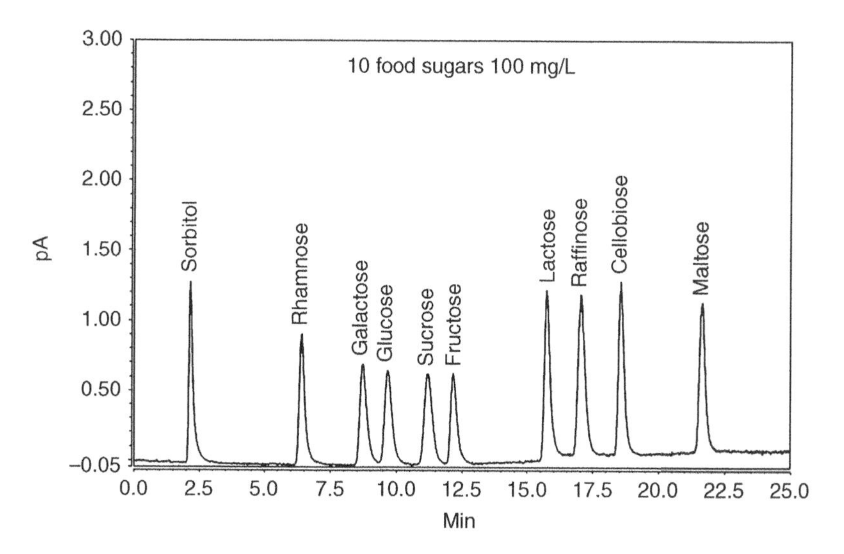

Figure 2.2 Separation of simple carbohydrates by ion-exchange chromatography (CarboPac PA20 Column—Thermo Scientific Dionex) using a sodium hydroxide gradient. Carbohydrates were measured by CAD following mobile phase desalting by a CMD 300 (Carbohydrate Membrane Desalter 300) with an RFC-10 Reagent-Free™ Controller (Thermo Scientific Dionex).

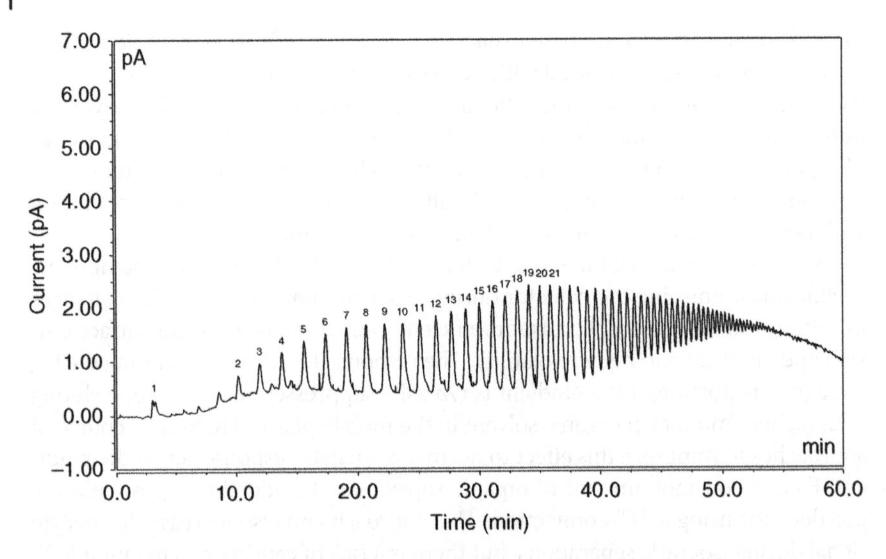

Figure 2.3 Separation of dahlia inulins by ion-exchange chromatography (CarboPac PA100 Column—Thermo Scientific Dionex) using sodium acetate–sodium hydroxide gradient. Carbohydrates were measured by CAD following mobile phase desalting by a CMD 300 (Carbohydrate Membrane Desalter 300), with a RFC-10 Reagent-Free™ Controller (Thermo Scientific Dionex).

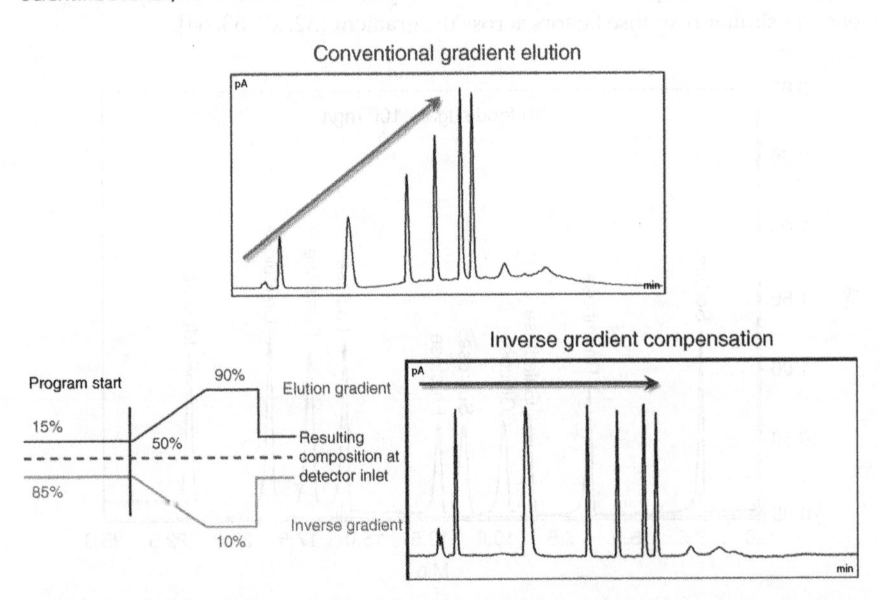

Figure 2.4 Analyte response factors change during conventional gradient elution due to altered nebulization efficiency. Inverse gradient compensation ensures that the mobile phase composition entering the nebulizer is consistent, so that analyte response factors are similar independent of elution time. *Source:* Reproduced with permission of Thermo Fisher Scientific Inc.

2.3 Application Areas

Charged aerosol detection is typically used to measure compounds that lack a chromophore or are not capable of being ionized. In this section we have chosen three groups of compounds—carbohydrates, lipids, and natural products, which have such characteristics—to illustrate how CAD is being used to address analytical challenges. We then discuss references regarding how CAD has been used to meet the analytical needs of different industries.

2.3.1 Carbohydrates

Carbohydrates are highly polar and generally exhibit weak UV absorption, making their analysis by reversed-phase HPLC-UV challenging. Anion-exchange chromatography with pulsed amperometric detection using basic conditions is frequently used to measure carbohydrates but typically requires a dedicated, metal-free ion chromatography system. Sodium hydroxide mobile phases readily absorb carbon dioxide forming carbonate that can adversely affect analyte retention time, requiring that mobile phase be freshly prepared or formed *in situ* using online eluent generation. Furthermore, as sodium hydroxide is not volatile, the mobile phase must undergo desalting if samples are to be analyzed using MS detection. More recently, with advances in column technology, HILIC-CAD approaches that predominantly use organic mobile phases are becoming more common. An overview of the publications using CAD for carbohydrate measurement is presented in Table 2.4. This topic is discussed in more detail in Chapter 7.

2.3.2 Lipids

Lipids are hydrophobic or amphiphilic small molecules showing a wide and diverse range of structures. For convenience, lipids can be classified into different categories, sometimes referred to as classes or families (e.g., fatty acids, phospholipids, sterols, triglycerides, polyketides, and carotenoids), some of which can be further refined (or speciated) based on their composition (e.g., although triglycerides are composed of one glycerol molecule and three fatty acids molecules, differences in fatty acid chain length, degree and position of unsaturation, and position on the glycerol backbone result in numerous triglyceride molecules). Numerous analytical approaches can be used to measure lipids including TLC, NMR, HPLC, MS, and GC approaches [81, 82]. GC-MS have been used for many years to measure a wide variety of lipids in many different sample matrices. Although GC-based approaches can offer phenomenal chromatographic resolution, sample preparation (including derivatization to render analytes volatile) and issues with measuring labile analytes can be a problem. HPLC separations of lipids tend to fall into two categories. Normal-phase HPLC is used to separate lipid classes, while reversed-phase HPLC is

Table 2.4 An overview of CAD publications relevant to carbohydrates.

Title	Application area	Overview	References
Characterization of an endoglucanase belonging to a new subfamily of glycoside hydrolase family 45 of the basidiomycete *Phanerochaete chrysosporium*. Applied and environmental microbiology	Agriculture	Cellulose decomposition. Cellulose is the most abundant biopolymer on earth, and its degradation is part of the natural carbon cycle. In nature, cellulose is degraded by microorganisms using a number of cellulose hydrolyzing enzymes, for example, cellobiohydrolases, endoglucanases, and β-glucosidases. Until this publication, the enzymes used by the wood decay fungus *Phanerochaete chrysosporium* were uncertain. HILIC-CAD was used to study the activity of the glycoside hydrolase family 45 endoglucanase, which the authors identified as a major hydrolytic enzyme of this fungus	[46]
Structural and biochemical analyses of glycoside hydrolase family 26 β-mannanase from a symbiotic protist of the termite *Reticulitermes speratus*	Agriculture	Cellulose decomposition. Symbiotic protists located in termite gut contribute to degradation of lignocellulosic biomass. A novel β-mannase is responsible for degradation of glucomannan, a major component of this biomass. The authors used HILIC-CAD to study carbohydrates released during the enzymatic degradation process to better understand the mechanism underlying heteropolysaccharide recognition by β-mannase	[65]
Two-step process for preparation of oligosaccharide propionates and acrylates using lipase and cyclodextrin glycosyl transferase (CGTase)	Biotechnology	Carbohydrate esters. These are used as surfactants, hydrogels, and other materials, but enzymatic acylation of carbohydrates longer than disaccharides is problematic. The authors overcame this issue by using a two-step enzymatic process using lipase and CGTase. Reversed-phase high-performance liquid chromatography (HPLC)-CAD was used for the analysis of reaction mixtures from the transglycosylation reactions	[36]
Differential selectivity of the *Escherichia coli* cell membrane shifts the equilibrium for the enzyme-catalyzed isomerization of galactose to tagatose	Biotechnology	Artificial sweetener. Tagatose, a naturally occurring monosaccharide, is very similar in texture and sweetness to sucrose but with only 38% of the calories. The biotransformation of galactose to tagatose using bacterial L-arabinose isomerase is an active area of research. The authors used a HILIC-CAD method to measure tagatose yields in gene knockout strains and how changes in the selectivity of the cell membrane can be used to manipulate the isomerase reaction	[47]

Appearance and distribution of regioisomers in metallo-, serine-, and protease-catalyzed acylation of sucrose in *N,N*-dimethylformamide	Biotechnology	Surfactants. Sugar fatty acids esters are nonionic surfactants that are typically synthesized by conventional chemical processes at high temperatures resulting in low regiospecificity and side-chain reactions. This publication explores the use of biocatalysis as a simpler alternative with improved activity and selectivity, without the need of the protection and deprotection steps often required in chiral and regioselective organic synthesis. Samples were analyzed using reversed-phase HPLC with CAD	[35]
Characterization of cyclodextrin glycosyltransferases (CGTases) and their application for synthesis of alkyl glycosides with oligomeric head group	Biotechnology	Surfactants. Alkyl glycosides are used as nonionic surfactants by a number of industries. Current commercial and enzymatic methods used to manufacture alkyl glycosides are limited to products with just one carbohydrate residue. There is interest in producing alkyl glycosides with an elongated hydrophilic group as these are milder to cells and tissues. The authors used HPLC-CAD to study the activities of different bacterial cyclodextrin glucosyltransferases capable of producing elongated alkyl glycosides	[66]
1,2-α-L-Fucosynthase: a glycosynthase derived from an inverting α-glycosidase with an unusual reaction mechanism	Biotechnology	The α-fucosyl residue found on sugar chains of glycoconjugates play an important role in inflammation, development, and signal transduction, as well as acting as a site for pathogens to bind to host cells. Fucosyloligosaccharides also have great therapeutic potential. The authors used glycosynthase technology to modify 1,2-α-L-fucosidase thereby creating a novel route for synthesizing a Fucα1,2-Gal linkage. Glycosynthase activity was monitored using HPLC-CAD	[67]
Determination of levoglucosan in atmospheric aerosols using HPLC with aerosol charge detection	Environmental	Biomass combustion. An HPLC-aerosol charge detection approach was used to measure levoglucosan and other monosaccharide anhydrides in biomass combustion smoke	[68]
Chromatography for foods and beverages: carbohydrates analysis applications notebook	Food/beverage	Review. This application booklet presents HPLC- and IC-based application literature for measurement of carbohydrates in foods, beverages, and supplements	[69]

(*Continued*)

Table 2.4 (Continued)

Title	Application area	Overview	References
Simultaneous separation and determination of erythritol, xylitol, sorbitol, mannitol, maltitol, fructose, glucose, sucrose, and maltose in food products by HPLC coupled to charged aerosol detector	Food/beverage	Sugar analysis. A simple, sensitive, and accurate method is described for the simultaneous determination of numerous simple sugars and sugar alcohols using gradient HILIC-CAD. Levels of sugars found in fruit juices, nectars, and syrups are presented	[43]
Sugar content in the sap of birches, hornbeams, and maples in southeastern Poland	Food/beverage	Sugar analysis. Sap from different trees was analyzed for glucose, fructose, and sucrose using HILIC-CAD	[45]
Comparison of two evaporative universal detectors for the determination of sugars in food samples by liquid chromatography	Food/beverage	Sugar analysis. Fructose, glucose, maltose, lactose, and maltotriose in different sauces, confectionary, and dairy products were detected using either CAD or ELSD	[3]
Carbohydrate analysis in beverages and foods using pulsed amperometric detection or charged aerosol detection	Food/beverage	Sugar analysis. HPLC-pulsed amperometric detection and HPLC-CAD methods were evaluated for measurement of simple carbohydrates in different beverages. HPLC-CAD was much simpler to use, did not require a metal-free system, and was unaffected by changes in analyte retention due to carbonate formation in the mobile phase	[70]
Simple assay of trehalose in industrial yeast	Food/beverage	Trehalose. A HILIC-CAD method was developed for analysis of trehalose in industrial yeasts. This method was more sensitive than conventional chromatography with UV absorbance detection. The method can be used in the baking industry to control quality of yeast products and to assess biotechnological significance of yeast strains	[44]

Determination of inulin-type fructooligosaccharides in edible plants by HPLC with charged aerosol detector	Food/beverage	Dietary fiber. Fructooligosaccharides are classified as dietary fiber and have a number of purported health benefits. The authors used a microwave-assisted extraction method and HPLC-CAD to analyze 11 inulin-type fructooligosaccharides with degree of polymerization from 3 to 13, in different plant extracts	[71]
Practical preparation of lacto-*N*-biose I, a candidate for the bifidus factor in human milk	Food/beverage	Bifidus factor. This is a compound that promotes the growth of beneficial bifidobacteria in a product or animal's intestines. The authors have developed a "one-pot" enzymatic approach to generate kg quantities of lacto-*N*-biose 1 (LNB), a compound proposed as the bifidus factor for humans. In order to study reaction mechanisms, HILIC-CAD was used to measure levels of LNB, GlcNAc, sucrose, fructose, and glucose	[72]
Distribution of *in vitro* fermentation ability of lacto-*N*-biose I, a major building block of human milk oligosaccharides, in bifidobacterial strains	Food/beverage	Bifidus factor. Lacto-*N*-biose I (LNB) is a building block for human milk oligosaccharides, which may be a factor for selective growth of beneficial bifidobacteria. This study investigated the utilization of LNB by different strains of bifidobacteria. LNB levels were determined by HILIC-CAD	[73]
Carbohydrate analysis using HPLC with PAD, FLD, CAD, and MS detectors	General methodology	A number of HPLC-based approaches for the measurement of simple sugars, modified sugars, and glycans liberated from glycoproteins were presented. HPLC-CAD enabled with direct measurement of carbohydrates without the need to use derivatization or a metal-free system	[74]
Selectivity issues in targeted metabolomics: separation of phosphorylated carbohydrate isomers by mixed-mode hydrophilic interaction/weak anion-exchange chromatography	General methodology	The authors developed a HILIC approach using a mixed-mode (reversed-phase/weak anion exchange) column along with CAD for the measurement of numerous phosphorylated carbohydrates. Acidic conditions combined with low temperatures allowed complete separation of α- and β-anomers. Dynamic HPLC enabled the investigation of mutorotation, interconversion kinetics, and measurement of energy barriers for interconversion	[49]

(*Continued*)

Table 2.4 (Continued)

Title	Application area	Overview	References
Investigation of polar organic solvents compatible with Corona- charged aerosol detection and their use for the determination of sugars by hydrophilic interaction liquid chromatography	General methodology	Due to the world shortage of acetonitrile in 2008, the authors investigated the effects of different organic solvents on chromatographic resolution and detection of numerous simple carbohydrates by HILIC-CAD. Acetone was used successfully as an alternative to acetonitrile. Examples include the measurement of carbohydrates in a beer sample and the analysis of a partially digested dextran sample	[75]
Composition of structural carbohydrates in biomass: precision of a liquid chromatography method using a neutral detergent extraction and a charged aerosol detector	Industrial	Biofuels. Lignocellulosic biomass is a feedstock that can be used for production of biofuels and chemicals. Compositional characterization of biomass is important for optimal biofuel production. The authors used neutral detergent extraction of feedstock followed by sulfuric acid hydrolysis prior to analysis by ligand-exchange chromatography-CAD. In addition to measuring cellulose and hemicellulose content, this method, unlike the reference Van Soest method, also measured xylan, arabinan, mannan, and galactan for better characterization	[76]
Efficient separation of oxidized cello-oligosaccharides generated by cellulose degrading lytic polysaccharide monooxygenases	Industrial	Biofuels. Enzymatic conversion of biomass to useful components such as glucose is fundamental to the biofuel industry. Unfortunately, enzymatic hydrolysis tends to be inefficient and slow. The recent discovery of novel oxidative enzymes (lytic polysaccharide monooxygenases) that can potentiate traditional hydrolytic enzymes is a major breakthrough. To better understand enzyme reaction mechanisms, the authors evaluated different chromatographic approaches—high-performance anion-exchange chromatography with pulsed amperometric detection (HPAEC-PAD), HILIC-CAD, and porous graphitized carbon liquid chromatography-CAD (PGC-LC-CAD). While HPAEC-PAD showed good resolution, sensitivity, and throughput, it is not compatible with LC-MS. The authors concluded that the LC-CAD//MS platform can better characterize samples	[77]

Enzymatic saccharification of soda pulp from sago starch waste using sago lignin-based amphipathic derivatives	Industrial	Biofuels. The authors developed a repetitive and efficient process capable of converting biomass (sago waste) into glucose that can be used for biofuel (alcohol) production as well as a food source. Sugars were analyzed using ion-exclusion chromatography-CAD	[78]
Hydrophilic interaction liquid chromatography—a potential alternative for the analysis of dextran-1	Pharmaceutical/ biopharmaceutical	Dextran. The pharmacopeial test for the average relative molecular mass is used as a quality property to define and control the overall distribution of glucose units present in the polymer length. The pharmacopeial method uses size-exclusion chromatography (SEC) and has a run time in excess of 500 min. The authors describe an improved gradient method using HILIC-CAD capable of resolving up to 15-Glu in under 13 min	[79]
Direct detection method of oligosaccharides by high-performance liquid chromatography with charged aerosol detection	Pharmaceutical/ biopharmaceutical	Sialic acids. A HILIC-CAD approach is described for the direct measurement of sialylglycoprotein from egg yolk. The method was five times more sensitive than UV absorbance detection. It was not as sensitive as fluorescence approaches but did not require derivatization. The method could also be used to measure monosiaolo-oligosaccharides, asialo-oligosaccharides, and free sialic acid released from sialylglycoprotein following acid hydrolysis	[80]

used to measure the different species within a particular class. Methods can also be targeted for the measurement of specific lipids, or global (untargeted), for studying the lipidome. Many lipids lack a chromophore making their determination by HPLC-UV challenging. HPLC-CAD is now routinely used to analyze all but the most volatile lipids (e.g., short-chain fatty acids and fatty acid esters). An overview of the publications using CAD for lipid measurement is presented in Table 2.5.

2.3.3 Natural Products

A natural product can be defined as any compound produced by a living organism. This definition is far too broad, so natural products are often classified according to their biological function, biosynthetic pathway, or source. We are using a more restrictive definition here and define a natural product as being a low molecular weight compound, found in terrestrial (herbal or animal) or marine organisms that has a purported biological activity, be it beneficial or harmful (e.g., a toxin). Natural products are useful to the pharmaceutical industry, providing novel chemical scaffolds that can be modified to generate new druggable chemical candidates.

Natural products are often consumed as supplements to promote health or to prevent or treat disease. Supplements may consist of the purified natural product or the natural product contained within the original matrix (e.g., dried herb, extract, tincture, etc.). Supplements are a multibillion dollar business, and there is growing concern regarding the authenticity, adulteration, and contamination (e.g., pesticides, heavy metals, fungal toxins, and pharmaceutical drugs) of products purchased by consumers.

Analysis of natural products can be challenging—they show a wide diversity of chemical structure and can have vastly different physicochemical properties, and for herbal supplements there is a high probability of analyte interferences due to the complex nature of the sample matrix. Many natural products lack a chromophore or ionize poorly, making them ideal candidates for measurement by HPLC-CAD. Numerous publications report the use of HPLC-CAD to characterize natural products, authenticate supplements, and help identify possible adulteration or contamination (Table 2.6).

2.3.4 Pharmaceutical and Biopharmaceutical Analysis

As presented in Table 2.7, CAD is being used throughout the pharmaceutical industry from discovery (research), through development (synthesis and formulation), and to manufacturing (QC of API and formulation). Although it is not possible to discuss each area in detail here, there are a couple of novel uses of CAD that are worth highlighting. First, as with natural products, CAD is being used with DAD and MS (DAD-CAD//MS) to generate orthogonal data in an attempt to measure all analytes in a sample—this is incredibly important

Table 2.5 An overview of CAD publications relevant to lipids.

Title	Application area	Overview	References
Determination of long-chain alcohols using high-performance liquid chromatography (HPLC) with charged aerosol detection	Food/beverage	Fatty alcohols. Long chain alcohols, typically found not only as a side product of cane sugar production but also in vegetable oils, bees wax, and in many epiticular waxes, exert their health benefits by lowering plasma LDL-C. This article describes an HPLC-CAD method for the measurement of linear long-chain alcohols (C_{20} to C_{34}) in food supplements, cane sugar, vegetable oils, and epiticular plant waxes	[83]
The analysis of lipids via HPLC with a charged aerosol detector	Food/beverage	General methods. Several HPLC-CAD methods were developed and evaluated for measurement of different lipid classes including normal-phase methods for either nonpolar lipids (triglycerides and lipid esters) or polar lipids (phospholipids), a normal-phase method for tocopherols and tocotrienols, and a reversed-phase method for lipid molecular species (cholesterol, fatty acids, esters, and triglycerides)	[38]
Effect of temperature toward lipid oxidation and nonenzymatic browning reactions in krill oil upon storage	Food/beverage	Global lipids. The authors showed that exposure of krill oil to elevated temperature for prolonged periods increased lipid peroxidation and nonenzymatic browning reactions—the latter due to the reaction of carbonyl compounds with amino acids or ammonia. Lipid classes were determined using HPLC-CAD	[84]
Quantitation of triacylglycerols from plant oils using charged aerosol detection with gradient compensation	Food/beverage	Glycerides. A nonaqueous reversed-phase HPLC-CAD method was developed for the quantitative analysis of triacylglycerols (TGs) in plant and animal oils. The authors used an inverse gradient approach to overcome the effects of changes in mobile phase composition on analyte signal. Variation in response between different TGs was <5% marked improvement over APCI-MS	[37]

(*Continued*)

Table 2.5 (Continued)

Title	Application area	Overview	References
Determination of olive oil adulteration by principal component analysis with HPLC–charged aerosol detector data	Food/beverage	Glycerides. An HPLC-CAD method was used to generate patterns of triglycerides found in oil samples. Chemometric modeling was then used to determine pure extra virgin olive oil samples from olive oil blended with corn oil, hazelnut oil, or pomace	[85]
Discriminating olive and non-olive oils using HPLC-CAD and chemometrics	Food/beverage	Glycerides. Triacylglyceride profiles were used to discriminate olive oil from other vegetable oils. The Triaclyglyceride (TAG) profiles generated by HPLC-CAD were analyzed using chemometric modeling	[86]
Olive oil quantification of edible vegetable oil blends using triacylglycerols chromatographic fingerprints and chemometric tools	Food/beverage	Glycerides. Triacylglyceride profiles were used to quantify olive oil in blends with vegetable oils. The TAG profiles generated by HPLC-CAD were analyzed using chemometric modeling	[87]
Authentication of geographical origin of palm oil by chromatographic fingerprinting of triacylglycerols and partial least square-discriminant analysis	Food/beverage	Glycerides. The effectiveness of triglyceride chromatographic fingerprinting for the characterization of palm oil from various geographical origins was evaluated using HPLC-CAD and GC-MS. A chemometric model could then be used to predict the provenance of palm oil samples	[88]
Compositional and thermal characteristics of palm olein-based diacylglycerol (DAG) in blends with palm super olein	Food/beverage	Glycerides. DAGs are promoted as a healthier form of oil. They are digested and metabolized differently from other oils, which significantly reduces the risk of obesity. The intent of this paper is to provide the palm oil industry with essential information on the physical and chemical characteristics of palm olein-based DAG blends and identify which has the most desirable physicochemical properties for the food industry. Acylglycerol (MAG, DAG, and TAG) composition was determined by HPLC-CAD	[89]

Determination of polymerized triglycerides by high-pressure liquid chromatography and Corona Veo-charged aerosol detector	Food/beverage	Glycerides. TAGs, which form the bulk of vegetable and animal oils, undergo polymerization reactions when heated, creating high boiling point, higher viscosity, and insoluble materials. These can affect product quality, properties, and nutritional content. HPLC-CAD was used to characterize polymerized TAGs	[90]
Quadruple parallel mass spectrometry (MS) for analysis of vitamin D and triacylglycerols in a dietary supplement	Food/beverage	Glycerides. The "dilute and shoot" method for the simultaneous measurement of vitamin D and triglycerides in dietary supplements described in Ref. [91] was extended to include a fourth MS in parallel	[92]
Characterization of used cooking oils by HPLC and Corona- charged aerosol detection	Food/beverage	Gutter oils. This is a term used in China and Taiwan to describe illicit cooking oil that has been recycled from waste oil collected from sources such as restaurant fryers, sewers, drains, grease traps, and slaughterhouse waste. Following rudimentary reprocessing it is packaged and resold as a cheaper alternative to normal cooking oil. HPLC-CAD was used to measure lipid profiles, which could be used to differentiate between fresh and used cooking oils (gutter oils)	[42]
Simple and efficient profiling of phospholipids in phospholipase D-modified soy lecithin by HPLC with charged aerosol detection	Food/beverage	Phospholipids. The authors developed an HPLC-CAD method to measure six different classes of phospholipids and used it to study the effects of phospholipase D on lecithin samples	[11]
A new liquid chromatography method with charge aerosol detector (CAD) for the determination of phospholipid classes. Application to milk phospholipids	Food/beverage	Phospholipids. Milk samples were prepared using SPE prior to the analysis of five classes of phospholipids using normal-phase HPLC-CAD	[40]

(*Continued*)

Table 2.5 (Continued)

Title	Application area	Overview	References
Development of analytical procedures to study changes in the composition of meat phospholipids caused by induced oxidation	Food/beverage	Rancidity. The authors developed a model useful for evaluating the oxidation level of meat. An accelerated solvent extraction HPLC-CAD method was used to study the effects of peroxidation conditions on a number of phospholipids including phosphatidyl species and cerebrosides	[93]
The effect of dietary antioxidants on iron-mediated lipid peroxidation in marine emulsions studied by measurement of dissolved oxygen consumption	Food/beverage	Rancidity. The behavior of several food antioxidants in the presence of low molecular weight iron in fish oil emulsions stabilized with marine phospholipids was studied. Interaction with iron-converted antioxidants into oxidizing species associated with lipid peroxidation. Phospholipid classes were measured using normal-phase HPLC-CAD	[41]
Chromatography for foods and beverages: fats and oils analysis applications notebook	Food/beverage	Review. HPLC-CAD application literature for measurement of lipids in foods, beverages, and supplements is reviewed	[94]
Simultaneous determination of five bile acids in pulvis fellis suis, pulvis billis bovis, pulvis fellis caprinus, and pulvis fellis galli by HPLC-charged aerosol detector	Food/beverage	Supplements. The authors developed an HPLC-CAD method for the simultaneous measurement of five bile acids in different commercially available gall bladder powder supplements. The authors noted that their method was three times more sensitive and had a wider linear dynamic range than an HPLC-ELSD method	[95]
"Dilute-and-shoot" triple parallel MS method for analysis of vitamin D and triacylglycerols in dietary supplements	General methodology	Glycerides. A "dilute-and-shoot" method for the simultaneous measurement of vitamin D and triglycerides in dietary supplements is described. Following chromatographic separation analytes were detected using DAD followed by parallel CAD, ELSD, and three mass spectrometers operating in different ionization modes	[91]

Composition analysis of positional isomers of phosphatidylinositol (PI) by HPLC	General methodology	Inositides. Myoinositol has six nonequivalent hydroxyl groups, and therefore there are six possible positional isomers when PI is formed enzymatically. As these isomers may have differing biological roles, it is important to be able to measure their levels independently. Normal-phase HPLC-CAD enabled the direct measurement of PI isomers	[39]
The use of charged aerosol detection with HPLC for the measurement of lipids	General methodology	Lipidomics. The authors report on how HPLC-CAD can be used for targeting specific lipids (e.g., fatty acids, fatty alcohols, fat soluble vitamins, and antioxidants) or for profiling all lipids contained in a sample	[81]
From lipid analysis toward lipidomics, a new challenge for the analytical chemistry of the twenty-first century. Part I: Modern lipid analysis	General methodology	Lipidomics. This review describes the different analytical procedures, including HPLC-CAD, used to measure lipids and how they can be used to explore the lipidome	[96]
Comparison of universal detectors for high-temperature micro-liquid chromatography	General methodology	Method comparison. The performance of CAD, ELSD, APCI-MS, and ESI-MS when used with micro-high-temperature liquid chromatography was evaluated using squalene, cholesterol, and ceramide as test compounds. CAD was more sensitive than ELSD (similar to MS), had a wider linear dynamic range compared to ELSD, and was the only detector capable of measuring all three compounds	[97]
Comparison between charged aerosol detection and light scattering detection for the analysis of *Leishmania* membrane phospholipids	General methodology	Phospholipids. Leishmaniasis is a parasitic disease widespread throughout the world. One treatment, miltefosine, appears to exert its effects through disruption of membrane phospholipid distribution. Phospholipid classes from *Leishmania* membranes were measured using HPLC-CAD	[10]

(*Continued*)

Table 2.5 (Continued)

Title	Application area	Overview	References
Rapid quantification of yeast lipid using microwave-assisted total lipid extraction and HPLC-CAD	General methodology	Polar lipids. The authors created a simple and rapid method for fast screening of lipids in yeast. Following microwave-assisted extraction neutral and polar lipids were measured using HPLC-CAD	[98]
Analysis of fatty acid samples by hydrophilic interaction liquid chromatography and charged aerosol detector	General methodology	Prostaglandins. Different HILIC-CAD approaches were evaluated for the measurement of prostacyclin (prostaglandin PG12), 6-keto-prostaglandin $F_{1\alpha}$, arachidonic acid, aleuritic acid, and 12-hydroxydecanoic acid	[99]
Extraction and analysis of food lipids	General methodology	Review. A number of different extraction and analytical methods for the analysis of lipids from foods are discussed	[100]
Advanced MS methods for analysis of lipids from photosynthetic organisms	General methodology	Review. Overview of different HPLC, MS, and LC-MS approaches for the measurement of different lipid classes	[59]
Lipid analysis via HPLC with a charged aerosol detector	General methodology	Review. This brief review discusses the use of HPLC-CAD for the measurement of different nonvolatile lipid species	[58]
Batch production of FAEE-biodiesel using a liquid lipase formulation	Industrial	Biofuel. The authors explored the use of liquid lipase, as a viable option to the use of immobilized enzyme, for the conversion of oils and fats to biodiesel. HPLC-CAD was used to quantify free fatty acids, fatty acid ethyl esters, monoglycerides, diglycerides, and triglycerides in oil phase samples	[101]
Azide improves triglyceride yield in microalgae	Industrial	Biofuel. Microalgae are an important source of lipids used in biofuel production. Azide was shown to significantly increase the amount of triglyceride production in 15 of 17 microalgae species tested. HPLC-CAD was used to measure changes in levels and the relative abundance of different triglycerides following exposure of microalgae to azide	[102]

Characterization of castor oil by HPLC and charged aerosol detection	Industrial	Castor oil. Castor oil has many uses including personal care products, chemical manufacturing, and industrial applications. The authors present an HPLC-CAD method capable of measuring different triglycerides in castor oil including the unique triglyceride, glyceryl triricinoleate	[103]
Analysis of oil stain on paper by charged aerosol detector	Industrial	Stains. The author used an HPLC-CAD approach to profile lipid components of different oil stains on paper samples and differentiate between stains caused by mineral, animal, and plant oils	[104]
Effects on immunogenicity by formulations of emulsion-based adjuvants for malaria vaccines	Pharmaceutical/biopharmaceutical	Adjuvants. Squaline-based stable emulsions adjuvant dose effects on humoral and cellular immune responses to a novel antimalarial vaccine were evaluated. Glucopyranosyl lipid adjuvant (GLA) levels were monitored using HPLC-CAD	[105]
Squalene emulsions for parenteral vaccine and drug delivery	Pharmaceutical/biopharmaceutical	Adjuvants. Squalene is a linear terpene used as a major component of parenteral emulsions for drug and vaccine delivery. This article reviews the physicochemical and biological properties of such emulsions and analytical techniques (including HPLC-CAD) for their characterization	[106]
Monitoring the effects of component structure and source on formulation stability and adjuvant activity of oil-in-water emulsions	Pharmaceutical/biopharmaceutical	Adjuvants. Oil-in-water emulsions have shown promise in adjuvant formulations for vaccines. This study evaluates the physicochemical properties and biological efficacies of emulsions consisting of oil and detergent components from animal, plant, and synthetic sources. HPLC-CAD was used for lipid quantification	[107]
Determination of phospholipid and its degradation products in liposomes for injection by HPLC-charged aerosol detection (CAD)	Pharmaceutical/biopharmaceutical	Drug delivery. HPLC-CAD was used to characterize phosphatidylcholine, lysophosphatidylcholine, distearoyl-glycero-3-phospho-glycerol, and distearoyl-glycero-3-phosphoethanolamine-polyethylene glycol (PEG) from liposomes	[108]

(*Continued*)

Table 2.5 (Continued)

Title	Application area	Overview	References
Analysis of cationic lipids used as transfection agents for siRNA with charged aerosol detection	Pharmaceutical/ biopharmaceutical	Drug delivery. Short segments of RNA, called small interference RNA (siRNA), are being investigated as novel approaches to treat numerous conditions, including cancer, AIDS, diabetes, age-related macular degeneration, and hepatitis. In order to effectively penetrate the cell membrane, the RNA complex is encapsulated with transfection reagents to provide a net positive charge. Suitable transfection reagents include the use of cationic liposomes, PEG pegylated nanocarrier complexes, polymeric systems, dendrimers, polyplexes, and natural polymers and cell-penetrating peptides. As clinical trials proceed to final stages, quality control measurements of purity and quantity of these cationic lipid delivery agents are required. An HPLC-CAD method enabling purity assessment, analyte quantitation, and measurement of stability of commonly used cationic lipids is described	[109]
Quantification of pegylated phospholipids decorating polymeric microcapsules of perfluorooctyl bromide by reversed-phase HPLC with a charged aerosol detector	Pharmaceutical/ biopharmaceutical	Drug delivery. Ultrasound contrast agents (UCAs) generally consist of gaseous perfluorocarbons encapsulated by polymer microcapsules. Unfortunately, these are rapidly cleared from the circulatory system. Pegylation of the particle's surface can delay such clearance. An HPLC-CAD method enabled the measurement a typical pegylated phospholipid—DSPE-PEG2000. The method was applicable to quantification of DSPE-PEG associated to microcapsules, free in the surfactant solution, or to the whole suspension	[110]
Phospholipid decoration of microcapsules containing perfluorooctyl bromide used as UCAs	Pharmaceutical/ biopharmaceutical	Drug delivery. This paper discusses an easy method for the modification of the surface chemistry of polymeric microcapsules used as UCAs. DSPE-PEG associated with the microcapsules was determined by HPLC-CAD	[111]

A nanoliposome delivery system to synergistically triggers TLR4 AND TLR7	Pharmaceutical/ biopharmaceutical	Drug delivery. The authors developed a manufacturable nanoliposome delivery system containing two synergistic ligands (TLR4 [GLA] and TLR7) made with pharmaceutically acceptable excipients and agonists. Formulations containing GLA were characterized by HPLC-CAD	[112]
The performance of PEGylated nanocapsules of perfluorooctyl bromide as an UCA	Pharmaceutical/ biopharmaceutical	Drug delivery. The surface of polymeric nanocapsules used as UCAs was modified with PEGylated phospholipids—PEG-2000-grafted istearoylphosphatidylethanolamine (DSPE-PEG). This process allows them to escape recognition and clearance by the mononuclear phagocyte system so as to achieve passive tumor targeting. DSPE-PEG associated to nanoparticles was quantified using HPLC-CAD	[113]
Optimizing manufacturing and composition of a TLR4 nanosuspension: physicochemical stability and vaccine adjuvant activity	Pharmaceutical/ biopharmaceutical	Drug delivery. Nanosuspensions are a class of delivery system for drugs and adjuvants. One nanosuspension, consisting of synthetic TLR4 ligand GLA and DPPC, is a clinical vaccine adjuvant called GLA-AF. The authors used HPLC-CAD to study the purity of DPPC from different suppliers	[114]
Amino alcohol cationic lipids for nucleotide delivery	Pharmaceutical/ biopharmaceutical	Drug delivery. This US patent describes the use of novel cationic lipids that can be used with other components such as cholesterol and PEG-lipids to form lipid nanoparticles with oligonucleotides and act as siRNA delivery vehicles. HPLC-CAD was used to measure individual lipid concentrations	[115]
Simple and precise detection of lipid compounds present within liposomal formulations using a charged aerosol detector	Pharmaceutical/ biopharmaceutical	Drug delivery. An HPLC-CAD method capable of determining cholesterol, α-tocopherol, phosphatidylcholine, and mPEG-2000-DSPE in liposomal formulations was developed and evaluated	[116]
An LC method for the analysis of phosphatidylcholine hydrolysis products and its application to the monitoring of the acyl migration process	Pharmaceutical/ biopharmaceutical	Drug delivery. Acyl migration in phospholipids limits the shelf life of liposome-based pharmaceuticals. Although determination of lysophospholipid regioisomers is a long-standing problem in phospholipid chemistry, the authors successfully addressed this issue using HPLC-CAD	[117]

(Continued)

Table 2.5 (Continued)

Title	Application area	Overview	References
Charged aerosol detection to characterize components of dispersed-phase formulations	Pharmaceutical/ biopharmaceutical	Drug delivery. This review discusses the use of CAD for the analysis of components found in dispersed-phase (or colloidal) formulations	[118]
Aerosol-based detectors for the investigation of phospholipid hydrolysis in a pharmaceutical suspension formulation	Pharmaceutical/ biopharmaceutical	Drug delivery. An HPLC method was developed to quantify free fatty acids in pharmaceutical suspension formulated with phospholipids as stabilizing agents	[20]
Determination of Impurities in 17β-estradiol reagent by HPLC with charged aerosol detector	Pharmaceutical/ biopharmaceutical	Impurity testing. HPLC-CAD was used to measure six low-level impurities in a 17β-estradiol reagent	[119]
Interactions between parenteral lipid emulsions and container surfaces	Pharmaceutical/ biopharmaceutical	Nutrition. Triglyceride emulsions are used to meet the nutritional needs of critically ill patients. Emulsion instability leading to formation of larger size lipid globules is a health concern. The contribution of different container materials to emulsion instability was evaluated using HPLC-CAD	[120]
Quantitative study of the stratum corneum lipid classes by normal-phase liquid chromatography: comparison between two universal detectors	Research	Lipid analysis. Fatty acids, ceramides, and cholesterol were measured in extracts of human skin using HPLC-CAD	[9]
Integrated analysis, transcriptome–lipidome, reveals the effects of INO-level (INO2 and INO4) on lipid metabolism in yeast	Research	Lipid analysis. INO transcription factors through their regulation of genes containing UAS_{INO} control lipid biosynthesis in yeast. The authors postulated that the expression level of INO genes may be altered by nutrient level, which in turn affects lipid metabolism. Lipid class separation, identification, and quantification used HPLC-CAD	[121]

Origin of β-carotene-rich plastoglobuli in *Dunaliella bardawil*	Research	Lipid analysis. Under conditions of nitrogen deprivation, the halotolerant microalgae *Dunaliella bardawil* accumulates two types of lipid droplets: plastoglobuli rich in β-carotene and cytoplasmic lipid droplets (CLD) rich in triglycerides (TGs). The authors were interested in the origin of these lipid pools and showed that CLD lipids were derived from the endoplasmic reticulum, whereas plastoglobuli lipids came from hydrolysis of chloroplast membrane lipids and transfer of TGs and fatty acids from CLD. TG profiles were determined using HPLC-CAD	[122]
Structure/function relationships of adipose phospholipase A_2 containing a cys–his–his catalytic triad	Research	Lipid analysis. Adipose tissue phospholipase A_2 (AdPLA) plays an important role in obesity onset by suppressing adipose tissue lipolysis. Based on enzyme modeling, and the generation of structural and biochemical data, the authors contend that the current model contingent on the release of arachidonic acid may be wrong. AdPLA activity was determined by the release of fatty acids whose levels were determined by HPLC-CAD	[123]
Mice deleted for glycerol-3-phosphate acyltransferase 3 (GPAT3) have reduced GPAT activity in white adipose tissue and altered energy and cholesterol homeostasis in diet-induced obesity	Research	Lipid analysis. Over the years cloning and knockout experiments have advanced the understanding of the role of triglyceride synthesizing enzymes in health and disease. Using these approaches, the authors evaluated the role of GPAT3 in the maintenance of glucose and lipid homeostasis. Free and esterified cholesterol, triglycerides, and diglycerides were simultaneously determined using normal-phase HPLC with CAD	[124]
Engineering of acetyl-CoA metabolism for the improved production of polyhydroxybutyrate in *Saccharomyces cerevisiae*	Research	Lipid analysis. Metabolic engineering of microorganisms can be used to generate new products in high yield. The authors modified yeast to overproduce the bacterial energy storage material, polyhydroxybutyrate, while minimizing formation of side products. A gradient HPLC-CAD method was used to profile lipids including triacylglycerides, free fatty acids, ergosterol, and individual phospholipids	[125]

(*Continued*)

Table 2.5 (Continued)

Title	Application area	Overview	References
Lovastatin decreases acute mucosal inflammation via 15-epi-lipoxin A_4	Research	Statins. These are used to lower cholesterol levels, possess pleiotropic anti-inflammatory properties. It appears that lovastatin triggers the biosynthesis of the anti-inflammatory and pro-resolving mediator 15-epi-lipoxin A_4, as well as increased 14,15-epoxyeicosatrienoic acid (14,15-EET) generation that may explain statin-mediated tissue protection. Levels of statin, 14,15-EET, and 11,12-EET were determined using HPLC-CAD	[126]
Determination of intraluminal individual bile acids by HPLC with charged aerosol detection	Research	Bile acids. Intraluminal bile acids are essential for lipid absorption and may affect absorption of drugs administered in solid dosage forms. Current methods used to measure bile acids in upper gastrointestinal lumen aspirates lack sensitivity. The authors developed an HPLC-CAD method capable of measuring seven bile acids in human gastric aspirates	[127]

Table 2.6 An overview of CAD publications relevant to natural products/supplements.

Title	Application area/ analyte family	Overview	References
Application of high-performance liquid chromatography (HPLC) with charged aerosol detection for universal quantitation of undeclared phosphodiesterase-5 (PDE-5) inhibitors in herbal dietary supplements	Adulteration/ contamination	Pharmaceutical agents. Many cases of illicit supplements in Europe involve herbal aphrodisiacs adulterated with PDE-5 inhibitors. Consumption can result in serious medical incidents, especially in patients with atherosclerosis, hypertension, or diabetes. The authors developed an isocratic HPLC-CAD method for the determination of analogs of PDE-5 inhibitors without the need for reference standards. PDE-5 inhibitors and 10 analogs were found in 22 illicit dietary supplements and two bulk powdered herbal materials	[128]
Determination of flibanserin and tadalafil in supplements for women sexual desire enhancement using HPLC with tandem mass spectrometer, diode array detector, and charged aerosol detector	Adulteration/ contamination	Pharmaceutical agents. Filbanserin is a synthetic adulterant in counterfeit herbal supplements that are used to stimulate sexual drive in women. Along with filbanserin, tadalafil, a PDE-5 inhibitor used to treat erectile dysfunction, has also been found. A validated HPLC-DAD-CAD//ES-TOF-mass spectrometry (MS) method for the simultaneous measurement of filbanserin and tadalafil is presented and its application to the analysis of adulterated herbal supplements is discussed	[129]
Rapid purification method for fumonisin B1 using centrifugal partition chromatography	Adulteration/ contamination	Toxins. Fumonisins are mycotoxins produced mainly by *Fusarium* species that often contaminate maize products. In order to produce large amounts of pure fumonisin B1, rice cultured with *Fusarium verticilloides* was subjected to purification by centrifugal partition chromatography. Relative abundance of the mycotoxin was determined using LC-MS and HPLC-CAD	[130]

(*Continued*)

Table 2.6 (Continued)

Title	Application area/ analyte family	Overview	References
Fumonisin measurement from maize samples by HPLC coupled with Corona-charged aerosol detector	Adulteration/ contamination	Toxins. Of the different fumonisins mycotoxins produced by *Fusarium* species, FB1, FB2, and FB3 are the most abundant and toxic constituents. An HPLC-CAD method is described for the direct measurement of B-series fumonisins without the need for time-consuming derivatization steps. The method was capable of measuring the maximum residue levels specified in Europe and the United States for maize and maize-based products	[131]
A biosynthetic pathway for BE-7585A, a 2-thiosugar-containing angucycline-type natural product	Angucyclines	Angucyclines. Angucyclines show both antibacterial and anticancer activities. BE-7585A produced by *Amycolatopsis orientalis* sub *vinearia* is a thymidialte synthase inhibitor containing an unusual 2-thioglucose moiety. HPLC-CAD was used during the characterization of the biosynthetic pathway	[132]
Chemotaxonomic differentiation between *Cortinarius infractus* and *Cortinarius subtortus* by supercritical fluid chromatography connected to a multi-detection system	β-Carboline alkaloids	Alkaloids. Differences in β-carboline profiles (β-carboline-1-propionic acid, 6-hydroxy-β-carboline-1-propionic acid, and infractopicrine), generated using SFC followed by HPLC-parallel CAD and MS, were used to distinguish between two *Cortinarius* species	[133]
Photostability of rebaudioside A and stevioside in beverages	Diterpene glycosides	Stevia. Extracts of the leaves of *Stevia rebaudiana* have been popular for many years due to their sweet taste. The sweetness is due to two diterpene glycosides of steviol, rebaudioside A (also called rebiana), and stevioside. Prior to commercialization, rebiana had to go through extensive stability testing including exposure to light. The authors used HPLC-CAD to study product stability and showed that an earlier publication reporting that rebiana was photosensitive was incorrect	[134]

Structural characterization of the degradation products of a minor natural sweet diterpene glycoside rebaudioside M under acidic conditions	Diterpene glycosides	Stevia. The diterpene glycoside rebaudioside M is a minor component of *Stevia rebaudiana* that is 160–500 times sweeter than glucose. The authors used acid hydrolysis at elevated temperature to generate three minor degradation products from rebaudioside M that were characterized using NMR, LC-MS, and HPLC-CAD	[135]
Dereplication of microbial extracts and related analytical technologies	General methodology	Dereplication. This is a process in screening natural products that analyzes the extracts of microbial fermentation broths or plant samples. In this review article the authors discuss (i) direct detection of microbial colonies, (ii) UHPLC-MS profiling of library constructs, (iii) micro-fractionation to identify active peaks, (iv) quantification of low abundance compounds, and (v) analyte structural identification in limited samples. HPLC-CAD is used for absolute analyte quantification	[136]
Decoding glycome of *Astragalus membranaceus* based on pressurized liquid extraction, microwave-assisted hydrolysis, and chromatographic analysis	General methodology	Glycomics. Although carbohydrates are well recognized as an energy source or structural materials, their biological roles are often overlooked. Many herbal carbohydrates show bioactivity and can be exploited as therapeutics, diagnostics, and food additives. Addressing the glycome is crucial for quality control of herbal medicines and function foods. In this paper, the authors present a number of approaches, including HPLC-CAD, to measure different components of the glycome, using the traditional Chinese medicine, *Astragalus membranaceus*, as the test example. Chemometric modeling (hierarchical clustering and principal components analysis) was used to differentiate plants from different geographical regions	[137]
Supplement analysis applications notebook: from raw materials to extracts and natural products	General methodology	Review. Numerous examples of HPLC-CAD analysis methods are described for ashwagandha, bacopa, black cohosh, caralluma, falcarinols, ginkgo, ginseng, gotu kola, hoodia, mangostins, milk thistle, phytoestrogens, and ursolic acid	[138]

(*Continued*)

Table 2.6 (Continued)

Title	Application area/ analyte family	Overview	References
Production of surfactin and iturin by *Bacillus licheniformis* N1 responsible for plant disease control activity	Lipopeptides	Biofungicide. *Bacillus lichenformis* N1, developed as a biofungicide formulation N1E, is used to control gray mold disease in plants. HPLC-CAD and LC-MS approaches were used to show that biofungicide activity primarily results from two lipopeptides, surfactin and iturin	[139]
Assessment of microcystin purity using charged aerosol detection	Microcystin	Toxins. Microcystins are a large group of non-proteinogenic amino acid-containing hepatotoxic cyclic heptapeptides. They are produced in large quantities during cyanobacteria blooms and pose a major threat to drinking water. Although several validated monitoring procedures exist, they are hampered by the availability of high-quality calibration standards. The authors used HPLC-CAD to supplement HPLC-UV and ESI-TIC MS approaches for the assessment of standard purity and showed that HPLC-CAD could measure impurities missed by the other procedures	[140]
HPLC in natural product analysis: the detection issue	Natural products	Review. The author reviews the various HPLC-based approaches used to analyze natural products. The potential limitations and new trends in HPLC hyphenation are discussed	[141]
Development and validation of HPLC-DAD-CAD–MS 3 method for qualitative and quantitative standardization of polyphenols in *Agrimoniae eupatoriae herba* (Ph. Eur)	Phenolics/ polyphenols	Agrimony. Agrimony is traditionally used as a mild astringent and anti-inflammatory for superficial skin damage and is taken internally to treat acute diarrhea and inflammation of the oral mucosa. The authors developed a validated HPLC-DAD-CAD method for the quantitative and qualitative standardization of the pharmacopeial plant material, *Agrimoniae eupatoriae herba*. Fourteen major polyphenols were quantified	[142]

Quantitative and qualitative investigations of pharmacopeial plant material *Polygoni Avicularis* Herba by UHPLC-CAD and UHPLC-ESI-MS methods	Phenolics/ polyphenols	Knotgrass. *Polygonum aviculare* L., known as common knotgrass, is used in the form of an infusion in the treatment of renal diseases, urinary bladder inflammation, and as an expectorant. The author evaluated UHPLC-MS and UHPLC-CAD approaches to quantify flavonoid glucuronides, the major constituents of common knotgrass that can be used for routine standardization of this herbal	[143]
Determination of C-glucosidic Ellagitannins in *Lythri herba* by ultra-high-performance liquid chromatography coupled with charged aerosol detector: method development and validation	Phenolics/ polyphenols	Loosestrife. *Lythri herba*, known as purple loosestrife, is a pharmacopeial plant material that is used in traditional medicine as an astringent and to treat inflammation of the mucosa and acute diarrhea. The authors developed a validated UHPLC-CAD method for determination of the four major ellagotannins (vescalagin, castalagin, and salicarinins A and B) in aqueous extracts of *Lythri herba*	[144]
Quantification of individual phenolic compounds' contribution to antioxidant capacity in apple: a novel analytical tool based on liquid chromatography with diode array, electrochemical, and charged aerosol detection	Phenolics/ polyphenols	Nutrition. Polyphenols are purported to have a number of health benefits (e.g., anticaricnogenic, antibacterial, and anti-inflammatory effects) that may be due, in part, to their antioxidant activities. The total antioxidant capacity of a food attempts to relate its antioxidant content to its possible health benefits. The authors used a novel HPLC-DAD-ECD-CAD method to examine the contribution of individual phenolic compounds found in the extract to the total antioxidant capacity of extracts of different apples	[145]
Qualitative and quantitative analyses of secondary metabolites in aerial and subaerial of *Scorzonera hispanica* L. (black salsify)	Phenolics/ polyphenols	Salsify. The roots of black salsify are used as a vegetable, as a coffee substitute, and as a diuretic and to enhance digestion. Although the subaerial parts are known to contain a number of sesquiterpenoids and lignans, very little is known about the chemical composition of the aerial parts. The authors developed a validated HPLC-DAD-CAD method to quantify the major phenolic constituents of both subaerial and aerial parts of *Scorzonera hispanica*	[146]

(*Continued*)

Table 2.6 (Continued)

Title	Application area/ analyte family	Overview	References
Polyketide analysis using MS, evaporative light scattering, and charged aerosol detector systems	Polyketides	Antibiotics. Polyketides are a large and diverse group of natural products that include antibiotics, antifungal, and anticancer agents. They play a significant role in drug development and discovery processes. The authors compared MS, ELSD, and CAD for the measurement of a representative polyketide, 6-deoxyerythronolide B, and concluded that with its sensitivity and wide dynamic range, CAD is a viable and more cost efficient alternative to MS	[15]
Linear aglycones are the substrates for glycosyltransferase DesVII in methymycin biosynthesis: analysis and implications	Polyketides	Antibiotics. The authors evaluated the role of glucosyltransferase DesVII in the biosynthesis of the macrolide antibiotic methymycin. HPLC-CAD was used to study reactions in the synthetic pathway	[147]
Comparison of ultraviolet detection and charged aerosol detection methods for liquid chromatographic determination of protoescigenin	Saponins	Ecsin. This is a complex mixture of pentacyclic triterpene saponins from the horse chestnut, *Aesculus hippocastanum*. It is used as a traditional herbal preparation of the treatment of chronic venous insufficiency and capillary blood vessel leakage. Escin can be chemically converted to protoescigenin, a starting material for synthesis of various saponin mimetics. An UHPLC-UV-CAD method for the quantification of protoescigenin was developed and evaluated	[13]
A new application of charged aerosol detection in liquid chromatography for the simultaneous determination of polar and less polar ginsenosides in ginseng products	Saponins	Ginsenosides. The HPLC-CAD method enabled the measurement of six polar ginsenosides (Rg_1, Re, Rb_1, Rc, Rb, and Rd) and eight less polar ginsenosides (Rg_6, F_4, Rk_3, Rh_4, $Rg_3(S)$, $Rg_3(R)$, Rk_1, and Rg_5) in one homemade red ginseng and 13 commercial ginseng products (liquid and solid samples)	[148]

Isolation and analysis of ginseng: advances and challenges	Saponins	Ginsenosides. This in-depth review discusses the various isolation and chromatographic approaches used to study ginseng	[149]
Certification of a pure reference material for the ginsenoside Rg_1	Saponins	Ginsenosides. A pure certified reference material for ginsenoside Rg_1 was prepared from roots of *Panax ginseng*. HPLC-CAD was used during testing for homogeneity and long-term stability	[150]
Recent methodology in ginseng analysis	Saponins	Ginsenosides. This paper reviews recent advances in analytical approaches used in the measurement of ginseng	[151]
Ginseng total saponins reverse corticosterone-induced changes in depression-like behavior and hippocampal plasticity-related proteins by interfering with GSK-3β-CREB signaling pathway	Saponins	Ginsenosides. The authors investigated the antidepressant mechanisms of ginseng total saponins in the corticosterone-induced mouse depression model and showed that it may be mediated partly through interfering with the GSK-3β-CREB signaling pathway. For quality control, HPLC-CAD was used to quantify ginseng marker components (Rd, Re, and Rg_1)	[152]
Determination of total ginsenosides in ginseng extracts using charged aerosol detection with post-column compensation of the gradient	Saponins	Ginsenosides. The authors note that quality control of plant products is limited by the restricted availability of reference standards. They developed a gradient UHPLC-CAD method with post-column gradient compensation and showed that the determination of each component based on a single marker peak was practical due to CAD's uniform inter-analyte response	[153]
Performance evaluation of charged aerosol and evaporative light scattering detection for the determination of ginsenosides by LC	Saponins	Ginsenosides. Comparison of HPLC-CAD and HPLC-ELSD methods for the analysis of the botanical, *Panax ginseng*	[14]
Sensitive determination of saponins in *Radix et Rhizoma Notoginseng* by charged aerosol detector coupled with HPLC	Saponins	Ginsenosides. The authors used an HPLC-CAD method to measure seven saponins, notoginsenoside R_1 and ginsenosides Rg_1, Re, Rb_1, Rg_2, Rh_1, and R_d in 30 batches of Sanqi samples. The method showed superior sensitivity compared with HPLC-ELSD and HPLC-UV and, unlike HPLC-UV, was not affected by changes in the gradient profile	[154]

(*Continued*)

Table 2.6 (Continued)

Title	Application area/ analyte family	Overview	References
Preparation and quality assessment of high-purity ginseng total saponins by ion-exchange resin combined with macroporous adsorption resin separation	Saponins	Ginsenosides. Total saponins were purified by dynamic anion–cation exchange following the removal of hydrophilic impurities by macroporous resin. For quality control marker components were qualified using UHPLC-CAD	[155]
HPLC analysis of plant saponins: an update 2005–2010	Saponins	Review. This publication briefly discusses the different HPLC-based approaches used for the determination of plant saponins	[156]
Optimization of pressurized liquid extraction for spicatoside A in *Liriope platyphylla*	Saponins	Steroidal saponins. *Liriope platyphylla*, also called wide-leaf monkey grass is used as a tonic, antitussive, and expectorant in Korea. The steroidal saponin, spicatoside A, a major secondary metabolite, is reported to stimulate secretion of growth hormone and to possess other biological effects. Following pressurized liquid extraction, HPLC-CAD was used to determine levels of spicatoside A	[157]
Simultaneous determination of triterpenoid saponins from *Pulsatilla koreana* using HPLC coupled with a charged aerosol detector (HPLC-CAD)	Saponins	Triterpene saponins. The dried root of the Korean pasque flower (*Pulsatilla koreana*) is used to treat a number of health issues such as dysentery and an antiparasitic and anti-inflammatory agent. The active components are the triterpenoid saponin lupanes (pulsatilloside E, anemoside B_4, and cussosaponin C) and oleanane (*Pulsatilla saponin* H). The authors developed a validated gradient HPLC-CAD method for measuring all four compounds and applied it to quality evaluation, quality control, and monitoring of *P. koreana*	[158]

Comparison between evaporative light scattering detection and charged aerosol detection for the analysis of saikosaponins	Saponins	Triterpene saponins. *Bupleuri radix*, the dried root of *Bupleurum falcatum*, also known as Chinese thoroughwax, is used as a Chinese multiherbal remedy to treat fever, inflammation, liver, and viral diseases. A number of triterpene saponins, the saikosaponins, appear to have biological activity. The paper describes an HPLC-CAD method capable of resolving 10 saikosaponins at 2–6× greater sensitivity than the previous HPLC-ELSD method	[12]
Profiling hoodia extracts by HPLC with charged aerosol detection and electrochemical array detection and pattern recognition	Steriodal glycosides	Oxypregnane steroidal glycosides. Eight of these glycosides isolated from dried plant material were measured using gradient HPLC-CAD. HPLC-CAD and HPLC-DAD-ECD electrochemical array methods generated metabolite patterns that could be interrogated by chemometric software to differentiate samples based on plant species, geographical origin, and processing methods	[159, 160]
Analysis of terpene lactones in a ginkgo leaf extract by HPLC using charged aerosol detection	Terpene lactones	Ginkgolides and bilobalide. Different terpene lactones from ginkgo leaf extracts were characterized using HPLC-CAD. The method showed that analyte levels varied widely in different health products obtained in Japan	[161]

Table 2.7 An overview of CAD publications relevant to pharmaceutical and biopharmaceutical industries.

Title	Application area	Overview	References
Novel analytical method to verify effectiveness of cleaning processes	Cleaning validation	Cleaning agent analysis. A gradient UHPLC-CAD method with mobile phase compensation and a mixed-mode column was evaluated for the measurement of actives, counterions, and the components found in several commercially available cleaning products used by the pharmaceutical industry. Two approaches were developed. Selective analysis enabled the simultaneous measurement of acetaminophen, PEG400, sodium, naproxen, chloride, dodecylsulfate, diclofenac, phosphoric acid, and citric acid. Fast screening by flow-injection analysis provides information about the approximate amount of material present	[50]
High-performance liquid chromatography (HPLC) with charged aerosol detection for pharmaceutical cleaning validation	Cleaning validation	Cleaning agent analysis. Ensuring that the production equipment is clean is a critical step in the pharmaceutical production process and there are strict requirements for the validation of equipment cleaning set up by the EMA and the FDA. HPLC-UV is typically used to measure actives, impurities, and formulants, but problems can occur when these compounds lack a chromophore. The authors used HPLC-CAD to measure test compounds using different solvents typically used in equipment cleaning. CAD showed equivalent or slightly better performance than UV for most compounds tested but surpassed UV for detection of lactose. Furthermore, compared with UV, CAD showed less interference from cleaning solvents. Taken together CAD is complementary to UV and can expand the range of analytes measured for cleaning validation studies	[162]
A mechanism enhancing macromolecule transport through paracellular spaces induced by poly-L-arginine: poly-L-arginine induces the internalization of tight junction proteins via clathrin-mediated endocytosis	Drug delivery	Polycationic materials. Poly-L-arginine (PLA) can promote the transmucosal delivery of macromolecules without inducing severe epithelial toxicity. In this paper, the mechanisms by which PLA can affect the permeability model Caco-2 cell monolayers were evaluated using erythritol as the permeation marker. Erythritol was quantified using HPLC-CAD	[163]

Quantitative analysis of polyethylene glycol (PEG)-functionalized colloidal gold nanoparticles using charged aerosol detection	Drug delivery	Nanomedicines. One major advantage of nonomedicines is their potential to more effectively deliver drugs. The biggest challenge, however, is avoiding the reticuloendothelial system that is responsible for their elimination from the circulatory system. One approach to overcome this problem is to modify the surface of the nanoparticle with PEG. In this article the authors describe an HPLC-CAD method capable of quantitating both free unbound PEG and PEG bound to gold nanoparticles. This approach can be used to assess purification efficiency, stability, and batch-to-batch consistency	[164]
Capture and exploration of sample quality data to inform and improve the management of a screening collection	Drug discovery	Library screening. The authors describe the design, development, and implementation of a data-base-oriented system for assessing the integrity of AstraZeneca's screening collection. HPLC-CAD is one approach used for the routine QC/QA of the corporate compound collection	[165]
Evaluation of CAD as a complementary technique for high-throughput LC-mass spectrometry (MS)-UV-ELSD analysis of drug discovery screening libraries	Drug discovery	Library screening. High-throughput screening used in drug discovery research has driven high-throughput organic synthesis to expand the library and the need for high-throughput analytical approaches (typically LC-MS) for compound library quality assessment. Rapid evaluation of both the purity and quantity of a compound present is challenging due to the absence of fully characterized reference materials. The authors describe a complementary rapid gradient HPLC-CAD approach that makes use of both the universal nature and similar inter-analyte response factors of CAD to quantify analytes using generic calibration curves	[26]
Implementation of charged aerosol detection in routine reversed-phase liquid chromatography methods	Drug discovery	Drug purity. The use of a walk-up open access HPLC-UV-CAD/MS system for drug purity monitoring is described. The authors discuss the importance of the purity of acetonitrile for optimal system performance	[34]
Applications of the charged aerosol detector in compound management	Drug discovery	Compound management. The authors describe how HPLC-CAD can benefit various aspects of compound management including the measurement of contaminants leaching from different storage microplates, monitoring the concentration of solubilized compound stock, solid dissolution processes, evaluation of synthetic purification processes, quantification of difficult-to-weigh solids, and "just-in-time" monitoring of solution concentrations for secondary screening destinations	[166]

(*Continued*)

Table 2.7 (Continued)

Title	Application area	Overview	References
Combined application of dispersive liquid–liquid microextraction based on the solidification of floating organic droplets and charged aerosol detection for the simple and sensitive quantification of macrolide antibiotics in human urine	Drug measurement	Antibiotics. The authors used dispersive liquid–liquid microextraction based on the solidification of floating organic droplets (DLLME-SFO) in combination with HPLC-CAD for the analysis of macrolide antibiotics (azithromycin, clarithromycin, dirithromycin, erythromycin, and roxithromycin) in human urine. The authors concluded that DLLME-SFO was more sensitive, specific, rapid, and environmentally friendly than organic solvent precipitation methods	[167]
Biosynthesis of fosfomycin, reexamination and reconfirmation of a unique Fe (II)- and NAD (P) H-dependent epoxidation reaction	Drug measurement	Antibiotics. HPLC-CAD was used to determine levels of the antibiotic fosfomycin during elucidation of its biosynthetic pathway	[168]
In vitro characterization of LmbK and LmbO identification of GDP-D-erythro-α-D-gluco-octose as a key intermediate in lincomycin A biosynthesis	Drug measurement	Antibiotics. HPLC-CAD was used to measure levels of intermediates in the lincomycin A biosynthetic pathway	[169]
Comparison of ultraviolet detection, evaporative light scattering detection, and charged aerosol detection methods for liquid chromatographic determination of antidiabetic drugs	Drug measurement	Antidiabetic drugs. A validated HPLC-CAD method was developed for the measurement of glipizide, gliclazide, glibenclamide, and glimepiride in drug and supplement tablets and capsules. Even though all compounds possess a chromophore, HPLC-CAD showed superior analytical performance to HPLC-UV	[17]

Comparison of two aerosol-based detectors for the analysis of gabapentin in pharmaceutical formulations by hydrophilic interaction chromatography	Drug measurement	Gabapentin. This hydrophilic antiepileptic drug lacks a chromophore and is often measured using UV or FL detection following derivatization. This paper describes an HILIC-CAD method for the direct measurement in commercial samples of this drug. During method development the authors evaluated four different HILIC columns	[16]
Non-derivatization method for the determination of gabapentin in pharmaceutical formulations, rat serum, and rat urine using HPLC coupled with charged aerosol detection	Drug measurement	Gabapentin. A validated reversed-phase HPLC-CAD method for the direct measurement of gabapentin in a pharmaceutical formulation and rat serum and urine following drug administration is described	[170]
The relationship between plasma concentration of metoprolol and CYP2D6 genotype in patients with ischemic heart disease	Drug measurement	Metoprolol. This is a common β-blocker used to treat ischemic heart disease. It is extensively metabolized in the liver by the CYP2D6 isoenzyme of cytochrome P450. Genetic polymorphism of CYP2D6 has a major impact on drug efficacy and safety. The authors developed an HPLC-DAD-CAD method for the measurement of metoprolol in plasma and related the level of drug to the activity of CYP2D6 in patients with ischemic heart disease	[171]
Photodiode array to charged aerosol detector response ratio enables comprehensive quantitative monitoring of basic drugs in blood by ultraHPLC	Drug measurement	Screening. Quantitative screening for a wide variety of drugs in blood is needed to assess drug abuse and poisoning. The authors describe a validated UHPLC-DAD-CAD approach for identification and quantification of 161 basic drugs from major drug classes. The response ratio of DAD/CAD provides additional analyte-specific information beyond retention times and UV spectra. Due to the stability of response, an historic one point calibration was utilized, resulting in a performance similar to that obtained with ordinary calibrated methods. The method was simple to use and did not suffer from ion suppression or enhancement and required less workload and costs associated with MS methods	[172]
Comparison of UV and charged aerosol detection approach in pharmaceutical analysis of statins	Drug measurement	Statins. A validated HPLC-CAD method is described for the analysis of three statins—simvastatin, lovastatin, and atorvastatin—in tablets	[173]

(*Continued*)

Table 2.7 (Continued)

Title	Application area	Overview	References
A new approach to threshold evaluation and quantitation of unknown extractables and leachables using HPLC/CAD	Extractables/ leachables	Contaminants. Extractables are compounds that can be extracted from the components of a container/closure system, while leachables leach into the drug product. Measurement of extractables and leachables are required by various regulatory agencies because of safety concerns. Measurement by HPLC-UV or HPLC-MS can be problematic as response factors can vary considerably between compounds and using a surrogate standard for quantitation can give highly uncertain results. The authors made use of the fact that CAD response is relatively constant across a broad range of compounds minimizing the need for reference standards. The HPLC-CAD method described and measured a wide variety of relevant compounds and was used to evaluate an extract of an elastomeric packaging component	[174]
Direct analysis of multicomponent adjuvants by HPLC with charged aerosol detection	Formulation	Adjuvants. A vaccine adjuvant is any substance that helps promote the effectiveness of a vaccine by reducing the amount or frequency of the required dose, by prolonging the duration of immunological memory, or by modulating the involvement of humoral or cellular responses. This encompasses a very diverse group of substances whose chemical structures and mechanisms of action vary widely. Adjuvants are typically subjected to rigorous standards of analysis including quantification of strength, purity, stability, and degradation behavior. Complicating such analysis, many adjuvants contain components that are not readily analyzed by traditional HPLC with UV detection. These include various mixtures of lipids, fatty acids, and glycosides that lack suitable UV chromophores. HPLC-CAD methods are presented that enable the measurement of components in a variety of adjuvants including AbISCO-100, AddaVax, a squaline-tocopherol-PS80 mixture, and synthetic MPLA	[175]

Hydrophilic interaction liquid chromatography–charged aerosol detection as a straightforward solution for simultaneous analysis of ascorbic acid (AA) and dehydroascorbic acid (DHA)	Formulation	Antioxidants. AA and DHA. AA, vitamin C, is a small polar water-soluble antioxidant forming DHA when it is oxidized. The AA/DHA ratio is sometimes used as an indicator of oxidative stress. The authors developed a validated HILIC-CAD for the simultaneous determination of AA and DHA in pharmaceutical tablets and evaluated the chromatographic performance of different commercially available HILIC columns	[176]
Simultaneous determination of positive and negative pharmaceutical counterions using mixed-mode chromatography coupled with charged aerosol detector	Formulation	Counterions. Pharmaceutical counterions are typically measured using IC with suppressed conductivity detection. However, this approach requires specialized equipment and separate columns, mobile phases, and suppressors for either cation or anion measurement. The conditioning and changeover time for IC is lengthy or otherwise duplicate systems are required. The authors describe an HPLC-CAD approach using a mixed-mode column that can measure inorganic and organic cations and anions, simultaneously. 25 commonly used pharmaceutical ions were separated in 20 min in a single method. This method was also able to simultaneously measure an API and associated counterion	[48]
Determination of inorganic pharmaceutical counterions using hydrophilic interaction chromatography coupled with a Corona® CAD detector	Formulation	Counterions. A simple generic HPLC-CAD approach was evaluated for the determination of inorganic pharmaceutical counterions using a polymer-based zwitterionic stationary phase operated in HILIC mode. When compared with IC-conductivity, the authors noted that their method could (i) measure both anions and cations simultaneously and did not require separate column/eluents; (ii) function with organic mobile phases enabling detection of counterions of poorly water-soluble drugs; and (iii) require minimal training of the operating analyst	[177]
Evaluation of methods for the simultaneous analysis of cations and anions using HPLC with charged aerosol detection and a zwitterionic stationary phase	Formulation	Counterions. A validated gradient HPLC-CAD method using a polymeric zwitterionic column for the analysis of inorganic ions as salts and counterions of both APIs and other compounds is described. Twelve inorganic cations and anions were resolved in <23 min. The method could simultaneously measure the API and its counterion and impurities at 0.1%	[178]

(*Continued*)

Table 2.7 (Continued)

Title	Application area	Overview	References
Comprehensive approaches for measurement of active pharmaceutical ingredients, counterions, and excipients using HPLC with charged aerosol detection	Formulation	Counterions. HPLC-CAD methods are described that enable (i) measurement of salts, (ii) simultaneous determination of inorganic and organic cations and anions, (iii) simultaneous measurement of APIs and counterion(s), (iv) use of HPLC-DAD-CAD//MS for measurement and identification of API and counterion(s), and (v) simultaneous measurement of an excipient, counterion(s), and impurities	[179]
Performance of charged aerosol detection with hydrophilic interaction chromatography	Formulation	Counterions. In this paper the performance of HILIC-CAD for the measurement of a diverse set of 29 solutes including acids, bases, and neutrals was evaluated and the role of organic modifier and buffer components in the mobile phase on analyte response was discussed	[180]
API and counterions in Adderall® using multimode liquid chromatography with charged aerosol detection	Formulation	Counterions. Adderall is used to treat attention deficit hyperactivity disorder (ADHD) and narcolepsy. It is a formulation of dextroamphetamine sulfate, dextroamphetamine saccharate, racemic amphetamine sulfate, and racemic amphetamine aspartate monohydrate. It is impossible to assay all components within the same analysis on any RP, ion exchange, or HILIC column. An HPLC-UV-CAD method using a mixed-mode column was developed allowing the simultaneous measurement of aspartate, sodium, saccharin, amphetamine, and sulfate	[53]
Simultaneous determination of metformin and its chloride counterion using multimode liquid chromatography with charged aerosol detection	Formulation	Counterions. Metformin is an oral antidiabetic drug, often formulated in its hydrogen chloride salt form. Because of the highly hydrophilic nature of both API and counterion, it is impossible to assay both components within the same analysis on any RP, ion exchange, or HILIC column. An HPLC-CAD method using a mixed-mode column was developed allowing the simultaneous measurement of the metformin and its counterion chloride	[52]

Simultaneous determination of tartaric acid and tolterodine in tolterodine tartrate	Formulation	Counterions. Tolterodine tartrate is a medication that is used to treat urinary urgency, thereby reducing the frequency of passing urine and urinary incontinence. It is important for pharmaceutical companies to ensure that a drug product contains the appropriate amount of its API. A gradient HPLC-CAD method using a mixed-mode column was developed allowing the simultaneous measurement of the API tolterodine and its counterion tartrate in drug capsules	[51]
Multimodal analyte detection of cyclodextrin and ketoprofen inclusion complex using UV and CAD on an integrated UHPLC platform	Formulation	Drug delivery. An UHPLC-UV-CAD method was developed for the simultaneous measurement of cyclodextrin (a non-chromophore) and the nonsteroidal anti-inflammatory, ketoprofen on the Vanquish™ UHPLC system	[181]
Chromatographic methods for characterization of poly (ethylene glycol)-modified polyamidoamine dendrimers	Formulation	Drug delivery. Dendrimers are monodisperse and globular macromolecules that can be used as drug delivery systems with poly(amidoamine) dendrimers (PAMAMs) being the most widely studies. PAMAMs can be used to either encapsulate drugs or have their terminal groups functionalized with drugs, targeting moieties or fluorescent agents. Unfortunately, the cytotoxicity and hemolytic toxicity associated with PAMAMs limits their use in biomedical applications. PEGylation of PAMAMs appears to be one approach to reduce such toxicity. The authors developed an HPLC-UV-CAD method to study PEG-PAMAMs. UV provides information about the degree of PEGylation, while CAD is used to characterize PEG-PAMAMs	[182]
A new approach for quantitative determination of γ-cyclodextrin in aqueous solutions: application in aggregate determinations and solubility in hydrocortisone/γ-cyclodextrin inclusion complex	Formulation	Drug delivery. A fast and simple HPLC-CAD method was developed for the measurement of γ-cyclodextrin in aqueous solutions and was successfully applied to permeation and phase-solubility studies	[183]

(*Continued*)

Table 2.7 (Continued)

Title	Application area	Overview	References
Investigating the stability of the nonionic surfactants tocopheryl PEG succinate and sucrose laurate by HPLC–MS, DAD, and CAD	Formulation	Excipients. The authors developed an HPLC-DAD-CAD-MS method for quality control and stability testing of two nonionic surfactants—D-α-tocopheryl PEG succinate and sucrose laurate. The method can be used to examine their pH stability, digestion by metabolic and digestive enzymes, and formation of degradation products	[184]
Characterization of hydroxypropylmethylcellulose (HPMC) using comprehensive two-dimensional liquid chromatography	Formulation	Excipients. HPMC is widely used as a tablet binder, film coating, thickening agent, and capsule shell and in extended-release tablet matrix. Modification of cellulose polymers by methylation and hydroxypropylation results in several types of HPMC differing in the degree of substitution and molecular weight distribution. The authors developed a direct 2D-HPLC method to directly characterize HPMC batches from different manufacturers. In the first dimension, reversed-phase HPLC was used to analyze the degree of substitution. Molecular weight was assessed by size-exclusion chromatography in the second dimension. Detection was accomplished by CAD. In addition this approach provided information about the temperature at which half of the polymer is gelated and this could be correlated with the "cloud-point temperature"	[185]
Simple and rapid HPLC method for the determination of polidocanol as bulk product and in pharmaceutical polymer matrices using charged aerosol detection	Formulation	Excipients. Polidocanol is heterogeneous consisting of ethers of lauryl alcohol and PEG of various chain lengths. Despite the fact that it acts as an emulsifier and as an active ingredient (due to its anesthetic and antipruritic activities), neither the European nor the US Pharmacopeia provide a method for its determination. The authors describe a fully validated HPLC-CAD method for the analysis of polidocanol and its application to the measurement of polidocanol released from a pharmaceutical polymer matrix	[186]

Quantitative comparison of a Corona-charged aerosol detector and an evaporative light scattering detector for the analysis of a synthetic polymer by supercritical fluid chromatography	Formulation	Excipients. The performance of supercritical fluid chromatography with either ELSD or CAD was evaluated using PEG-certified reference material and a well-defined equimass mixture of uniform PEG oligomers. CAD was ten times more sensitive than ELSD. ELSD underestimated molecular mass distribution of the certified value of PEG 1000 by 4.6%	[21]
Size-exclusion chromatography with Corona-charged aerosol detector for the analysis of PEG polymer	Formulation	Excipients. PEG is used as an emulsifier and surfactant. It is also used in making PEGylated drugs and biologics that improve their solubility, stability, bioavailability, and efficacy. PEG reagent analysis entails not only determination of purity (e.g., contamination with dimers and other oligomers) but also determination of its polydispersity. In this paper a size=-exclusion chromatographic-CAD approach was validated using a PEG reference standard of known polydispersity	[22]
Characterization and stability study of polysorbate 20 in therapeutic monoclonal antibody (MAb) formulation by multidimensional ultrahigh-performance liquid chromatography–charged aerosol detection–MS	Formulation	Excipients. Polysorbate 20 is a nonionic surfactant used in the formulation of therapeutic antibodies to prevent protein denaturation and aggregation. It is important to understand the molecular heterogeneity and stability of polysorbate 20 in formulations. The authors used a 2D HPLC-CAD-MS approach to characterize and study stability of polysorbate 20 in a MAb formulation. In the first dimension a mixed-mode column was used to separate polysorbate from protein in the formulation sample, while polysorbate esters were trapped online then separated using reversed-phase UHPLC in the second dimension. Online MS was then used to further resolve and identify polysorbate 20 ester subspecies. A second 2D method used cation exchange in the first dimension (the second dimension was kept the same). The authors noted that these 2D methods showed that different polysorbate 20 esters degraded at different rates and that some of these degradation rates are different in the protein formulation compared with a placebo that lacks protein	[187]

(*Continued*)

Table 2.7 (Continued)

Title	Application area	Overview	References
Fast and sensitive determination of polysorbate 80 in solutions containing proteins. Journal of pharmaceutical and biomedical analysis	Formulation	Excipients. Polysorbate 80 (Tween 80) is a nonionic surfactant added to many protein formulations to maintain their biological activity by stabilizing the protein's three-dimensional conformation during storage, transportation, and delivery to patients. In liquid formulations, surfactants help prevent adsorption of proteins to surfaces of containers and syringes and reduce interfacial surface tension, in order to decrease protein denaturation that can lead to aggregation. A fast and sensitive gradient HPLC-CAD method using a superficially porous column is described capable of simultaneously measuring Polysorbate 80, excipients, and proteins in a single analysis	[188]
Fatty acid composition analysis in polysorbate 80 with HPLC coupled to charged aerosol detection	Formulation	Excipients. Polysorbate 80, a nonionic surfactant used in pharmaceutical formulations, is composed of fatty acid esters of sorbitol. Although polysorbate 80 is primarily esterified with oleic acid, other fatty acids (C14 to C18) are present. European pharmacopoeial and US pharmacopeial methods require that at least 58% oleic acid be present in commercially used polysorbate 80. The authors developed a validated HPLC-CAD method that could determine fatty acid profiles released by alkaline hydrolysis and used it to characterize 16 different batches of polysorbate 80. This approach also measure unique fatty acids including petroselinic acid and an oxidized fatty acid, 11-hydroxy-9-actdecanoic acid (both identified using LC-MS/MS)	[189]
Quantitative determination of nonionic surfactants with CAD	Formulation	Excipients. An HPLC-CAD method is described that is capable of profiling Tween 80 and Span 85, two nonionic surfactants widely used in pharmaceutical formulations	[23]
A highly sensitive method for the quantitation of polysorbate 20 and 80 to study the compatibility between polysorbates and m-cresol in the peptide formulation	Formulation	Excipients. A highly sensitive method was developed for the quantitation of polysorbate 20 (PS20) and 80 (PS80) in therapeutic peptide formulations. A mixed-mode HPLC column was used to separate polysorbates from the peptide and other excipients, and CAD was used for the detection. The method was used to study the compatibility between polysorbates and m-cresol in the peptide formulation. It was found that both PS20 and PS80 are compatible with m-cresol when their levels were not greater than 20 ppm. Significant losses of polysorbates were observed when PS20 and PS80 concentrations were above 50 ppm. Furthermore, the agitation study demonstrated that even trace levels of PS20 and PS80 (e.g., 20 ppm) could stabilize the peptide against fibrillation and aggregation	[190]

Analysis of ionic surfactants by HPLC with evaporative light scattering detection and charged aerosol detection	Formulation	Excipients. The separation of five cationic and seven anionic surfactants was carried out by gradient HPLC using two analytical columns and mobile phases varying in amount of organic modifier, buffer salts, and buffer pH. CAD showed better linearity and greater sensitivity than ELSD	[19]
Design of experiments and multivariate analysis for evaluation of reversed-phase HPLC with charged aerosol detection of sucrose caprate regioisomers	Formulation	Excipients. Sugar fatty acid esters are nonionic surfactants used as additives by the pharmaceutical, cosmetic, and food industry. In this paper, different step-down gradient elution profiles were investigated during the development of a quantitative reversed-phase HPLC-CAD method	[191]
Elution strategies for reversed-phase HPLC analysis of sucrose alkanoate regioisomers with charged aerosol detection	Formulation	Excipients. Further to their original paper [60], Lie and Pedersen systematically investigated a broad range of elution strategies for reversed-phase-HPLC analysis of sucrose alkanoate regioisomers with CAD. Methods were investigated using design-of-experiments methodology and analyzed by analysis of variance (ANOVA) and regression modeling	[192]
Material Identification by HPLC with charged aerosol detection	Formulation	Material identification. A generic approach for material identification is presented through HPLC-CAD and a mixed-mode column. Current industry standards in the pharmaceutical industry rely on USP191 for material identification. However, the HPLC-CAD approach is a rapid and concise way for replacing many sections of USP191 for material identification. In addition to identification, the technique can also simultaneously perform quantitation and detect impurities	[193]
HPLC-CAD surfactants and emulsifiers applications notebook	Formulation	Review. Numerous examples of HPLC-CAD methods are described for the measuremkent of cationic, anionic, and nonionic surfactants	[194]
Charged aerosol detection: factors for consideration in its use as a generic quantitative detector	General methodology	Compound libraries. The authors conclude that the CAD is a powerful tool in the analysis of diverse compounds contained in compound libraries. They noted the need to use an inverse gradient makeup flow when using gradient chromatography. They also suggest using a buffer that contains the smallest possible counterions to overcome possible quantification errors due to unpredicted ion pairing with drug compounds	[63]

(*Continued*)

Table 2.7 (Continued)

Title	Application area	Overview	References
Corona-charged aerosol detection in supercritical fluid chromatography for pharmaceutical analysis. Analytical chemistry	General methodology	Impurity testing. A packed column supercritical fluid chromatography (pSFC)-CAD approach for the measurement of selected pharmaceutical compounds and impurities at 0.05% versus the main component is described. The authors discuss how CAD is coupled to pSFC column by way of a back pressure regulator and the advantage of mobile phase compensation to improve CAD response uniformity	[56]
Serial coupling of reversed-phase and hydrophilic interaction liquid chromatography to broaden the elution window for the analysis of pharmaceutical compounds	General methodology	Method development. The authors explored a novel approach to extend the range of analytes measured in a pharmaceutical sample by coupling gradient reversed-phase HPLC with a HILIC separation. A t-piece prior to the HILIC column ensured that sufficient organic modifier was present for effective HILIC separation. Analytes were measured using UV-CAD//MS	[195]
Use of suppressors for signal enhancement of weakly-acidic analytes in ion chromatography with universal detection methods	General methodology	Mobile phase suppression. This review evaluates the use of suppressors as eluent desalting devices to enable the coupling of IC to universal detectors (MS, ELSD, and CAD)	[60]
Determination of pharmaceutically related compounds by suppressed ion chromatography: IV. Interfacing ion chromatography with universal detectors	General methodology	Mobile phase suppression. The authors evaluated the use of suppressed IC with different universal detectors (MS, ELSD, and CAD) for the measurement of 10 weakly anionic pharmaceuticals. Mobile phase nonvolatile ions were removed by a suppressor prior to detection. A prototype electrolytic suppressor performed better than a chemical suppressor, showing wider linear response ranges	[61]

Increased process understanding for Quality by Design (QbD) by introducing universal detection at several stages of the pharmaceutical process	General methodology	QbD. An initiative mandated by the FDA is to improve product quality and increase process understanding during development and production. QbD is an approach that places increased emphasis on stages defined as being risk points in the process, in order to reduce the emphasis on final product testing. This paper evaluates how HPLC-CAD can be implemented in the QbD process including: lot-to-lot variability, excipient measurement, salt selection, reaction monitoring, degradation studies, and measurement of extractables and leachables	[196]
Charged aerosol detection in pharmaceutical analysis	General methodology	Review. This publication gives a background to charged aerosol detection and how HPLC-CAD can be used to measure impurities, assay drugs, and measure formulants	[62]
Validating analytical methods with charged aerosol detection	General methodology	The authors discuss how to address the various parameters for method validation using an HPLC-CAD method for the simultaneous measurement of an API and its counterion	[197]
Assessment of the complementarity of temperature and flow rate for response normalization of aerosol-based detectors	General methodology	In this paper the authors empirically investigate the use of temperature and flow-rate gradients to overcome the effects of solvent composition on CAD response. The applicability of these approaches was evaluated using a typical pharmaceutical impurity profiling application	[198]
Control of impurities in L-aspartic acid and L-alanine by HPLC coupled with a Corona-charged aerosol detector	Impurity testing	Amino acids. This fully validated HPLC-CAD method, suitable for pharmaceutical quality control, was used to measure L-aspartic acid and related substances. With slight modification the method could be applied to L-alanine. The LOQ for impurities was 0.03%. Malic acid and alanine were the major impurities in aspartic acid; aspartic acid and glutamic acid were the major impurities in alanine	[24]
Alternatives to amino acid analysis for the purity control of pharmaceutical grade L-alanine	Impurity testing	Amino acids. The authors evaluated a number of detectors (qNMR, ELSD, MS, NQAD, and CAD) for the measurement of pharmaceutical grade L-alanine purity. qNMR enabled qualitative and quantitative determination of major and minor components, had the ability to assign structure, but lacked sensitivity. ELSD was too insensitive and had a nonlinear response. CAD was superior to NQAD in terms of repeatability and sensitivity	[199]

(Continued)

Table 2.7 (Continued)

Title	Application area	Overview	References
Development and validation of a RP-HPLC method for the determination of gentamicin sulfate and its related substances in a pharmaceutical cream using a short pentafluorophenyl column and a charged aerosol detector	Impurity testing	Antibiotics. A validated HPLC-CAD method for the direct measurement of gentamicin and its impurities (gentamicin C1, C1a, C2, C2a, and C2b, deoxystreptamine, garamine, and sisomicin) in a pharmaceutical cream was developed. The method was validated by two analysts from two different laboratories and was shown to have good accuracy, linearity, precision, reproducibility, specificity, and robustness. The authors conclude that their method "can be used in quality control labs for stability studies and also for routine analysis of gentamicin in commercial lots of celestoderm-V cream including release testing"	[200]
Determination of gentamicin sulfate composition and related substances in pharmaceutical preparations by LC with CAD	Impurity testing	Antibiotics. This paper describes a validated HPLC-CAD/ESI/MS/MS-TOF method for the measurement of gentamicin and related substances (gentamicin C1, C1a, C2, C2a, and C2b, garamine, and sisomicin) in three pharmaceutical formulations from two manufacturers. CAD provided quantitation without the need for labeled standards. MS was used for analyte identification. Slight differences in impurity profiles were noted for the different products but these were in agreement to European Pharmacopoeial requirements	[201]
Gentamicin sulfate assay by HPLC with charged aerosol detection	Impurity testing	Antibiotics. An HPLC-CAD method was developed and evaluated for measurement of gentamicin sulfate and its impurities in ointments, solutions, and creams	[202]
Development of HPLC methods with charged aerosol detection for the determination of lincomycin, spectinomycin. and its impurities in pharmaceutical products	Impurity testing	Antibiotics. This paper describes a validated HPLC-CAD/ESI/MS/MS-TOF method for the measurement of lincomycin, spectinomycin, and related substances in commercially available veterinary products. CAD provided quantitation without the need for labeled standards. MS was used for analyte identification. The level of actives agreed with the product label. Levels of impurities (e.g., spectinomycin impurities A, C, D, E, and F) complied with European Pharmacopoeial requirements	[203]

Determination of neomycin and related substances in pharmaceutical preparations by reversed-phase HPLC with MS and charged aerosol detection	Impurity testing	Antibiotics. This paper describes a validated HPLC-CAD/ESI/MS/MS-TOF method for the measurement of neomycin and related substances (neomycin A, B, C, D, E, and F; neomycin LP-A, LP-B, and LP-C; ribostamycin, paromamine, neamine, paromomycin I and II) in nine commercially available pharmaceutical formulations. CAD provided quantitation without the need for labeled standards. MS was used for analyte identification. Only one sample failed to meet European Pharmacopoeial criteria	[204]
Development and validation of a RP-HPLC method for the estimation of netilmicin sulfate and its related substances using charged aerosol detection	Impurity testing	Antibiotics. A validated HPLC-CAD method for the direct measurement of netilmicin and related substances was developed. The approach can be used to a) determine product composition and b) as a stability-indicating method	[205]
Identification and control of impurities in streptomycin sulfate by HPLC coupled with mass detection and Corona-charged aerosol detection	Impurity testing	Antibiotics. A reversed-phase (volatile) ion pair HPLC-CAD method capable of measuring streptomycin and 21 impurities at an LOQ of at least 0.1% was developed. Impurity identification and possible structures were determined using parallel MS/MS. Analysis of 12 samples obtained from different manufacturers showed impurity levels between 4.6 and 16%. The authors note that the current European Pharmacopoeia (EP) monograph for streptomycin sulfate limits streptomycin B by a TLC test to 3% and stress the importance of in introducing a state-of-the-art test for control of impurities in the monograph	[206]
SEC assay for polyvinylsulfonic (PVS) impurities in 2-(*N*-morpholino) ethanesulfonic acid using a charged aerosol detector	Impurity testing	Biological buffer. 2-(*N*-Morpholino) ethanesulfonic acid (MES) is a biological buffer commonly used in enzymatic studies and protein purification. PVS acid is an impurity in MES that can interfere with enzyme activity and cause protein precipitation. This paper describes an SEC-CAD method capable of simultaneously measuring MES and PVS	[207]

(*Continued*)

Table 2.7 (Continued)

Title	Application area	Overview	References
Impurity profiling of carbocisteine by HPLC-CAD, qNMR, and UV/Vis spectroscopy	Impurity testing	Cabocisteine. This drug is an anti-inflammatory mucolytic agent used to treat chronic obstructive lung disease. A fully validated HPLC-CAD method using a mixed-mode column and post-column addition of acetonitrile was used to measure carbocisteine, sodium, ammonium, and numerous impurities. Volatile chloroacetic acid was determined by qNMR and unstable cysteine by UV/Vis spectroscopy following derivatization with Ellman's reagent. The authors conclude that HPLC-CAD is a suitable alternative to amino acid analysis and HPLC-UV detection and in addition can be used to measure related process and degradation impurities (e.g., cabocisteine lactam). The universal nature of HPLC-CAD enables quantification without external standards. HPLC-CAD is a suitable alternative to the current TLC test	[208]
Development of a reversed-phase HPLC impurity method for a UV variable isomeric mixture of a CRF drug substance intermediate with the assistance of Corona CAD	Impurity testing	CRF drug substance intermediate. This publication describes an HPLC-UV-CAD method that could determine an antagonist of the corticotropin releasing factor receptor 1 and related substances. The different compounds had different UV response factors and, as reference standards, were unavailable and quantitation was difficult. The combination of UV and CAD could be used to determine the relative amount of substances in the sample, which subsequently permitted the use of conventional UV detection for routine analysis	[209]
Analysis of pharmaceutical impurities using multi-heartcutting 2D LC coupled with UV-charged aerosol MS detection	Impurity testing	General method. This HPLC-DAD-CAD//MS method used a single analytical column in the first dimension and six orthogonal columns in the second dimension for the ultimate in flexibility and selectivity. A wide variety of analytes can be determined including those with and without a chromophore, organics, and inorganics. It also provides structural information for impurity identification. This approach was used to reveal a minor degradation product that co-eluted with the drug main peak, thus enabling a more accurate stability-indicating method to be developed	[210]

Sensitive and direct determination of lithium by mixed-mode chromatography and charged aerosol detection	Impurity testing	Lithium. Lithium carbonate is prescription drug used to treat mania and bipolar disorders. It has a very narrow therapeutic window, so it is important to monitor and maintain circulating lithium levels below toxic levels. Lithium can also occur as an impurity in pharmaceutical products resulting from lithium-based reagents used in drug synthesis and, thus, could cause problems for patients taking lithium carbonate medication. The authors describe a validated HPLC-CAD method using a mixed-mode column for the direct analysis of lithium, potential interfering ions, and actives in pharmaceutical matrices	[211]
Simultaneous determination of Maillard reaction impurities in memantine tablets using HPLC with charged aerosol detector	Impurity testing	Memantine. The reaction between an amino compound and a reducing sugar is known as the Maillard reaction and is probably most familiar with browning seen in cooking. This reaction can take place in tablets, where amine actives react with sugar formulants such as lactose, leading to the formation of unwanted impurities. This paper describes the use of HPLC-CAD for the measurement of impurities formed between memantine and lactose, galactose, glucose, and glycine. The approach could measure impurities at 0.02% of the level of memantine and is suitable for long-term stability studies	[212]
Metoprolol and select impurities analysis using a hydrophilic interaction chromatography method with combined UV and charged aerosol detection	Impurity testing	Metoprolol. The drug metoprolol succinate USP is a selective beta1-adrenoreceptor antagonist that reduces chest pain and lowers high blood pressure. Several pharmacopeias (US Pharmacopeia, EP, and British Pharmacopoeia) have indicated acceptable levels of impurities allowed by drug manufacturers. Impurity profiling of a drug substance is important since the presence of unwanted chemicals, even at small amounts, may influence the efficacy and safety of the pharmaceutical product. Quantification of metoprolol and metoprolol EP impurity A by HPLC with UV detection has been described but some impurities do not possess a detectable UV chromophore. The EP indicates that some impurities can be analyzed by thin layer chromatography, but this is not a suitable technique for producing reliable quantitative data at lower concentrations. The USP monograph modernization program indicates that a liquid chromatographic method is more desirable. The author describes a simple, rapid, sensitive, accurate, and direct HILIC-DAD-CAD method for the simultaneous quantification of metoprolol and its impurities A, M, and N	[213]

(*Continued*)

Table 2.7 (Continued)

Title	Application area	Overview	References
Determination of atracurium, cisatracurium, and mivacurium with their impurities in pharmaceutical preparations by liquid chromatography with charged aerosol detection	Impurity testing	Neuromuscular blocking drugs. Atracurium, cisatracurium, and mivacurium are non-depolarizing neuromuscular blocking agents. Atracurium and its more potent isomer cisatracurium undergo Hofmann elimination, a nonenzymatic process dependent upon pH and temperature, yielding laudanosine and quaternary monoacrylate. In acidic solution, atracurium but not cisatracurium is degraded by ester hydrolysis. The authors developed a gradient HPLC-CAD method capable of measuring actives and impurities in standards and pharmaceutical preparations. Peaks were identified using LC-ESI-TOF-MS	[214]
Comprehensive impurity profiling of nutritional infusion solutions by multidimensional offline reversed-phase liquid chromatography × hydrophilic interaction chromatography–ion trap MS and charged aerosol detection with universal calibration	Impurity testing	Nutritional infusion. A comprehensive qualitative and quantitative impurity profile of a stressed multi-constituent nutritional infusion solution (composed of amino acids and dipeptides) was developed and validated. The approach used a multidimensional analysis making use of an offline two-dimensional reversed-phase liquid chromatography (RPLC) × hydrophilic interaction chromatography (HILIC) separation, combination of complementary detection involving ion trap MS (IT-MS), and CAD. A universal calibration function was set up with a set of standards and was then used to quantify unknown impurities. Impurities detected comprised di-, tri-, and tetrapeptides, cyclic dipeptides (diketopiperazines), pyroglutamic acid derivatives, and their condensation products	[215]
Determination of relative response factors of impurities in paclitaxel with HPLC equipped with ultraviolet and charged aerosol detectors	Impurity testing	Paclitaxel. This compound is a complex tetracyclic substance with 11 stereoisomers. First isolated from the bark of the Pacific yew tree, it is a widely used in anticancer therapy. In this paper, the relative response factors (RRFs) for paclitaxel-related impurities were determined by HPLC-CAD and compared with those obtained using HPLC-UV. CAD provided almost consistent RRFs for paclitaxel and nine of its impurities. Using UV, RRFs varied significantly from one compound to another and indicated that significant errors in quantification can if the response factor of an unknown compound is assumed to be one	[216]

Development of a purity control strategy for pemetrexed disodium and validation of associated analytical methodology	Impurity testing	Pemetrexed. This synthetic compound is used in the treatment of cancers. It is formulated as a lyophilized sterile powder of pemetrexed disodium that is reconstituted prior to intravenous administration. Understanding the process and formation of degradation impurities is key to patient safety. The authors developed a purity control strategy and a validated analytical method using HPLC-UV in tandem with either CAD or NMR detection	[217]
Forced degradation and impurity profiling: recent trends in analytical perspectives. Journal of pharmaceutical and biomedical	Impurity testing	Review. This publication describes the various analytical approaches used for studying forced degradation and impurity profiling during 2008–2012	[218]
Validated HPLC method for the quantitative analysis of a 4-methanesulfonyl-piperidine hydrochloride salt	Impurity testing	Starting materials. Raw materials and intermediates used in early stage development of drug substances need to be pure and free from contaminants/impurities. This paper describes a validated HPLC-CAD method for analysis of a typical starting material, 4-Methanesulfonyl-piperidine hydrochloride salt	[219]
Sensitive analysis of underivatized amino acids using UHPLC with charged aerosol detection	Protein characterization	Tryptic digests. The separation and measurement of free amino acids is challenging due to their structural similarity and, for most, the lack of a chromophore. Consequently, free amino acids are typically measured following derivatization of a suitable chromophore or fluorophore. Presented in this paper are two methods for the direct measurement of unlabeled amino acids. The first method used gradient HPLC-CAD for the direct determination of 18 amino acids in under 9 min, with low ng sensitivity. The second method used a longer gradient HPLC-UV-CAD with an inverse compensation gradient for the characterization of protein tryptic digests. This automated approach could be used to study the time course of tryptic digestion and was found to offer a more comprehensive profile of the sample than when using UV (at 214 and 254 nm) detection alone	[64]

(Continued)

Table 2.7 (Continued)

Title	Application area	Overview	References
Label-free analysis by UHPLC with charged aerosol detection of glycans separated by charge, size, and isomeric structure	Protein characterization	Glycan analysis. Glycoproteins of biological, diagnostic, or therapeutic interest owe key aspects of their normal function to the oligosaccharides attached to the protein backbone. Changes in the number, type, composition, or linkage pattern of these glycans may serve as a biomarker of disease or influence the efficacy of a biotherapeutic product. For this reason, the ability to correctly identify and measure these glycans is of scientific interest, and to do so reliably, quickly, and inexpensively is of practical benefit. The authors describe a UHPLC-CAD method for the direct measurement of key *N*-linked glycans released from glycoproteins using PNGase-F. The method is simple, accurate, and precise and, unlike the commonly used fluorescence approach, does not require labeling	[220, 221]
Label-free profiling of *O*-linked glycans by HPLC with charged aerosol detection	Protein characterization	Glycan analysis. An HPLC-CAD method for the quantitative profiling of *O*-linked glycans released from glycoproteins by reductive beta-elimination is presented. Reduction helps prevent peeling side reactions; however, the resulting *O*-linked glycan alditols cannot be derivatized easily with a fluorescent label. Charged aerosol detection does not require a fluorophore or chromophore for sensitive, accurate quantification, and so HPLC-CAD provides a simple, direct approach to separate and quantify native glycans. Examples discussed included bovine fetuin, submaxillary mucin, hemocyanin, and IgG	[222]
Evaluation of charged aerosol detector for purity assessment of protein	Protein characterization	Purity. Commercially available protein reference standard materials are widely used for the quantitation of intact proteins in biopharmaceuticals. Unfortunately, their purity is often assumed to be 100%, or they may be assigned an inaccurate value because the methods used to measure their purity often lack specificity and accuracy. This paper uses HPLC-CAD to evaluate reference standard purity and showed that UV and MS approaches can lead to significant errors in purity measurement	[223]

Direct measurement of sialic acids released from glycoproteins, by HPLC and charged aerosol detection	Protein characterization	Sialic acid analysis. Sialic acid plays a number of vital physiological functions and is involved with neural transmission, synaptogenesis, and immunity. Sialylation can also improve the qualities of therapeutic glycoproteins, such as circulatory half-life, biological activity, and solubility. Therefore, it is important to determine the sialic acid content of such proteins when assaying for pharmaceutical therapeutic function and efficacy. Two important sialic acids are *N*-acetyl-neuraminic acid (Neu5Ac or NANA) and *N*-glycolyl-neuraminic acid (Neu5Gc), and these can be found in a number of biochemically relevant molecules including glycoproteins, proteoglycans, and glycolipids. Presented in this paper is an HPLC-CAD method for the direct measurement of NANA and Neu5Gc in neuraminidase digest of human and bovine transferrin	[224]
PEGylation of cholecystokinin prolongs its anorectic effect in rats	Protein conjugates	PEGylation. Many peptides showing useful biological activity are rapidly inactivated *in vivo*. PEGylation conjugation is an approach used to prolong a peptide's activity by preventing its renal clearance and inhibiting the activity of proteolytic enzymes. The octapeptide cholecystokinin-8 (CCK8) is known to induce meal termination and delay gastric emptying and induce satiety behavior. In order to study possible biological mechanisms of action a stable analog, PEG-CCK-9 was synthesized. A gradient HPLC-CAD method was used to study the CCK-9 PEGylation process	[225]
Analytical methods to qualify and quantify PEG and PEGylated biopharmaceuticals	Protein conjugates	PEGylation. An online 2D HPLC-CAD approach for the determination of PEGylated protein and free PEG was developed and used to study manual preparation of PEGylated IgG and the online preparation of PEGylated BSA	[226]
Measurement of stability and purity of cell-penetrating peptides (CPP) used for siRNA delivery	Stability testing	Drug delivery. CPP are often enriched with cationic groups (e.g., arginine, lysine residues). These charged amino acids impart a positive charge to the carrier that helps form the peptide/siRNA complex and facilitates penetration of the cell membrane and delivery of siRNA molecules. Analysis of CPP and fragments can be complicated, as amino acids have variable responses using HPLC-UV detection due to varying extinction coefficients. HPLC-CAD was used to quantify a number of cationic peptides along with their stability and purity	[227]

(*Continued*)

Table 2.7 (Continued)

Title	Application area	Overview	References
Direct stability-indicating method development and validation for analysis of etidronate disodium using a mixed-mode column and charged aerosol detector	Stability testing	Etidronate. Editronate disodium, a bisphosphonate, is a potent inhibitor of bone resorption used in the treatment of cancer hypercalcemia and metastatic bone disease. Analysis is challenging as etidronate is a highly polar, strong chelator and lacks a chromophore. The authors developed a validated HPLC-CAD method using a mixed-mode column. The method can be used for release and stability testing and has applicability to other bisphosphonate compounds	[228]
Novel MS solutions inspired by MIST	Safety testing	MIST. The metabolites in safety testing (MIST) guidelines were released in 2008 by the USDA, FDA, and CDER. The guidelines are a suggested, FDA-endorsed list of practices for safety testing of new compounds before clinical trials. The guidelines take into account differences between human and animal drug metabolism, the possible generation of phase 1 reactive intermediates, and address their detection prior to clinical trials, so that they do not show up as a surprise during the trials. This review paper discusses novel MS solutions to address MIST and how the universal response of CAD can also benefit this area	[229]
Use of charged aerosol detection as an orthogonal quantification technique for drug metabolites in safety testing (MIST)	Safety testing	MIST. HPLC-CAD applications to the area of MIST, including metabolism of buspirone and erythromycin are discussed	[230]
Metabolites in safety testing: metabolite identification strategies in discovery and development	Safety testing	MIST. In this review on MIST, the author evaluates a number of analytical methods used for metabolite quantitation including HPLC-CAD	[231]

for mass balance studies, degradant, and impurity measurement where uncharacterized analytes might not possess a chromophore or may poorly ionize [34, 179, 195]. Second, a number of papers studying drug formulation describe the use of CAD to measure the API, counter ion(s), and other excipients simultaneously on an HPLC system without the need for separate assays or analytical systems [48, 177–179].

Although there is some overlap between the analytical needs of pharmaceutical and biopharmaceutical industries, the latter has some unique challenges that can be addressed using CAD. Selected examples include *label-free* glycan measurement without the need for derivatization and fluorescence detection or anion-exchange PAD approaches [74, 220–222] and the characterization of different drug delivery vehicles including liposomes [108, 109, 112, 117].

2.3.5 Other Application Areas

Table 2.8 presents how CAD is being applied to other application areas including the analysis of environmental samples (e.g., antibiotic contamination of soils and ground water resulting from their use in chicken feed), characterization of biofuels, and measurement of artificial sweeteners in food.

2.4 Conclusions

As can be seen by the nearly 300 references, CAD has been utilized in a wide variety of application areas in multiple industries. Although it is most popular for analytes that do not contain chromophores, its universal (chemically independent) quantitative response has made it an ideal complementary technology to other HPLC detectors (e.g., UV and MS). Furthermore, relative to the closest alternative technology (ELSD), CAD typically demonstrates improved sensitivity, wider dynamic range, and more universal response. CAD is often the method of choice for detecting lipids and surfactants and is a label-free alternative to fluorescence and MS when high sensitivity or chemical identification is not required. However, for all its advantages, CAD is not without limitations and, similar to other aerosol-based detectors, requires the use of volatile mobile phases and shows a gradient effect on analyte response. Nonetheless, with continued evolution of CAD technology, one would expect its adoption to broaden while addressing ever more challenging applications.

Acknowledgements

The authors are grateful to Rainer Bauder and Brent Morrison for their editorial review.

Table 2.8 Miscellaneous CAD publications.

Title	Application area	Overview	References
Support of academic synthetic chemistry using separation technologies from the pharmaceutical industry	Academic	Review. This mini-review article gives an overview of how state-of-the-art separation tools used by the pharmaceutical industry can be applied by those involved with academic synthetic chemistry. One example shows the use of chiral SFC-DAD-CAD-MS for the separation and identification of a complex mixture of six major isomers and other impurities from a reaction mixture	[57]
Determination of parabens in cosmetic products using multi-walled carbon nanotubes as solid-phase extraction sorbent and Corona-charged aerosol detection system	Cosmetics	Preservatives. Parabens are used as preservatives in cosmetics, foods, and pharmaceuticals to prevent microbial and fungal damage. The authors investigated the use of carbon nanotubes for the solid-phase extraction of parabens in cosmetic products. Four parabens were determined using a validated high-performance liquid chromatography (HPLC)-CAD method	[232]
Quantitation of pluronics by HPLC and Corona-charged aerosol detection	Cosmetics	Surfactants. Pluronics are surfactants used in a variety of consumer products such as fluoride-containing mouthwash. Two methods are discussed—an HPLC-CAD method capable of quantifying different pluronics standards (single peak measurement) including pluronics F68, L64, P85, P123, and F127 and a method for measurement of F127 in mouthwash	[233]
The effect of composting on the degradation of a veterinary pharmaceutical	Environmental	Antibiotics. Composting is a valuable means of lessening the environmental impact of antibiotics in manure. The authors studied the impact of different parameters (pH, temperature, microbial enzymes, and microorganisms) on the degradation of salinomycin in poultry manure. Salinomycin levels were determined by HPLC-CAD	[234]

Effect of soil pH on sorption of salinomycin in clay and sandy soils	Environmental	Antibiotics. Salinomycin is an antibiotic used in the poultry industry to prevent infection and to promote growth. A large amount of the antibiotic is excreted unchanged and may result in the contamination of agricultural land. How the drug reacts with soil and the environmental consequences are the topic of this paper. The authors studied the drug's sorption–desorption characteristics in four different agricultural soils at different pH levels. They concluded that salinomycin in soil poses a significant contamination threat to both shallow ground water and surface water bodies. Salinomycin levels were determined by HPLC-CAD	[235]
Determination of erythrocin in chicken manure by HPLC-Corona-charged aerosol detection coupled with online solid-phase extraction	Environmental	Antibiotics. A novel HPLC-CAD method with online solid-phase extraction was used to measure erythromycin (erythrocin) in chicken manure	[236]
Enhancing bioethanol production from delactosed whey permeate by upstream desalination techniques	Environmental	Food. Industrial cheese production results in tons of whey that cannot be disposed of via the local sanitation system due to its organic and inorganic content. Whey is usually processed to remove valuable compounds such as lactose and proteins, but such procedures are inefficient and the resulting delactosed whey permeate (LWP) may still contain lactose (around 10% of the original content). One approach to this problem is to convert the remaining lactose to ethanol using yeast. However, high salt content can deactivate yeast. The authors evaluated three approaches to lower salt levels: (i) simple dilution, (ii) nanofiltration, and (iii) electrodialysis. Lactose levels were determined by HILIC-CAD	[237]
Enzymatic reaction coupled with flow-injection analysis with charged aerosol, coulometric, or amperometric detection for estimation of contamination of the environment by pesticides	Environmental	Pesticides. The authors developed a method to indirectly measure chlorinated pesticides. The enzyme haloalkane deyhdrogenase was used to cleave the test compound, 1-chlorocyclohexane, and the chloride formed was then determined using either HPLC-electrochemical detector (ECD) or HPLC-CAD	[238]

(*Continued*)

Table 2.8 (Continued)

Title	Application area	Overview	References
Characterization of dispersants by reversed-phase high-pressure liquid chromatography (LC) and charged aerosol detection	Environmental	Oil spills. Dispersants (surfactants) are used to treat oils spills. This paper describes an HPLC-CAD method for that can be used to characterize different dispersants including Span® 20, 83, and 85, as well as the different components of Corexit® 9500 (including Aerosol® OT, Spans 60 and 80, and Tween® 80 and 85)	[239]
Determination of perfluorinated carboxylic acids (PFCAs) in water using LC coupled to a Corona-charged aerosol detector	Environmental	PFCAs. PFCAs are employed in lubricants, paints, cosmetics, and fire-fighting foams and have become a common environmental pollutant of river water, surface water, groundwater, and sewage. The US Environmental Protection Agency has classified PFCAs as likely human carcinogens, thus monitoring of their environmental levels is essential. This paper describes an HPLC-CAD method for monitoring PFCAs in wastewater samples	[240]
Acetone as a greener alternative to acetonitrile in liquid chromatographic fingerprinting	Environmental	Solvents. A considerable amount of organic solvent waste is generated from LC analysis. One solvent typically used is acetonitrile, which has favorable physicochemical properties but unwelcome environmental impact. The authors investigated the possibility of substituting acetonitrile with acetone, using HPLC-CAD methods, and showed that acetone is a viable alternate solvent	[241]
A review of separation methods for the determination of estrogens and plastics-derived estrogen mimics from aqueous systems	Environmental	Xenoestrogens. This paper reviews the analytical approaches, including HPLC-CAD, that can be used to measure xenoestrogens in aqueous systems	[242]
A new approach to the simultaneous analysis of underivatized ionophoric antibiotics using LC and charged aerosol detection	Food	Antibiotics. A gradient UHPLC-CAD method was developed for the simultaneous measurement of the antibiotics lasolacid A, onensin A and B, narasin, and salinomycin. Spiked chicken muscle extract was used as the example	[243]

Simultaneous determination of aspartame, acesulfame-K, saccharin, citric acid, and sodium benzoate in various food products using HPLC-CAD-UV/DAD	Food	Arificial sweeteners. A validated HPLC-DAD-CAD method was developed capable of simultaneously measuring aspartame, acesulfame-K, saccharin, citric acid, and sodium benzoate in various food products including table sweeteners, sports drinks, and dietary supplements	[244]
A fast and efficient method to assess 2D-HPLC column and method combinations for food metabolomics studies	Food	Metabolomics. Understanding metabolism of distinct food ingredients is an important prerequisite in sensory science. Measuring the various components of the food metabolome is challenging due to the diversity in chemical structure, differences in relative abundance, and complexity of food matrices. The authors describe a straightforward 2D-LC method development approach, avoiding full 2D-LC experiments and enables exploration of all possible combinations of fast and easy methods. Compounds identification was achieved by MS/MS, prequantification by CAD and exact quantification by MS with stable isotope-labeled internal standards	[245]
Analysis of emulsifiers in foods by HPLC and Corona-charged aerosol detection	Food	Surfactants. Emulsifiers, typically surfactants, are used by the food industry to maintain consistency, for example, in margarine and mayonnaise, as dough conditioners and as thickening agents, for example, in ice cream. Presented here are HPLC-CAD methods for the measurement of soy lecithin in granola bars and hydroxypropylmethylcelluose in ice cream and popsicles	[246]
Determination of water-soluble vitamins in infant milk and dietary supplement using a LC online coupled to a Corona-charged aerosol detector	Food	Vitamins. A simple and rapid HPLC-CAD method for the simultaneous determination of seven water-soluble vitamins (thiamine, folic acid, nicotinic acid, ascorbic acid, pantothenic acid, pyridoxine, and biotin) is presented. The optimized method was used to analyze different infant milk samples and dietary supplements. The results were in good agreement with the declared values	[247]

(*Continued*)

Table 2.8 (Continued)

Title	Application area	Overview	References
A single method for the direct determination of total glycerols in all biodiesels using LC and charged aerosol detection	Industrial	Biofuel. Biodiesel provides a clean and renewable liquid fuel that can be used in current diesel engines and oil burners without significant modifications. Natural oils, such as virgin and waste cooking oils and algal oils, are used as feedstock and are esterified to form biodiesel. The simplest approach uses a basic esterification reaction with methanol, sodium hydroxide, and heat. The reaction esterifies the fatty acids of the oil, producing fatty acid methyl esters (FAMEs), which is the biodiesel fuel. Harmful impurities, such as unreacted acylated and free glycerols, must be removed to avoid fuel system damage, for example, fuel filter clogging or fuel injector damage. The simple, normal-phase HPLC-CAD method described in this article provides a measurement of all acylated and free glycerols. Any biodiesel sample, in-process, finished, or blended, is diluted and analyzed directly in <25 min. The method also provides the necessary sensitivity to quantify total glycerols to the current ASTM specifications	[248]
Determination of residual acylglycerols in biodiesel	Industrial	Biofuel. Many methods have been proposed to determine acylglycerols in biodiesel samples. This analysis is commonly performed by gas chromatography/flame ionization detection (GC/FID) methods such as ASTM D6584 and EN14105. However, GC methods often require sample derivatization, and high inlet temperatures can cause thermal degradation and underestimation of the sample amount. HPLC-CAD is emerging as the preferred method for biodiesel analysis. Presented here is a rapid separation LC (RSLC)-CAD method for analysis of acylglycerols in biodiesel fuel in under 10 min	[249]
An improved global method for the quantitation and characterization of lipids by high-pressure LC and charged aerosol detection	Industrial	Biofuel. A gradient HPLC-CAD method was developed capable of resolving and detecting various classes of lipids including esters, acylglycerols, fatty alcohols, fatty acids, and paraffins. The method was used to measure the lipid profiles of algal oil and emu oil	[250]

Determination of polyacrylic acid (PAA) in boiler water using size-exclusion chromatography with charged aerosol detection	Industrial	Electricity generation. Boiler scale is formed when impurities in water precipitate onto heat transfer processes and can lead to decreased efficiency and boiler damage. PAA is an anti-scaling additive typically used in conventional and nuclear steam generators. A SEC-CAD method is described for routine measurement of PAA in boiler water	[251]
Quantitation of hindered amine light stabilizers (HALS) by LC and charged aerosol detection	Industrial	HALS. Plastics, especially those developed for outside use, require additives such as HALS to provide polymeric stability from the effects of temperature and light. HALS consist of a wide array of compounds, designed to absorb light, provide increased mechanical strength, and/or improve thermal stability. Their measurement can be challenging due to their complexity and physicochemical properties. HPLC-CAD methods using either solid core or porous shell columns were developed. The solid core approach was used to profile different HALS including Hostavin® N30, Irgafos® 168, and Tinuvin 622 and 770; the porous shell approach was used for Cyasorb UV3529 and Sabo® stab UV119 (also now known as Chimassorb® 119 FL)	[252]
Analysis of quaternary ammonium and phosphonium ionic liquids (IL) by reversed-phase HPLC with charged aerosol detection and unified calibration	Industrial	Ionic liquids. An IL is a salt in the liquid state. Their unique properties (e.g., negligible vapor pressure, good thermal stability, tunable miscibility) have been utilized in numerous application including chemical synthesis, gas handling, pharmaceutical development, cellulose processing, waste recycling, and battery development. This paper describes a fully validated HPLC-DAD-CAD method that can be used to measure a new suite of IL derived from either tricaprylmethylammonium chloride or trihexyltetradecylphosphonium chloride as precursors. The method used a unified calibration function as no single component calibrant was available	[253]

(*Continued*)

Table 2.8 (Continued)

Title	Application area	Overview	References
Evaluation of column bleed by using an ultraviolet and a charged aerosol detector coupled to a high-temperature liquid chromatographic system	Industrial	Manufacturing. Stability of guard and analytical columns is a key performance characteristic for companies manufacturing HPLC columns. The authors describe a simple method to help characterize column stability using HPLC-DAD-CAD. Five columns were heated to 200°C, and the detector response was used to indicate column performance. The authors concluded that silica-based C18 column had the highest bleed, and carbon-clad titanium dioxide columns (typically used for extreme pH applications) had the lowest bleed. Measurement was independent of detection technique or wavelength used	[254]
Evaluation of analytical methods for determination of kinetic hydrate inhibitor (KHI) in produced waters	Industrial	Oil/gas. Hydrates (ice-like crystals) can form under the high pressures and low temperatures found in pipelines of oil and gas fields and can cause blockage, hindering process operations and causing economic loss. The addition of KHIs, such as hyperbranched poly-ester-amide polymers, is used to mitigate this problem. The authors developed a size-exclusion method coupled to either a refractive index detector or CAD and evaluated their method for the measurement of KHIs in produced water	[55]
Effect of temperature on the analysis of asphaltenes by the on-column filtration/redissolution method	Industrial	Oil/gas. Asphaltenes are organic molecular substances that are found in crude oil. In the form of asphalt or bitumen products, they are used as paving materials on roads, shingles for roofs, and waterproof coatings on building foundations. Asphaltenes impart high viscosity to crude oils and can negatively impact production, through fouling in the heat exchangers of the crude oil distillation preheat train. They are present within micelles in crude oil, which can be broken down by reaction with paraffins under high temperature. Once the protective micelle is destroyed, polar asphaltenes agglomerate and are transported to the tube walls, where they can stick and form a foulant layer. The authors combined an on-column filtration/redisolution method along with HPLC-CAD to study the effects of temperature on asphaltene analysis and used it to study asphaltenes from different geographical locations. Mass balance for virgin vacuum residue and processed n-heptane asphaltenes were effectively carried out and the presence of adsorbed species was found	[255]

A comparison of scale inhibitor return concentrations obtained with a novel analytical method and current commercial techniques	Industrial	Oil/gas. Wells that produce water are likely to develop deposits of inorganic scales that can coat key components (e.g., valves and pumps) that, if allowed to proceed, will limit production and eventually requiring abandonment of the well. In order to prevent this from occurring, scale inhibitor chemicals are pumped into the well, for example, the "inhibitor squeeze." Here an inhibitor-containing solution is forced into the formation. The inhibitor then resides on the rock surface and slowly leaches back into the produced-water phase at or above the critical concentration needed to prevent scaling. The authors developed an HPLC-CAD method for the determination of polymer-based scale inhibitors. The method was used to evaluate inhibitor content in samples obtained from several commercial squeeze programs performed on various North Sea fields. The method was able to generate reliable data where current commercial techniques fail	[256]
Quantitation and characterization of copper-plating bath additives by LC with charged aerosol detection and electrochemical detection	Industrial	Plating baths. Copper plating using an acid bath is a process widely used in the manufacture of a variety of products. Different additives can be used to control different aspects of the copper-plating process and must be accurately quantified. In this article, an approach using a dual LC system with an ECD and a charged aerosol detector (CAD) is described for the quantitation of three additives typically used. The methods are precise and sensitive for the determination of all additives, giving a quantitative measure of suppressor and suppressor degradation. Calibration curves and sample analysis results are reported for all additives. Both analyses can be run using the same sample preparation and on the same system	[257, 258]
Measurement and control of copper additives in electroplating baths by HPLC	Industrial	Plating baths. A comprehensive analytical approach to measurement of key additives in neutralized plating bath solution. One approach, HPLC-ECD, is used to simultaneously measure accelerator and leveler, while a second approach HPLC-CAD is used to simultaneously measure accelerator and suppressor. This approach was shown to provide higher data quality than the typical cyclic voltammetric stripping method used to measure levels of plating bath additives	[259]

(*Continued*)

Table 2.8 (Continued)

Title	Application area	Overview	References
Quantitation and characterization of copper- and nickel-plating bath additives by LC	Industrial	Plating baths. A comprehensive analytical approach to measurement of key additives in neutralized copper-plating bath solutions using one analytical system is presented. HPLC-ECD was used to simultaneously measure accelerator and leveler, while a HPLC-CAD was used to simultaneously measure accelerator and suppressor. The use of HPLC-UV-CAD for measuring saccharin and sodium alkyl sulfate in nickel plating bath solutions is also presented	[260]
Analysis of silicone oils by HPLC and Corona-charged aerosol detection	Industrial	Silicone oil. Two HPLC-CAD methods were developed to either qualify (measure relative abundance of individual components) or quantify (measure levels of analyte as a single peak) heating bath silicone oil or silicone oils present in shampoo, hair conditioner, or gas relief products	[261]
Quantitation of surfactants in samples by HPLC and Corona-charged aerosol detection	Industrial	Surfactants. The use of HPLC-CAD for the measurement of surfactants in metal working oils, fluoride mouthwash, hair conditioner, and laundry detergents is discussed	[262, 263]
Unusual temperature-induced retention behavior of constrained β-amino acid enantiomers on the zwitterionic chiral stationary phases ZWIX (+) and ZWIX (−)	Research	Chirality. The effects of temperature on the chiral recognition of cyclic β-amino acids enantiomers on zwitterionic chiral stationary phases were investigated using HPLC-CAD	[264]

References

1 Cohen RD, Liu Y. Advances in aerosol-based detectors. Advances in Chromatography, 2014; 52; 1–53.
2 Thomas D, Bailey B, Plante M, Acworth IN. Charged aerosol detection and evaporative light scattering detection—fundamental differences affecting analytical performance. Thermo Scientific. PN70990; 2015.
3 Márquez-Sillero I, Cárdenas S, Valcárcel M. Comparison of two evaporative universal detectors for the determination of sugars in food samples by liquid chromatography. Microchemical Journal, 2013; 110; 629–635.
4 Thomas D, Bailey B, Plante M, Acworth IN. Charged aerosol detection and evaporative light scattering detection: Fundamental differences affecting analytical performance. Thermo Scientific. PN70990; 2014.
5 Cohen RD, Liu Y, Gong X. Analysis of volatile bases by high performance liquid chromatography with aerosol-based detection. Journal of Chromatography A, 2012; 1229; 172–179.
6 Wipf P, Werner S, Twining LA, Kendall C. HPLC determinations of enantiomeric ratios. Chirality, 2007; 19; 5–9.
7 Hutchinson JP, Li J, Farrell W, Groeber E, Szucs R, Dicinoski G, Haddad PR. Comparison of the response of four aerosol detectors used with ultra-high pressure liquid chromatography. Journal of Chromatography A, 2011; 1218; 1646–1655.
8 He Y, Cook KS, Littlepage E, Cundy J, Mangalathillam R, Jones MT. Ion-pair reversed phase liquid chromatography with ultraviolet detection for analysis of ultraviolet transparent cations. Journal of Chromatography A, 2015; 1408; 261–266.
9 Merle C, Laugel C, Chaminade P, Baillet-Guffroy A. Quantitative study of the stratum corneum lipid classes by normal phase liquid chromatography: Comparison between two universal detectors. Journal of Liquid Chromatography, 2010; 33; 629–644.
10 Ramos RG, Libong D, Rakotomanga M, Gaudin K, Loiseau PM, Chaminade P. Comparison between charged aerosol detection and light scattering detection for the analysis of Leishmania membrane phospholipids. Journal of Chromatography A, 2008; 1209; 88–94.
11 Damnjanović J, Nakano H, Iwasaki Y. Simple and efficient profiling of phospholipids in phospholipase D-modified soy lecithin by HPLC with charged aerosol detection. Journal of the American Oil Chemists' Society, 2013; 90; 951–957.
12 Eom HY, Park SY, Kim MK, Suh JH, Yeom H, Min JW, Kim U, Lee J, Youm J-R, Han SB. Comparison between evaporative light scattering detection and charged aerosol detection for the analysis of saikosaponins. Journal of Chromatography A, 2010; 1217; 4347–4354.

13 Filip K, Grynkiewicz G, Gruza M, Jatczak K, Zagrodzki B. Comparison of ultraviolet detection and charged aerosol detection methods for liquid-chromatographic determination of protoescigenin. Acta Poloniae Pharmaceutica, 2014; 71; 933–938.
14 Wang L, He WS, Yan HX, Jiang Y, Bi KS, Tu PF. Performance evaluation of charged aerosol and evaporative light scattering detection for the determination of ginsenosides by LC. Chromatographia, 2009; 70; 603–608.
15 Pistorino M, Pfeifer BA. Polyketide analysis using mass spectrometry, evaporative light scattering, and charged aerosol detector systems. Analytical and Bioanalytical Chemistry, 2008; 390; 1189–1193.
16 Jia S, Park JH, Lee J, Kwon SW. Comparison of two aerosol-based detectors for the analysis of gabapentin in pharmaceutical formulations by hydrophilic interaction chromatography. Talanta, 2011; 85; 2301–2306.
17 Shaodong J, Lee WJ, Ee JW, Park JH, Kwon SW, Lee J. Comparison of ultraviolet detection, evaporative light scattering detection and charged aerosol detection methods for liquid-chromatographic determination of anti-diabetic drugs. Journal of Pharmaceutical and Biomedical Analysis, 2010; 51; 973–978.
18 Cintrón JM, Risley, DS. Hydrophilic interaction chromatography with aerosol-based detectors (ELSD, CAD, NQAD) for polar compounds lacking a UV chromophore in an intravenous formulation. Journal of Pharmaceutical and Biomedical Analysis, 2013; 78; 14–18.
19 Kim BH, Jang JB, Moon DC. Analysis of ionic surfactants by HPLC with evaporative light scattering detection and charged aerosol detection. Journal of Liquid Chromatography and Related Technologies, 2013; 36; 1000–1012.
20 Nair LM, Werling JO. Aerosol based detectors for the investigation of phospholipid hydrolysis in a pharmaceutical suspension formulation. Journal of Pharmaceutical and Biomedical Analysis, 2009; 49; 95–99.
21 Takahashi K, Kinugasa S, Senda M, Kimizuka K, Fukushima K, Matsumoto T, Christensen J. Quantitative comparison of a corona-charged aerosol detector and an evaporative light-scattering detector for the analysis of a synthetic polymer by supercritical fluid chromatography. Journal of Chromatography A, 2008; 1193; 151–155.
22 Kou D, Manius G, Zhan S, Chokshi HP. Size exclusion chromatography with Corona charged aerosol detector for the analysis of polyethylene glycol polymer. Journal of Chromatography A, 2009; 1216; 5424–5428.
23 Lobback C, Backensfeld T, Funke A, Weitschies W. Quantitative determination of nonionic surfactants with CAD. Chromatography Techniques, 2007; November, 18–20.
24 Holzgrabe U, Nap CJ, Almeling S. Control of impurities in L-aspartic acid and L-alanine by high-performance liquid chromatography coupled with a corona charged aerosol detector. Journal of Chromatography A, 2010; 1217; 294–301.

25 Vervoort N, Daemen D, Török G. Performance evaluation of evaporative light scattering detection and charged aerosol detection in reversed phase liquid chromatography. Journal of Chromatography A, 2008; 1189; 92–100.
26 Loughlin J, Phan H, Wan M, Guo S, May K, Lin B. Evaluation of charged aerosol detection (CAD) as a complementary technique for high-throughput LC-MS-UV-ELSD analysis of drug discovery screening libraries. American Laboratory, 2007; 39; 24–27.
27 Kaufman SL (2003). Evaporative Electrical Detector. US Patent # US 6,568,245 B2.
28 Dixon RW, Peterson DS. Development and testing of a detection method for liquid chromatography based on aerosol charging. Analytical Chemistry, 2002; 74; 2930–2937.
29 Vehovec T, Obreza A. Review of operating principle and applications of the charged aerosol detector. Journal of Chromatography A, 2010; 1217; 1549–1556.
30 Magnusson LE, Risley DS, Koropchak JA. Aerosol-base detectors for liquid chromatography. Journal of Chromatography A, 2015; 1421; 68–81.
31 Chaminade P. (2011). Evaporative light scattering and charged aerosol detector. In: Hyphenated and Alternative Methods of Detection in Chromatography. Chap 5. 146–159. Shalliker AR (editor). CRC Press, Boca Raton.
32 Gorecki T, Lynen F, Szucs R, Sandra P. Universal response in liquid chromatography using charged aerosol detection. Analytical Chemistry, 2006; 78; 3186–3192.
33 Mitchell CR, Bao Y, Benz NJ, Zhang S. Comparison of the sensitivity of evaporative universal detectors and LC/MS in the HILIC and the reversed-phase HPLC modes. Journal of Chromatography B, 2009; 877; 4133–4139.
34 Reilly J, Everatt B, Aldcroft C. Implementation of charged aerosol detection in routine reversed phase liquid chromatography methods. Journal of Liquid Chromatography and Related Technologies, 2008; 31; 3132–3142.
35 Lie A, Meyer AS, Pedersen LH. Appearance and distribution of regioisomers in metallo-and serine-protease-catalyzed acylation of sucrose in *N,N*-dimethylformamide. Journal of Molecular Catalysis B: Enzymatic, 2014; 106; 26–31.
36 Ayres BT, Valença GP, Franco TT, Adlercreutz P. Two-step process for preparation of oligosaccharide propionates and acrylates using lipase and Cyclodextrin Glycosyl Transferase (CGTase). Sustainable Chemical Processes, 2014; 2; 6; http://www.biomedcentral.com/content/pdf/2043-7129-2-6.pdf (accessed January 28, 2017).
37 Lisa M, Lynen F, Holčapek M, Sandra P. Quantitation of triacylglycerols from plant oils using charged aerosol detection with gradient compensation. Journal of Chromatography A, 2007; 1176; 135–142.

38 Moreau RA. The analysis of lipids via HPLC with a charged aerosol detector. Lipids, 2006; 41; 727–734.
39 Iwasaki Y, Masayama A, Mori A, Ikeda C, Nakano H. Composition analysis of positional isomers of phosphatidylinositol by high-performance liquid chromatography. Journal of Chromatography A, 2009; 1216; 6077–6080.
40 Kiełbowicz G, Micek P, Wawrzeńczyk C. A new liquid chromatography method with charge aerosol detector (CAD) for the determination of phospholipid classes. Application to milk phospholipids. Talanta, 2013; 105; 28–33.
41 Kristinova V, Aaneby J, Mozuraityte R, Storrø I, Rustad T. The effect of dietary antioxidants on iron-mediated lipid peroxidation in marine emulsions studied by measurement of dissolved oxygen consumption. European Journal of Lipid Science and Technology, 2014; 116; 857–871.
42 Plante M, Bailey B, Acworth IN. Characterization of used cooking oils by high performance liquid chromatography and corona charged aerosol detection. Thermo Scientific. PN70536; 2013.
43 Grembecka M, Lebiedzińska A, Szefer P. Simultaneous separation and determination of erythritol, xylitol, sorbitol, mannitol, maltitol, fructose, glucose, sucrose and maltose in food products by high performance liquid chromatography coupled to charged aerosol detector. Microchemical Journal, 2014; 117; 77–82.
44 Kus-Liśkiewicz M, Górka A, Gonchar M. Simple assay of trehalose in industrial yeast. Food Chemistry, 2014; 158; 335–339.
45 Łuczaj Ł, Bilek M, Stawarczyk K. Sugar content in the sap of birches, hornbeams and maples in southeastern Poland. Central European Journal of Biology, 2014; 9; 410–416.
46 Igarashi K, Ishida T, Hori C, Samejima M. Characterization of an endoglucanase belonging to a new subfamily of glycoside hydrolase family 45 of the basidiomycete *Phanerochaete chrysosporium*. Applied and Environmental Microbiology, 2008; 74; 5628–5634.
47 Kim JH, Lim BC, Yeom SJ, Kim YS, Kim HJ, Lee JK, Oh DK. Differential selectivity of the *Escherichia coli* cell membrane shifts the equilibrium for the enzyme-catalyzed isomerization of galactose to tagatose. Applied and Environmental Microbiology, 2008; 74; 2307–2313.
48 Zhang K, Dai L, Chetwyn NP. Simultaneous determination of positive and negative pharmaceutical counterions using mixed-mode chromatography coupled with charged aerosol detector. Journal of Chromatography A, 2010; 1217; 5776–5784.
49 Hinterwirth H, Lämmerhofer M, Preinerstorfer B, Gargano A, Reischl R, Bicker W, Lindner W. Selectivity issues in targeted metabolomics: Separation of phosphorylated carbohydrate isomers by mixed-mode hydrophilic interaction/weak anion exchange chromatography. Journal of Separation Science, 2010; 33; 3273–3282.

50 Crafts C, Plante M, Bailey B, Acworth IN. Novel analytical method to verify effectiveness of cleaning processes. Thermo Scientific. 1820-4; 2012.
51 Chantarasukon C, Tukkeeree S, Rohrer J. Simultaneous determination of tartaric acid and tolterodine in tolterodine tartrate. Thermo Scientific. AN1047; 2013.
52 Liu X, Tracy M. Simultaneous determination of metformin and its chloride counterion using multi-mode liquid chromatography with charged aerosol detection. Thermo Scientific. AN20868; 2013.
53 Tracy M, Liu X. API and counterions in Adderall® using multi-mode liquid chromatography with charged aerosol detection. Thermo Scientific. AN20870; 2013.
54 He Y, Friese OV, Schlittler MR, Wang Q, Yang X, Bass LA, Jones MT. On-line coupling of size exclusion chromatography with mixed-mode liquid chromatography for comprehensive profiling of biopharmaceutical drug product. Journal of Chromatography A, 2012; 1262; 122–129.
55 Turkmen IR, Upadhyay N, Adham S, Gharfeh S. Evaluation of analytical methods for determination of kinetic hydrate inhibitor (KHI) in produced waters. Journal of Petroleum Science and Engineering, 2015; 126; 63–68.
56 Brunelli C, Gorecki T, Zhao Y, Sandra P. Corona-charged aerosol detection in supercritical fluid chromatography for pharmaceutical analysis. Analytical Chemistry, 2007; 79; 2472–2482.
57 Regalado EL, Kozlowski MC, Curto JM, Ritter T, Campbell MG, Mazzotti AR, Hamper B, Spilling C, Mannino M, Wan L, Yu J-Q, Liu J, Welch CJ. Support of academic synthetic chemistry using separation technologies from the pharmaceutical industry. Organic and Biomolecular Chemistry, 2014; 12; 2161–2166.
58 Moreau RA. Lipid analysis via HPLC with a charged aerosol detector. Lipid Technology, 2009; 21; 191–194.
59 Seiwert B, Giavalisco P, Willmitzer L. (2009). Advanced mass spectrometry methods for analysis of lipids from photosynthetic organisms. In: Lipids in Photosynthesis. 445–461. Wada H, Murata N (editors). Springer, Dordrecht.
60 Karu N, Dicinoski GW, Haddad PR. Use of suppressors for signal enhancement of weakly-acidic analytes in ion chromatography with universal detection methods. TrAC Trends in Analytical Chemistry, 2012; 40; 119–132.
61 Karu N, Hutchinson JP, Dicinoski GW, Hanna-Brown M, Srinivasan K, Pohl CA, Haddad PR. Determination of pharmaceutically related compounds by suppressed ion chromatography: IV. Interfacing ion chromatography with universal detectors. Journal of Chromatography A, 2012; 1253; 44–51.
62 Almeling S, Ilko D, Holzgrabe U. Charged aerosol detection in pharmaceutical analysis. Journal of Pharmaceutical and Biomedical Analysis, 2012; 69; 50–63.
63 Sinclair I, Gallagher R. Charged aerosol detection: Factors for consideration in its use as a generic quantitative detector. Chromatography Today, 2008; 1; 5–9.

64 Crafts C, Plante B, Bailey B, Acworth IN. Sensitive analysis of underivatized amino acids using UHPLC with charged aerosol detection. Thermo Fisher. PN70038; 2012.

65 Tsukagoshi H, Nakamura A, Ishida T, Touhara KK, Otagiri M, Moriya S, Arioka M. Structural and biochemical analyses of glycoside hydrolase family 26 β-mannanase from a symbiotic protist of the termite *Reticulitermes speratus*. The Journal of Biological Chemistry, 2014; 289; 10843–10852.

66 Rather MY, Ara KZG, Karlsson EN, Adlercreutz P. Characterization of cyclodextrin glycosyltransferases (CGTases) and their application for synthesis of alkyl glycosides with oligomeric head group. Process Biochemistry, 2015; 50; 722–728.

67 Wada J, Honda Y, Nagae M, Kato R, Wakatsuki S, Katayama T, Yamamoto K. 1, 2-α-L-Fucosynthase: A glycosynthase derived from an inverting α-glycosidase with an unusual reaction mechanism. FEBS Letters, 2008; 582; 3739–3743.

68 Dixon RW, Baltzell G. Determination of levoglucosan in atmospheric aerosols using high performance liquid chromatography with aerosol charge detection. Journal of Chromatography A, 2006; 1109; 214–221.

69 Acworth IN. Chromatography for foods and beverages: Carbohydrates analysis applications notebook. Thermo Scientific. AI71469; 2015.

70 Zhang Q, Hvizd M, Bailey B, Thomas D, Plante M, Acworth IN. Carbohydrate analysis in beverages and foods using pulsed amperometric detection or charged aerosol detection. Thermo Scientific. PN71433; 2014.

71 Li J, Hu D, Zong W, Lv G, Zhao J, Li S. Determination of inulin-type fructooligosaccharides in edible plants by high-performance liquid chromatography with charged aerosol detector. Journal of Agricultural and Food Chemistry, 2014; 62; 7707–7713.

72 Nishimoto M, Kitaoka M. Practical preparation of lacto-*N*-biose I, a candidate for the bifidus factor in human milk. Bioscience, Biotechnology, and Biochemistry, 2007; 71; 2101–2104.

73 Xiao JZ, Takahashi S, Nishimoto M, Odamaki T, Yaeshima T, Iwatsuki K, Kitaoka M. Distribution of *in vitro* fermentation ability of lacto-*N*-biose I, a major building block of human milk oligosaccharides, in bifidobacterial strains. Applied and Environmental Microbiology, 2010; 76; 54–59.

74 Bailey B, Ullucci P, Bauder R, Plante M, Crafts C, Acworth IN. Carbohydrate analysis using HPLC with PAD, FLD, CAD and MS detectors. Thermo Scientific. PN70026; 2012.

75 Hutchinson JP, Remenyi T, Nesterenko P, Farrell W, Groeber E, Szucs R, Dicinoski G, Haddad PR. Investigation of polar organic solvents compatible with Corona Charged Aerosol Detection and their use for the determination of sugars by hydrophilic interaction liquid chromatography. Analytica Chimica Acta, 2012; 750; 199–206.

76 Godin B, Agneessens R, Gerin PA, Delcarte J. Composition of structural carbohydrates in biomass: Precision of a liquid chromatography method using a neutral detergent extraction and a charged aerosol detector. Talanta, 2011; 85; 2014–2026.

77 Westereng B, Agger JW, Horn SJ, Vaaje-Kolstad G, Aachmann FL, Stenstrøm YH, Eijsink VG. Efficient separation of oxidized cello-oligosaccharides generated by cellulose degrading lytic polysaccharide monooxygenases. Journal of Chromatography A, 2013; 1271; 144–152.

78 Winarni I, Koda K, Waluyo TK, Pari G, Uraki Y. Enzymatic saccharification of soda pulp from sago starch waste using sago lignin-based amphipathic derivatives. Journal of Wood Chemistry and Technology, 2014; 34; 157–168.

79 Barber M, Hammersley E. Hydrophilic interaction liquid chromatography: A potential alternative for the analysis of dextran-1. Chromatography Today, May/June, 2011.

80 Inagaki S, Min JZ, Toyo'oka T. Direct detection method of oligosaccharides by high-performance liquid chromatography with charged aerosol detection. Biomedical Chromatography, 2007; 21; 338–342.

81 Plante M, Bailey B, Acworth IN. (2009). The use of charged aerosol detection with HPLC for the measurement of lipids. In: Lipidomics. Volume 1: Methods and Protocols. 269–482. Armstrong D (editor). Springer Protocols. Humana Press, New York.

82 Wenk MR. The emerging field of lipidomics. Nature, 2005; 4; 594–610.

83 León-Tamariz F, Cokelaere M, Van Boven M. Determination of long-chain alcohols using HPLC with charged aerosol detection. In Abstracts of the 4th European Fed Lipid Congress: Oils, Fats and Lipids for a Healthier Future, the Need for Interdisciplinary Approaches. October 1–4, 2006, Madrid, Spain.

84 Lu FSH, Bruheim I, Haugsgjerd BO, Jacobsen C. Effect of temperature towards lipid oxidation and non-enzymatic browning reactions in krill oil upon storage. Food Chemistry, 2014; 157; 398–407.

85 Plante M, Bailey B, Acworth IN. Determination of olive oil adulteration by principal component analysis with HPLC–charged aerosol detector data. Thermo Scientific. PN70689; 2013.

86 De la Mata-Espinosa P, Bosque-Sendra JM, Bro R, Cuadros-Rodriguez L. Discriminating olive and non-olive oils using HPLC-CAD and chemometrics. Analytical and Bioanalytical Chemistry, 2011; 399; 2083–2092.

87 De la Mata-Espinosa P, Bosque-Sendra JM, Bro R, Cuadros-Rodriguez L. Olive oil quantification of edible vegetable oil blends using triacylglycerols chromatographic fingerprints and chemometric tools. Talanta, 2011; 85; 177–182.

88 Ruiz-Samblás C, Arrebola-Pascual C, Tres A, van Ruth S, Cuadros-Rodríguez L. Authentication of geographical origin of palm oil by chromatographic fingerprinting of triacylglycerols and partial least square-discriminant analysis. Talanta, 2013; 116; 788–793.

89 Ng SP, Lai OM, Abas F, Lim HK, Beh BK, Ling TC, Tan CP. Compositional and thermal characteristics of palm olein-based diacylglycerol in blends with palm super olein. Food Research International, 2014; 55; 62–69.

90 Plante M, Bailey B, Thomas D, Acworth IN. Determination of polymerized triglycerides by high pressure liquid chromatography and Corona Veo charged aerosol detect. Thermo Scientific. PN71561; 2015.

91 Byrdwell WC. "Dilute-and-shoot" triple parallel mass spectrometry method for analysis of vitamin D and triacylglycerols in dietary supplements. Analytical and Bioanalytical Chemistry, 2011; 401; 3317–3334.

92 Byrdwell WC. Quadruple parallel mass spectrometry for analysis of vitamin D and triacylglycerols in a dietary supplement. Journal of Chromatography A, 2013; 1320; 48–65.

93 Cascone A, Eerola S, Ritieni A, Rizzo A. Development of analytical procedures to study changes in the composition of meat phospholipids caused by induced oxidation. Journal of Chromatography A, 2006; 1120; 211–220.

94 Acworth IN. Chromatography for foods and beverages: Fats and oils analysis applications notebook. Thermo Scientific. AI71471; 2015.

95 Yi-Bo W, Chun-Yu W, Fan-Na Q, Li-Ying Z, Li-Na L, Yan-Hai Y, Yuan-Yuan P. Simultaneous determination of five bile acids in pulvis fellis suis, pulvis billis bovis, pulvis fellis caprinus and pulvis fellis galli by high performance liquid chromatography-charged aerosol detector. Chinese Journal of Analytical Chemistry, 2014; 42; 109–112.

96 Carrasco-Pancorbo A, Navas-Iglesias N, Cuadros-Rodriguez L. From lipid analysis towards lipidomics, a new challenge for the analytical chemistry of the 21st century. Part I: Modern lipid analysis. TrAC Trends in Analytical Chemistry, 2009; 28; 263–278.

97 Hazotte A, Libong D, Matoga M, Chaminade P. Comparison of universal detectors for high-temperature micro liquid chromatography. Journal of Chromatography A, 2007; 1170; 52–61.

98 Khoomrung S, Chumnanpuen P, Jansa-Ard S, Ståhlman M, Nookaew I, Borén J, Nielsen J. Rapid quantification of yeast lipid using microwave-assisted total lipid extraction and HPLC-CAD. Analytical Chemistry, 2013; 85; 4912–4919.

99 Roy CE, Kauss T, Prevot S, Barthelemy P, Gaudin K. Analysis of fatty acid samples by hydrophilic interaction liquid chromatography and charged aerosol detector. Journal of Chromatography A, 2015; 1383; 121–126.

100 Moreau RA, Winkler Moser JK. (2011). Extraction and analysis of food lipids. In: Methods of Analysis of Food Components and Additives, 2nd Edition. Chap 6. 115–134. Otles S (editor). CRC Press, Boca Raton.

101 Pedersen AT, Nordblad M, Nielsen PM, Woodley JM. Batch production of FAEE-biodiesel using a liquid lipase formulation. Journal of Molecular Catalysis B: Enzymatic, 2014; 105; 89–94.

102 Zalogin TR, Pick U. Azide improves triglyceride yield in microalgae. Algal Research, 2014; 3; 8–16.

103 Plante M, Crafts C, Bailey B, Acworth IN. Characterization of castor oil by HPLC and charged aerosol detection. Thermo Scientific. PN2822; 2011.

104 Omatsu M. Analysis of oil stain on paper by charged aerosol detector. Japan TAPPI Journal, 2008; 62; 94–100.

105 Fox CB, Baldwin SL, Vedvick TS, Angov E, Reed SG. Effects on immunogenicity by formulations of emulsion-based adjuvants for malaria vaccines. Clinical and Vaccine Immunology, 2012; 19; 1633–1640.

106 Fox CB. Squalene emulsions for parenteral vaccine and drug delivery. Molecules, 2009; 14; 3286–3312.

107 Fox CB, Anderson RC, Dutill TS, Goto Y, Reed SG, Vedvick TS. Monitoring the effects of component structure and source on formulation stability and adjuvant activity of oil-in-water emulsions. Colloids and Surfaces B: Biointerfaces, 2008; 65; 98–105.

108 Jiang Q-W, Yang R, Mei X-G. Determination of phospholipid and its degradation products in liposomes for injection by HPLC-charged aerosol detection (CAD). Chinese Pharmaceutical Journal, 2007; 2; 1794–1796.

109 Gendeh G, Plante M, Bailey B, Crafts C, Acworth IN. Analysis of cationic lipids used as transfection agents for siRNA with charged aerosol detection. Thermo Scientific. PN2828; 2011.

110 Diaz-Lopez R, Libong D, Tsapis N, Fattal E, Chaminade P. Quantification of pegylated phospholipids decorating polymeric microcapsules of perfluorooctyl bromide by reverse phase HPLC with a charged aerosol detector. Journal of Pharmaceutical and Biomedical Analysis, 2008; 48; 702–707.

111 Díaz-López R, Tsapis N, Libong D, Chaminade P, Connan C, Chehimi MM, Fattal E. Phospholipid decoration of microcapsules containing perfluorooctyl bromide used as ultrasound contrast agents. Biomaterials, 2009; 30; 1462–1472.

112 Fox CB, Sivananthan SJ, Duthie MS, Vergara J, Guderian JA, Moon E, Coblen R, Carter D. A nanoliposome delivery system to synergistically trigger TLR4 AND TLR7. Journal of Nanobiotechnology, 2014; 12; 17–22.

113 Díaz-López R, Tsapis N, Santin M, Bridal SL, Nicolas V, Jaillard D, Libong D, Chaminade P, Marsaud V, Vauthier C, Fattal E. The performance of PEGylated nanocapsules of perfluorooctyl bromide as an ultrasound contrast agent. Biomaterials, 2010; 31; 1723–1731.

114 Fung HM, Mikasa TJ, Vergara J, Sivananthan SJ, Guderian JA, Duthie MS, Vedvick TS, Fox CB. Optimizing manufacturing and composition of a TLR4 nanosuspension: Physicochemical stability and vaccine adjuvant activity. Journal of Nanobiotechnology, 2013; 11; 43–48.

115 Budzik BW, Colletti SL, Seifried DD, Stanton MG, Tian L. Amino alcohol cationic lipids for nucleotide delivery. U.S. Patent No. 8,802,863. Washington, DC: U.S. Patent and Trademark Office; 2014.

116 Schönherr C, Touchene S, Wilser G, Peschka-Süss R, Francese G. Simple and precise detection of lipid compounds present within liposomal formulations using a charged aerosol detector. Journal of Chromatography A, 2009; 1216; 781–786.

117 Kiełbowicz G, Smuga D, Gładkowski W, Chojnacka A, Wawrzeńczyk C. An LC method for the analysis of phosphatidylcholine hydrolysis products and its application to the monitoring of the acyl migration process. Talanta, 2012; 94; 22–29.

118 Fox CB, Sivananthan SJ, Mikasa TJ, Lin S, Parker SC. Charged aerosol detection to characterize components of dispersed-phase formulations. Advances in Colloid and Interface Science, 2013; 199; 59–65.

119 Yamazaki T, Ihara T, Nakamura S, Kato K. Determination of impurities in 17 beta-estradiol reagent by HPLC with charged aerosol detector. Bunseki Kagaku, 2010; 59; 219–224.

120 Gonyon T, Tomaso AE, Kotha P, Owen H, Patel D, Carter PW, Cronin, H, Green JBD. Interactions between parenteral lipid emulsions and container surfaces. PDA Journal of Pharmaceutical Science and Technology, 2013; 67; 247–254.

121 Chumnanpuen P, Nookaew I, Nielsen J. Integrated analysis, transcriptome-lipidome, reveals the effects of INO-level (INO2 and INO4) on lipid metabolism in yeast. BMC Systems Biology, 2013; 7; S7–S12.

122 Davidi L, Shimoni E, Khozin-Goldberg I, Zamir A, Pick U. Origin of β-carotene-rich plastoglobuli in *Dunaliella bardawil*. Plant Physiology, 2014; 164; 2139–2156.

123 Pang XY, Cao J, Addington L, Lovell S, Battaile KP, Zhang N, Rao D, Moise AR. Structure/function relationships of adipose phospholipase A2 containing a cys-his-his catalytic triad. Journal of Biological Chemistry, 2012; 287; 35260–35274.

124 Cao J, Perez S, Goodwin B, Lin Q, Peng H, Qadri A, Gimeno RE. Mice deleted for GPAT3 have reduced GPAT activity in white adipose tissue and altered energy and cholesterol homeostasis in diet-induced obesity. American Journal of Physiology. Endocrinology and Metabolism, 2014; 306, E1176–E1187.

125 Kocharin K, Chen Y, Siewers V, Nielsen J. Engineering of acetyl-CoA metabolism for the improved production of polyhydroxybutyrate in *Saccharomyces cerevisiae*. AMB Express, 2012; 2; 52–57.

126 Planagumà A, Pfeffer MA, Rubin G, Croze R, Uddin M, Serhan CN, Levy BD. Lovastatin decreases acute mucosal inflammation via 15-epi-lipoxin A4. Mucosal Immunology, 2010; http://www.nature.com/mi/journal/vaop/ncurrent/full/mi2009141a.html (accessed January 28, 2017).

127 Vertzoni M, Archontaki H, Reppas C. Determination of intralumenal individual bile acids by HPLC with charged aerosol detection. Journal of Lipid Research, 2008; 49; 2690–2695.

128 Poplawska M, Blazewicz A, Bukowinska K, Fijalek Z. Application of high-performance liquid chromatography with charged aerosol detection for universal quantitation of undeclared phosphodiesterase-5 inhibitors in herbal dietary supplements. Journal of Pharmaceutical and Biomedical Analysis, 2013; 84; 232–243.

129 Poplawska M, Blazewicz A, Zolek P, Fijalek Z. Determination of flibanserin and tadalafil in supplements for women sexual desire enhancement using high-performance liquid chromatography with tandem mass spectrometer, diode array detector and charged aerosol detector. Journal of Pharmaceutical and Biomedical Analysis, 2014; 94; 45–53.

130 Szekeres A, Lorántfy L, Bencsik O, Kecskeméti A, Szécsi Á, Mesterházy Á, Vágvölgyi C. Rapid purification method for fumonisin B1 using centrifugal partition chromatography. Food Additives and Contaminants. Part A, 2013; 30; 147–155.

131 Szekeres A, Budai A, Bencsik O, Németh L, Bartók T, Szécsi Á, Vágvölgyi C. Fumonisin measurement from maize samples by high-performance liquid chromatography coupled with corona charged aerosol detector. Journal of Chromatographic Science, 2014; 52; 1181–1185.

132 Sasaki E, Ogasawara Y, Liu HW. A biosynthetic pathway for BE-7585A, a 2-thiosugar-containing angucycline-type natural product. Journal of the American Chemical Society, 2010; 132; 7405–7417.

133 Brondz I, Høiland K. Chemotaxonomic differentiation between *Cortinarius infractus* and *Cortinarius subtortus* by supercritical fluid chromatography connected to a multi-detection system. Trends in Chromatography, 2008; 4; 79–87.

134 Clos JF, DuBois GE, Prakash I. Photostability of rebaudioside A and stevioside in beverages. Journal of Agriculture and Food Chemistry, 2008; 56; 8507–8513.

135 Prakash I, Chaturvedula VSP, Markosyan A. Structural characterization of the degradation products of a minor natural sweet diterpene glycoside rebaudioside M under acidic conditions. International Journal of Molecular Sciences, 2014; 15; 1014–1025.

136 Ito T, Masubuchi M. Dereplication of microbial extracts and related analytical technologies. The Journal of Antibiotics, 2014; 67; 353–360.

137 Lv GP, Hu DJ, Cheong KL, Li ZY, Qing XM, Zhao J, Li SP. Decoding glycome of *Astragalus membranaceus* based on pressurized liquid extraction, microwave-assisted hydrolysis and chromatographic analysis. Journal of Chromatography A, 2015; 1409; 19–29.

138 Acworth IN. Supplements analysis applications notebook: From raw materials to extracts and natural products. Thermo Scientific. AI71473; 2015.

139 Kong HG, Kim JC, Choi GJ, Lee KY, Kim HJ, Hwang EC, Lee SW. Production of surfactin and iturin by *Bacillus licheniformis* N1 responsible for plant disease control activity. The Plant Pathology Journal, 2010; 26; 170–177.

140 Edwards C, Lawton LA. Assessment of microcystin purity using charged aerosol detection. Journal of Chromatography A, 2010; 1217; 5233–5238.

141 Wolfender JL. HPLC in natural product analysis: The detection issue. Planta Medica, 2009; 75; 719–734.

142 Granica S, Krupa K, Kłębowska A, Kiss AK. Development and validation of HPLC-DAD-CAD–MS 3 method for qualitative and quantitative standardization of polyphenols in *Agrimoniae eupatoriae herba* (Ph. Eur). Journal of Pharmaceutical and Biomedical Analysis, 2013; 86; 112–122.

143 Granica S. Quantitative and qualitative investigations of pharmacopoeial plant material *Polygoni avicularis* herba by UHPLC-CAD and UHPLC-ESI-MS methods. Phytochemical Analysis, 2015; 26(5); 374–382; http://online library.wiley.com/doi/10.1002/pca.2572/abstract;jsessionid=96086F684C 27864D35F38A18CD781A70.f03t01?userIsAuthenticated=false&denied AccessCustomisedMessage= (accessed January 28, 2017).

144 Granica S, Piwowarski JP, Kiss AK. Determination of C-glucosidic Ellagitannins in *Lythri salicariaeherba* by ultra-high performance liquid chromatography coupled with charged aerosol detector: Method development and validation. Phytochemical Analysis, 2014, 25, 201–206.

145 Plaza M, Kariuki J, Turner C. Quantification of individual phenolic compounds' contribution to antioxidant capacity in apple: A novel analytical tool based on liquid chromatography with diode array, electrochemical, and charged aerosol detection. Journal of Agricultural and Food Chemistry, 2014; 62; 409–418.

146 Granica S, Lohwasser U, Jöhrer K, Zidorn C. Qualitative and quantitative analyses of secondary metabolites in aerial and subaerial of *Scorzonera hispanica* (black salsify). Food Chemistry, 2015; 173; 321–331.

147 Kao CL, Borisova SA, Kim HJ, Liu HW. Linear aglycones are the substrates for glycosyltransferase DesVII in methymycin biosynthesis: Analysis and implications. Journal of the American Chemical Society, 2006; 128, 5606–5607.

148 Jia S, Li J, Yunusova N, Park JH, Kwon SW, Lee J. A new application of charged aerosol detection in liquid chromatography for the simultaneous determination of polar and less polar ginsenosides in ginseng products. Phytochemical Analysis, 2013; 24; 374–380.

149 Qi LW, Wang CZ, Yuan CS. Isolation and analysis of ginseng: Advances and challenges. Natural Product Reports, 2011; 28; 467–495.

150 Kim DH, Chang JK, Sohn HJ, Cho BG, Ko SR, Nho KB, Lee SM. Certification of a pure reference material for the ginsenoside Rg_1. Accreditation and Quality Assurance, 2010; 15; 81–87.

151 Baek SH, Bae ON, Park JH. Recent methodology in ginseng analysis. Journal of Ginseng Research, 2012; 36; 119–123.

152 Chen L, Dai J, Wang Z, Zhang H, Huang Y, Zhao Y. Ginseng total saponins reverse corticosterone-induced changes in depression-like behavior and hippocampal plasticity-related proteins by interfering with GSK-3β-CREB signaling pathway. Evidence-Based Complement and Alternative Medicine, 2014; http://www.hindawi.com/journals/ecam/2014/506735/abs/ (accessed January 28, 2017).
153 Ouyang LF, Wang Z-L, Dai J-G, Chen L, Zhao Y-N. Determination of total ginsenosides in ginseng extracts using charged aerosol detection with post-column compensation of the gradient. Chinese Journal of Natural Medicines, 2014; 12; 857–868.
154 Bai CC, Han SY, Chai XY, Jiang Y, Li P, Tu PF. Sensitive determination of saponins in *Radix et Rhizoma Notoginseng* by charged aerosol detector coupled with HPLC. Journal of Liquid Chromatography and Related Technologies, 2008; 32; 242–260.
155 Zhao Y-N, Wang Z-L, Dai J-G, Chen L, Huang Y-F. Preparation and quality assessment of high-purity ginseng total saponins by ion exchange resin combined with macroporous adsorption resin separation. Chinese Journal of Natural Medicines, 2014; 12; 382–392.
156 Negi JS, Singh P, Pant GJN, Rawat MSM. High-performance liquid chromatography analysis of plant saponins: An update 2005–2010. Pharmacognosy Reviews, 2011; 5; 155–164.
157 Kim SH, Kim HK, Yang ES, Lee KY, Du Kim S, Kim YC, Sung SH. Optimization of pressurized liquid extraction for spicatoside A in *Liriope platyphylla*. Separation and Purification Technology, 2010; 71; 168–172.
158 Yeom H, Suh JH, Youm JR, Han SB. Simultaneous determination of triterpenoid saponins from *Pulsatilla koreana* using high performance liquid chromatography coupled with a charged aerosol detector (HPLC-CAD). Bulletin of the Korean Chemical Society, 2010; 31; 1159–1164.
159 Acworth IN, Zhang Q, Thomas D. Profiling hoodia extracts by HPLC with charged aerosol detection and electrochemical array detection and pattern recognition. Planta Medica, 2013; 79; 127.
160 Acworth IN, Bailey B, Plante M, Zhang Q, Thomas D. Profiling hoodia extracts by HPLC with charged aerosol detection, electrochemical array detection and principal component analysis. Thermo Scientific. PN70540; 2013.
161 Kakigi Y, Mochizuki N, Icho T, Hakamatsuka T, Goda Y. Analysis of terpene lactones in a ginkgo leaf extract by high-performance liquid chromatography using charged aerosol detection. Bioscience, Biotechnology, and Biochemistry, 2010; 74; 590–594.
162 Forsat, B, Snow NH. HPLC with charged aerosol detection for pharmaceutical cleaning validation. LCGC North America, 2007; 25; 960–964.

163 Yamaki T, Kamiya Y, Ohtake K, Uchida M, Seki T, Ueda H, Kobayashi J, Morimoto Y, Natsume H. A mechanism enhancing macromolecule transport through paracellular spaces induced by poly-L-arginine: Poly-L-arginine induces the internalization of tight junction proteins via clathrin-mediated endocytosis. Pharmaceutical Research, 2014; 31; 2287–2296.

164 Smith MC, Crist RM, Clogston JD, McNeil SE. Quantitative analysis of PEG-functionalized colloidal gold nanoparticles using charged aerosol detection. Analytical and Bioanalytical Chemistry, 2015; 407; 3705–3710.

165 Charles I, Sinclair I, Addison DH. Capture and exploration of sample quality data to inform and improve the management of a screening collection. Journal of Laboratory Automation, 2013; http://jla.sagepub.com/content/early/2013/08/22/2211068213499758.abstract (accessed January 28, 2017).

166 Sinclair I, Charles I. Applications of the charged aerosol detector in compound management. Journal of Biomolecular Screening, 2009; 14; 531–537.

167 Jia S, Li J, Park SR, Ryu Y, Park IH, Park JH, Lee J. Combined application of dispersive liquid–liquid microextraction based on the solidification of floating organic droplets and charged aerosol detection for the simple and sensitive quantification of macrolide antibiotics in human urine. Journal of Pharmaceutical and Biomedical Analysis, 2013; 86; 204–213.

168 Yan F, Munos JW, Liu P, Liu HW. Biosynthesis of fosfomycin, re-examination and re-confirmation of a unique Fe (II)-and NAD (P) H-dependent epoxidation reaction. Biochemistry, 2006; 45; 11473–11481.

169 Lin CI, Sasaki E, Zhong A, Liu HW. In vitro characterization of LmbK and LmbO: Identification of GDP-D-erythro-α-D-gluco-octose as a key intermediate in lincomycin A biosynthesis. Journal of the American Chemical Society, 2014; 136; 906–909.

170 Jia S, Lee HS, Choi MJ, Hyun Sung SB, Han SH, Park J, Lee J. Non-derivatization method for the determination of gabapentin in pharmaceutical formulations, rat serum and rat urine using high performance liquid chromatography coupled with charged aerosol detection. Current Analytical Chemistry, 2012; 8; 159–167.

171 Wojtczak A, Wojtczak M, Skrętkowicz J. The relationship between plasma concentration of metoprolol and CYP2D6 genotype in patients with ischemic heart disease. Pharmacological Reports, 2014; 66; 511–514.

172 Viinamäki J, Ojanperä I. Photodiode array to charged aerosol detector response ratio enables comprehensive quantitative monitoring of basic drugs in blood by ultra-high performance liquid chromatography. Analytica Chimica Acta, 2015; 865; 1–7.

173 Nováková L, Lopéz SA, Solichová D, Šatínský D, Kulichová B, Horna A, Solich P. Comparison of UV and charged aerosol detection approach in pharmaceutical analysis of statins. Talanta, 2009; 78; 834–839.

174 Yu X, Zdravkovic S, Wood D, Li C, Cheng Y, Ding X. A new approach to threshold evaluation and quantitation of unknown extractables and leachables using HPLC-CAD. Drug Delivery Technology, 2009; 9; 50–55.
175 Thomas D, Acworth IN, Bailey B, Plante M. Direct analysis of multicomponent adjuvants by HPLC with charged aerosol detection. Thermo Scientific. PN70333; 2012.
176 Nováková L, Solichová D, Solich P. Hydrophilic interaction liquid chromatography–charged aerosol detection as a straightforward solution for simultaneous analysis of ascorbic acid and dehydroascorbic acid. Journal of Chromatography A, 2009; 1216; 4574–4581.
177 Huang Z, Richards MA, Zha Y, Francis R, Lozano R, Ruan J. Determination of inorganic pharmaceutical counterions using hydrophilic interaction chromatography coupled with a Corona® CAD detector. Journal of Pharmaceutical and Biomedical Analysis, 2009; 50; 809–814.
178 Crafts C, Bailey B, Plante M, Acworth IN. Evaluation of methods for the simultaneous analysis of cations and anions using HPLC with charged aerosol detection and a zwitterionic stationary phase. Journal of Chromatographic Science, 2009; 47; 534–539.
179 Crafts C, Bailey B, Gamache P, Liu X, Acworth IN. (2012). Comprehensive approaches for measurement of active pharmaceutical ingredients, counter-ions, and excipients using HPLC with charged aerosol detection. In: Applications of Ion Chromatography in the Analysis of Pharmaceutical and Biological Products. 221–236. Bhattacharyya L, Rohrer JS (editors). John Wiley & Sons, Inc., Hoboken, NJ.
180 Russell JJ, Heaton JC, Underwood T, Boughtflower R, McCalley DV. Performance of charged aerosol detection with hydrophilic interaction chromatography. Journal of Chromatography A, 2015; 1405; 72–84.
181 Plante M, Bailey B, Acworth IN, Sneekes E-J, Steiner F. Multi-modal analyte detection of cyclodextrin and ketoprofen inclusion complex using UV and CAD on an integrated UHPLC platform. Thermo Scientific. PN71690; 2015.
182 Park EJ, Cho H, Kim SW, Na DH. Chromatographic methods for characterization of poly (ethylene glycol)-modified polyamidoamine dendrimers. Analytical Biochemistry, 2014; 449; 42–44.
183 Saokham P, Loftsson T. A new approach for quantitative determination of γ-cyclodextrin in aqueous solutions: Application in aggregate determinations and solubility in hydrocortisone/γ-cyclodextrin inclusion complex. Journal of Pharmaceutical Sciences, 2015; 104; 3925–3933.
184 Christiansen A, Backensfeld T, Kühn S, Weitschies W. Investigating the stability of the nonionic surfactants tocopheryl polyethylene glycol succinate and sucrose laurate by HPLC–MS, DAD, and CAD. Journal of Pharmaceutical Sciences, 2011; 100; 1773–1782.

185 Greiderer A, Steeneken L, Aalbers T, Vivó-Truyols G, Schoenmakers P. Characterization of hydroxypropylmethylcellulose (HPMC) using comprehensive two-dimensional liquid chromatography. Journal of Chromatography A, 2011; 1218; 5787–5793.

186 Ilko D, Puhl S, Meinel L, Germershaus O, Holzgrabe U. Simple and rapid high performance liquid chromatography method for the determination of polidocanol as bulk product and in pharmaceutical polymer matrices using charged aerosol detection. Journal of Pharmaceutical and Biomedical Analysis, 2015; 104; 17–20.

187 Li Y, Hewitt D, Lentz YK, Ji JA, Zhang TY, Zhang K. Characterization and stability study of polysorbate 20 in therapeutic monoclonal antibody formulation by multidimensional ultrahigh-performance liquid chromatography–charged aerosol detection–mass spectrometry. Analytical Chemistry, 2014; 86; 5150–5157.

188 Fekete S, Ganzler K, Fekete J. Fast and sensitive determination of polysorbate 80 in solutions containing proteins. Journal of Pharmaceutical and Biomedical Analysis, 2010; 52; 672–679.

189 Ilko D, Braun A, Germershaus O, Meinel L, Holzgrabe U. Fatty acid composition analysis in polysorbate 80 with high performance liquid chromatography coupled to charged aerosol detection. European Journal of Pharmaceutics and Biopharmaceutics, 2015; 94; 569–574.

190 Shi S, Chen Z, Rizzo JM, Semple A, Mittal S. A highly sensitive method for the quantitation of polysorbate 20 and 80 to study the compatibility between polysorbates and m-cresol in the peptide formulation. Journal of Analytical and Bioanalytical Techniques, 2015; 6; 2–8.

191 Lie A, Wimmer R, Pedersen LH. Design of experiments and multivariate analysis for evaluation of reversed-phase high-performance liquid chromatography with charged aerosol detection of sucrose caprate regioisomers. Journal of Chromatography A, 2013; 1281; 67–72.

192 Lie A, Pedersen LH. Elution strategies for reversed-phase high-performance liquid chromatography analysis of sucrose alkanoate regioisomers with charged aerosol detection. Journal of Chromatography A, 2013; 1311; 127–133.

193 Scott B, Zhang K, Wigman L. Material identification by HPLC with charged aerosol detection. LCGC North America, 2013; 31; 564–569.

194 Acworth IN. HPLC-CAD Surfactants and emulsifiers applications notebook. Thermo Scientific. AN71104; 2014.

195 Louw S, Pereira AS, Lynen F, Hanna-Brown M, Sandra P. Serial coupling of reversed-phase and hydrophilic interaction liquid chromatography to broaden the elution window for the analysis of pharmaceutical compounds. Journal of Chromatography A, 2008; 1208; 90–94.

196 Crafts C, Plante M, Acworth IN, Bailey B, Waraska J, Gamache P. Increased process understanding for QbD by introducing universal detection at several stages of the pharmaceutical process. Thermo Scientific. PN2438; 2010.

197 Crafts C, Bailey B, Plante M, Acworth IN. Validating analytical methods with charged aerosol detection. Thermo Scientific. PN2949; 2011.

198 Khandagale MM, Hilder EF, Shellie RA, Haddad PR. Assessment of the complementarity of temperature and flow-rate for response normalization of aerosol-based detectors. Journal of Chromatography A, 2014; 1356; 180–187.

199 Holzgrabe U, Nap CJ, Beyer T, Almeling S. Alternatives to amino acid analysis for the purity control of pharmaceutical grade L-alanine. Journal of Separation Science, 2010; 33; 2402–2410.

200 Joseph A, Rustum A. Development and validation of a RP-HPLC method for the determination of gentamicin sulfate and its related substances in a pharmaceutical cream using a short pentafluorophenyl column and a charged aerosol detector. Journal of Pharmaceutical and Biomedical Analysis, 2010; 51; 521–531.

201 Stypulkowska K, Blazewicz A, Fijalek Z, Sarna K. Determination of gentamicin sulphate composition and related substances in pharmaceutical preparations by LC with charged aerosol detection. Chromatographia, 2010; 72; 1225–1229.

202 Li R, Hurum D, Wang J, Rohrer J. Gentamicin sulfate assay by HPLC with charged aerosol detection. Thermo Scientific. AN70016; 2012.

203 Stypulkowska K, Blazewicz A, Brudzikowska A, Warowna-Grzeskiewicz M, Sarna K, Fijalek Z. Development of high performance liquid chromatography methods with charged aerosol detection for the determination of lincomycin, spectinomycin and its impurities in pharmaceutical products. Journal of Pharmaceutical and Biomedical Analysis, 2015; 112; 8–14.

204 Stypulkowska K, Blazewicz A, Fijalek Z, Warowna-Grzeskiewicz M, Srebrzynska K. Determination of neomycin and related substances in pharmaceutical preparations by reversed-phase high performance liquid chromatography with mass spectrometry and charged aerosol detection. Journal of Pharmaceutical and Biomedical Analysis, 2013; 76; 207–214.

205 Joseph A, Patel S, Rustum A. Development and validation of a RP-HPLC method for the estimation of netilmicin sulfate and its related substances using charged aerosol detection. Journal of Chromatographic Science, 2010; 48; 607–612.

206 Holzgrabe U, Nap CJ, Kunz N, Almeling S. Identification and control of impurities in streptomycin sulfate by high-performance liquid chromatography coupled with mass detection and corona charged-aerosol detection. Journal of Pharmaceutical and Biomedical Analysis, 2011; 56; 271–279.

207 Zhang T, Hewitt D, Kao YH. SEC assay for polyvinylsulfonic impurities in 2-(N-morpholino) ethanesulfonic acid using a charged aerosol detector. Chromatographia, 2010; 72; 145–149.

208 Wahl O, Holzgrabe U. Impurity profiling of carbocisteine by HPLC-CAD, qNMR and UV/vis spectroscopy. Journal of Pharmaceutical and Biomedical Analysis, 2014; 95; 1–10.

209 Huang Z, Neverovitch M, Lozano, R, Tattersall P, Ruan J. Development of a reversed-phase HPLC impurity method for a UV variable isomeric mixture of a CRF drug substance intermediate with the assistance of corona CAD. Journal of Pharmaceutical Innovation, 2011; 6; 115–123.

210 Zhang K, Li Y, Tsang M, Chetwyn NP. Analysis of pharmaceutical impurities using multi-heartcutting 2D LC coupled with UV-charged aerosol MS detection. Journal of Separation Science, 2013; 36; 2986–2992.

211 Dai L, Wigman L, Zhang K. Sensitive and direct determination of lithium by mixed-mode chromatography and charged aerosol detection. Journal of Chromatography A, 2015; 1408; 87–92.

212 Rystov L, Chadwick R, Krock K, Wang T. Simultaneous determination of Maillard reaction impurities in memantine tablets using HPLC with charged aerosol detector. Journal of Pharmaceutical and Biomedical Analysis, 2011; 56; 887–894.

213 Bailey B. Metoprolol and select impurities analysis using a hydrophilic interaction chromatography method with combined UV and charged aerosol detection. Thermo Scientific. AN1126; 2015.

214 Blazewicz A, Fijalek Z, Warowna-Grzeskiewicz M, Jadach, M. Determination of atracurium, cisatracurium and mivacurium with their impurities in pharmaceutical preparations by liquid chromatography with charged aerosol detection. Journal of Chromatography A, 2010; 1217; 1266–1272.

215 Schiesel S, Lämmerhofer M, Lindner W. Comprehensive impurity profiling of nutritional infusion solutions by multidimensional off-line reversed-phase liquid chromatographyhydrophilic interaction chromatography–ion trap mass-spectrometry and charged aerosol detection with universal calibration. Journal of Chromatography A, 2012; 1259; 100–110.

216 Sun P, Wang X, Alquier L, Maryanoff CA. Determination of relative response factors of impurities in paclitaxel with high performance liquid chromatography equipped with ultraviolet and charged aerosol detectors. Journal of Chromatography A, 2008; 1177; 87–91.

217 Warner A, Piraner I, Weimer H, White K. Development of a purity control strategy for pemetrexed disodium and validation of associated analytical methodology. Journal of Pharmaceutical and Biomedical Analysis, 2015; 105; 46–54.

218 Jain D, Basniwal PK. Forced degradation and impurity profiling: Recent trends in analytical perspectives. Journal of Pharmaceutical and Biomedical Analysis, 2013; 86; 11–35.

219 Soman A, Jerfy M, Swanek F. Validated HPLC method for the quantitative analysis of a 4-methanesulfonyl-piperidine hydrochloride salt. Journal of Liquid Chromatography and Related Technologies, 2009; 37; 1000–1009.

220 Thomas D, Acworth IN. Label-free analysis by UHPLC with charged aerosol detection of glycans separated by charge, size, and isomeric structure. Thermo Scientific. AN1127; 2015.

221 Thomas D, Acworth IN. Label-free analysis by UHPLC with charged aerosol detection of glycans separated by charge, size, and isomeric structure. Thermo Scientific. PO71734; 2015.

222 Thomas D, Acworth IN, Bauder R, Kast L. Label-free profiling of O-linked glycans by HPLC with charged aerosol detection. Thermo Scientific. PO71733; 2015.

223 Wang R, Wang X, Paulino J, Alquier L. Evaluation of charged aerosol detector for purity assessment of protein. Journal of Chromatography A, 2013; 1283; 116–121.

224 Zhang Q, Acworth IN. Direct measurement of sialic acids released from glycoproteins, by high performance liquid chromatography and charged aerosol detection. Thermo Scientific. PN71726; 2015.

225 León-Tamariz F, Verbaeys I, Van Boven M, De Cuyper M, Buyse J, Clynen E, Cokelaere M. PEGylation of cholecystokinin prolongs its anorectic effect in rats. Peptides, 2007; 28; 1003–1011.

226 Crafts C, Bailey B, Plante M, Acworth IN. Analytical methods to qualify and quantify PEG and PEGylated biopharmaceuticals. Thermo Scientific. AN70160; 2012.

227 Courtemanche K, Bailey B, Crafts C, Plante M, Waraska J, Acworth IN, Swartz M. Measurement of stability and purity of cell-penetrating peptides used for siRNA delivery. Thermo Scientific. PN2829; 2011.

228 Liu XK, Fang JB, Cauchon N, Zhou P. Direct stability-indicating method development and validation for analysis of etidronate disodium using a mixed-mode column and charged aerosol detector. Journal of Pharmaceutical and Biomedical Analysis, 2008; 46; 639–644.

229 Ramanathan R, Josephs JL, Jemal M, Arnold M, Humphreys WG. Novel MS solutions inspired by MIST. Bioanalysis, 2010; 2; 1291–1313.

230 Malek G, Crafts C, Plante M, Neely M, Bailey B. Use of charged aerosol detection as an orthogonal quantification technique for drug metabolites in safety testing (MIST). Thermo Scientific. PN2953; 2011.

231 Nedderman AN. Metabolites in safety testing: Metabolite identification strategies in discovery and development. Biopharmaceutics and Drug Disposition, 2009; 30; 153–162.

232 Márquez-Sillero I, Aguilera-Herrador E, Cárdenas S, Valcárcel M. Determination of parabens in cosmetic products using multi-walled carbon nanotubes as solid phase extraction sorbent and corona-charged aerosol detection system. Journal of Chromatography A, 2010; 1217; 1–6.

233 Plante M, Bailey B, Acworth IN. Quantitation of pluronics by high performance liquid chromatography and corona charged aerosol detection. Thermo Scientific. PN70535; 2013.

234 Ramaswamy J, Prasher SO, Patel RM, Hussain SA, Barrington SF. The effect of composting on the degradation of a veterinary pharmaceutical. Bioresource Technology, 2010; 101; 2294–2299.

235 Jayashree R, Prasher SO, Kaur R, Patel RM. Effect of soil pH on sorption of salinomycin in clay and sandy soils. African Journal of Environmental Science and Technology, 2011; 5; 661–667.
236 Zhou Q, Chen M, Zhu L, Yao-Bin D. Determination of erythrocin in chicken manure by High Performance Liquid Chromatography-Corona-Charged Aerosol Detection coupled with on-line solid phase extraction. Chinese Journal of Analytical Chemistry, 2014; 42; 1838–1841.
237 Wagner C, Benecke C, Buchholz H, Beutel S. Enhancing bioethanol production from delactosed whey permeate by upstream desalination techniques. Engineering in Life Sciences, 2014; 14; 520–529.
238 Mikelova R, Prokop Z, Stejskal K, Adam V, Beklova M, Trnkova L, Kizek R. Enzymatic reaction coupled with flow-injection analysis with charged aerosol, coulometric, or amperometric detection for estimation of contamination of the environment by pesticides. Chromatographia, 2008; 67; 47–53.
239 Plante M, Bailey B, Acworth IN, Neeley M. Characterization of dispersants by reversed-phase high pressure liquid chromatography and charged aerosol detection. Thermo Scientific. PN2737; 2011.
240 Zhou Q, Chen M, Zhu L, Tang H. Determination of perfluorinated carboxylic acids in water using liquid chromatography coupled to a corona-charged aerosol detector. Talanta, 2015; 136; 35–41.
241 Funari CS, Carneiro RL, Khandagale MM, Cavalheiro AJ, Hilder EF. Acetone as a greener alternative to acetonitrile in liquid chromatographic fingerprinting. Journal of Separation Science, 2015; 38; 1458–1465.
242 LaFleur AD, Schug KA. A review of separation methods for the determination of estrogens and plastics-derived estrogen mimics from aqueous systems. Analytica Chimica Acta, 2011; 696; 6–26.
243 Plante M, Bailey B, Acworth IN, Crafts C. A new approach to the simultaneous analysis of underivatized ionophoric antibiotics using liquid chromatography and charged aerosol detection. Thermo Scientific. PN70054; 2012.
244 Grembecka M, Baran P, Błażewicz A, Fijałek Z, Szefer P. Simultaneous determination of aspartame, acesulfame-K, saccharin, citric acid and sodium benzoate in various food products using HPLC–CAD–UV/DAD. European Food Research and Technology, 2014; 238; 357–365.
245 Steiner F, Grubner M, Dunkel A, Hofmann T. A fast and efficient method to access 2D-HPLC column and method combinations for food metabolomics studies. Thermo Fisher. PN71130; 2014.
246 Plante M, Bailey B, Acworth IN. Analysis of emulsifiers in foods by high performance liquid chromatography and corona charged aerosol detection. Thermo Scientific. PN70995; 2014.
247 Márquez-Sillero I, Cárdenas S, Valcárcel M. Determination of water-soluble vitamins in infant milk and dietary supplement using a liquid

chromatography on-line coupled to a corona-charged aerosol detector. Journal of Chromatography A, 2013; 1313; 253–258.
248 Plante M, Bailey B, Acworth IN, Crafts C. A single method for the direct determination of total glycerols in all biodiesels using liquid chromatography and charged aerosol detection. Thermo Scientific. AN1049; 2012.
249 Hurrum D, Rohrer J. Determination of residual acylglycerols in biodiesel. Thermo Scientific. AB70486; 2013.
250 Plante M, Bailey B, Acworth IN. An improved global method for the quantitation and characterization of lipids by high performance liquid chromatography and charged aerosol detection. Thermo Scientific. PN70533; 2013.
251 Tracy M, Liu X, Acworth IN. Determination of polyacrylic acid in boiler water using size-exclusion chromatography with charged aerosol detection. Thermo Scientific. AN20984; 2014.
252 Plante M, Bailey B, Acworth IN. Quantitation of hindered amine light stabilizers (HALS) by liquid chromatography and charged aerosol detection. Thermo Scientific. PN70022; 2012.
253 Stojanovic A, Lämmerhofer M, Kogelnig D, Schiesel S, Sturm M, Galanski M, Lindner W. Analysis of quaternary ammonium and phosphonium ionic liquids by reversed-phase high-performance liquid chromatography with charged aerosol detection and unified calibration. Journal of Chromatography A, 2008; 1209; 179–187.
254 Teutenberg T, Tuerk J, Holzhauser M, Kiffmeyer TK. Evaluation of column bleed by using an ultraviolet and a charged aerosol detector coupled to a high-temperature liquid chromatographic system. Journal of Chromatography A, 2006; 1119; 197–201.
255 Ovalles C, Rogel E, Moir ME, Morazan H. Effect of temperature on the analysis of asphaltenes by the on-column filtration/redissolution method. Fuel, 2015; 146; 20–27.
256 Thompson A, Gangstad A, Kotlar HK. Oil field data/return analysis: A comparison of scale inhibitor return concentrations obtained with a novel analytical method and current commercial techniques. In: SPE International Oilfield Scale Conference (SPE 114049. 2008), May 28–29, 2017, Aberdeen, UK.
257 Acworth IN, Bailey B, Plante M. Quantitation and characterization of copper plating bath additives by liquid chromatography with charged aerosol detection and electrochemical detection. LC-GC The Magazine of Separation Science, 2014; 10(7); 10–16.
258 Plante M, Bailey B, Acworth IN. Quantitation and characterization of copper plating bath additives by liquid chromatography with charged aerosol detection and electrochemical detection. Thermo Scientific. PN70008; 2012.
259 Plante M, Fairlie S, Bailey B, Acworth IN. Measurement and control of copper additives in electroplating baths by HPLC. Thermo Scientific. WP71211; 2014.

260 Plante M, Bailey B, Acworth IN. Quantitation and characterization of copper and nickel plating bath additives by liquid chromatography. Thermo Scientific. PN71179; 2014.
261 Plante M, Bailey B, Acworth IN. Analysis of silicone oils by high performance liquid chromatography and corona charged aerosol detection. Thermo Scientific. PN70538; 2013.
262 Plante M, Bailey B, Acworth IN. Quantitation of surfactants in samples by high performance liquid chromatography and corona charged aerosol detection. Thermo Scientific. PN70539; 2013.
263 Steiner F, Plante M, Bailey B, Acworth IN. An easy way to a fast universal method for surfactant analysis. The Column, LCGC, 2012; 8(12); 2–9.
264 Ilisz I, Pataj Z, Gecse Z, Szakonyi Z, Fülöp F, Lindner W, Péter A. Unusual temperature-induced retention behavior of constrained β-amino acid enantiomers on the zwitterionic chiral stationary phases ZWIX (+) and ZWIX (–). Chirality, 2014; 26; 385–393.

3

Practical Use of CAD

Achieving Optimal Performance

Bruce Bailey, Marc Plante, David Thomas, Chris Crafts, and Paul H. Gamache

Thermo Fisher Scientific, Chelmsford, MA, USA

CHAPTER MENU

3.1 Summary

This chapter discusses practical approaches to optimize performance of charged aerosol detection (CAD) for routine quantitative analysis with a main focus on newer instrument designs. Key differences with newer designs are the capability to operate with lower eluent flow rates, production of aerosols with a relatively larger size distribution and number concentration (particles/cm^3 gas), and use of temperature-controlled evaporation to optimize performance. As described in Chapter 1, the aerosol characteristics of the new design have the effect of simplifying the overall response curve, enabling lower detection limits, improving detection of semivolatile analytes, and increasing the quasi-linear response range.

A primary factor affecting the performance of CAD is the concentration of nonvolatile plus semivolatile impurities in the column eluent. Impurity concentration can be minimized through use of mobile phase solvents with

Charged Aerosol Detection for Liquid Chromatography and Related Separation Techniques, First Edition. Edited by Paul H. Gamache.

low residue after evaporation, columns that exhibit low bleed under conditions of use and exposure, LC systems and operational practices that minimize contamination, and volatile mobile phase additives (pH modifiers, buffers, ion-pairing reagents) at the lowest required concentration. Choice of mobile phase additive and concentration requires consideration of both the desired chromatographic separation and the possible influence on CAD response of salt formation with ionic solutes (analyte or impurity). Strategies are described for method transfer from older to newer models and to exploit newer capabilities to improve performance and extend the technology to additional applications.

3.2 Introduction

Charged aerosol detection (CAD) can be used with a range of chromatographic techniques and in many configurations where several variables can influence performance. This chapter provides practical information to help analysts achieve optimal results with CAD for quantitative analysis. This includes consideration of chromatographic variables (e.g., mobile phase composition, liquid flow rate, solvent gradients, columns, and separation modes), analyte properties, detector settings, and various system configurations.

CAD theory is described in Chapter 1. Briefly (Figure 3.1) CAD works on the principle of (i) nebulizing an eluent, (ii) drying it to leave an aerosol residue, (iii) charging the residue, and (iv) measuring aggregate charge. CAD behaves primarily as a mass flow-dependent device where response is proportional to the mass of nonvolatile solute (impurity + analyte) reaching the detector per unit time.

While CAD and other evaporative aerosol (EA) techniques (evaporative light scattering detection (ELSD) and condensation nucleation light scattering detection (CNLSD)) are considered to be nonselective (i.e., universal), a certain minimum level of selectivity is needed to distinguish analytes from other components of an *eluent*—defined here as the liquid that is nebulized and therefore includes solvents, additives, impurities, and any analyte present. In all cases, selectivity is based on relative partitioning of species between the two phases of an aerosol—gas phase and condensed phase. While it can be generally stated that analytes must be nonvolatile and other eluent components must be volatile, several factors that pertain to the specific process of aerosol evaporation and to mobile phase composition should be taken into account when predicting detectability and sensitivity for a given analyte. These factors are further discussed in this chapter and in Chapter 1.

A main variable affecting the performance of all EA detectors is the mass concentration of nonvolatile impurities in the column eluent [1–4]. These impurities usually exist as solutes that, like analytes, contribute to the

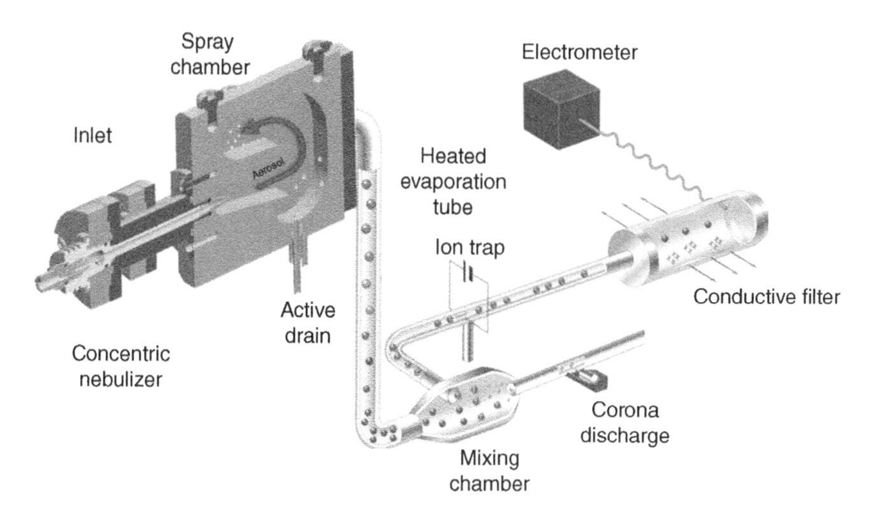

Figure 3.1 CAD involves pneumatic nebulization of liquid column eluent, removal of larger droplets within a spray chamber, solvent evaporation, diffusion charging of aerosol residue within a mixing chamber by an opposing ion jet formed via corona discharge, removal of excess ions and high mobility charged particles in an ion trap, and measurement of aggregate charge of aerosol particles with a filter/electrometer. *Source*: Reproduced with permission of Thermo Fisher Scientific Inc.

size of each dried aerosol particle in proportion to their mass concentration or, more precisely, volume concentration after a minor density correction (Equation 1.5). Since CAD signal is directly related to particle size, higher concentrations of impurities produce higher baseline signal, drift, and noise (e.g., Figure 1.22), thus limiting performance. This chapter will therefore discuss likely sources of impurities, recommendations to minimize their concentration, and practical use of instrumental settings to optimize detection selectivity.

3.2.1 First- and Second-Generation Instrument Designs

As reviewed in Chapter 2, most literature references are, *at present*, based on the original design of commercial CAD instruments, which was replaced in 2013 starting with the introduction of the Corona™ Veo™ detector (basic design is depicted in Figure 3.1). The main differences are summarized in Table 3.1 and some practical implications are briefly discussed in the following text.

3.2.2 Liquid Flow Range

A key change with newer-generation instruments is the use of a concentric rather than cross-flow nebulizer. With the cross-flow design, operation at lower liquid flow rates often results in erratic baselines. This is typically

Table 3.1 Main differences between first- and second-generation CAD instruments.

Generation	1	2
Representative model	Corona® Ultra® RS	Corona Veo RS
Nebulizer	Cross-flow	Concentric
Relative impactor distance	Proximal	Distal
Eluent flow rate (mL/min)	0.20–2.0	0.01–2.0
Evaporation T (°C)	Ambient	(Ambient + 5) to 100
Nebulizer T (°C)	5–35	Ambient
Excess liquid removal	Gravity with "closed" waste system	Actively pumped to "open" waste

observed at ~0.2 mL/min with aqueous eluents or at even higher flow rates with solvents having lower viscosity and surface tension. This is attributed to an unstable spray due to siphoning of liquid by the gas flow at the nebulizer tip. An advantage of the concentric design, as implemented in newer instruments, is its ability to produce a stable aerosol spray under lower flow conditions, thus extending the detector's operating range (e.g., Figure 3.2). For some applications, this may be further optimized by altering nebulizer gas flow.

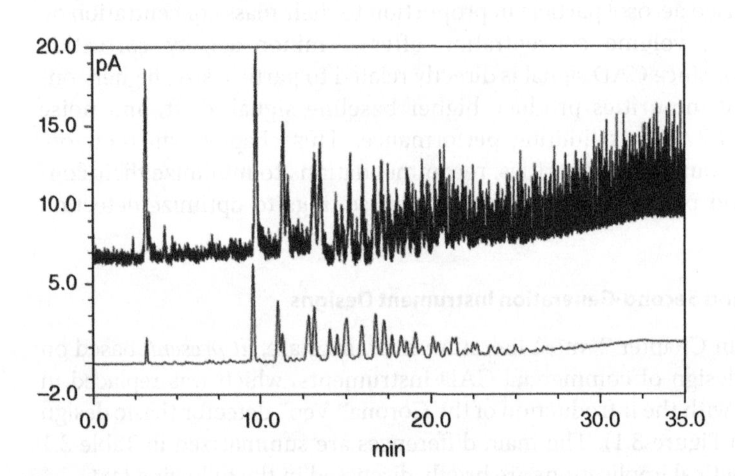

Figure 3.2 Baseline noise observed with older CAD design (Corona® ultra, top trace) attributed to unstable nebulization at low liquid flow rates and/or solvents with low viscosity and surface tension. Bottom trace (Corona Veo). Conditions: 0.30 mL/min, isocratic 59:40:1, acetonitrile: ethyl acetate: 0.1 M ammonium acetate buffer, pH 5.5 for 1 min; then 27 min linear gradient to and 10 min hold at 19:80:1, acetonitrile: ethyl acetate: 0.1 M ammonium acetate buffer, pH 5.5.

3.2.3 Excess Liquid Removal

Older and newer CAD designs also differ significantly in the handling of excess liquid. To maintain a stable spray, the pressure within the spray chamber must be relatively constant and any excess liquid must be removed smoothly and efficiently. In the older design, the spray chamber and waste bottle were included in a closed loop whose pressure was slightly above atmospheric. Intermittent baseline disturbances, sometimes observed with the older design, are mainly attributed to uneven drainage and corresponding pressure spikes within the spray chamber. This issue has been mitigated with the new design by actively pumping excess liquid to an "open" waste. This configuration also adds significant operational convenience.

3.2.4 Temperature Control

The nebulizer temperature control capability of the first-generation design was mainly intended to prevent freezing of water due to rapid evaporative cooling that occurs with highly volatile eluents. To our knowledge this has only been observed with the cross-flow nebulizer–impactor design and only with tetrahydrofuran (THF)–water solvent gradients. For most other applications, evaporation temperature (T_e) has been found to have a much greater influence on performance than nebulizer temperature (T_n)—as also observed with ELSD [5]. The relatively minor effect of T_n may be due to limited heat transfer to the aerosol during its very short residence time within the nebulizer. Since T_e has been found to be a more robust control variable for method optimization, the newer design operates at ambient T_n while controlling T_e at a user-specified value.

3.2.5 Aerosol Creation and Transport

As described in Chapter 1, for a given solvent and flow rate, the size distribution of the aerosol that is transported to the evaporation tube with the newer CAD design has a significantly larger (e.g., median) droplet size and a higher number concentration (e.g., droplets/cm^3 of gas). Temperature-controlled (heated) evaporation is then used to accommodate a range of eluent compositions and flow rates and can be a useful variable to optimize "selectivity" between analyte and non-analyte eluent components. For a given method, the different aerosol characteristics produced by the newer design have the effect of simplifying response curves, extending the quasi-linear and overall dynamic response range, and increasing sensitivity to both low concentrations of nonvolatile analytes and of semivolatile analytes. Because of these changes, detector settings of the newer design may need to be optimized when transferring methods from older models. This chapter will therefore provide guidance for method transfer and for exploiting the added capabilities of the newer design.

3.3 Factors Influencing CAD Performance

This section will discuss the main factors that influence CAD response and best practices to optimize performance. In most cases, these concepts apply similarly to all types of EA detectors.

3.3.1 Analyte Properties

For any chromatographic method, understanding the relationships between analyte properties and detector response is of great interest, for example, in method development and troubleshooting. This information is especially important for universal detectors since they are often geared toward quantitative analysis in the absence of individual standards. The main analyte-specific properties that influence EA detector response can be separated into two general categories: (i) those that determine the extent of partitioning of solutes (analytes or impurities) into an aerosol residue particle (i.e., volatility) and (ii) for a given particle size distribution, those properties that influence the response of the downstream aerosol detector. In principle, the first category applies very similarly to all EA detectors, which explains observations of both similar detection scope (i.e., nonvolatiles and some semivolatile) and similar eluent requirements (i.e., volatile solvents and additives). The second category represents the inherent capability of a given type of EA detector to provide uniform response for a diverse range of analytes. These two categories are further discussed in the following text.

3.3.1.1 Formation of Aerosol Residue Particles

As discussed in Chapter 1, EA detectors respond to chemicals that, under a given set of conditions, are able to form stable residue particles. Briefly, "spray drying" involves the creation of droplets that rapidly shrink, which increases the concentration of the less volatile components and dynamically changes droplet solvent composition. This enhances solute–solute interactions, precipitation, and other effects producing a condensed phase—a solid or nonvolatile liquid. Several studies have shown that solute (i.e., analyte or impurity) properties such as vapor pressure (P_s) at a reference temperature, boiling point (B_p) at a reference pressure, enthalpy of vaporization (ΔH_v), or molecular weight (M) can describe an approximate "volatility limit" above, or below, which they are reliably detected and therefore considered nonvolatile (e.g., $B_p > \sim 400°C$, [6, 7] or $M > 350\,g/mol + \Delta H_v > 65\,kJ/mol$ [8]). A main practical consideration is the evaporation temperature setting of the instrument (T_e), which largely determines the value of this volatility limit for a given instrument design and method. It is well recognized, however, that the aforementioned parameters, alone or in combination, are not fully predictive of detectability or sensitivity. Additional factors are discussed in Chapter 1. Among these, acidity and basicity are perhaps the most significant since ionic interactions between analyte and mobile phase additives can also influence response. Another factor is analyte density, which weakly influences residue particle diameter and therefore the response of all EA detectors. If density is known, then this can be accounted for as described in Chapter 1.

3.3.1.2 Inherent Response of Downstream Aerosol Detector

A main goal for universal detection is to achieve uniform response, which requires equivalent sensitivity for a wide diversity of analytes, independent of their physicochemical properties. The extent to which a given type of EA detector can achieve uniform response is mainly related to the inherent response dependency of the downstream aerosol measurement technique on analyte/particle material. As described in Chapter 1, the more uniform response that is widely observed with CAD can be attributed to its dependence on particle dielectric constant, which is relatively minor compared with both the dependency of ELSD on particle refractive index and the apparent dependency of CNLSD on particle wettability and solubility.

3.3.1.3 Summary of Analyte Properties

The most significant analyte properties that influence CAD response are those that determine analyte volatility as it relates to the specific process and conditions of aerosol evaporation. For ionizable analytes, detectability and sensitivity may also be influenced by interactions with ionic mobile phase components. Analyte density also has a minor influence. These relationships are effectively shared by all EA detectors, and the described factors affect impurities as well as analytes. Among these detectors, CAD is otherwise the least dependent on analyte properties and therefore inherently provides the most uniform response.

As described previously, T_e is a useful parameter to alter the volatility or relative partitioning of solutes between the two phases of an aerosol. This leads to the following practical recommendations for a given method:

- Use the lowest T_e that consistently produces the required sensitivity limits. This should provide the best reproducibility and most uniform response among analytes.
- Use small incremental changes in T_e (e.g., ≤5°C) that investigate method robustness and relative temperature stability of each analyte focusing on levels near the limit of detection (LOD).

3.3.2 Eluent Properties and Composition

3.3.2.1 Mass Transport

Key considerations for solvents used with CAD are the physical properties that influence nebulization and evaporation. As widely described for evaporative detectors, mass transport efficiency to the downstream aerosol detector is higher for eluents (e.g., solvent mixtures) with lower viscosity and surface tension (e.g., high % CH_3CN) [9]. As a result, sensitivity (i.e., signal per unit mass) is higher when compared with eluents with higher viscosity and surface tension (e.g., high % water). These transport effects are assumed to be independent of solute properties or solute concentration, which is evident in Figure 4.5. Here the response for several analytes are of similar shape (mostly parallel) and increase with increasing % CH_3CN due its higher transport efficiency. Techniques to address changes in mass transport during solvent gradients to help achieve universal quantitation are discussed later in this chapter and also in Chapter 4.

It should be emphasized that eluent-dependent differences in aerosol mass transport apply not only to analytes but also to nonvolatile impurities. Therefore, eluents that are more efficiently transported can produce higher baseline current and noise and, especially during solvent gradients, baseline drift. Common sources of impurities and practices to minimize their concentration are further discussed in Section 3.3.2.2.

3.3.2.2 Eluent Purity

A main factor affecting CAD performance is the mass concentration of impurities (nonvolatile + semivolatile) in the column eluent, which contributes to the baseline (background) signal, noise, and, especially with solvent gradients, drift. These impurities often exist as solutes that cannot be removed by filtration. For some high-purity organic solvents (e.g., HPLC grade or better), an additional filtration step can contaminate the solvent and increase impurity concentration. It is therefore common to filter only the aqueous buffer portion of a mobile phase. After droplet evaporation, less volatile impurities contribute to the size of each dried aerosol particle in proportion to their mass concentration. Recommended practices to minimize impurity concentration are similar to LC-MS and are further discussed in the following text.

Water is a very common LC solvent and can be a main source of less volatile impurities. The highest quality water for EA detection and LC-MS is most often obtained from a well-maintained point-of-use purification system. Commercially available HPLC-grade bottled water is often found to produce a significantly higher background level than that obtained from point-of-use systems. Typical ultrapure water specifications are 18.2 MΩ·cm resistivity at 25°C and <5 parts per billion total organic carbon. Since impurities can accumulate upon storage and in a stagnant system, it is recommended to obtain freshwater from a thoroughly flushed purification system.

The most relevant specification for an organic solvent is its residue after evaporation. This is especially important since the properties of many organic solvents lead to much higher mass transport than water. As expected, solvents with lower residue after evaporation typically produce lower background and noise. There are some reports of better sensitivity limits with CAD using HILIC compared with RPLC, which may be attributed to improved analyte mass transport efficiency with high-purity acetonitrile [5, 10, 11]. For gradient methods, it should also be considered that impurities in the weaker solvent (e.g., water in reversed phase) can be retained and concentrated within the stationary phase only to elute later as "ghost" peaks or baseline artifacts.

3.3.2.3 Mobile Phase Additives

The choice of mobile phase additive type and concentration should take into account not only the requirements of the chromatographic separation but also the detection technique. Common additives include pH modifiers (i.e., acids and bases), buffers (i.e., weak acid + conjugate base or weak base + conjugate acid), and ion-pairing reagents (e.g., for reversed-phase chromatography an acid or base with hydrophobic tail). A fundamental requirement for EA detectors is that additives be sufficiently volatile to ensure an effectively complete partitioning into the dried aerosol's gas phase. Otherwise, if they significantly contribute to residue particle size, then high detector background signal and noise will be observed. Additive concentration is frequently expressed in units of molarity or volume %. However, given the mass flow-dependent behavior of EA detectors, it is also useful to consider, as with analytes, the mass concentration of additives. Table 3.2 lists common volatile

Table 3.2 Volatile additives and typical concentrations.

Acidic additive	Typical conc.	mM	mg/mL	pK_a	pH	
Acetic acid	**0.1% v/v**	17.4	1.04	4.8	3.27	
Formic acid	**0.1% v/v**	23.6	1.09	3.8	2.7	
Trifluoroacetic acid	**0.1% v/v**	13	1.48	0.0	1.9	
Buffer					**~Buffer range**	**Buffer species**
Ammonium acetate	**10 mM**	10	0.77	4.8	3.8–5.8	Acetic acid–acetate
Ammonium formate	**10 mM**	10	0.63	3.8	2.8–4.8	Formic acid–formate
Ammonium carbonate	**10 mM**	10	0.96	10.3	7–11	Ammonia–ammonium
				9.3		Hydrogen carbonate–carbonate
				7.8		Carbonic acid–hydrogen carbonate

additives and typical concentrations used for both EA detection and LC-MS. It should be noted that the indicated 10 mM buffer concentrations are simply based on the molar mass of each salt. However, it is very common to prepare a 10 mM HPLC buffer by adding a concentrated acid (e.g., formic acid (FA)) to a 10 mM salt solution (e.g., ammonium formate) until the desired pH is obtained as measured with a pH electrode. With this method, the actual mass and molar concentration can therefore be much higher than stated. By comparison, for a typical chromatographic peak volume (V) of 0.1 mL and normally distributed analyte, an injection of >20 µg of analyte (m_{inj}) would be required to achieve a comparable >0.5 mg/mL instantaneous concentration at peak apex (C_{max}) based on

$$C_{max} = \frac{m_{inj}}{V} \times (2\pi)^{0.5}$$

This illustrates that, for many analyses, additive mass concentration is several orders of magnitude higher than instantaneous analyte mass concentration, which clearly highlights the need to use only volatile additives. It should also be emphasized that the previously described pH titration method should only be performed with aqueous solutions since pH electrode measurement is often not accurate or reproducible when organic solvent is present.

It is important to note that most mobile phase additives are ionizable and, as mentioned earlier, are often present at relatively high concentration in each aerosol droplet. These factors and the rapid increase in concentration that occurs during droplet evaporation improve the likelihood that less volatile salts will form. As described in Chapter 1, volatile additives can extend the detection scope of CAD by enhancing sensitivity of volatile ionizable analytes. They can also, to a lesser extent, increase the sensitivity of non-volatile ionizable analytes, which adversely affects response uniformity. In addition, ionizable additives can combine with impurities, sample matrix components, and other additives of opposite charge. Practical implications of these types of ionic interactions are discussed in more detail in the following text.

Evidence suggests that the likelihood that ionizable additives will contribute to the mass of a stable aerosol particle increases with both concentration and extent of ionization (i.e., related to solution pH and strength of acidic and/or basic additives) and decreases with the volatility of the resulting salt [12–15]. These factors are especially important to consider when using buffers and ion-pairing reagents. For example, Petritis *et al.* [13] showed that a mixture of ammonia ($\geq$100 mM = 1.7 mg/mL) and FA ($\geq$100 mM = 4.6 mg/mL) produced a much lower ELSD background signal than a more dilute mixture of ammonia ($\leq$5 mM = 0.085 mg/mL) and TFA ($\leq$5 mM = 0.57 mg/mL). In this case, the stronger acid, TFA ($pK_a = 0$), is more fully ionized than FA ($pK_a = 3.75$), and a corresponding TFA-ammonium salt may be less volatile than a FA-ammonium salt. Importantly, this illustrates that when using volatile anionic ion-pairing reagents such as TFA, use of a basic additive (e.g., ammonia) should be done with caution. Likewise, as indicated in studies by Cohen *et al.* [15], caution should also be taken if adding an acid to a mobile phase containing a cationic ion-pairing reagent (e.g., triethylamine).

For many RPLC methods it is very common to use only an acidic modifier. A typical starting point when using CAD or LC-MS is a silica-based C18 stationary phase and water–acetonitrile eluents containing 0.1% v/v FA or acetic acid. These acids are sufficiently volatile, and the low pH (e.g., 2.7 for FA in water) suppresses ionization of both a majority of acidic analytes and of weakly acidic un-bonded silanol groups on the silica particle surface. This facilitates retention of acidic analytes and minimizes peak tailing for basic analytes. However, retention of polar analytes (e.g., basic analytes ionized at this low pH) is not always sufficient nor is the selectivity for a given analysis. A common alternative is to use TFA in place of, or in addition to, these weaker acids. With 0.1% v/v TFA, the lower pH (1.9) further suppresses solution-phase ionization of acids (not to be confused with its unwanted suppression of electrospray ionization) while forming ion pairs with basic analytes to increase their retention. However, in addition to the previously mentioned potentially adverse effects of TFA, there are several other things to consider when using TFA or other higher molar mass ion-pairing reagents such as other perfluorocarboxylic acids (e.g., heptafluorobutyric acid (HFBA)). One possible issue is the formation of non-volatile salts with cationic eluent impurities, leading to an increase in background. These anionic ion-pairing reagents may also interact with cationic sample matrix components, which if present at high concentration can produce a large solvent front, "ghost peaks," or other baseline artifacts. TFA has also been reported to be unstable, and therefore single-use vials are preferred for mobile phase preparation. These issues and cautions also apply to the use of basic additives where salt formation may likewise occur with impurities and sample matrix components (e.g., Cl^-) [16]. Several mixed-mode stationary phases, for example, with reversed-phase + ion-exchange retention characteristics or ion-exchange + normal phase (via hydrophilic interaction) characteristics, have become available to address the many challenging separations faced by chromatographers. These columns may avoid several problems associated with use of ion-pairing reagents. In particular, mixed-mode columns that exhibit low column bleed (see Section 3.3.2.5) and provide good separation with minimal additive concentration can provide an ideal complement to the previously described RPLC technique for use with CAD and LC-MS.

3.3.2.4 Additional Sources of Eluent Impurities

Eluent impurities may originate not only from the solvents and additives but also from columns, other system components, and general laboratory equipment. Common practices to minimize these sources of eluent and sample contamination are discussed in the following text.

3.3.2.5 Column Bleed

For EA detectors, column bleed refers to the addition of nonvolatile impurities into the eluent by the column itself, which increases background signal and noise. This is commonly attributed to degradation (i.e., cleavage of bonded phase, dissolution of support material) but may also include slow dissolution of impurities from the stationary phase. Impurities may originate from the support itself (e.g., metals in lower purity silica) or from prior exposure. For example, it is often recommended to avoid using columns with LC-MS or CAD that have been previously exposed to nonvolatile buffers (e.g., sodium phosphate) or to nonvolatile ion-pairing reagents (e.g., octanesulfonic acid). A simple approach to screen columns for use with CAD is to compare the background signal with, and without, a column. Columns are tested after sufficient flushing to waste and, once connected to the detector, after equilibration. For gradient methods, the full range of solvent strengths and compositions (e.g., pH) should be tested. Some columns (e.g., amino, cyano) have been observed by CAD to exhibit significant bleed throughout their operating range [6]. However, it should be emphasized that, for many columns, significant bleed is mainly observed when operated near the extremes of their compatibility range. Also, some columns may exhibit significant bleed while still maintaining fairly stable retention times and peak shapes. Most reports of column bleed with silica-based columns appear to be for methods operating toward either the lower or higher end of the pH compatibility range (typically 2–8). Column bleed may therefore occur with 0.1% TFA (pH 1.9) and be more pronounced at higher concentrations (e.g., pH of 0.5% TFA = 1.2). There are several examples with HILIC using neutral or slightly basic eluents where silica-based stationary phases (e.g., with amino, diol, or zwitterionic functional groups) exhibit significant column bleed. In these cases, alternative columns developed for wider pH compatibility can be used including those with polymeric, protected silica, or other more pH-stable supports [17]. As observed by Teutenberg *et al.* [18], columns designed for wide pH compatibility also seem to exhibit low bleed when operated at high T.

3.3.2.6 Basic Eluents

Some investigators have observed high background current and noise with CAD when using higher pH eluents and reported general incompatibility with ammonium carbonate buffers [6]. From the previous discussion, likely sources of impurities in higher pH eluents include additives used to achieve the desired pH and their potential to form salts and column bleed, which is especially pronounced at high pH with silica-based supports. Our own

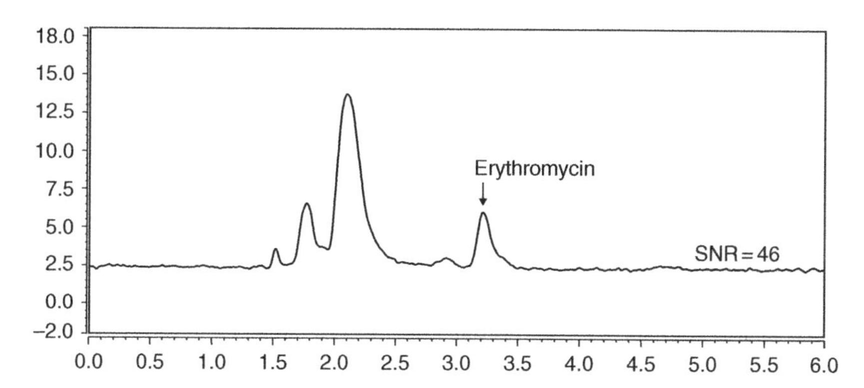

Figure 3.3 Use of CAD with higher pH eluents. 40 ng erythromycin; polymer-encapsulated C18 column, 5.0 μm, 4.6 × 150 mm; 30% 10 mM ammonium carbonate, pH 9.0; 70% acetonitrile. Isocratic 0.8 mL/min at 70% B; Corona Veo RS T_e = 75°C.

studies using earlier model CAD instruments and no column showed very high background signal and noise with aqueous ammonium carbonate buffers (e.g., >3 pA noise for 1.0 mM NH_4HCO_3). Lower background and noise (<0.01 pA), however, was observed with freshly prepared ammonium–ammonia buffers with acetate as counterion at similar pH and concentration, but this increased over time. One potential explanation is that CO_2 absorbed from the atmosphere may lead to the formation of bicarbonate and the divalent anion carbonate. The latter, which is favored at higher pH, may accumulate over time [19] and during droplet evaporation may form a less volatile salt $(NH_4)_2CO_3$. This might explain both the increase in noise over time with an ammonium buffer and the high noise with ammonium carbonate buffers. Our studies with newer CAD models demonstrate that increasing the T_e is very effective at reducing background signal and noise with ammonium carbonate buffers and higher pH eluents. For example, with the same 1.0 mM NH_4HCO_3 buffer, noise with Corona Veo, while very high at the default T_e of 35°C, was reduced to <0.01 pA at 50°C. Figure 3.3 shows a further example where a signal-to-noise ratio (SNR) of 46 was obtained for 40 ng of erythromycin with a 3 mM ammonium bicarbonate buffer and a column designed for higher pH stability.

3.3.2.7 System Components and Laboratory Equipment

Possible sources of contamination, which increase the mass concentration of nonvolatile impurities in eluents and samples, are briefly discussed here. As with columns, many of the issues with components in an LC system's flow path can be attributed to "memory effects" after extended exposure to nonvolatile solutes (e.g., phosphate buffers). In some cases, the component can be flushed with a suitable solvent, but in others, it should be replaced with a part that is then dedicated for use with CAD and LC-MS. As with columns, it is useful to compare the background signal and noise with, and without, the component or with a suitable replacement. Commonly reported sources of long-term contamination after extended exposure to nonvolatile solutes include in-line filters (e.g., solvent reservoir filters), online degassers (membrane with vacuum), and pulse dampers (membrane with solvent reservoir). The decision whether to flush or replace depends on different factors including ease of replacement and effectiveness of cleaning, which may be evaluated as described previously.

Additional sources of contamination include pH electrodes, liquid handling, and storage materials and the environment. A pH electrode used to adjust mobile phase buffer pH can contribute a significant amount of nonvolatile impurity since it is typically calibrated and stored using highly concentrated, nonvolatile buffers. For mobile phase preparation it is therefore recommended to only immerse a pH electrode in a separate aliquot of the buffer and to avoid any pH adjustment of solutions containing an organic solvent [20]. These practices are particularly important with CAD and LC-MS since they avoid both contamination and an often unwarranted addition of high concentrations of acidic or basic modifier. A common observation with CAD is a measurable peak response from a mobile phase blank with flow injection (no column) or in the solvent front (with column). The most likely cause is contamination from the sample vial or well either through leaching from the material (e.g., glass) or prior exposure. Likewise, contamination of sample or mobile phase can arise from various sources such as glassware and liquid handling materials. As mentioned earlier, the source should be identified and the issue minimized, for example, by prior rinsing of materials with appropriate solvent or solvents.

3.3.2.8 Summary

From the previous discussion, the following general practices are recommended:

- Use mobile phase solvents with low residue after evaporation and ultrapure water fresh from a purifier and not from a stored bottle.
- Use high-purity volatile mobile phase additives at lowest concentration needed for desired separation.
- For nonvolatile analytes, the most uniform response is obtained with low molar mass additives.
- Minimize column bleed through proper choice of column and avoid using columns near the extremes of their operating range (e.g., pH, T, pressure, solvent strength).
- Avoid sources of contamination (e.g., LC system components, labware, pH electrode buffers, etc.).

3.4 System Configurations

This section discusses some practical aspects of different system configurations including micro-flow LC, post-column addition, and flow splitting.

3.4.1 Microscale LC

Newer CAD designs allow routine operation with eluent flow rates as low as 10 μL/min. Advantages of microscale LC include reduction of both sample and solvent consumption. Practical use of CAD at lower flow liquid flow rates may be best understood by considering that CAD behaves primarily as a mass flow-dependent device (see Chapter 1). For a true mass flow-dependent device and a given analyte mass, when volumetric flow rate is changed, peak area remains constant while peak height changes. The approximate mass flow-dependent behavior of CAD is demonstrated in Chapter 1 (Figure 1.21) where similar peak area was obtained for a given mass when using a smaller internal diameter column at the same flow velocity. At the lower flow, baseline noise was reduced due to lower mass flow of impurity per unit time, which resulted in higher SNR. While mass sensitivity is improved at the lower flow, concentration sensitivity is unchanged if injection volume is scaled in direct proportion to column volume. Significant advantages may however be achieved if the column has sufficient mass load capacity, which is often true with gradient elution.

3.4.2 Post-column Addition

There are several examples using post-column addition with EA detectors. The most common technique is called inverse gradient compensation first described for use with CAD by Gorecki *et al.* in 2006 [21]. The approach employs two gradient pumps, where a second pump delivers a solvent gradient that is the exact inverse of that for the analytical separation. The solvent streams are combined post-column typically using a simple T-connector. This delivers a constant solvent composition to the detector (i.e., midpoint of gradient) at twice the liquid flow. A practical requirement is to match the gradient timing and therefore the volume of the two flow paths. Since the analytical flow includes a column, a typical approach is to include an identical column within the compensation flow as shown in Figure 3.4. An alternative is to use an appropriate volume of tubing. The overall goal of this technique is to compensate for the eluent-dependent, analyte-independent, change in mass transport efficiency during gradients and therefore achieve more uniform response. Since CAD behaves as a mass flow-dependent device (i.e., peak area is preserved), a twofold dilution is compensated by a twofold increase in flow velocity. A further advantage of this technique is that a more stable (less drifting)

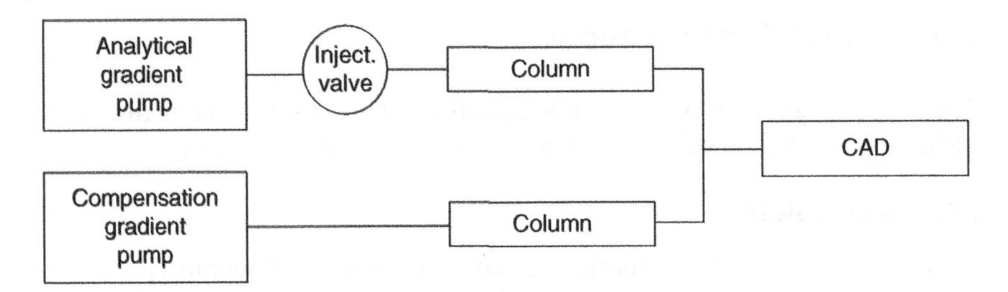

Figure 3.4 Inverse gradient compensation configuration, which includes two identical columns in order to match the volumes of both flow paths.

baseline can be obtained with a constant eluent. Post-column addition has also been used to increase mass transport and therefore sensitivity by adding, isocratically, a lower viscosity/surface tension solvent. Since mass transport of impurities is also increased, it is important to choose a solvent that has low residue after evaporation. A practical consideration is to ensure that the solvents are completely miscible and that analytes, matrix components, and additives remain fully soluble. A further consideration with any post-column addition is the flow rate capacity of the instrument, since this involves an increase in solvent load. A common symptom of inadequate solvent evaporation is baseline noise that is characteristically irregular.

3.4.3 Multi-detector Configurations

There are many applications using CAD combined with one or more complementary detectors. The use of a UV (diode array detector (DAD)) flow cell upstream, in series, with CAD is now commonplace. The ever-increasing need to rapidly identify and quantify many diverse species in complex samples has furthermore led to significant growth in use of multi-detector systems with CAD for universal detection and comprehensive analysis. A very common configuration involves liquid splitting after a DAD flow cell to parallel CAD and MS devices. Universal detection and global analysis is the focus of Chapter 4 and includes discussions of multi-detector configurations, orthogonal separation techniques (e.g., HPLC and supercritical fluid chromatography (SFC)), and multidimensional separation techniques such as heart-cutting and comprehensive 2D LC. Further examples are included in Chapter 4 and in studies by Poplawska *et al.* (LC-DAD, CAD, MS) [22], Li *et al.* (2D-LC-CAD, MS) [23], Zhang *et al.* (2D-LC-DAD, CAD, MS) [24], Schiesel *et al.* (2D-LC-CAD, MS) [25], and Brondz and Høiland (SFC-CAD, MS) [26]. Here we describe main practical aspects of various configurations.

Primary concerns when combining CAD with other detectors include eluent and flow rate compatibility, the destructive nature of nebulizer-based detectors, dead volume, and flow cell back pressure rating. As noted earlier, CAD and MS have very similar eluent requirements. In some instances, analyte-specific eluent requirements for MS (e.g., pH for electrospray ionization) may influence CAD response. For example, use of an acidic modifier to promote positive electrospray ionization may influence CAD response for volatile bases. In most cases, eluent requirements for MS are more stringent than CAD and can be accommodated within the previously described guidelines.

CAD and MS are both nebulizer based and therefore must be configured in parallel at the end of separate flow paths after a flow split. CAD behaves as a mass flow-dependent device, and any split reduces the mass of analyte and impurity seen by the detector. Signal from both analyte and impurity should thus decrease and SNR should, in most cases, be similar. For many methods, the optimal flow to MS is ≤0.2 mL/min, and, with standard LC flow rates (e.g., 0.5–2.0 mL/min), the remaining flow is typically directed to CAD. For most applications, it is not critical to establish an exact split ratio. A passive flow splitter such as a simple Y-connector is often used where the split ratio depends on the relative flow restriction between the two paths. This can be adjusted by using different tubing lengths and internal diameters. Adjustable passive flow splitters are also available to simplify this step. The ratio is measured by collecting the liquid from the outlet of the detector's inlet tubing or, to account for further flow restriction, from the nebulizer outlet. Since it is more accurate to take measurements from the lower flow path, this is usually done with a low volume precision measuring device. With gradient methods, a change in solvent viscosity will often change the split ratio of passive flow splitters. It is therefore important to use a starting split ratio that is well within the requirements of both devices. Some splitters can maintain a constant ratio during gradients (e.g., ASI, Richmond, CA, USA) and can be very useful for these applications. As described earlier, inverse gradient compensation is often used with CAD to provide uniform response and may also serve to normalize the performance of an MS ionization source. It may therefore be beneficial to combine the compensation and analytical solvent streams before the flow split since this would also maintain the split ratio independent of solvent composition.

All of the previous concerns must take into account the dead volume and back pressure introduced by tubing and splitters and especially the dead volume and back pressure rating of upstream flow cells. In most cases a UV flow cell is upstream before any post-column addition or flow split and its volume, back pressure rating, and the back pressure from downstream restrictions must be considered. This in turn depends on the flow rate, solvent viscosity, and temperature. Use of upstream low volume flow cells with a high back pressure rating is often a best choice to use in these configurations. For some detectors (e.g., RI, Fl), the available flow cells contribute significant volume and/or have low pressure rating, which poses significant constraints for this type of application.

3.5 Method Transfer

This section provides a basic approach to transfer a method from first-to second-generation CAD instruments. The logic should also generally apply to method transfer from ELSD to CAD. A main assumption in that regard is that the default conditions of most instruments are based around RPLC and common liquid flow rates (e.g., 0.5–2.0 mL/min). Here the requirement to quickly and completely evaporate water imposes some constraints on the measured aerosol size distributions, the range of which may be largely encompassed by different CAD designs. For so-called "type A ELSD instruments [3], this should correspond to an "impactor on" setting if available. In all cases, evaporation *T* is then a primary variable to optimize performance.

The basic steps for method transfer are as follows:

1) Start with default settings of the newer CAD instrument. This is also a good starting point for new method development. While some earlier CAD models include a nebulizer *T* setting, its influence is considered relatively minor (Section 3.2.1) and should be ignored.
2) After equilibration (e.g., 0.5–1 h), perform a trial run of mobile phase and calibrator injections (triplicate) across the full mass range of interest including the desired lower sensitivity limits. A good approach is to include calibration levels in 1, 2, 5, and 10 increments across each order of magnitude.

   Note: Downstream aerosol detector sensitivity drops exponentially for small particles (≤10 nm for CAD; ≤~50 nm for ELSD). Particles of this size are typically predominant for lower concentrations of eluent impurities and, for low analyte mass levels, impurity + analyte. Initial performance comparison based on SNR for higher level standards or solely on baseline noise can be very misleading if response is assumed to be linear. Also, the aerosol mass transported to the downstream detector differs between CAD designs, and so the absolute peak response and shape of the response curve may be different. Before changing any variable such as evaporation *T*, it is first recommended to evaluate the shape of the response curve, as described in Section 3.6.
3) Using the previous method's calibration model (e.g., linear, quadratic, double log), compare goodness of fit for the data obtained with the different instruments. Some CAD instruments include a power function (PF) setting, which may be used to "linearize" the response output of the instrument. If a setting other than the default was used for older instruments, then it is likely that a similar value may be warranted. Guidance for evaluating goodness of fit and choosing a PF value other than the default is provided in Section 3.6:
4) Optimize T_e. As previously discussed, a best practice is to use the lowest T_e setting that consistently produces the required sensitivity limits. Given the

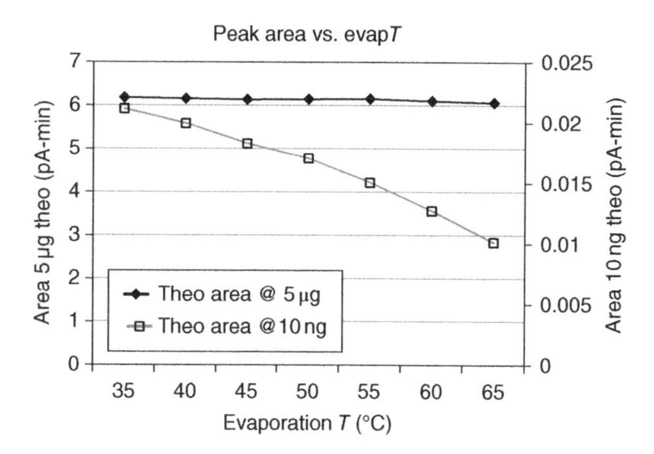

Figure 3.5 Peak area response for low and high amounts of theophylline as a function of T_e.

previously mentioned exponential signal drop that occurs with ELSD and, to a lesser degree, with CAD (see Chapter 1), sensitivity limits should only be assessed from SNR for lower level calibrators (e.g., near target LOD). The importance of this when optimizing T_e is shown in Figure 3.5 where the effect is more pronounced for low analyte levels. A best practice is to optimize T_e by testing small increments (±5°C). An increase in T_e may reduce noise but possibly at the expense of response for semivolatile analytes. Conversely, an increase in T_e should increase signal for semivolatile analytes often at the expense of higher baseline noise.

5) Optimize digital filter settings, which are typically based on a time constant. A smaller (for fast peaks) or larger (for broad peaks) time constant may provide better SNR. Specific guidance can be obtained from the appropriate operating manual.

After completing the previous steps, a more thorough evaluation of method performance (e.g., reproducibility and robustness) can be made with additional optimization as needed based on the concepts and best practices described in this chapter.

When comparing chromatographic data between newer and older CAD instruments, using the default conditions for both, it is fairly common to see a somewhat higher baseline level, noise, and drift with the newer design. This is typically due to a nonlinear "drop-off" in sensitivity of the earlier design at the extreme low end of the dynamic range. This "drop-off" in sensitivity can mislead the user to think that the achievable lower limits of detection or limit of quantitation (LOQ) with newer instruments are poorer than that of earlier models. However, this is more likely due to the better sensitivity (relative absence of signal drop-off) of the newer design to very low levels of nonvolatile residue

3.6 Calibration and Sensitivity Limits

This section will discuss calibration approaches and determination of sensitivity limits with CAD. This includes use of an optional PF setting on some CAD instruments to help "linearize" the signal output of the detector. As described in Chapter 1, CAD response is typically quasi-linear over a range of ~10^2, but over a wider range is nonlinear. Briefly, response can be described by a power law equation, as shown in the following, where peak area (A) is equal to the injected mass (m_{inj}) raised to an exponent (b) and multiplied by a sensitivity coefficient (a):

$$A = a\left(m_{inj}\right)^b$$

The power law exponent (b) describes the shape of the curve, which is linear when $b = 1.0$, sublinear when $b < 1$, and supralinear when $b > 1$. With linear response, the slope of A/m_{inj} or "response factor" remains constant over the investigated range. When $b \neq 1$, the response factor changes. The more b differs from 1, in either direction, the greater the degree of curvature or deviation from linear response. Importantly, over the dynamic range of CAD not only does the response factor change ($b \neq 1$), but so does the value of b or degree of curvature. For a nonvolatile analyte and an instrument default PF setting of 1.0, b is expected to approach a maximum of ~1.1 near the lower LOD; gradually decrease throughout its >10^4 dynamic range, which includes a quasi-linear segment of ~10^2; and reach a minimum of ~0.6 at the high end. By comparison, ELSD response shows a directionally similar trend with more dramatic changes (~2.0 to ~0.67) over smaller (~10^3) dynamic and (~10^1) quasi-linear ranges. For either detector, response will be more variable among semivolatile analytes, and the observed exponents will be higher and supralinear, especially near the lower mass range.

For any quantitative method it is important to generate and evaluate calibration data and choose an appropriate curve fitting model. This is especially important when measuring over a wide analyte mass range since a linear regression model may not provide an adequate fit. Shown in the following text are practical suggestions based on the response characteristics of CAD for generating and evaluating calibration models for "quality of fit." The following references form a basis for this discussion [27–31] and are recommended for additional guidance. A key consideration is the number and spacing of levels that should be analyzed. Given the nature of CAD response, more levels are typically needed than, for example, with UV. As described previously for method transfer, a good approach is to include calibration levels ($n = 3$ each level) in 1, 2, 5, and 10, or similar, increments within each order of magnitude across the full mass range of interest, making sure to include points around the desired lower sensitivity limits.

A good general practice is to use the simplest curve fitting model that adequately describes the response–amount relationship over the required range of interest [30]. A linear curve fit can be used for initial evaluation and is often adequate for many CAD methods that require quantitation over smaller mass ranges ($\sim 10^2$). The curve should be fit to the individual data points as opposed to first averaging the response between replicates at each level. A first step in evaluating a calibration model is to determine whether to force the curve through zero, which is only appropriate if the Y-intercept is less than or equal to standard error of the Y-intercept [27]. Once determined, a very common practice is to use correlation coefficient (r) or coefficient of determination (r^2) from least squares regression as the sole metric to evaluate the curve fit. However, this metric is based on an assumption that data are homoscedastic (i.e., equal absolute error across the range). For most HPLC analyses, it is common to see somewhat higher absolute error (e.g., standard deviation (SD) of replicates) at higher amounts, which brings this assumption into question. For such data, the higher levels may exert too much influence on a least squares regression line. As shown by Kiser and Dolan [30], a curve fit with an r^2 of 0.9990 may still be poor, with largest error typically near the low end. In general, it is highly recommended to more closely examine the curve fit especially in the lower mass range and consider using weighted regression (e.g., $1/x^{0.5}$, $1/x$, $1/x^2$) to counteract this influence [30, 32]. Weighted curve fitting options are available in most chromatography software applications. Quality of fit near the low end can be visually examined by zooming in on the signal versus amount plot. Another effective way to more closely evaluate the quality of fit is to plot the residuals, expressed as % recovery or % error, against the analyte amount as shown in Figure 3.6. These plots can be generated by most chromatography software applications or by spreadsheet software using calculations described by Kiser and Dolan [29, 30]. This allows better visualization of the data where the best fit is obtained when the points are evenly scattered closely above and below a 100% recovery or 0% error line.

If a linear curve fit model is found to be inadequate, then additional options available in most chromatography data systems (e.g., log–log, point to point, quadratic (i.e., second-order polynomial)) may be used and evaluated in the same manner as before. Also, if the problems are mainly near the low end, then it may be appropriate where possible to use higher injection volumes. Among the available curve fitting models, log–log is commonly used with ELSD and CAD. As noted in Chapter 1, this is a simple transformation of the previously described power law equation where the exponent b now becomes the slope of the line. With this option and a plot of log area versus log mass, the data points usually appear very close to the fit line, and the r^2 value is usually very close to 1. Here it is especially important to more closely examine the quality of fit since these data can be very misleading. An example of this is shown by the residual plot in Figure 3.6 (left). In this case, it is clear that a poor fit was obtained.

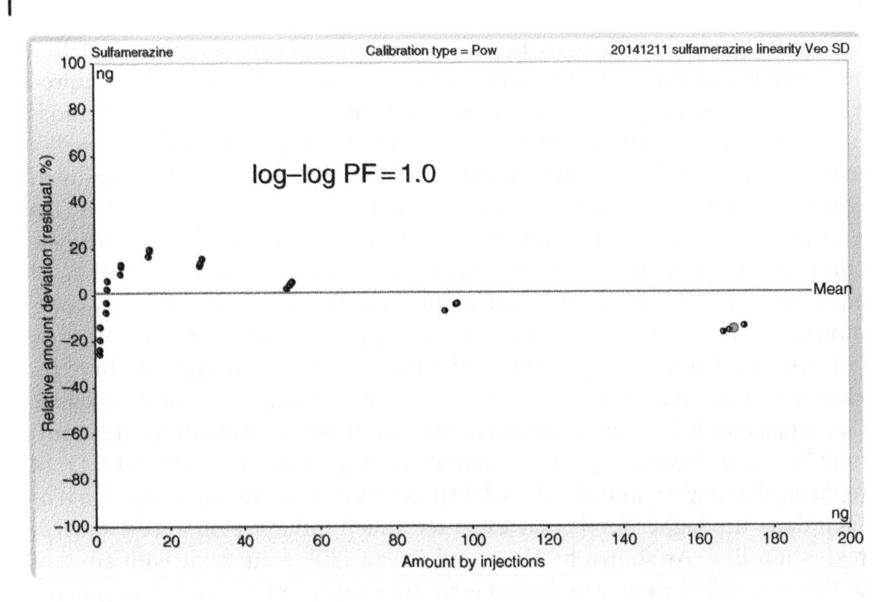

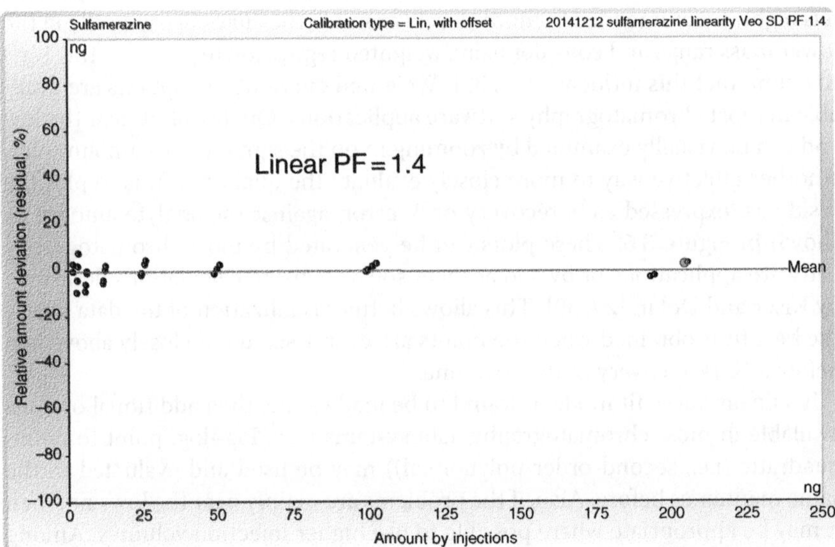

Figure 3.6 Residual plots showing % amount deviation (error) from regression curve versus mass injected using either linear fit to log (area) versus log (mass) for data acquired with PF = 1.0 (left) or linear fit to area versus mass for data acquired with a PF = 1.4.

A quadratic fit is another model that is commonly used for CAD when the mass range of interest is >10^2. For some methods, a best fit has been found with an inverse quadratic model where the signal and mass axes are swapped prior to curve fitting. As with any approach, the best model to use depends on the results and the method requirements (e.g., mass range, sensitivity limits, etc.).

3.6.1 Power Function

The PF is an instrumental setting that may be used to help "linearize" the signal output of CAD. Chapter 1 describes the PF setting as effectively a multiplication factor to the power law exponent b across the dynamic range. For example, if b is observed to range from ~1.1 to 0.6 for data acquired with a PF = 1.0, then a setting of 1.5 would be expected to produce results whose b values range from ~1.65 at the low end to ~0.9 at the high end. The main effects are therefore to shift the ~10^2 m_{inj} range over which response is approximately linear and to produce a supralinear response (i.e., exponential signal drop) at the low end. This approach can be very useful for lower sensitivity methods where the mass range of interest is relatively high and for fast gradient/high separation efficiency methods where analytes often elute in more concentrated solute bands. In both cases, the response curve is more likely to be sublinear over the range of interest. Figure 3.6 compares residual plots for two curve fitting models applied to data obtained using a fast gradient method. The plot on the left used a log–log curve type to fit data acquired using the default PF of 1.0. The plot on the right used a linear fit to data acquired with a PF of 1.4. In this case, the PF value of 1.4 was derived from the results obtained with PF = 1.0 and using a feature of Chromeleon™ software (Thermo Fisher Scientific). Another way to estimate a PF value is to determine b from the slope of a log–log plot keeping in mind the caveats mentioned previously. The appropriate PF to use is then the reciprocal of b. By either approach, the PF setting should be verified experimentally. The results in Figure 3.6 show a significant benefit to using a PF of 1.4 in terms of quality of fit.

As described in Chapter 1 with nonlinear response, it is common to be misled by SNR obtained from higher level standards and by peak distortion, which for sublinear response causes artificial peak broadening and for supralinear response causes peak shaving. Besides the better curve fit obtained in this example, use of an instrumental PF setting to better linearize the signal output therefore may help to avoid these common pitfalls. It is particularly recommended to consider use of an optional PF setting for methods that rely heavily on relative response between peaks for quantitation (e.g., profiling mass distribution of polymers). It should be emphasized, however, that a PF setting should only be chosen to linearize response rather than to artificially sharpen peaks or to exaggerate the SNR values for high level standards.

One of the most challenging aspects of any quantitative method is determining the lower sensitivity limits. Most commonly, an SNR value of 3 is used for the LOD and an SNR value of 10 is used for the LOQ. Since response may be nonlinear especially toward these limits, an accurate estimation of LOD and LOQ usually requires actual analysis of appropriately low analyte levels rather than estimation from higher analyte levels. As previously mentioned, the latter can lead to very misleading results especially when assuming linear response. In general, it is better to perform a more statistically rigorous estimate of sensitivity limits based on sufficient levels and replicates over the course of several days. This is often facilitated by tools within chromatography software (e.g., Chromeleon™). As described by Dolan [28], it is useful to check or confirm these limits by also measuring imprecision at these levels. A good rule of thumb is that the % relative SD should be ~17% at an SNR of 3 and ~5% at an SNR of 10.

3.6.2 Summary of Calibration and Sensitivity Limits

From the previous discussion, the following general practices are recommended:

- Use the simplest calibration model that adequately describes the response curve.
- Evaluate quality of fit using a sufficient number of levels and replicates over the entire range with special consideration to the upper and lower limits.
- Do not use r^2 as the only metric for assessing quality of fit and consider using weighted regression to achieve a better fit.
- Consider using a PF setting to help linearize signal output but avoid using too high a value as it could create or exaggerate a "drop-off" in response near the lower limit of the dynamic range.
- Make sure to determine LOD and LOQ from actual analysis of standards around the typical metrics of SNR 3 and 10, respectively.
- Do not use SNR for higher level standards as a metric to assess the quality of results.

References

1 Young, C. S. and J. W. Dolan (2003). "Success with evaporative light-scattering detection." LC-GC The Magazine of Separation Science 21(2): 120–128.
2 Aruda, W. O., *et al.* (2008). "Review and optimization of linearity and precision in quantitative HPLC-ELSD with chemometrics." LC-GC North America 26(10): 1032.
3 Magnusson, L.-E., *et al.* (2015). "Aerosol-based detectors for liquid chromatography." Journal of Chromatography A 1421: 68–81.

4 Reilly, J., *et al.* (2008). "Implementation of charged aerosol detection in routine reversed phase liquid chromatography methods." Journal of Liquid Chromatography & Related Technologies 31(20): 3132–3142.
5 Hutchinson, J. P., *et al.* (2011). "Comparison of the response of four aerosol detectors used with ultra high pressure liquid chromatography." Journal of Chromatography A 1218(12): 1646–1655.
6 Cohen, R. D. and Y. Liu (2014). "Advances in aerosol-based detectors." Advances in Chromatography 52: 1.
7 Hutchinson, J. P., *et al.* (2010). "Universal response model for a corona charged aerosol detector." Journal of Chromatography A 1217(47): 7418–7427.
8 Squibb, A. W., *et al.* (2008). "Application of parallel gradient high performance liquid chromatography with ultra-violet, evaporative light scattering and electrospray mass spectrometric detection for the quantitative quality control of the compound file to support pharmaceutical discovery." Journal of Chromatography A 1189(1): 101–108.
9 Cobb, Z., *et al.* (2001). "Evaporative light-scattering detection coupled to microcolumn liquid chromatography for the analysis of underivatized amino acids: sensitivity, linearity of response and comparisons with UV absorbance detection." Journal of Microcolumn Separations 13(4): 169–175.
10 Hutchinson, J. P., *et al.* (2012). "Investigation of polar organic solvents compatible with Corona Charged Aerosol Detection and their use for the determination of sugars by hydrophilic interaction liquid chromatography." Analytica Chimica Acta 750: 199–206.
11 Russell, J. J., *et al.* (2015). "Performance of charged aerosol detection with hydrophilic interaction chromatography." Journal of Chromatography A 1405: 72–84.
12 Megoulas, N. C. and M. A. Koupparis (2004). "Enhancement of evaporative light scattering detection in high-performance liquid chromatographic determination of neomycin based on highly volatile mobile phase, high-molecular-mass ion-pairing reagents and controlled peak shape." Journal of Chromatography A 1057(1): 125–131.
13 Petritis, K., *et al.* (2002). "Volatility evaluation of mobile phase/electrolyte additives for mass spectrometry." LC-GC Europe 15(2): 98–103.
14 Deschamps, F. S., *et al.* (2002). "Mechanism of response enhancement in evaporative light scattering detection with the addition of triethylamine and formic acid." Analyst 127(1): 35–41.
15 Cohen, R. D., *et al.* (2012). "Analysis of volatile bases by high performance liquid chromatography with aerosol-based detection." Journal of Chromatography A 1229: 172–179.
16 Lantz, M. D., *et al.* (1997). "Simultaneous resolution and detection of a drug substance, impurities, and counter ion using a mixed-mode HPLC column with evaporative light scattering detection." Journal of Liquid Chromatography & Related Technologies 20(9): 1409–1422.

17 Huang, Z., *et al.* (2009). "Determination of inorganic pharmaceutical counterions using hydrophilic interaction chromatography coupled with a Corona® CAD detector." Journal of Pharmaceutical and Biomedical Analysis 50(5): 809–814.

18 Teutenberg, T., *et al.* (2006). "Evaluation of column bleed by using an ultraviolet and a charged aerosol detector coupled to a high-temperature liquid chromatographic system." Journal of Chromatography A 1119(1): 197–201.

19 Cataldi, T. R., *et al.* (1999). "Isocratic separations of closely-related mono-and disaccharides by high-performance anion-exchange chromatography with pulsed amperometric detection using dilute alkaline spiked with barium acetate." Journal of Chromatography A 855(2): 539–550.

20 Subirats, X., *et al.* (2009). "Buffer considerations for LC and LC-MS." LC-GC North America 27(11): 1000–1004.

21 Gorecki, T., *et al.* (2006). "Universal response in liquid chromatography using charged aerosol detection." Analytical Chemistry 78(9): 3186–3192.

22 Poplawska, M., *et al.* (2014). "Determination of flibanserin and tadalafil in supplements for women sexual desire enhancement using high-performance liquid chromatography with tandem mass spectrometer, diode array detector and charged aerosol detector." Journal of Pharmaceutical and Biomedical Analysis 94: 45–53.

23 Li, Y., *et al.* (2014). "Characterization and stability study of polysorbate 20 in therapeutic monoclonal antibody formulation by multidimensional ultrahigh-performance liquid chromatography–charged aerosol detection–mass spectrometry." Analytical Chemistry 86(10): 5150–5157.

24 Zhang, K., *et al.* (2013). "Analysis of pharmaceutical impurities using multi-heartcutting 2D LC coupled with UV-charged aerosol MS detection." Journal of Separation Science 36(18): 2986–2992.

25 Schiesel, S., *et al.* (2012). "Comprehensive impurity profiling of nutritional infusion solutions by multidimensional off-line reversed-phase liquid chromatography × hydrophilic interaction chromatography–ion trap mass-spectrometry and charged aerosol detection with universal calibration." Journal of Chromatography A 1259: 100–110.

26 Brondz, I. and K. Høiland (2008). "Chemotaxonomic differentiation between *Cortinarius infractus* and *Cortinarius subtortus* by supercritical fluid chromatography connected to a multi-detection system." Trends in Chromatography 4: 79–87.

27 Dolan, J. W. (2009). "Calibration curves, part I: to b or not to b?" LC-GC North America 27(3): 224.

28 Dolan, J. W. (2009). "Calibration curves, part II: what are the limits?" LC-GC North America 27(4): 306–312.

29 Dolan, J. W. (2009). "Calibration curves, part 3: a different view." LC-GC North America 27(5): 392–400.

30 Kiser, M. M. and J. W. Dolan (2004). "Selecting the best curve fit." LC-GC North America 22(2): 112–117.

31 Hinshaw, J. V. (2002). "Nonlinear calibration." LC-GC North America 20(4): 350–355.

32 Almeida, A. M., *et al.* (2002). "Linear regression for calibration lines revisited: weighting schemes for bioanalytical methods." Journal of Chromatography B 774(2): 215–222.

4

Aerosol-Based Detectors in Liquid Chromatography

Approaches Toward Universal Detection and to Global Analysis

Joseph P. Hutchinson, Greg W. Dicinoski, and Paul R. Haddad

Australian Centre for Research on Separation Science (ACROSS), School of Chemistry, Faculty of Science, Engineering and Technology, University of Tasmania, Hobart, Tasmania, Australia

CHAPTER MENU

4.1 Summary

There is an increasing need in the pharmaceutical, nutraceutical, and natural product areas to separate and identify components present in complex samples. Universal detection remains a goal for the discipline of liquid chromatography (LC), and the major attribute of a universal detector is that all analytes are detected sensitively with uniform response factors, regardless of their physicochemical properties. Such a detector should be widely compatible with the methods and techniques routinely used by chromatographers. Aerosol-based detectors display some of the desirable attributes of a universal detector.

Charged Aerosol Detection for Liquid Chromatography and Related Separation Techniques, First Edition. Edited by Paul H. Gamache.

These detectors nebulize the separated sample and the resultant aerosol is dried, leaving analyte particles that can be detected. The Corona Charged Aerosol Detector (Corona CAD) is an example of an aerosol detector that transfers a charge to the dried particles using a countercurrent stream of gas that has passed a corona discharge. These charged particles can then be sensitively detected with uniform response factors being observed for nonvolatile analytes. Aerosol detectors are therefore able to detect a wide range of nonvolatile compounds and have been particularly useful in the analysis of analytes that do not possess a chromophore and hence are not suitable for UV/visible detection. An advantage of the uniform response factors displayed by aerosol detectors is that they allow unknown species to be quantified on the basis of the response of a well-characterized standard. This characteristic is particularly useful in the pharmaceutical industry.

This chapter will discuss the various types of universal detection techniques that can be used in LC. The advantages and disadvantages of aerosol detectors will be discussed with particular reference to the factors that affect analyte response in these detectors and efforts that have been made to characterize these effects. A major challenge for aerosol detection is the nonlinear change in response exhibited when mobile phase gradients are introduced into the detector. Normalizing this effect will greatly improve the acceptance of these detectors. Furthermore, due to their inherent mode of operation, aerosol detectors are compatible with a wide range of volatile solvents consistent with the principles of green chemistry and allow these detectors to be hyphenated to a wide range of liquid chromatographic methods, including high temperature water and supercritical carbon dioxide chromatographic separations. Finally, aerosol detectors will be discussed in the context of global analysis and their use in combination with multidimensional separations. By understanding the fundamentals that affect analyte response in these aerosol-based detectors, method and detector design can be optimized to suit the separation needs of today along with those of the future.

4.2 Introduction

Analytical separation science is concerned with separating complex mixtures, identifying components, quantitative measurement of the amounts of these components that are present in the mixture, and providing chemical structural information on the components. It finds extensive applications across a wide range of sciences, such as environmental science; agricultural science; biomedical, clinical, and pharmaceutical science; forensics; oceanography; materials science; mining; and geology. The role of the analytical chemist is becoming increasingly more challenging due to the complexity of samples requiring separation and the ever-increasing need to detect smaller amounts of components within a sample due to mandated regulatory requirements. In particular, the advent of the "omic" studies such as genes (genomics), proteins (proteomics), lipids (lipidomics), and metabolites (metabolomics) require the separation, detection, and measurement of large numbers of analytes to find correlations with biological endpoints, such as disease, toxicity, or potential drug candidates. It is important in such time-consuming and resource-intensive studies to use technology that is able to detect all components present so that potential causal factors are not overlooked. An interrelated requirement is that the detector should be able to detect analytes at trace levels such that components of the sample present at low concentration, which may be a vital parameter in the study being performed, are not overlooked. The ideal characteristics of a detector are that it shows universal response (i.e., all analyte types can be detected at low levels with uniform response factors) and that the detector can be interfaced to a separation technique that employs a wide variety of mobile phases/separation media for the purpose of achieving efficient separations. It is also advantageous for a universal detector to be of low cost and to be suitable for use by personnel with minimal training. Unfortunately, such an ideal universal detector currently does not exist.

Universal detection techniques have been available in conjunction with gas chromatography (GC) for some time. Mass spectrometry (MS) is a popular choice due to the additional mass and fragmentation pattern information it provides. Flame ionization detection (FID) has also been used routinely in conjunction with GC as it exhibits high sensitivity, wide linear range, and more universal response factors for organic compounds in comparison with MS. FID is capable of detecting most organic species as the response is related to the number of pyrolyzed carbon atoms (ions) reaching the detector per unit time. However, not all samples are suitable for separation in the gas phase, and there is a growing need for liquid chromatography (LC) to separate nonvolatile and thermally labile species. This is evidenced by the global LC market increasing by greater than 5% per annum and is projected to reach US$4.13 billion by 2021 [1].

LC systems in combination with acetonitrile or methanol gradient separations allow a wide variety of organic species to be separated based on their hydrophobicity and are used to analyze, identify, purify, and quantify compounds in many industries. Acetonitrile is the solvent of choice for reversed-phase liquid chromatography (RPLC) due to its favorable physicochemical properties (e.g., water miscibility, low boiling temperature, and ability to solvate many analytes), especially when combined with detectors that nebulize the sample, such as MS and the emerging class of aerosol detectors, including the Corona CAD, Nano Quantity Analyte Detector (NQAD), and evaporative light scattering detector (ELSD). MS is by far the most universally applicable detection technique used in conjunction with LC, due to its ability to detect a wide range of compounds and the extra separation dimension it provides based on mass spectral information. However, MS does have drawbacks such as its high cost, a requirement for skilled operators, and variable response factors for different analytes resulting from differing ionization efficiencies. The use of FID as a detector hyphenated to LC separations has been investigated [2, 3], but this approach suffers from practical considerations, such as the potential for the mobile phase to extinguish the flame and an increase in baseline noise when organic modifiers are used in the mobile phase.

While universal detection methods compatible with liquid-phase separations will benefit all users of LC, they are of particular interest to the pharmaceutical industry, which is approaching one trillion dollars globally in terms of market capitalization. The pharmaceutical industry is subject to a variety of laws and regulations regarding patenting, testing, and ensuring drug safety and efficacy. A universal detector providing a uniform response toward all analytes would allow well-characterized reference standards to be used to obtain purity results for unknown compounds in a sample. While MS has achieved prominence as an important tool in drug discovery for activities such as high-throughput screening, combinatorial synthesis, and *in vitro/in vivo* metabolic studies [4], many analyte species are not ionizable or exhibit variable ionization. Similarly, while photometric detection (UV/visible) is still widely used in routine assays in the later development activities, such as formulations and batch reproducibility testing, UV detection requires the analyte to contain a suitable chromophore.

This chapter will provide an overview of the aerosol detection methods suitable for use with LC and highlight the challenges that must be overcome for these detectors to be considered truly universal in their response. The discussion will include the factors affecting analyte response in these detectors, efforts to overcome these challenges, and some insights into the future of charged aerosol detection with regard to orthogonal separation techniques and global analysis.

4.3 Universal Detection Methods

A range of different detection techniques have been used in conjunction with LC. These can be broadly classified as destructive or nondestructive. Nondestructive detectors can be used to increase the universality of an analytical system when they are placed in-line before another form of detection. Alternatively, if two or more destructive detectors are required to detect all components in a sample, the mobile phase can be split post-separation and a portion introduced into the respective detectors. This approach leads to reduced detection limits as not all the analyte is introduced to a given detector, and determination of the exact split ratio can be dependent on the back pressures of the detectors used, complicating the quantification of analytes.

Several (quasi) universal detection techniques have been described in the literature for liquid-phase separations [5–7]. However, none are truly universal nor do they display uniform response factors when the composition of the mobile phase is changing during a separation or when used with different separation conditions. Of these detectors, the most popular include photometric detectors (UV/visible) at low wavelengths [8] and MS detectors [9] such as time of flight, quadrupole, ion trap, particle beam, and ion cyclotron resonance mass spectrometers, which utilize a range of ionization techniques (e.g., matrix-assisted laser desorption ionization, electrospray ionization, thermospray ionization, ionspray ionization, atmospheric pressure ionization, and chemical ionization).

MS has long been the universal detection system of choice in combination with LC separations due to its high sensitivity and the structural information provided [10]. However, it is not feasible to place such expensive instrumentation on every benchtop for the routine screening of pharmaceuticals as these detectors require specialist skills for their operation. Furthermore, mass spectrometers are known to suffer from variable response factors that can be attributed to variation in their ability to ionize different compounds based on the chemical structure of the analyte and the surrounding ionization environment in the MS [11–13]. Refractive index (RI) detectors are a more affordable alternative, and these measure the difference between the RI of the analyte band and the mobile phase as it passes through a flow cell [6]. An inherent problem with this type of detector is that analytes that closely match the RI of the mobile phase suffer from poor sensitivity or even appear to be invisible to the detector. Gradient elution poses a further problem as the background RI is constantly changing throughout the separation. Another alternative is the mass-specific chemiluminescence nitrogen detector (CLND). Some limitations of this technique are that the compound of interest must contain nitrogen and the mobile phase must be kept free of nitrogen-containing components (which excludes the use of acetonitrile as the organic modifier in the LC eluent) [14, 15].

Another class of detectors that provides a response directly proportional to the mass of analyte in the sample is the aerosol-based detectors. Aerosol-based detectors nebulize the LC column effluent, which is subsequently dried, leaving analyte particles that are detected, for example, through their ability, to scatter light. This process accommodates a large variety of compound classes, provided that the compounds to be detected are substantially less volatile than the mobile phase. These dried particles are then detected optically in the case of the ELSD [16] and the more sensitive condensation nucleation light scattering detector (CNLSD) [17–19] or by charge transfer in the case of the Corona CAD [20]. ELSDs have been around for the last 20 years [16, 21], but they have not been implemented widely due to their relatively poor sensitivity in comparison with UV detectors [22].

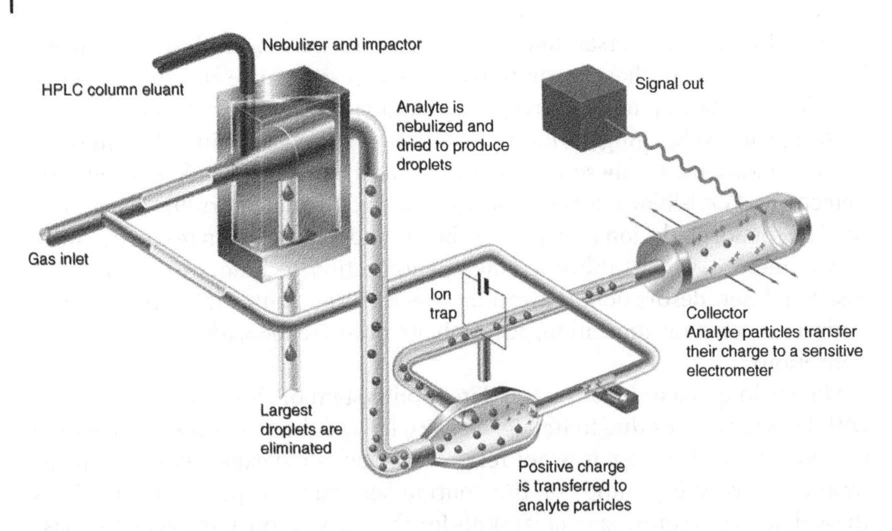

Figure 4.1 Schematic diagram of the Corona CAD. *Source*: Reproduced with permission of Thermo Fisher Scientific Inc.

The Corona CAD was first commercialized in late 2004 by ESA Biosciences. The Corona CAD nebulizes and dries the column effluent, leaving dried analyte particles. A countercurrent flow of gas is then passed over a high voltage corona needle that transfers a charge to the dried particles. A response is detected when the charged particles come into contact with a highly sensitive electrometer [23]. Figure 4.1 provides a schematic diagram showing the important features of this detector. The Corona CAD exhibits a wide dynamic range of approximately four orders of magnitude, ranging from low nanogram to high microgram amounts of analyte on-column. The response does not rely on the optical properties of analytes nor on the ability of analytes to be ionized in the gas phase [23]. Like any aerosol detector, the Corona CAD exhibits reduced response if the analyte is volatile or not all if the analyte forms particles for detection purposes. However, the Corona CAD is capable of detecting all nonvolatile analytes and, with reduced response, most semi-volatile analytes and has been used in combination with a variety of different separation modes (isocratic and gradient RPLC), ion chromatography (IC), hydrophilic interaction liquid chromatography (HILIC), supercritical fluid chromatography (SFC), and size exclusion chromatography (SEC) in normal and narrow-bore column formats, for a wide range of analytes [23]. Applications utilizing the Corona CAD include the analysis of synthetic polymers [24], inorganic ions [25], extractables and leachables from storage containers [26], lipids [27], and ionic liquids [28]; determination of enantiomeric ratios [29]; cleaning validation in the pharmaceutical industry [30]; evaluating column bleed in high temperature LC [31]; and analysis of pharmaceuticals [32–34], including assessment of their purity [11, 35]. The Corona CAD has been compared with other detectors, such as the ELSD, RI, UV, and MS detectors, on several occasions. The Corona CAD has been stated to be more sensitive and reproducible than ELSD [34, 36, 37] and to exhibit more uniform response factors [24, 29]. UV detection displayed poorer sensitivity than the Corona CAD for some analytes [27, 33] with the organic modifier interfering with the analyte signal [23] and did not exhibit the uniform relative response factors displayed by the Corona CAD [38]. Furthermore, the Corona CAD was capable of detecting a greater range of analytes than the UV detector due to its ability to produce a response for nonchromophoric compounds [30]. In comparison with MS, the Corona CAD provided a more uniform response for particular analytes due to poor electrospray ionization of these analytes in the MS [11]. Pistorino and Pfeifer [12] compared the Corona CAD with MS for the analysis of erythromycin and its precursor and found that the Corona CAD was slightly more sensitive and exhibited better precision and also greater accuracy over the measured dynamic range. Hazotte *et al.* [13] further compared the Corona CAD to an MS with interchangeable APCI and ESI ionization sources and found that the Corona CAD could universally detect all compounds of interest, whereas the MS required both ionization sources in order to detect all analytes. The Corona CAD was three to nine times more sensitive than the MS, and it was suggested that the Corona CAD response should be used to supplement the mass spectral information of the MS due to its low cost, precision, similar dynamic range, and excellent precision and accuracy of measurement.

A barrier to the widespread implementation of the aerosol detectors has been that they exhibit nonlinear calibration curves. The response of the Corona CAD [20, 23] and ELSD [13] has been described by the following equation:

$$Y = Am^b \tag{4.1}$$

where Y is the output signal from the detector (peak area or height), m is the mass injected, and A and b are constants (A represents the response intensity and b represents the response shape).

The Corona CAD has been found to yield a lower response as the particle diameter increases, and this explains the nonlinear calibration curves obtained [20]. If linear calibration curves are desired, Equation 4.1 can be converted to a linear relationship by taking the logarithm of both sides to give Equation 4.2:

$$\log Y = b \log m + \log A \tag{4.2}$$

The uniform response factors exhibited by the Corona CAD under isocratic conditions make it a potential alternative to UV detection, particularly for analytes that do not contain a chromophore. This characteristic potentially allows the use of a single universal standard for calibrating all other compounds, including those whose identities are unknown. This is particularly useful for complex real-world samples, and this utility will gain widespread attention if the aerosol detectors can be coupled with gradient separations capable of resolving a wide range of compounds in a single analysis.

4.4 Factors Affecting the Response in Charged Aerosol Detection

There are several factors that affect the response of aerosol detectors, such as the Corona CAD. The relationship between the signal produced and the amount of analyte is nonlinear, as described by Equation 4.1, but Equation 4.2 provides a linear relationship between peak area and quantity of analyte and allows accurate quantification using two or three point calibration curves. The Corona CAD is a mass-dependent detector and under constant experimental conditions is expected to display only minor differences in response to identical amounts of different analytes. Analyte response on the Corona CAD does not depend on spectral or physicochemical properties, as is the case for UV detectors, which exhibit concentration-dependent response. Gamache *et al.* [39] have tested the response for 17 chemically different compounds under isocratic conditions and found a 7% relative standard deviation in the Corona CAD response for these compounds. Given that the precision is approximately 2% for individual replicates on the Corona CAD, this indicates that the physicochemical properties of analytes play only a small role in the detector response. This was further investigated by Hutchinson *et al.* [40] who measured the response of 23 chemically different analytes under isocratic conditions. Once analytes that exhibited significant volatility were dismissed from the sample set, the average absolute relative error was found to be approximately 12%. Figure 4.2 shows the relative response of 18 (of the original 23) compounds of varied physicochemical properties under identical conditions in the Corona CAD. Further investigations are required on a larger sample set to determine any significant correlations between particular physicochemical properties and deviations in analyte response in this detector.

The design and operation of the nebulizer significantly affect the size distribution of the resultant aerosol and thereby the instrument performance. Factors such as the identity of the nebulizer gas, the gas flow rate, the flow rate of the liquid entering the nebulizer, and the temperature of the various components all contribute to the nature of the polydisperse aerosol created, which typically has droplets ranging in diameter from low nm to 100 μm [41]. Several empirical models have been proposed to predict the characteristics of nebulized aerosols. The Nukiyama and Tanasawa [42] (N-T) model was developed to predict the diameter of the primary aerosol produced by a pneumatic concentric nebulizer. It predicts the Sauter mean diameter of the

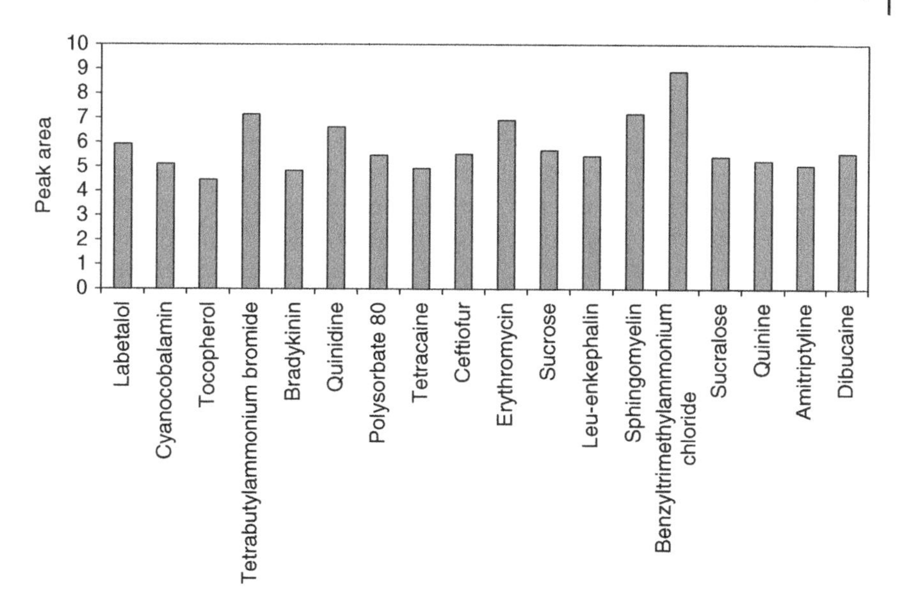

Figure 4.2 Response of 18 nonvolatile analytes of varied physicochemical properties on the Corona CAD. A mobile phase containing 50% acetonitrile and 50% formic acid (0.1% in Milli-Q) was used, and analyte concentration was kept constant at 0.01 mg/mL (25 μL injection).

droplets (arithmetic mean of the aerosol surface distribution, $D_{3,2}$) in relation to the surface tension (σ), viscosity (η), and density (ρ) of the liquid, in addition to nebulization parameters such as the gas (Q_g) and liquid (Q_l) flow rates and the difference between the axial velocity of the gas and liquid, V. The N-T model is shown as Equation 4.3:

$$D_{3,2} = \frac{585}{V}\left(\frac{\sigma}{\rho}\right)^{0.5} + 597\left(\frac{\eta}{(\sigma\rho)^{0.5}}\right)^{0.45}\left(\frac{1000Q_l}{Q_g}\right)^{1.5} \tag{4.3}$$

However, it has been noted that this equation generally overestimates the droplet size [43]. An alternative model that has been developed to estimate the droplet size distribution is the Rizk–Lefebvre (R-L) model [44] shown by Equation 4.4, where d_l is the solution capillary i.d. and the subscripts "l" and "g" refer to liquid and gas properties, respectively:

$$D_{3,2} = 0.48d_l\left(\frac{\sigma_l}{\rho_g V^2 d_l}\right)^{0.40}\left(1+\frac{Q_l\rho_l}{Q_g\rho_g}\right)^{0.40} + 0.15d_l\left(\frac{\eta_l^2}{\sigma_l\rho_l d_l}\right)^{0.5}\left(1+\frac{Q_l\rho_l}{Q_g\rho_g}\right) \tag{4.4}$$

This model was developed to be used in a mechanical engineering context and needs to be experimentally validated for its applicability to the nebulization process in the Corona CAD. Kahen and coworkers [45] developed a modified N-T model for microconcentric nebulizers that provided a more accurate prediction of Sauter mean diameter than the original N-T model, and this was measured using a Phase Doppler Particle Analyzer (PDPA). They similarly developed a modified R-L model that compensated for the observed underestimation of Sauter mean diameter by the original model. While the modified N-T model was shown to be applicable to a different nebulizer design, the modified R-L model was specific to the nebulizer on which the model was developed and may not be widely applicable.

After nebulization, the Corona CAD utilizes an impactor placed close to the point of nebulization to remove larger droplets and to ensure that a smaller subset of droplets enters the drift tube to form dried particles, thereby reducing the solvent load during the evaporation process. The droplets are then evaporated to form dried particles. The size of the dried particles d_p can be predicted using Equation 4.5, where d_d is the droplet diameter, C is the concentration, and ρ_p is the density of the particle, which is given by the density of the analyte [46]:

$$d_\mathrm{p} = d_\mathrm{d}\left(\frac{C}{\rho_\mathrm{p}}\right)^{1/3} \tag{4.5}$$

It should be noted that the use of such models in conjunction with Corona CAD is complicated by a number of factors. The models assume that there is a normal distribution of droplet sizes, which may be true after nebulization but is not necessarily the case once the particles are dried and reach the point of detection. Factors that contribute to the difficulty of applying physicochemical models to predict response in the Corona CAD include the loss of droplets to walls in the spray chamber; the possible charging of droplets from spray electrification; the use of an impactor after nebulization to remove larger droplets, droplet coagulation, and condensation; and the ion trap removing high mobility particles prior to detection [20].

Hutchinson *et al.* [19] investigated several of the parameters affecting the nebulization and evaporation procedures on a Varian 385 ELSD with regard to their effect on detector response. This detector was chosen due to the number of operational parameters able to be varied when compared to other commercially available aerosol detectors. While the nebulization and evaporation procedures remain conceptually the same for different aerosol detectors, caution must be taken in generalizing these results to other detectors as they all have differing designs that will affect the aerosol distribution and the resultant particles that reach the point of detection. The Varian 385 ELSD allowed the nebulizer temperature to be varied between 25 and 90°C, the evaporator temperature to be varied between 10 and 80°C, and the evaporator gas flow rate to be varied between 0.9 and 3.25 standard liters per minute (SLM), and the affect of manipulating these parameters on detector response is illustrated in Figure 4.3 [19]. It was found that changing the nebulizer temperature had little effect on detector response, and it is believed that this is due to the limited heat transfer that would be imparted on the eluent and gas from the short time they spend in the nebulization device. In accordance with Equations 4.3 and 4.4, it would be better to alter the temperature of the eluent and/or gas prior to nebulization as surface tension, density, and viscosity are all temperature dependent. On the other hand, the evaporator temperature and evaporator gas flow rate both had significant effects on detector response. Increasing the evaporator temperature leads to more efficient desolvation in the evaporator tube, providing greater particle transport to the point of detection. The effect of the evaporator gas flow rate was counterintuitive as it was initially thought that increasing the flow rate would increase the desolvation process. However, without knowledge of the specific design and engineering of this detector, it is difficult to infer the exact reasons for the observed results. The effect on detector response of the liquid flow rate entering into the aerosol detectors was also investigated by these authors. Typical of a mass-dependent detector, as the liquid flow rate increased, the Corona CAD response also increased. Figure 4.4 illustrates the relationship between the analyte response and the liquid flow rate entering the detector over the operating range of the Corona CAD.

Mobile phase additives are used frequently to improve separations in conjunction with aerosol detection. Organic acids such as formic, acetic, and trifluoroacetic acids are most commonly used volatile additives to control the pH of the mobile phase and to suppress silanol activity in the column in order to improve peak shape and resolution. When a mobile phase having neutral pH is required, volatile salts such as ammonium formate and acetate are used, while for higher pH, ammonium bicarbonate and triethylamine can be used. Vervoort *et al.* [34] investigated the effects of acetic acid, formic acid, and three different concentrations of ammonium acetate on the Corona CAD response. They found that the S/N ratio decreased as the concentration of ammonium acetate increased. However, it should be noted that when using the Corona CAD, the addition of extra components to the mobile phase increases the amount of background particles reaching the point of detection, hence increasing the background noise. A similar trend occurs with the inclusion of formic and acetic acids; however, the magnitude of the detector noise was a little lower than that observed when using ammonium acetate. Furthermore, using 0.1% acetic acid provided lower noise than using the same concentration of formic acid.

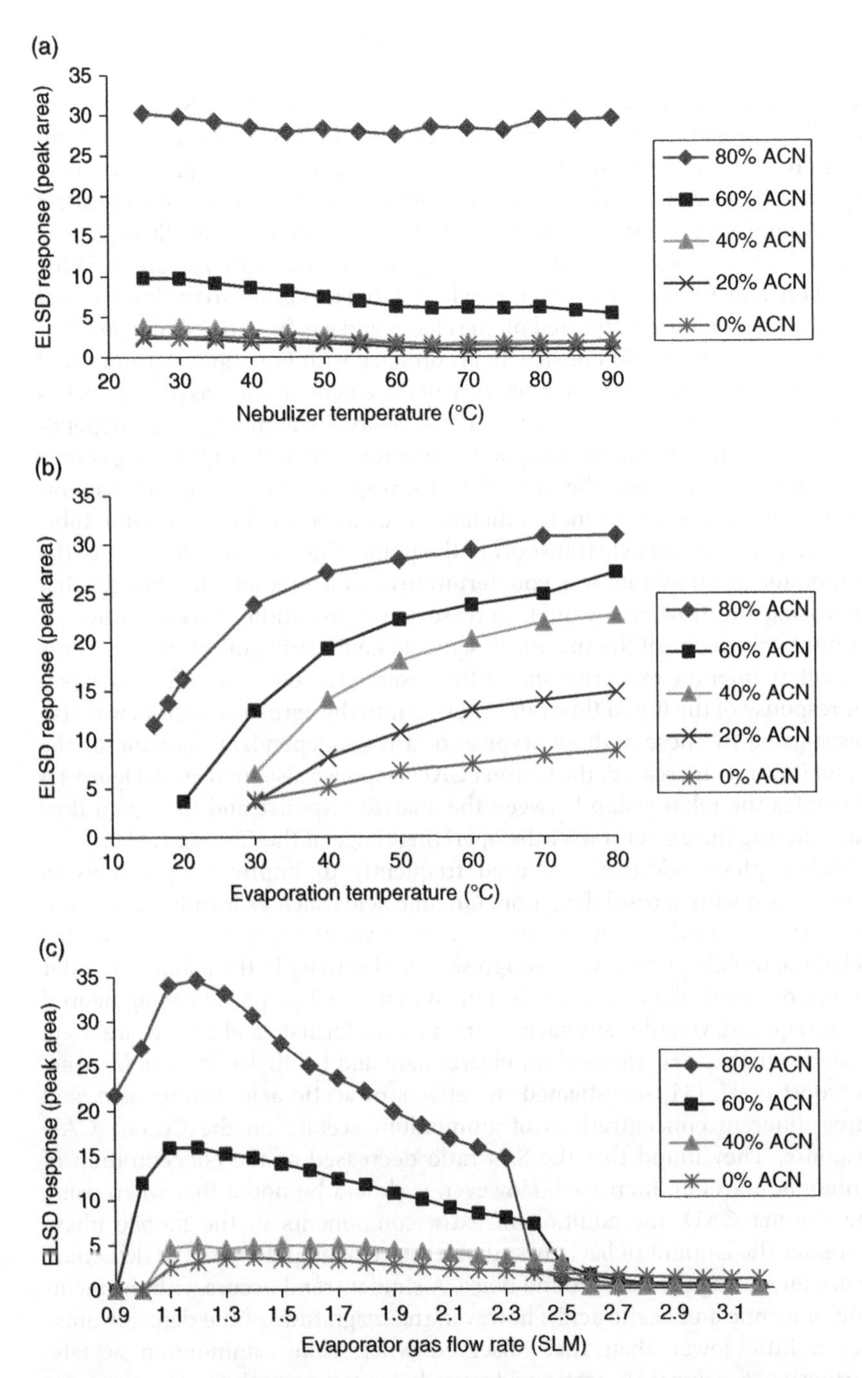

Figure 4.3 ELSD response when manipulating the (a) nebulizer temperature, (b) evaporator temperature, and (c) carrier gas flow-rate settings available on the detector and keeping all other conditions constant. *Source*: Hutchinson *et al.* [19]. Reproduced with permission of Elsevier.

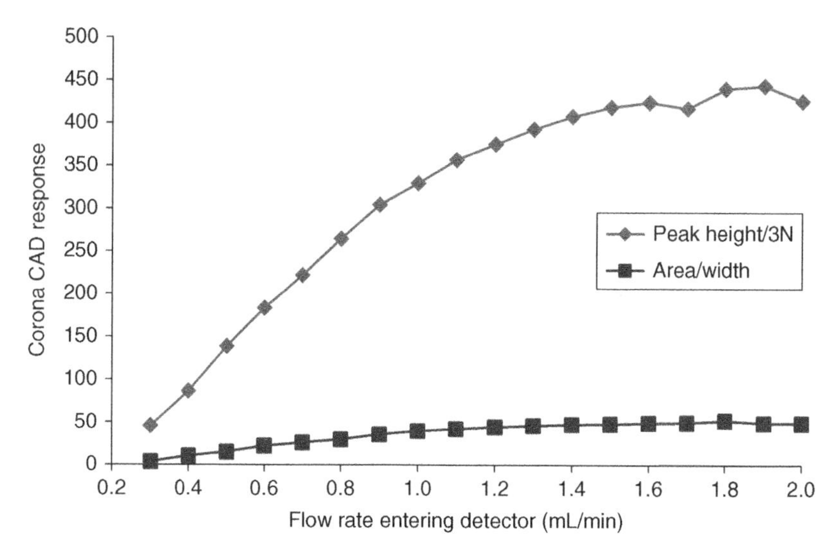

Figure 4.4 Effect of mobile phase flow rate on the response of a Corona CAD. *Source*: Hutchinson *et al.* [19]. Reproduced with permission of Elsevier.

For aerosol detectors to gain wider acceptance, it is also necessary to negate the "gradient effect" exhibited by these detectors. Organic modifier gradients are commonly used to achieve the desired separation conditions in RPLC, the result of which is that the mobile phase composition in which each analyte is eluted is different. Although the changing gradient is not seen visually as a change in the baseline of the detector output, the nebulization and droplet evaporation processes in the detector are affected by the changing composition of the mobile phase and can lead to a 5–10-fold increase in the response of an individual analyte due to an increase in the transport efficiency of droplets/particles within the detector [47]. The increase in analyte response on the Corona CAD with increasing content of organic modifier is nonlinear, and the effect of increasing the percentage of acetonitrile in the mobile phase on four analytes is shown in Figure 4.5. It can be seen from Figure 4.5 that the optimum response occurs using a mobile phase composition of approximately 70% acetonitrile, and it is best to use mobile phases of similar composition to achieve the lowest limits of detection during method development. Another observation was that the detector response becomes erratic when using mobile phases containing greater than 90% acetonitrile, which contributes to greater relative standard deviations when performing replicate analyses and reduces the reproducibility of the system.

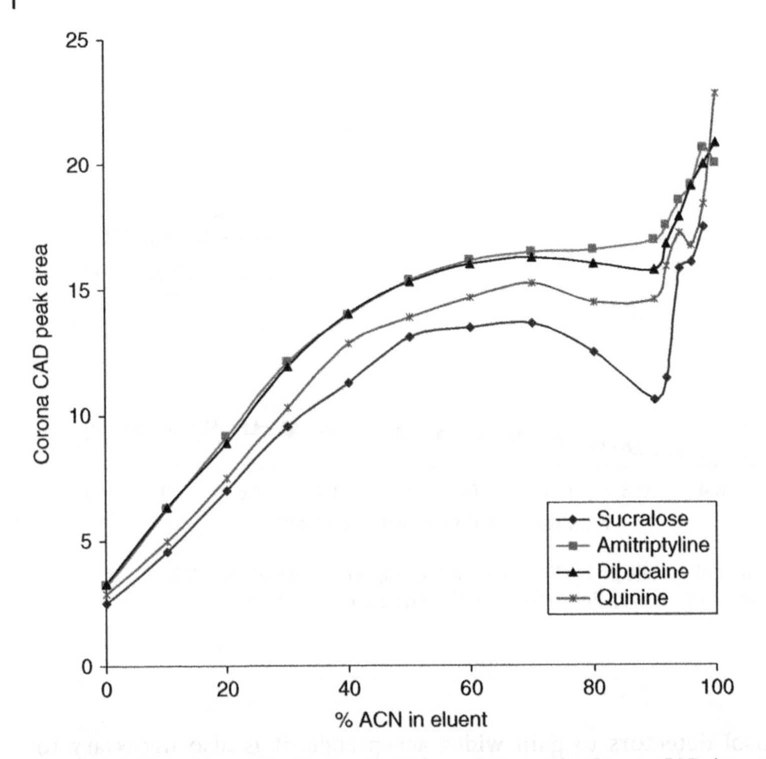

Figure 4.5 The response for four nonvolatile analytes on the Corona CAD detector with changing eluent composition (increasing the percentage of acetonitrile). *Source:* Hutchinson *et al.* [40]. Reproduced with permission of Elsevier.

4.5 Gradient Compensation

The nonlinear increase in response exhibited by the aerosol-based detectors when organic modifiers are used is related to the changes in the physicochemical properties of the mobile phase. This causes changes in the droplet size distribution after nebulization, along with the variations in the desolvation process when forming dried analyte particles. Important physicochemical properties related to these processes include the surface tension, viscosity, and enthalpy of vaporization. Efforts to mitigate the gradient effect have involved a gradient compensation approach whereby a second pump has been used to deliver a post-column inverse gradient, prior to the aerosol detector [35, 48]. This process ensured that the composition of the mobile phase entering the detector was constant at the midpoint of the gradient profile used. A drawback of this approach is that two identical systems of equivalent gradient-generating capabilities and equivalent back pressure are required to perform this task. Dionex Corporation have now commercialized a dual pump system for this purpose. Another disadvantage is that the midpoint of the optimized gradient to achieve the separation may not be the optimum composition to maximize sensitivity in the detector. This solution to the "gradient effect" has been criticized as being an instrumental solution external to the aerosol detector itself. This approach requires a dual pump system to be purchased for each detector, greatly increasing the associated purchase and running costs of using aerosol detectors. Despite these drawbacks, this approach provides a simple means of mitigating the gradient effect in aerosol detectors and has been used for both reversed-phase [35] and supercritical CO_2 chromatography separations [32].

4.6 Response Models

Another approach to overcoming the gradient effect is to develop empirical models that relate detector response to both the composition of the mobile phase and the amount of analyte injected. Such approaches do not attempt to explain the observed response in relation to the physicochemical properties and processes involved but to create a post-detection solution for calibration when two independent variables have a nonlinear effect on the observed response. The Corona CAD is known to exhibit uniform response factors under isocratic conditions, and different analytes show similar changes in response when the composition of the mobile phase is varied. This allows a generic response model to be used for quantifying a wide range of nonvolatile analytes, including unknown species where standards are not available for calibration purposes. A calibration routine, such as that performed by Mathews *et al.* [49] on an ELS detector, where a single, non-retained compound is injected at regular intervals during gradient analysis, is used to construct a three-dimensional calibration surface, which can be employed to account for variations in response caused by changes in mobile phase composition. Although the method suffered from the poor sensitivity generally attributed to ELS detectors, the benefits of universal detection allowed it to be successfully incorporated into in-house software used by Squibb *et al.* [50] for high-throughput analyses of compounds during the pharmaceutical discovery process. Further, a three-dimensional empirical model was developed by Hutchinson *et al.* [40] on the Corona CAD under flow-injection conditions using the average response of four nonvolatile analytes over a wide range of analyte concentrations and mobile phases containing between 0 and 80% acetonitrile. The resultant three-dimensional relationship is shown in Figure 4.6. Detector response was described by two second-order polynomial equations: one relating to detector response to the percentage of acetonitrile in the mobile phase and the other relating to detector response to analyte concentration. The model was used for quantifying analytes of varied physicochemical properties, separated using isocratic and gradient RPLC conditions. The overall error in this approach was around 13% for analytes that conformed to the model, and this level of error is satisfactory for the quantification of unknowns in the

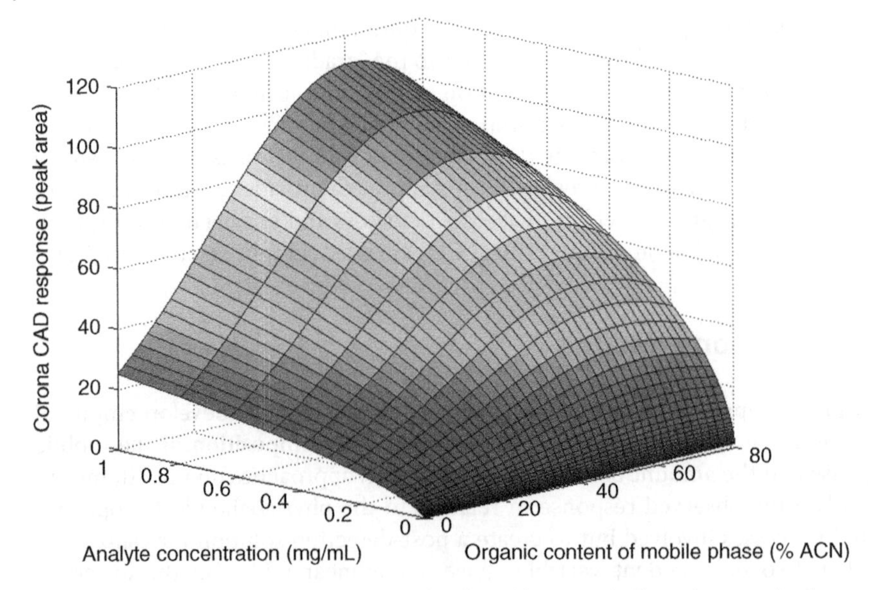

Figure 4.6 The three-dimensional relationship found for nonvolatile analytes on the Corona CAD relating the detector response, analyte concentration, and composition of the mobile phase. *Source*: Hutchinson *et al.* [40]. Reproduced with permission of Elsevier.

pharmaceutical industry. To put this into perspective, if analyte response is not corrected for the changing composition of the mobile phase, errors of up to 500% can result when using aerosol detection.

Ideally, it would be advantageous to generate a mathematical model that relates the detector response back to the physicochemical components of the system. While Equations 4.3–4.5 provide some insight into this process, creating such a model is quite challenging and involves a fundamental understanding and ability to measure in real time the variables that affect the nebulization, evaporation, and detection processes. This has not yet been achieved. Furthermore, the derivation of such models would involve constants that are instrumentation-specific, thereby limiting their widespread use.

4.7 Green Chemistry

Green chemistry can be defined by the 12 principles introduced by Anastas and Warner [51], which are further discussed in the context of separation science in a recent series of reviews by Sandra and coworkers [52, 53]. This overarching philosophy intends to eliminate pollution at its source to protect human health and, if this is not possible, then take steps to minimize pollution in the interim. Green chemistry is not only driven by ideals, but it also represents a way for companies to become more efficient in their processes and, in turn, reduce costs and save money. In RPLC, acetonitrile is the solvent of choice due to its favorable physicochemical properties (e.g., water miscibility, low boiling temperature, and ability to solvate many analytes), especially when combined with detectors that nebulize the sample, such as MS and the emerging class of aerosol detectors, including the Corona CAD. Unfortunately, acetonitrile usage is not in-line with green chemistry principles, and it is important to find alternative, nontoxic solvents that are compatible and provide comparable chromatographic performance.

Efforts to apply the concepts of green chemistry to reduce solvent usage in separation science have focused primarily on instrumental design. This has been performed by reducing internal volumes and using higher pressure pumps that are capable of performing separations on smaller internal diameter columns, packed with smaller stationary-phase particles. In practice, there are technological limitations to continue further reductions in column internal diameters and packed particle size. An alternative is to switch to environmentally friendly solvents such as ethanol, high temperature water, and supercritical CO_2, which can also provide the required separation performance for a given set of analytes. Hence, the ideal green approach is to use modern instrumental and column technology as best practice and to develop methods using environmentally friendly solvents to achieve performance comparable to acetonitrile-based systems.

Aerosol detectors provide an opportunity to investigate alternative green solvents. Unlike UV detection, the choice of solvent to be used in combination with aerosol detectors is not dictated by its UV transparency at low wavelengths so that a wider range of solvents can be considered. Furthermore, preliminary results from the literature have shown that aerosol detectors exhibit uniform response when used with high temperature water gradients [31] and gradient SFC [32]. Combining the benefits of green separation methods with universal detection techniques will provide a sustainable approach to chromatography and provide a more economical means of delivering highly efficient separations. Table 4.1 compares selected physical and chemical properties of possible alternatives to acetonitrile, which are compatible with aerosol detectors.

Methanol is the second most popular organic solvent used in RPLC. While methanol is less toxic than acetonitrile (see Table 4.1), like acetonitrile, aqueous waste streams containing methanol need to be disposed of as chemical waste [54].

Ethanol is a more favorable alternative that is less toxic and is also biodegradable [55], but it is seldom used in LC. Ethanol is also increasingly being used as a fuel that will ensure its availability, decrease its cost, and increase the quality of production. In addition, ethanol is considered to be a renewable resource as it is produced via the fermentation of starch-/sugar-based crops. Growing such crops reduces CO_2 from the atmosphere through the process of photosynthesis. Ethanol and methanol are less suitable for UV detection due to their higher UV

Table 4.1 Selected physical and chemical properties of acetonitrile and alternatives.

Property	Acetonitrile	Methanol	Ethanol	Acetone	Carbon dioxide	Water
Density (g/mL)	0.782	0.791	0.789	0.790	0.770 (56 atm, 20°C)	1.000
Viscosity (cP)	0.38	0.55	1.07	0.36	0.07 (−78°C)	
Boiling point (°C)	81.6	64.7	78.3	56.3	−57°C (5.117 atm)	100
Vapor pressure (hPa at 20°C)	118	33	90	246	57,226	23
Eluotropic strength on C18 (ε°)	3.1	1.0	3.1	8.8	3.02 (supercritical)	—
UV cutoff (nm)	190	205	210	330	UV transparent	UV transparent
LD_{50} (oral; rat) (mg/kg)	2,460	5,628	7,060	5,800	—	—
LD_{50} (inhalation; rat)	7,551 ppm (8 h)	64,000 ppm (4 h)	64,000 ppm (4 h)	50,100 mg/m^3 (8 h)	657,190 ppm (15 min)	—
LD_{50} (dermal) (mg/kg	2,000 (rabbit)	15,800 (rabbit)	Irritant	7,426 (guinea pig)	—	—

Source: Fritz *et al.* [54]. Reproduced with permission of John Wiley & Sons.

cutoffs, but this does not pose a problem for aerosol-based detectors where the solvent is evaporated. Furthermore, the higher viscosity of ethanol is now less of a concern with the advent of ultrahigh pressure LC instrumentation. Welch *et al.* [55] have compared the performance of acetonitrile, ethanol, and methanol and concluded that while acetonitrile delivers the best performance for gradient LC separations in combination with UV detection, greener alternatives such as ethanol performed well and may be substituted in many cases.

Acetone is another less toxic alternative to acetonitrile, which is not suitable for UV detection due to its UV cutoff of 330 nm. However this solvent has not received much attention in the literature as a solvent for reversed-phase separations [54]. Acetone exhibits higher eluotropic strength than methanol, ethanol, or acetonitrile, which can potentially further reduce analysis time and solvent usage. The viscosity of acetone is comparable to that of acetonitrile, which minimizes the associated change in back pressure of the system and allows increased flow rates and faster analyses. Acetone has a lower boiling point than acetonitrile, which aids the desolvation process in the aerosol-based detectors and MS. Fritz and coworkers [54] compared acetone to methanol and acetonitrile for LC-MS analysis of standard proteins from tryptic digests. They were able to identify the same number of peptides resulting from a tryptic digest using either acetonitrile or acetone, and they found that the use of acetone reduced the total analysis time by 20%.

4.8 Temperature Gradient Separations

Water is subcritical at temperatures and pressures lower than 374°C and 218 atm, where it is possible to manipulate its physicochemical properties such as dielectric constant, surface tension, viscosity, and dissociation constant by simply altering the temperature, provided adequate pressure is maintained to keep water in its liquid state [56]. Water acts like a polar organic solvent at elevated temperatures and is compatible with aerosol detectors. Therefore, under these conditions, pure water mobile phases can achieve the separation of hydrophobic species and, due to its volatility, does not generate a detector response in the Corona CAD. Columns capable of withstanding high temperatures are now commercially available, and only thermally stable analytes should be considered for this technique. Organic modifiers, similar to those used in conventional LC, can also be used to aid separation selectivity and improve the elution of highly retained analytes; however the amounts required are greatly reduced due to the eluotropic strength of high temperature water.

Apart from its green nature, subcritical water chromatography offers several advantages over traditional LC, including faster analysis times due to decreased viscosity of the mobile phase, differences in separation selectivity, and its compatibility with most gas- and liquid-phase detectors. Subcritical water chromatography has been demonstrated in conjunction with MS [57] and a Corona CAD [31]. Subcritical water chromatography requires specialized LC instrumentation, such as low volume mobile phase preheaters to ensure that thermal mismatch between the column temperature and the mobile phase introduced is minimized to reduce significant band broadening. Specialized column heaters that are capable of raising column temperatures up to 200°C are required to achieve the full benefits of high temperature water separations. After separation has occurred, the column effluent may also require post-column cooling before it enters the aerosol (or UV) detector as the spray distribution is sensitive to changes in temperature, due to variations in the viscosity and surface tension of the liquid to be nebulized. A further practical consideration when using high temperature water mobile phases is that a pressure restrictor is required downstream from the column to ensure that the water remains in the liquid phase while the separation is taking place. A relatively low back pressure of 15 bar is required to keep water in its liquid state at 200°C [58].

4.9 Supercritical CO_2 Separations

Supercritical fluids exhibit the properties of both a gas and a liquid, and because of their volatility they are also compatible with Corona CAD. Supercritical fluids offer the advantage that their density can be changed in a continuous manner by manipulating the pressure and temperature of the system. Diffusivities are approximately an order of magnitude higher than the corresponding liquid at room temperature and pressure, while viscosities are an order of magnitude lower. These properties allow supercritical fluids to have liquid-like solvating power with the mass transport characteristics of a gas. The solvent strength of CO_2 varies markedly with the density of the fluid [59]. Therefore, the use of CO_2 in the form of a supercritical fluid offers a substitute for an organic solvent. Supercritical CO_2 is suitable for many forms of chromatography, including normal- and reversed-phase formats. The critical properties of various solvents are shown in Table 4.2. CO_2 is a supercritical fluid under relatively mild conditions, and, in combination with its green nature (extracted from the atmosphere, hence carbon neutral), CO_2 is considered the ideal choice for use in SFC. Multiple gradients (pressure, temperature, and mobile phase modifiers) can be used to increase the separating power of the technique. SFC has been around for 50 years but has recently regained favor due to technological advances, as well as its green character. SFC can be used analytically or scaled up to preparative applications where the solvent merely evaporates under atmospheric conditions, which is particularly attractive for the pharmaceutical industry. It is also possible to recover the CO_2 and recycle this for further use [53], again adding to its green nature. SFC is compatible with a wide variety of detection

Table 4.2 Critical properties of various solvents.

Fluid	Critical temperature (K)	Critical pressure (bar)
Carbon dioxide	304.1	73.8
Ethane	305.4	48.8
Ethylene	282.4	50.4
Trifluoromethane	299.3	48.6
Ammonia	405.5	113.5
Water	647.3	221.2

Source: Reid *et al.* [60]. Reproduced with permission of McGraw Hill Book Co.

systems and has been shown to work with a Corona CAD [32] and MS [61]. Another advantage of SFC is its mild separation conditions. It operates at close to ambient temperature and is suitable for thermally labile analytes.

4.10 Capillary Separations

Corona CAD has typically been used coupled to analytical separations; however, one example exists in the literature of Corona CAD detection in conjunction with microseparations (i.e., columns smaller than 2 mm i.d.) and was performed by Hazotte *et al.* [13]. Hazotte *et al.* performed micro high temperature LC to separate lipids and coupled this to Corona CAD, ELSD, and MS for comparative purposes. They connected the microcolumn (150 × 0.53 mm i.d.), packed with 5 μm Hypercarb particles, to a fused silica capillary of appropriate dimensions (50 μm i.d., 375 μm o.d.), which was inserted directly into the nebulizer of the Corona CAD. Under these conditions, the Corona CAD exhibited a linear response over two orders of magnitude and approached the limits of detection of APCI-MS for certain analytes. One advantage of using the Corona CAD over MS was that all lipids gave a response, whereas on the MS, different ionization modes were required to detect particular lipids. ELSD has been also been used for this purpose, and this field has recently been reviewed by Lucena *et al.* [16].

There are several benefits of moving to miniaturized chromatographic separations. These include reduced consumption of solvents and samples, higher resolution separations, increased axial thermal dissipation, and increased sensitivity due to the smaller column volume reducing on-column sample dilution. This approach is again aligned with green chemistry principles and is useful for high temperature separations, which require rapid changes in column temperature to perform the separation. Furthermore, since the contribution to band broadening from ELSD is very small, this detector is well suited to high resolution microscale separations.

Coupling microscale separations to aerosol detectors typically requires micro-nebulization. Conventional nebulizers generally suffer from low transport efficiencies and large dead volumes when coupled with the μL/min flow rates required for capillary separations. However, there have been some examples in the literature of retrofitting capillaries into conventional nebulizers [62]. Micro-nebulizers increase the uniformity in droplet size distribution, which can lead to greater transport efficiency through the detector, thus increasing sensitivity and improving the linearity of response over conventional pneumatic nebulizer approaches.

Several commercial ELSDs are now available where the nebulizer can be changed to suit flow rates typically associated with analytical, narrow-bore, micro-bore, and capillary separations. Other options include directly inserting the capillary column into the drift tube or using a micro-nebulizer capable of generating a stable droplet distribution at low flow rates. A great deal of work has been performed in the literature investigating micro-nebulizers for their suitability in conjunction with ICP analysis [41]. A particular example is the monodisperse dried microparticulate injector (MDMI), which is capable of generating small, identical droplets of uniform velocity, along with reducing the solvent load in comparison with conventional nebulizers. These attributes should benefit signal response in the aerosol detectors [63].

4.11 Global Analysis and Multidimensional Separations

Research is increasingly trending toward miniaturization and high-throughput data collection methods that benefit from a global approach to data analysis. Microarrays and fast separations allow the screening of large numbers of samples, such as in chemical screens for biological activity or gene expression studies. The term *global analysis* has been used in different contexts and refers to a nonspecific (generic) analytical approach, which identifies all components of a complex sample. The resulting information is chemometrically surveyed to find meaningful relationships to the hypothesis being tested. Global analysis emphasizes the integration and analysis of data from *all* sources to form an overall picture [64]. An example of this is the Human Genome Project, for which the goal was to sequence, define, and understand all the genes specifying the making of a human being. Ultimately, this knowledge will improve the human condition through genetic diagnostics, gene therapy, and an improved understanding of biochemical and physiological processes. Over the past few years, the completion of genetic maps has been transitioning to physical mapping of genomes and to the characterization of their functional content. After genomics, other "omic" studies have followed (e.g., proteomics) to study and more fully understand biological systems. Proteomics is even more complicated than genomics, since, while an organism's genome is more or less constant, the protein expression from a given genome differs between cell types and conditions. The global analysis of protein levels can provide clues to cellular processes that occur during development and how cells respond to environmental conditions and whether tissues are performing normally or are present in a diseased state [65]. In turn, as the systems to be studied become more complex, this in turn challenges the analytical methods required to collect this information.

LC would greatly benefit from a universal detector that can be used for quantitative analysis without the need for an authentic standard. Indeed, a detector capable of universal detection is something that greatly interests pharmaceutical companies and other industries that require nonspecific analytical approaches capable of identifying all components in a complex sample. Universal detection systems would have an impact on biomarker research and "omic" studies where routine analytical measurement of complex samples is performed. Often, such studies do not attempt to determine the identity of the large number of analytes present but rather try to find patterns and quantitative differences that correlate to the hypothesis being tested.

Traditionally, protein abundance has been examined using two-dimensional gel electrophoresis (2D-GE) where complex protein mixtures can be separated into several thousand spots using orthogonal separation methods (pI and mass) and visualizing the proteins using protein-binding dyes. This leads to a semi-accurate determination of the mass of individual proteins present; however, complete protein resolution is often not achieved, and the identity of each spot remains unknown [65]. Off-line coupling of 2D-GE to MS overcomes these problems. Multidimensional LC techniques now provide a high resolution, automated alternative that can be directly coupled in-line to MS [66]. Comprehensive LC × LC using different (orthogonal) separation mechanisms maximizes the peak capacity of a system, which is the maximum possible number of separated compounds. By adding extra separation dimensions to the system, the overall theoretical peak capacity that can be achieved is the product of the peak capacity of each individual separation dimension, providing much greater resolving power than each individual separation on its own.

For example, peak capacity in LC can be increased to more than 400 peaks using columns packed with sub 2 μm particles [67]. In a single dimension, maximum peak capacities can be achieved by coupling columns together and using elevated temperatures to achieve peak capacities approaching 1000 [68]. However, such separations may be excessively long (in the order of tens of hours) due to back pressure limitations. Peak capacities of over 1000 can be achieved more efficiently using orthogonal, two-dimensional (2D) separations [69]. In practice however, the peak capacity required to resolve randomly distributed peaks must exceed the number of peaks by approximately 100-fold [70]. Furthermore, hyphenating LC × LC systems to MS provides an extra separation

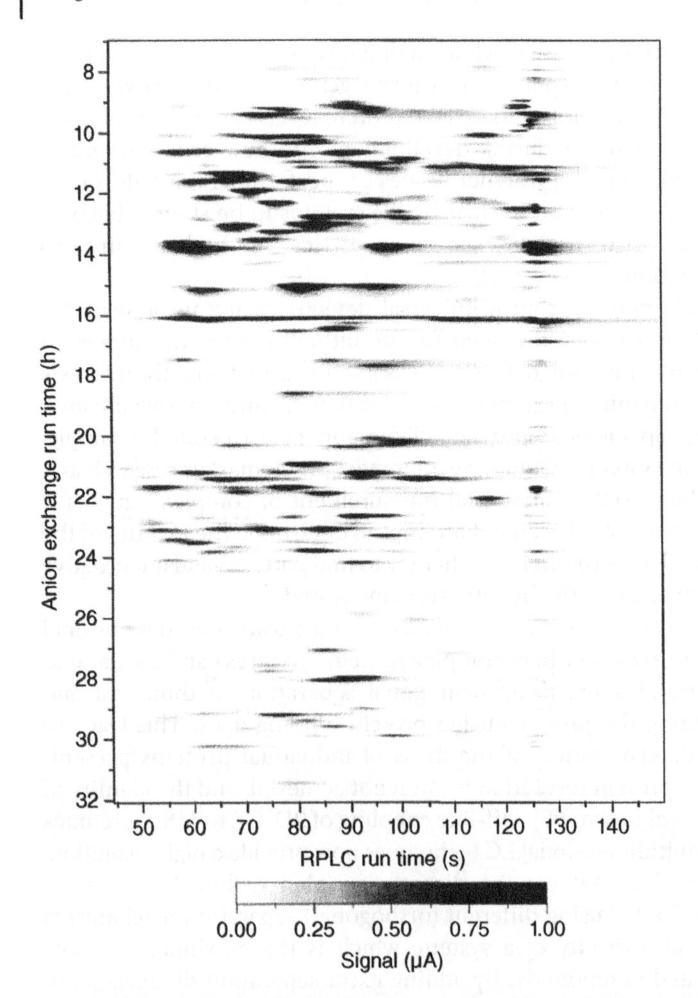

Figure 4.7 A two-dimensional separation showing approximately 150 peptide components obtained from a tryptic digest of reduced porcine thyroglobulin. *Source*: Holland and Jorgenson [71]. Reproduced with permission of American Chemical Society.

dimension that has been shown to be capable of separating 2000–8000 proteins from whole cell extracts [66]. Visualizing 2D separations require specialized software and two examples are shown in Figures 4.7 and 4.8, respectively. Figure 4.7 relates to a microcolumn LC separation system combining strong anion exchange (SAX) and reversed-phase dimensions to separate tagged peptides and cell lysate with laser-induced fluorescence (LIF) detection [71]. Figure 4.8 is a 2D separation of human urine metabolites using a strong cation exchange (SCX)–SAX separation in the first dimension and a monolithic octadecylsilyl (ODS) column in the second dimension, both with gradient elution.

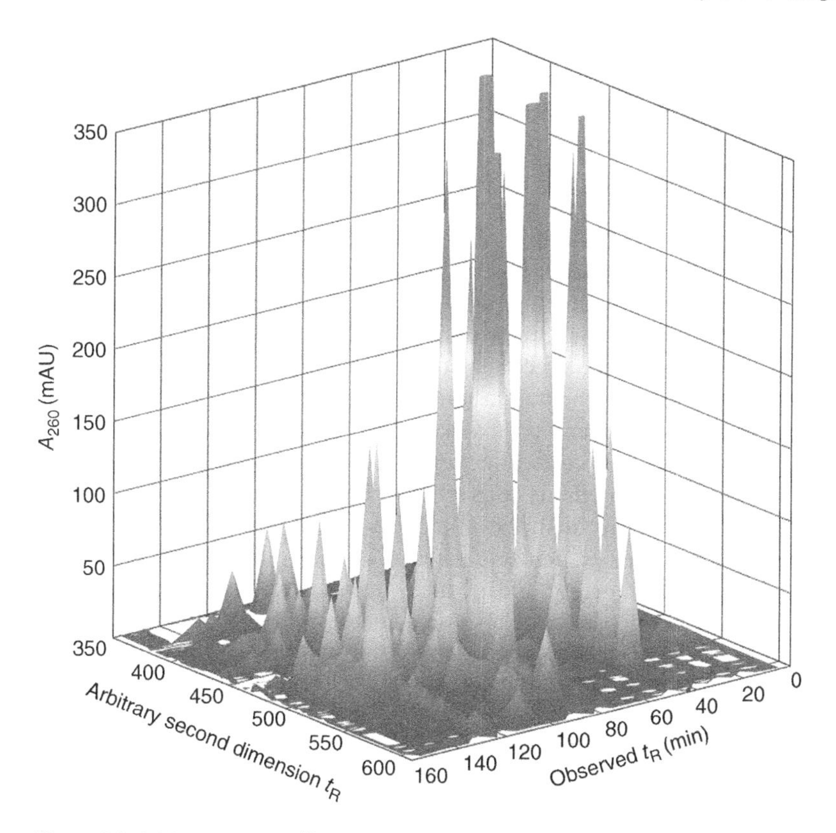

Figure 4.8 A 2D separation of human urine components using SCX–SAX columns in the first dimension and monolithic ODS columns in the second dimension. Both dimensions were gradient eluted. *Source*: Dixon *et al*. [70]. Reproduced with permission of John Wiley & Sons.

It was noted that the system was limited in that the detection method (UV detection at 260 nm) was not able to identify all compounds present in the sample [70]. While Corona CAD cannot provide the additional peak deconvolution, structural information, and identification that MS provides, its ability to provide a uniform response for nonvolatile species warrants its parallel inclusion in 2D systems to ensure that all analytes are detected and for quantitative purposes to determine relative abundance. This approach has yet to be explored.

4.12 Conclusions

LC would benefit from a universal detector capable of detecting all analytes with uniform response factors, regardless of their physicochemical properties. Ideally, such a detector would also be able to detect trace amounts of components within complex samples. While MS has become an important detection tool for LC, it has disadvantages, which include its high cost of purchase and operation, the requirement for a skilled operator, and nonuniform response factors for analytes of differing physicochemical properties. Although the emergent aerosol-based detectors are not ideal universal detectors, they offer a simple, low-cost detection alternative, providing relatively uniform response factors for a large number of nonvolatile analytes, regardless of their physicochemical properties. Due to these characteristics, aerosol detection is currently seen as a complementary detection technique to MS and UV detection. To date, aerosol detectors have not received the fundamental research required to fully understand the factors that affect analyte response. This research would lead to improvements in instrumental design to make the detectors more tolerant of changes in the mobile phase used and further reduce the limits of detection observed. The aerosol detectors offer several advantages, such as being compatible with separation methods consistent with the ideals of green chemistry and offering the ability to quantify unknowns due to the universal response factors observed. Therefore, aerosol detectors are a significant addition to the analyst's arsenal for detecting and quantifying a wide range of compounds separated by LC. It is predicted that these detectors will be particularly useful for the global analysis of complex samples and will be a suitable detection technique for multidimensional separations.

References

1 High-Performance Liquid Chromatography (HPLC) Market by Product (Instruments (Systems, Detectors), Consumables (Columns, Filters), and Accessories), Application (Clinical Research, Diagnostics, Forensics)—Analysis & Global Forecast to 2021. By marketsandmarkets.com. 2017. Report Code: AST 5029.
2 T.J. Causon, R.A. Shellie, E.F. Hilder, Analyst 134 (2009) 440.
3 R.M. Smith, J. Chromatogr. A 1184 (2008) 441.
4 X. Cheng, J. Hochlowski, Anal. Chem. 74 (2002) 2679.
5 W.R. Lacourse, Anal. Chem. 74 (2002) 2813.
6 B. Zhang, X. Li, B. Yan, Anal. Bioanal. Chem. 390 (2008) 299.
7 M. Swartz, LCGC N. Am. 28 (2010) 880.
8 Y. Chen, M. Manshouwer, W.L. Fitch, Pharm. Res. 24 (2007) 248.
9 M.P. Balogh, LCGC N. Am. 27 (2009) 130.
10 Y. Hsieh, E. Fukuda, J. Wingate, W.A. Korfmacher, Comb. Chem. High Throughput Screen. 9 (2006) 3.
11 J. Reilly, B. Everatt, C. Aldcroft, J. Liq. Chromatogr. Relat. Technol. 31 (2008) 3132.
12 M. Pistorino, B.A. Pfeifer, Anal. Bioanal. Chem. 390 (2008) 1189.

13 A. Hazotte, D. Libong, M. Matoga, P. Chaminade, J. Chromatogr. A 1170 (2007) 52.

14 S. Ojanpera, I. Rasanen, J. Sistonen, A. Pelander, E. Vuori, I. Ojanpera, Ther. Drug Monit. 29 (2007) 423.

15 E.M. Fujinari, L.O. Courthaudon, J. Chromatogr. A 592 (1992) 209.

16 R. Lucena, S. Cardenas, M. Valcarcel, Anal. Bioanal. Chem. 388 (2007) 1663.

17 J.A. Koropchak, C.L. Heenan, L.B. Allen, J. Chromatogr. A 736 (1996) 11.

18 J. Olsovska, Z. Kamenik, T. Cajthaml, J. Chromatogr. A 1216 (2009) 5774.

19 J.P. Hutchinson, J. Li, W. Farrell, E. Groeber, R. Szucs, G. Dicinoski, P.R. Haddad, J. Chromatogr. A 1218 (2011) 1646.

20 R.W. Dixon, D.S. Peterson, Anal. Chem. 74 (2002) 2930.

21 N.C. Megoulas, M.A. Koupparis, Crit. Rev. Anal. Chem. 35 (2005) 301.

22 C.S. Young, J.W. Dolan, LCGC N. Am. 21 (2003) 120.

23 P.H. Gamache, R.S. McCarthy, S.M. Freeto, D.J. Asa, M.J. Woodcock, K. Laws, R.O. Cole, LCGC Eur. 18 (2005) 345.

24 D. Kou, G. Manius, S. Zhan, H.P. Chokshi, J. Chromatogr. A 1216 (2009) 5424.

25 Z. Huang, M.A. Richards, Y. Zha, R. Francis, R. Lozano, J. Ruan, J. Pharm. Biomed. Anal. 50 (2009) 809.

26 X. Yu, S. Zdravkovic, D. Wood, C. Li, Y. Cheng, X. Ding, Drug Deliv. Technol. 9 (2009) 50.

27 C. Schonherr, S. Touchene, G. Wilser, R. Peschka-Suss, G. Francese, J. Chromatogr. A 1216 (2009) 781.

28 A. Stojanovic, M. Lammerhofer, D. Kogelnig, S. Schiesel, M. Sturm, M. Galanski, R. Krachler, B.K. Keppler, W. Lindner, J. Chromatogr. A 1209 (2008) 179.

29 P. Wipf, S. Werner, L.A. Twining, C. Kendall, Chirality 19 (2007) 5.

30 B. Forsatz, N.H. Snow, LCGC N. Am. 25 (2007) 960.

31 T. Teutenberg, J. Tuerk, M. Holzhauser, T.K. Kiffmeyer, J. Chromatogr. A 1119 (2006) 197.

32 C. Brunelli, T. Gorecki, Y. Zhao, P. Sandra, Anal. Chem. 79 (2007) 2472.

33 L. Novakova, S.A. Lopez, D. Solichova, D. Satinsky, B. Kulichova, A. Horna, P. Solich, Talanta 78 (2009) 834.

34 N. Vervoort, D. Daemen, G. Torok, J. Chromatogr. A 1189 (2008) 92.

35 T. Gorecki, F. Lynen, R. Szucs, P. Sandra, Anal. Chem. 78 (2006) 3186.

36 K. Takahashi, S. Kinugasa, M. Senda, K. Kimizuka, K. Fukushima, T. Matsumoto, Y. Shibata, J. Christensen, J. Chromatogr. A 1193 (2008) 151.

37 R.G. Ramos, D. Libong, M. Rakotomanga, K. Gaudin, P.M. Loiseau, P. Chaminade, J. Chromatogr. A 1209 (2008) 88.

38 P. Sun, X. Wang, L. Alquier, C.A. Maryanoff, J. Chromatogr. A 1177 (2008) 87.

39 P.H. Gamache, R.S. McCarthy, S.M. Freeto, D.J. Asa, M.J. Woodcock, K. Laws, R.O. Cole, LCGC N. Am. 23 (2005) 150.

40 J.P. Hutchinson, J. Li, W. Farrell, E. Groeber, R. Szucs, G. Dicinoski, P.R. Haddad, J. Chromatogr. A 1217 (2010) 7418.

41 J.A. McLean, M.G. Minnich, L.A. Iacone, H. Liu, A. Montaser, J. Anal. At. Spectrom 13 (1998) 829.
42 S. Nukiyama, Y. Tanasawa, in Defense Research Board, Department of National Defense, Trans. Soc. Mech. Eng. (Japan), Reports 4, 5, and 6, 1938–1940. Translated by E. Hope (Editor), Ottawa, Ontario, Canada, 1950.
43 A. Canals, V. Hernandis, J. Anal. At. Spectrom 5 (1990) 61.
44 N.K. Rizk, A.H. Lefebvre, J. Eng. Gas Turbines Power 106 (1984) 634.
45 K. Kahen, B.W. Acon, A. Montaser, J. Anal. At. Spectrom. 20 (2005) 631.
46 A. Stolyhwo, H. Colin, G. Guiochon, Anal. Chem. 57 (1985) 1342.
47 Z. Cobb, P. Shaw, L. Lloyd, N. Wrench, D. Barrett, J. Microcolumn Sep. 13 (2001) 169.
48 A. de Villiers, T. Gorecki, F. Lynen, R. Szucs, P. Sandra, J. Chromatogr. A 1161 (2007) 183.
49 B.T. Mathews, P.D. Higginson, R. Lyons, J.C. Mitchell, N.W. Sach, M.J. Snowden, M.R. Taylor, A.G. Wright, Chromatographia 60 (2004) 625.
50 A.W. Squibb, M.R. Taylor, B.L. Parnas, G. Williams, R. Girdler, P. Waghorn, A.G. Wright, F.S. Pullen, J. Chromatogr. A 1189 (2008) 101.
51 P.T. Anastas, J.C. Warner, Green Chemistry: Theory and Practice, Oxford University Press, New York, 1998.
52 P. Sandra, K. Sandra, A. Pereira, G. Vanhoenacker, F. David, LCGC Eur. 23 (2010) 242.
53 P. Sandra, A. Pereira, M. Dunkle, C. Brunelli, F. David, LCGC Eur. 23 (2010) 396.
54 R. Fritz, W. Ruth, U. Kragl, Rapid Commun. Mass Spectrom. 23 (2009) 2139.
55 C.J. Welch, T. Brkovic, W. Schafer, X. Gong, Green Chem. 11 (2009) 1232.
56 Y. Yang, J. Sep. Sci. 30 (2007) 1131.
57 R.M. Smith, O. Chienthavorn, I.D. Wilson, B. Wright, S.D. Taylor, Anal. Chem. 71 (1999) 4493.
58 C.V. McNeff, B. Yan, D.R. Stoll, R.A. Henry, J. Sep. Sci. 30 (2007) 1672.
59 D.R. Gere, Science 21 (1983) 253.
60 R.C. Reid, J.M. Prausnitz, B.E. Poling, The Properties of Gases and Liquids, McGraw-Hill, New York, 1987.
61 J.D. Pinkston, Eur. J. Mass Spectrom. 11 (2005) 189.
62 K. Gaudin, A. Baillet, P. Chaminade, J. Chromatogr. A 1051 (2004) 43.
63 J.W. Olesik, J.A. Kinzer, Spectrochim. Acta Part B 61 (2006) 696.
64 D.E. Root, B.P. Kelley, B.R. Stockwell, Curr. Opin. Drug Discov. Devel. 5 (2002) 355.
65 G.A. Michaud, M. Snyder, Biotechniques 33 (2002) 1308.
66 R. Matthiesen, L. Azevedo, A. Amorim, A.S. Carvalho, Proteomics 11 (2011) 604.
67 T.J. Causon, K. Broeckhoven, E.F. Hilder, R.A. Shellie, G. Desmet, S. Eeltink, J. Sep. Sci. 34 (2011) 877.

68 D. Guillarme, E. Grata, G. Glauser, J.-L. Wolfender, J.-L. Veuthey, S. Rudas, J. Chromatogr. A 1216 (2009) 3232.
69 F. David, G. Vanhoenacker, B. Tienpont, I. Francois, P. Sandra, LCGC Eur. 20 (2007) 154.
70 S.P. Dixon, I.D. Pitfield, D. Perrett, Biomed. Chromatogr. 20 (2006) 508.
71 L.A. Holland, J.W. Jorgenson, Anal. Chem. 67 (1995) 3275.

Section 2

Charged Aerosol Detection of Specific Analyte Classes

5

Lipid Analysis with the Corona CAD

Danielle Libong[1], *Sylvie Héron*[2], *Alain Tchapla*[2], *and Pierre Chaminade*[1]

[1] *Lip(Sys)*[2] *Lipids, Analytical and Biological Systems, Chimie Analytique Pharmaceutique, Université Paris-Sud, Université Paris-Saclay, Châtenay-Malabry, France*
[2] *Lip(Sys)*[2] *Lipids, Analytical and Biological Systems, LETIAM, Université Paris-Sud, Université Paris-Saclay, IUT d'Orsay, Orsay, France*

CHAPTER MENU

5.1 Introduction

Lipids had not been studied in detail until the nineteenth century, which can probably be explained by the difficulty in separating and identifying them. The definition and classification of lipids is complex. Basically, lipids are defined as products of natural origin, which have the common property of being soluble in nonpolar organic solvents such as chloroform ($CHCl_3$). From a structural but incomplete way, lipids can be grouped into compounds containing fatty acids (FAs) and to those that are similar in terms of biosynthesis and/or functional properties such as aliphatic ethers, sterols (St), cerides (or waxes: WE) and polyisoprénoides such as carotenoids (prenol lipids, PR). Lipids can be categorized into classes according to their chemically functional backbone, each class containing various numbers of molecular species, that is, unique molecular structures. Lipid nomenclature follows the International Union of

Charged Aerosol Detection for Liquid Chromatography and Related Separation Techniques, First Edition. Edited by Paul H. Gamache.

Pure and Applied Chemistry–International Union of Biochemistry and Molecular Biology (IUPAC-IUBMB) rules that can be retrieved at http://www.chem.qmul.ac.uk/iupac/bibliog/white.html#6.

Traditionally, lipids have been classified into simple or complex depending on the number of hydrolysis products obtained by saponification: two for simple lipids such as triacylglycerols (TAGs) where glycerol and FAs are retrieved and three or more for lipids such as phosphoglycerolipids (PL) or glycolipids. They have also been classified according to their polarity into nonpolar (also called neutral) and polar. Nonpolar lipids correspond to FAs, mono-, di-, and TAGs, waxes (WE), sterols (St), carotenoids (PR), and tocopherols, whereas polar lipids encompass PLs, glycerophospholipids (GP), and sphingolipids (SP) and their glyco and phospho derivatives and their lyso forms.

Current progress toward understanding the biological role of various lipid structures through lipidomic approaches necessitates the development of lipid-structure databases and the subsequent need of a widely accepted classification system. The current consensus [1, 2] is to split lipids into eight categories: fatty acids (FA), glycerolipids (GL), glycerophospholipids (GP), sphingolipids (SP), sterol (St), prenol lipids (PR), saccharolipids (SL), and polyketides (PK). Table 5.1 shows the structures of some representative lipids of each class. FAs have hydrocarbon chain lengths usually between 4 and 30 carbons, most often with an even number. This first source of variation, the chain length, is complemented by the presence of unsaturations with not only a Z stereochemistry (E are rare) but also cyclopropane and cyclopentane rings. A third source of structural variation is the occurrence of hydroxyl or epoxy groups. GL are abundant, the most well known being the TAG group. This group also contains mono- and diglycerides together with glycerolglycans where one or more sugar residues are attached to the glycerol group of a mono- or diglyceride. GP are key components of membrane bilayers divided in subclasses depending on the nature of the phospho head group. Phosphatidylcholine (PC), the major component of biological membranes, is represented in Table 5.1.

SP contain a fatty base called a sphingoid base, which may vary in alkyl chain length, number and position of double bonds, and presence of hydroxyl groups. The sphingoid base may be linked to a phosphate or glycosyl group. Ceramides (Cer) are a subclass of SP where an FA is linked to the sphingoid base by an amide link. St encompass various structures such as cholesterol, steroids, steroid conjugates and sterol esters (SE), vitamins D_2 and D_3, and bile acids and derivatives. PR are synthesized from C5 precursors to form the isoprenoid and polyterpene subclass, the ubiquinones and vitamins E and K subclass, and polyprenols. SL are compounds where FAs are directly linked to the sugar backbone. They can occur as glycan or phosphorylated derivatives such as lipid X illustrated in Table 5.1. The most familiar SLs are precursors of lipopolysaccharide components. Finally, PK are members of a very diverse class from either structural or biological activity point of view. PK are biosynthesized through malonyl-CoA condensation in a similar way to FA synthesis.

Table 5.1 Examples of lipid structures from various classes.

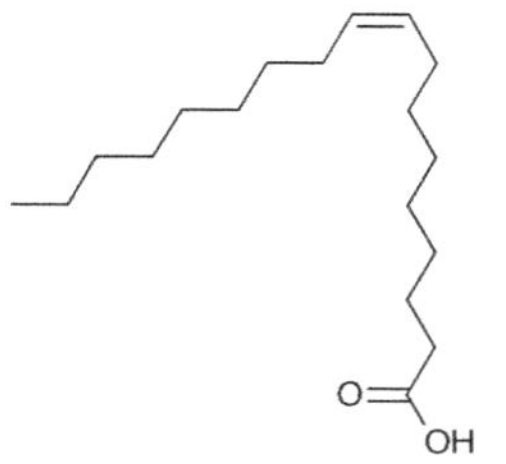	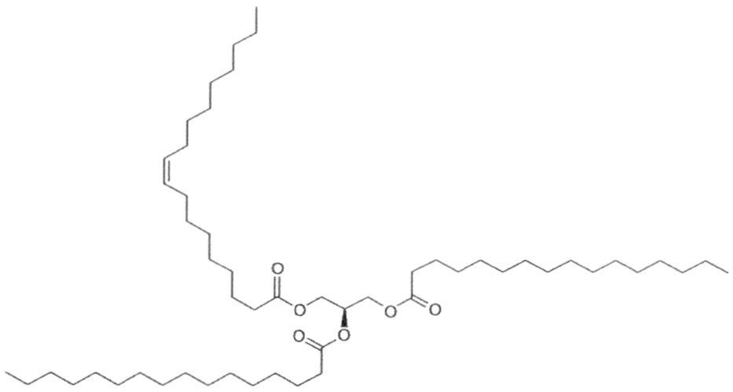	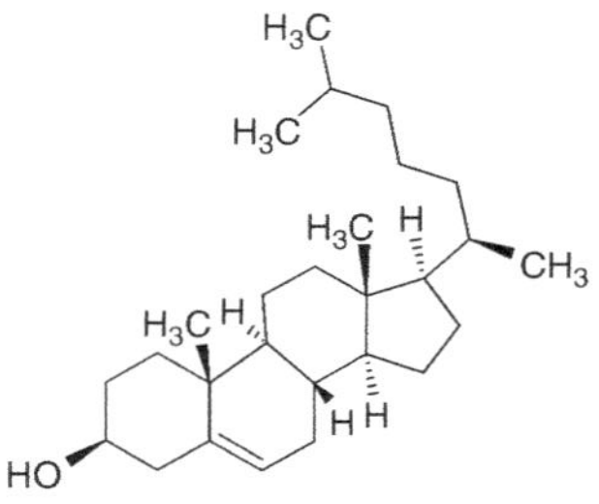
FA: oleic acid IUPAC name: 9Z-octadecenoic acid	TAG: TG(16:0/16:0/18:1(9Z)) IUPAC name: 1,2-dihexadecanoyl-3-(9Z-octadecenoyl)-sn-glycerol	St: cholesterol IUPAC name: (3β)-cholest-5-en-3-ol
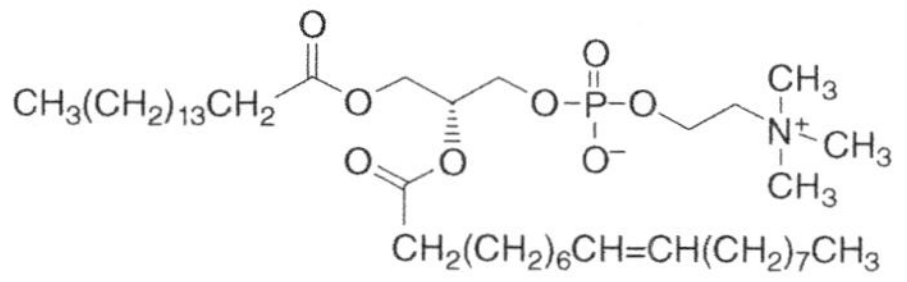	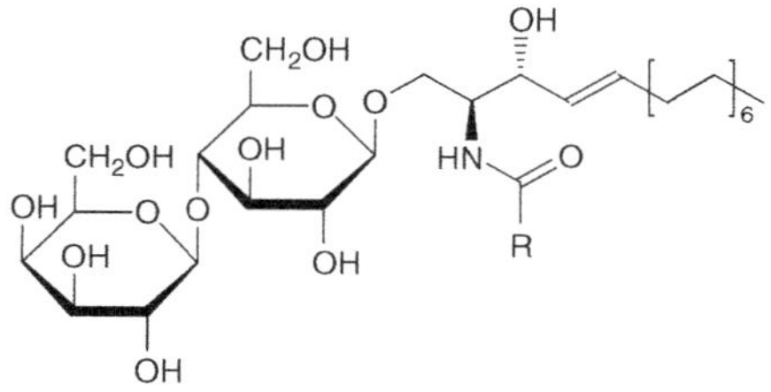	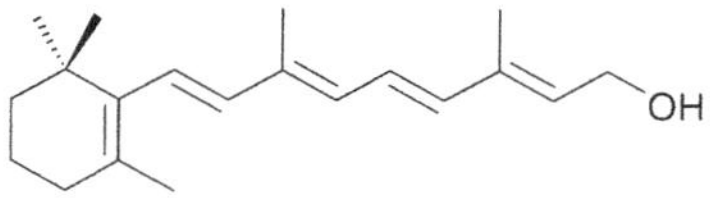

(Continued)

Table 5.1 (Continued)

PL(PC): 1-palmitoyl-2-oleoylphosphatidylcholine IUPAC name: 1-hexadecanoyl-2-(9Z-octadecenoyl)-sn-glycero-3-phosphocholine	SP: lactosylceramide IUPAC name: *N*-(dodecanoyl)-1-*b*-lactosyl-sphing-4-enine	PR: retinol (vitamin A) IUPAC name: (2E,4E,6E,8E)-3,7-dimethyl-9-(2,6,6-trimethylcyclohex-1-enyl)nona-2,4,6,8-tetraen-1-ol
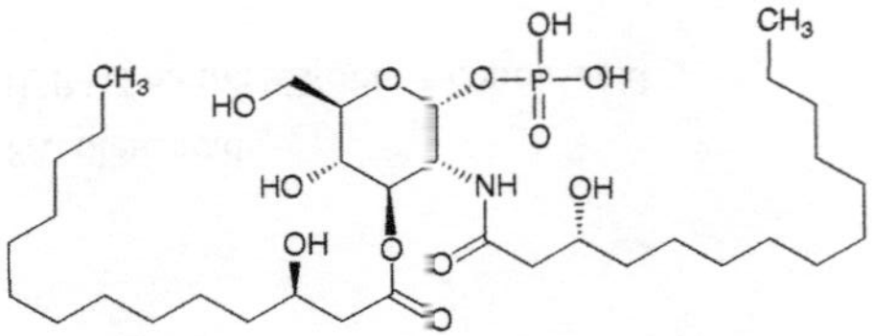	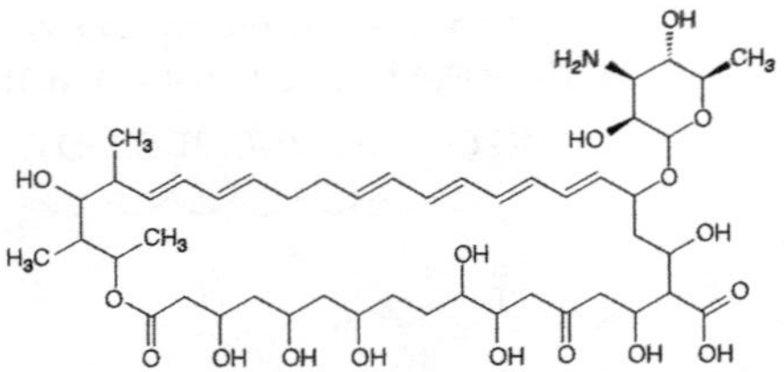	
SL: LipidX IUPAC name: 2,3-bis(3-hydroxytetradecanoyl)-alpha-D-glucosaminyl 1-phosphate	PK: nystatin IUPAC name: (21E,23E,25E,27E,31E,33E)-20-[(3S,4S,5S,6R)-4-amino-3,5-dihydroxy-6-methyloxan-2-yl]oxy-4,6,8,11,12,16,18,36-octahydroxy-35,37,38-trimethyl-2,14-dioxo-1-oxacyclooctatriaconta-21,23,25,27,31,33-hexaene-17-carboxylic acid	

Thus, in terms of separation science, the analysis of lipids has two challenges:

1) In lipid class analysis, separation of compounds corresponding to a wide range of polarity.
2) Separation of lipids within a class according to chain length and presence of unsaturations and/or hydroxyl group. Most lipids lack chromophoric groups exploitable by usual spectrophotometric detectors. Their detection requires near-universal detectors such as evaporative light scattering detector (ELSD) or Corona charged aerosol detector (CAD).

5.2 Principles of Chromatographic Separation of Lipids

5.2.1 Theory of Retention Mechanism in Reversed-Phase Liquid Chromatography

From a theoretical point of view, retention involves a process of solute transfer at infinite dilution from mobile phase (M) into or onto a stationary phase (S). The association of solute (A) with S can involve partitioning, adsorption, or both [3]. Another popular model of retention has been the "solvophobic theory" [4–6]. The distinction between the three models is that "partitioning" implies that the isolate solute is approximately fully embedded within S, "adsorption" implies that the isolate solute is in surface contact with S and is not fully embedded, and "pure solvophobic effect" implies that at least two isolate solutes come together in close contact with themselves giving a bimolecular complex fully embedded in M without accounting for transfer from one solvent to another. "Solvophobic effect applied to chromatography" implies that isolate solute is in close contact with semi-ordered S giving a bimolecular complex partially embedded in M [7]. In partitioning, adsorption, and solvophobic effects applied to chromatography, transfer is characterized by an exchange by neighboring mobile phase molecules and finally surrounded, fully or partially, by neighboring S.

The experimentally observed retention factor k is the product of the distribution constant K, which characterizes the equilibrium constant for solute transfer process between two immiscible phases multiplied by the phase ratio ϕ, the ratio of the volumes of stationary (S) and mobile (M) phases inside a chromatographic column. K can be expressed as the difference in standard-state chemical potentials μ_A° for the solute A, which is conveniently expressed in terms of the binary solution interaction parameters $\chi_{A\text{-solvent}}$ with the two bulk immiscible solvents in which it is distributed:

$$k = K\phi = K\left(\frac{V_S}{V_M}\right) \tag{5.1}$$

$$\ln K = -\frac{\left(\mu_{AS}^\circ - \mu_{AM}^\circ\right)}{RT} = \left(\chi_{AM} - \chi_{AS}\right) = -\frac{\Delta\Delta G_0}{RT} \tag{5.2}$$

Using the simple lattice model for liquids wherein every molecule is taken to be surrounded by nearest-neighboring molecules, Dill demonstrated that the transfer process involves the formation of bonds of type A–S and the breaking of bonds of type A–M whose number is not the same in partitioning, adsorption, and pure solvophobic effect [3, 8].

Depending on the process involved, this leads to the three following equations:

$$\ln k_{\mathrm{A}} = \frac{1}{6}\left[\chi_{\mathrm{MS}} - \chi_{\mathrm{AS}} + \chi_{\mathrm{AM}}\right] + \ln\left(\frac{n_{\mathrm{S}}}{n_{\mathrm{M}}}\right) \quad \text{if adsorption mechanism.} \tag{5.3}$$

$$\ln k_{\mathrm{A}} = \left[\chi_{\mathrm{AM}} - \chi_{\mathrm{AS}}\right] + \ln\left(\frac{n_{\mathrm{S}}}{n_{\mathrm{M}}}\right) \quad \text{if partition mechanism.} \tag{5.4}$$

$$\ln k_{\mathrm{A}} = \left[2\chi_{\mathrm{AS}} + \chi_{\mathrm{AM}}\right] + \ln\left(\frac{n_{\mathrm{S}}}{n_{\mathrm{M}}}\right) \tag{5.5}$$

if solvophobic mechanism applied to chromatography [7], where $n_{\mathrm{S}}/n_{\mathrm{M}}$ is the phase ratio expressed in mole number instead of volume.

For simple molecules for which the dominant interactions are due to the temporary moments (dispersion forces derived by London) and induced dipole moments (derived by Debye) [9], the Hildebrand solubility parameter (δ) concept [10] has been used to provide a useful additional simplification to this model for retention [11–14]. In these cases, the binary interaction parameter is approximated as a product of molar volume $\tilde{V}$ with unitary interaction constants, the δ_s being related to the partial molar enthalpy:

$$\chi_{\mathrm{XY}} = \frac{\tilde{V}}{RT}\left(\delta_{\mathrm{X}} - \delta_{\mathrm{Y}}\right)^2 \tag{5.6}$$

This factorization into unitary constants is often a poor approximation, particularly if forces other than dispersion interactions are involved. While this model is not adequate for a quantitative description or precise prediction of retention, it is convenient for understanding and choosing a chromatographic mode especially for separating nonpolar or weakly polar solutes such as lipids. It could explain many of the features of liquid chromatography (LC) in qualitative terms. Independent of the nature of three possible LC mechanisms, the retention of a solute A is dependent on a term that involves the binary solution interaction parameters and consequently the solubility parameters of the three partners (δ_{A}, δ_{M}, and δ_{S}) of the chromatographic equilibrium.

A lot of different studies were led to determine what kind of model occurs in reversed-phase liquid chromatography (RP-LC). A compilation of all results obtained during a period of 10 years around 1980–1990 was reported in a review that gave a general view of molecular interaction mechanisms in RP-LC

[15]. Valuable information came from studies about the influence of temperature (T) on retention [15–17].

Considering that lipids are more often analyzed with C_{18} (sometimes with C_{30})-bonded silicas and a mixture of organic modifier rich in solvent of low dielectric constant, in the following part of this chapter, we will develop theoretical considerations using Equation 5.4 relative to the partition model.

Combining Equation 5.4 with Equation 5.6 and neglecting the (small) entropy correction terms produces Equation 5.7, which could be rearranged to Equation 5.8 [13, 18, 19]:

$$\ln k_A = \frac{\tilde{V}}{RT}\left[(\delta_A - \delta_M)^2 - (\delta_A - \delta_S)^2\right] + \ln\left(\frac{n_S}{n_M}\right) \quad (5.7)$$

$$\ln k_A = \frac{\tilde{V}}{RT}\left[(\delta_M + \delta_S - 2\delta_A)(\delta_M - \delta_S)\right] + \ln\left(\frac{n_S}{n_M}\right) \quad (5.8)$$

Since the work of Hildebrand [10], the solubility parameter values of non-electrolyte were experimentally determined and different tables were published [20]. If not available, they are relatively easily calculable using attraction constants of structural units described by Small [21], [22], or following the calculation of Patton [23] after calculating the heat of vaporization L_V from boiling point T_E using the relation between L_V and T_E from Hildebrand. At last values of solubility parameters of some solid surfaces have been suggested in the literature [14], [24].

As developed by Schoenmakers in his book according to Equation 5.7, solute retention (k_A) varies exponentially with its polarity δ_A, and that as soon as $[(\delta_M + \delta_S - 2\delta_A)(\delta_M - \delta_S)]$ becomes highly positive or negative, k_A will be either too high or too low [18]. Although the retention factor of solutes is also dependent on the phase ratio value, it does not play major role in the following discussion. In fact for currently used RP-LC stationary phases, the phase ratio n_S/n_M falls around 10^{-3}. Recalling that in optimized isocratic conditions all solutes must fall in a range of k from 1 to 15 for understanding the chromatographic process, it is sufficient to look for experimental conditions where $[(\delta_M + \delta_S - 2\delta_A)(\delta_M - \delta_S)]$ is equal to 0 or is minimized. There are two possibilities for obtaining that:

1) Choosing M with the same solubility parameter as S. In such experimental conditions independent of their nature, all solutes are eluted together (not separated). This explains how to choose the nature of a solvent for cleaning a contaminated S.
2) Taking the polarity of solute to be roughly intermediate between the polarities of two other chromatographic partners (Equation 5.9),

$$\delta_A = \frac{\delta_S + \delta_M}{2} \quad (5.9)$$

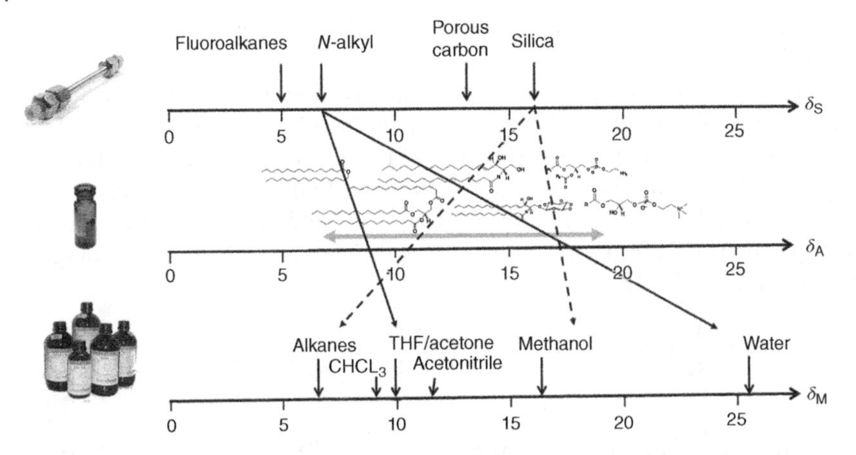

Figure 5.1 Solubility parameters of stationary phases, lipid solutes, and mobile phase solvents.

Thus Schoenmakers gives a very simple rule of thumb for selecting chromatographic conditions.

Adapted from the publication of Schoenmakers, three parallel axes could be drawn from top to bottom: the solubility parameters of S (δ_S), solute (δ_A), and M (δ_M) (Figure 5.1). On the middle axis the common values of some classes of solutes from Hansen data or derived from the Hansen system are reported, that is, from Crowley or Teas [20, 25]. As can be seen, the higher the polarity of a chemical family (or solute class), the higher is its solubility parameter. Independent of the nature of the polar group of a chemical family, the higher the alkyl chain length of the solute, the smaller is its solubility parameter. On the top axis the solubility parameters of the four considered S by Schoenmakers are reported: bonded perfluoroalkyl chain ($\delta = 5\,cal^{1/2}\,cm^{-3/2}$), bonded alkyl chain ($\delta = 7.2\,cal^{1/2}\,cm^{-3/2}$), porous carbon ($\delta = 14\,cal^{1/2}\,cm^{-3/2}$), and pure silica ($\delta = 16\,cal^{1/2}\,cm^{-3/2}$). On the bottom axis the solubility parameters of some common pure solvents M used in chromatography were reported: heptane ($\delta \sim 7\,cal^{1/2}\,cm^{-3/2}$), tetrahydrofuran (THF) ($\delta = 9.5\,cal^{1/2}\,cm^{-3/2}$) near both to methylene chloride ($\delta = 9.9\,cal^{1/2}\,cm^{-3/2}$) and to acetone [$(CH_3)_2CO$] ($\delta = 9.8\,cal^{1/2}\,cm^{-3/2}$), CH_3CN ($\delta = 11.8\,cal^{1/2}\,cm^{-3/2}$), methanol ($CH_3OH$) ($\delta = 14.35\,cal^{1/2}\,cm^{-3/2}$), and water ($\delta = 23.50\,cal^{1/2}\,cm^{-3/2}$).

When S is less polar than M, which is the case in RP-LC, if a sample molecule has a polarity similar to that of M, it will prefer M and k is ~0. If the polarity of the sample molecule is similar to that of S, k will be quite large. For some intermediate polarity of the sample molecule, the solutes will distribute more or less equally in both phases and k is around 1. The two solid arrows starting from the δ_S axis at the *N*-alkyl position to join THF or water on the δ_M axis set the polarity limits of S/M system in RP-LC. Solutes on the δ_A axis whose polarity is between the two arrows can be eluted in the favorable range of $1 \leq k \leq 15$ by adjusting the composition (and thus the polarity) of M. By using this concept, Schoenmakers derived Equation 5.7 into a very simple expression that expresses the selectivity of a phase system when two solutes A and B with similar molar volume have to been separated:

$$\ln \alpha_{AB} = \frac{2\tilde{V}}{RT}\left[(\delta_B - \delta_A)(\delta_M - \delta_S)\right] \tag{5.10}$$

Its examination leads to the following: the higher the difference ($\delta_M - \delta_S$), the higher will be α_{AB}. Thus it could also be easily understandable that using bonded alkyl silica (RP-LC), the mixture of THF and water permits to analyze solutes with a range of δ comprised between 8.1 and 15 $cal^{1/2}\,cm^{-3/2}$. Depending on the range of solute to be analyzed, the choice between gradient and isocratic mode becomes evident. If the entire range of δ is present in a complex mixture, gradient elution is the only solution. If the range of solutes to analyze is narrow, isocratic mode must be chosen, such that the solubility parameter of M would be a little less than the most polar analyte (A).

This diagram permits also to understand how to choose experimental conditions to fractionate different solutes using solid-phase extraction (SPE). For example, all analytes whose $\delta < 12\,cal^{1/2}\,cm^{-3/2}$ will be blocked and all analytes whose $\delta > 18\,cal^{1/2}\,cm^{-3/2}$ will be unretained on an alkyl-bonded silica SPE cartridge when using a solvent of high polarity such as water. It can be demonstrated that the same reasoning could be applied to normal-phase liquid chromatography (NP-LC) using pure silica as S. This is depicted by two dashed lines starting from silica to join alkane solvents and CH_3OH. The solutes whose δ is comprised between 8 and 15 $cal^{1/2}\,cm^{-3/2}$ could be eluted when changing M from pure heptane to pure CH_3OH. When using SPE cartridges, that is, disposable columns, water could be used as strong solvent, which is not the case in analytical NP-LC. In SPE all the analytes whose $\delta > 20\,cal^{1/2}\,cm^{-3/2}$ were blocked on the silica SPE cartridge when using water as solvent. In a general way, this reasoning could be adapted with any organic modifier (OM)–water (W) binary mixture or pure organic solvent mixtures. The solubility parameter concept provides a very simple rule for approximating the polarity of a mixture considering that the sum of the two volume fractions ϕ should equal 1 (see, e.g., Equation 5.11):

$$\delta_M = (1 - \phi_{OM})\delta_W + \phi_{OM}\delta_{OM} \tag{5.11}$$

5.2.2 Optimizing Selectivity

According to Equation 5.11, a given mixture of any organic modifier and water will have an intermediate polarity between the two pure solvents. The same solubility parameter of M may be obtained with several mixtures. To a first approximation it could be expected that mixtures of the same polarity yield the same retention factor. This very simple conclusion was verified with a huge number of experiments. However, due to specific solute–solvent interactions, the corresponding composition of organic modifier (i.e., CH_3OH, CH_3CN, or THF)–water mixtures used in partially aqueous reversed-phase liquid chromatography (PARP-LC) will not be exactly the same for all solutes accounting the polarity of solute and specific interactions they could develop with each given organic modifier. Consequently, when the isoeluotropic composition is calculated from the total solubility parameter theory, some solutes will elute later and some earlier, but approximately within the same range of k values with the change of CH_3OH by CH_3CN or THF in partially aqueous mobile phase. The relative differences may account to a factor of two for certain solutes. That is true also when considering a mixture of pure organic solvents used in nonaqueous LC (either nonaqueous reversed-phase liquid chromatography (NARP-LC) or NP-LC). This will be of fundamental importance to understand how to choose the nature of strong organic solvent (THF, $(CH_3)_2CO$, ethyl acetate, isopropanol (IPA), butanol, $CHCl_3$, methylene chloride (CH_2Cl_2) or methyl tertiobutyl ether (MTBE), trifluoro-1,1,2-trichloro-1,2,2-ethane (TTE)) and the weak solvent (CH_3OH or CH_3CN) in nonaqueous chromatography of nonpolar lipids ($\delta < 9$–$9.5\,cal^{1/2}\,cm^{-3/2}$). It is obvious that in NP-LC the weak–strong classification of solvents is inversed when compared with NARP-LC.

From a theoretical point of view, the differences described previously were rationalized by the introduction of the partial solubility parameter concept. In each phase the interaction energies are determined not only from the total solubility parameter, which is related to the overall polarity of a compound, but also by several different intermolecular forces: dispersion, dipole induction, dipole orientation, and hydrogen bonding [10]. The overall value δ_T is determined by the composite of the various interactions in the pure liquid. It is possible to split up the total solubility parameter of Hildebrand into components from these different interactions: δ_d for dispersion interactions (L and D), δ_o for dipole interactions, and $\delta_h = (2\delta_a\ \delta_b)^{1/2}$ for hydrogen-bonding interactions divided into hydrogen acceptor and proton donor bonding interactions. Each of these partial solubility parameters measures different aspects of solvent polarity. By calculating dispersion fraction polarity f_d, dipolar fraction polarity f_p, and hydrogen-bonding fraction polarity f_h following Equation 5.12, all the organic solvents could be classed into four families dispatched on the Teas triangle [20, 22, 26], which is very useful for choosing isoeluotropic mixtures in NARP-LC:

$$\delta_T = \left(\delta_d^2 + \delta_p^2 + \delta_h^2\right)^{1/2} \tag{5.12}$$

$$f_X = \frac{100\delta_X}{\delta_d + \delta_p + \delta_h}, \text{ where X represent d, p, or h} \tag{5.13}$$

Based on the same reasoning, Snyder classified possible organic solvents used in chromatography according to their relative dipole moment (polar interactions), basicity (hydrogen-bonding acceptor interactions), and acidity (hydrogen-bonding donor interactions) by means of the very well-known and widely used solvent selectivity triangle [27, 28] (Figure 5.2).

In PARP-LC, the organic modifier must be freely soluble in water. Thus Snyder's triangle allows the choice of three organic modifiers with different major partial polarities: CH_3OH, CH_3CN, and THF, which belong to group II, VI, and III, respectively. They provide different selectivities for specific pairs of solutes while keeping M composition in isoeluotropic conditions. As nonpolar solutes are insoluble in partially aqueous mobile phases, this triangle is not appropriate for selecting the organic solvents to be mixed for their analysis. To overcome this inconvenience, Heron and Tchapla suggested the use of the Teas triangle for selecting M composition for analysis of TAGs and cerides (WE) by NARP-LC [29]. The Teas approach permits to understand how to choose two different isoeluotropic mobile phases composed of two different organic modifiers whose dominant partial solubility parameter is not the same. Such modifiers could consequently lead to different selectivities. The weak solvent could belong to either C family (with both strong hydrogen-bonding and middle dispersion interactions) or D family (with both strong polar and middle dispersion interactions). The strong solvent could belong to either B family (with middle hydrogen-bonding, dipole, and dispersion interactions) or A family (with strong dispersion interactions). This explains why selectivities were found different in the analysis of apolar lipids ($\delta < 9–9.5\,cal^{1/2}\,cm^{-3/2}$) when using CH_3CN (D)–$(CH_3)_2CO$ (D) mixtures instead of CH_3CN (D)–dichloromethane (CH_2Cl_2) (B) or CH_3CN (D)–THF (B). As the partial dominant polarity of $(CH_3)_2CO$, CH_2Cl_2, or THF is not the same (or of the same magnitude), subtle difference in the interactions between some solutes and these solvents occurs, and consequently selectivity of some solute pairs could be strongly modified.

In a general way all the process of selectivity optimization described by Schoenmakers [18] is based on this principle independent of the chromatography mode: PARP-LC, NARP-LC, and NP-LC. It is possible to revisit this concept considering S. Considering the nature of the grafted silica, S with the same global polarity could have different partial solubility parameters and consequently could develop (with the same M composition) a very different separation for a given pair of solutes according to their relative polarities. This could explain why the separation of lipids by classes is not the same with aminopropyl (NH_2), cyanopropyl (CN), or glyceropropyl (diol) graftings. It is interesting to underline that in HILIC mode M is a mixture of CH_3CN and water that is very rich in organic modifier. As the true S is water that interacts with the support (e.g., native silica), it could be demonstrated that it becomes possible to analyze very polar solutes in these conditions (with $\delta \sim 20\,cal^{1/2}\,cm^{-3/2}$) that are

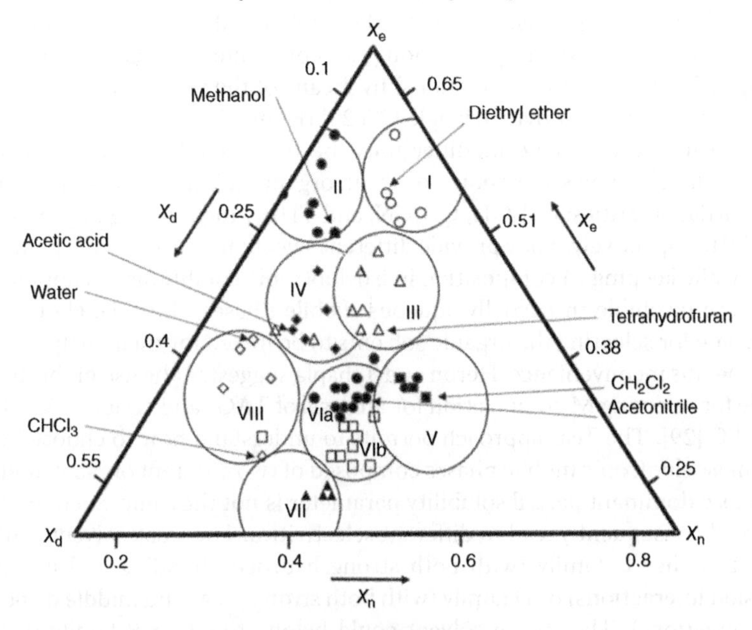

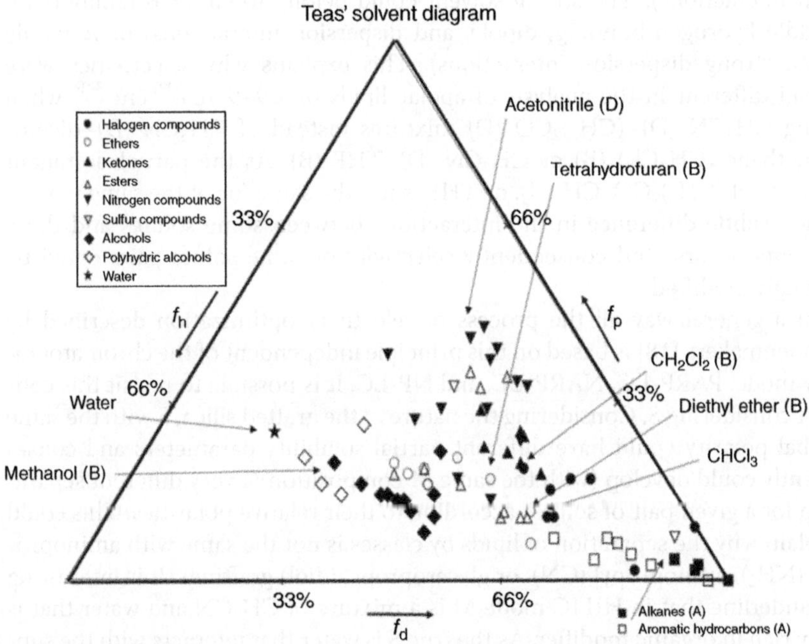

Figure 5.2 Snyder's and Teas' solvent selectivity triangles.

impossible with any support used in NP-LC. Another interest is the recent development of a lot of mixed bonded S that cover a large range of polarity between pure silica and pure C_{18}-bonded silica. For solutes of average polarity ($10\,cal^{1/2}cm^{-3/2} < \delta < 15\,cal^{1/2}cm^{-3/2}$), they present very different selectivities. This will be shown later with an application to the separation of phospholipids. At least we must underline that the solubility parameter of porous graphitized carbon (PGC) is different from the first native pyrocarbon-synthesized S. This is due to the very different process of synthesis of this type of material. From recent results it seems that δ of PGC falls nearest ($\delta \sim 9.5–10\,cal^{1/2}cm^{-3/2}$).

Considering the solubility parameters of different classes of lipids, now the published rules of thumb for separation of some classes of solutes become understandable and could be eventually modified to accommodate different separations.

5.2.3 Note on Using pH Modifiers for Selectivity Optimization

Generally speaking, lipids are not pH-sensitive solutes. The two noticeable exceptions are free fatty acids (FFA) and PLs. In both cases volatile organic acids (formic, acetic, and trifluoroacetic acids) or volatile buffers (usually ammonium acetate or formate, triethylamine: acetic or formic acid) are used to control the ionization of lipid classes. The basic principle is that any of pH modifiers or buffers usable for mass spectrometry (MS) is compatible with Corona CAD. Some organic ion pairs exhibit a proven effect on ELSD response by increasing the size of the solute particles in the detection cell [30]. Although a similar effect was not evidenced with Corona CAD, variations in signal-to-noise ratio were reported when using various amounts of ammonium acetate in the mobile phase [31].

5.3 Application: Strategy of Lipid Separation

As reported by Christie in 2003 [29], "literally thousands of papers have appeared over the last 40 years detailing the structures and compositions of lipids from particular tissues and species, as determined by modern chromatographic methods, but there appears to have been a little effort to collate and critically compare this data in any systematic way. I attempted a comprehensive survey of lipid compositions of ruminant tissues in 1981, but it would be a daunting task to update it or extend it to other species." Accounting for this, our present goal will be, by choosing some significant examples among this very abundant literature, to describe different complementary LC conditions for separating either different lipid classes or different congeners of a particular class. For a much more detailed and another description of strategy of lipid separation than the one developed in following part of this chapter, readers have to read the excellent books of Christie on the subject [29]. Our purpose is to help the comprehension of different ways that have been described for lipid separations using the solubility parameter theory. The following discussion is divided into three sections concerning successively the separation of individual lipid classes, molecular species of lipids (subclasses), and congeners belonging to a given class.

5.3.1 Separation of Individual Lipid Classes

In general, the lipidic extract from various plant or animal tissues covers a large range of classes from apolar lipids (such as cerides (WE) and carotenoids (PR)) to very polar lipids such as phospholipids (PL), in particular the PC. Lipid classes are in variable amounts in the samples depending on the biological origin. For example, the nature and weight percentages of MGDG, DGDG, TAG, and PL from different organs of different vegetables are given in Table 5.2, keeping in mind that under the denomination "other," a great number of different classes of apolar lipids are included. In Table 5.3 the composition of lipid classes of different organs of an animal is also reported.

The overall classes of lipids lie in a range of solubility parameters from 7.6 to ~19 $cal^{1/2} cm^{-3/2}$. Consequently, separation of all lipid classes in a single run could be only developed using gradient elution. As described previously, both pure silica and polar-ended alkyl-bonded supports allow analysis of solute mixtures in the range of solubility parameters (from 7 to 20 $cal^{1/2} cm^{-3/2}$). This implicates starting the analysis in normal mode using mixtures of nonaqueous solvents whose composition is continuously modified along all the analysis

Table 5.2 Lipid class composition of various plants (wt% of total lipids) [32].

Lipid class	Potato tuber	Apple fruit	Soybean seed	Clover leaves	Rye grass	Spinach chloroplast
Monogalactosyldiacylglycerol (MGDG)	6	1		46	39	36
Digalactosyldiacylglycerol (DGDG)	16	5		28	29	20
Sulfoquinosyldiacylglycerol (SQDG)	1	1		4	4	5
Triacylglycerol (TAG)	15	5	88			
Phosphatidylcholine (PC)	26	23	4	7	10	7
Phosphatidylethanolamine (PE)	13	11	2	5	5	3
Phosphatidylinositol (PI)	6	6	2	1	2	2
Phosphatidylglycerol (PG)	1	1		6	7	7
Other	15	42	5	3	4	

Source: Christie [32]. Reproduced with permission of Elsevier.

Table 5.3 Lipid classes of rat tissues (wt% of total lipids) [32].

Lipid class	Heart	Liver	Erythrocytes	Plasma
Cholesterol esters (CE)	Traces	2		16
Triacylglycerols (TAG)	4	7		49
Cholesterol (St(C))	4	5	30	6
Diacylglycerols (DAG)	1		Traces	Traces
Free fatty acids (FFA)		Traces		2
Diphosphatidyldiglycerol (DPG)	12	5		
Phosphatidylethanolamine (PE)	33	20	21	
Phosphatidylinositol (PI)	4	4	3	
Phosphatidylserine (PS)			3	
Phosphatidylcholine (PC)	39	55	32	24
Sphingomyelin (SPH)	2	2	8	2
Lysophosphatidylcholine (LPC)			1	1

Source: Christie [32]. Reproduced with permission of Elsevier.

process and finishing it in HILIC mode using a partially aqueous/predominantly organic M composition. In these conditions retention order is governed by the polarity of the lipid classes, which is directly correlated with their polar head group.

All these separations were only possible using evaporative detectors such as either the charge aerosol detectors (Corona CAD) or the oldest one: the ELSD and the ones derived from it (condensation nucleation light scattering detection (CNLSD)). When using these detectors, all solutes including non-UV-absorbing molecules could be detected using gradient elution (step or continuous). In spite of abrupt or continuous changes in M composition, no baseline disturbances were reported, and samples composed of the main simple lipid to phospholipid classes were generally adequately resolved. For example, using step gradient elution, Christie [32] separated lipidic components of rat kidney ranging in polarity from CE (unretained), TAG, S, C-MonoHexoïde + diphosphatidylglycerol (DPG), PE, PI, PS, and PC to SPH in 20 min on a 3 μm silica with absolutely steady baseline. Plant lipids present added difficulties; thus with an adaptation of ternary elution system, Moreau reported the separation of SE (unretained), TAG, St, FFA, acetylated sterol glycosides, MGDG, sterol glycosides SG, DGDG, cardiolipins (CL), PE, PG, PI, and PC in less than 50 min [33]. The higher the phase ratio, the higher is the selectivity of the separation. Since adsorbents with higher specific surface have higher phase ratios, the best selectivities could be gained using silica with ~400 m^2/g of specific area.

According to the composition of M, the selectivity of separation of SE, TAG, FFA, and St could be drastically modified: for instance, the order of retention of FFA and St could be inverted. For example, using a three-step gradient such as hexane/diethyl ether/formic acid 99/1/0.05 during 9 min followed by hexane/diethyl ether/formic acid 80/20/0.1 during 9 min and finishing by $(CH_3)_2CO$ 100% and then $CHCl_3/CH_3OH$/water 50/40/10 during 12 min, hydrocarbons (HC), cerides (WE) + SE, and FAME has been separated on the first step; TAG, FFA, St, and DAG were separated on the second step; and acetone-eluted polar lipids (AMPL) and phospholipids (PL) were separated on the final step [34].

Separative method of lipid classes from the *stratum corneum* was also developed with silica and subcritical $CH_3OH–CO_2$ mixture. The elution order was SE, TAG co-eluted with HC (squalene), FFA, St, Cer, and glycosylated ceramides (CG). The Cer eluted in several fractions where retention increased as a function of the number of hydroxyl groups, independent of unsaturations or alkyl chain. The Cer response with evaporative detector was improved by turning the influence of the solvent nature on the response to advantage [35].

Another option is the use of polyvinyl alcohol (PVA)-coated silica. This S whose total solubility parameter is less than pure silica also permitted the separation of overall lipid classes with some different selectivities. In particular, PG is less retained than PE—opposite to the retention order on pure silica [29, 36]. The great advantage of this S, compared with pure silica, is the independence of gained separation with the percentage of trace levels of water solubilized in organic solvent composing M. It was successfully used in NP-LC.

In NP-LC using S less polar than pure silica such as polar-ended alkyl-bonded silica, that is, diol, NH_2, or CN, the range of polarity of lipid classes that could be separated is reduced, because the total solubility parameter of M is less than 16 $cal^{1/2} cm^{-3/2}$, but best separations were gained in a restricted range of solubility parameters.

For example:

- With CN-bonded silica using an elution gradient from hexane to methyl tert-butyl ether, the following classes of lipids of total solubility parameters were perfectly separated: CE (unretained), TAG, FFA, St, 1-3 DAG, 1-2 DAG, and MAG (covering a range of 7.9–9.6 $cal^{1/2} cm^{-3/2}$) [37].
- With NH_2 stationary phase, lipid classes were separated with different selectivities than on CN phases. Thus FFA ($\delta \sim 8.5\,cal^{1/2} cm^{-3/2}$) are more retained than MAG ($\delta \sim 10\,cal^{1/2} cm^{-3/2}$) on amino phases. Conversely, they are eluted faster than St ($\delta \sim 10\,cal^{1/2} cm^{-3/2}$), DAG ($\delta \sim 9\,cal^{1/2} cm^{-3/2}$), and MAG. Even though reported separations were obtained with different M compositions, this could also be rationalized by the difference of the nature of preponderant partial solubility parameter, which is not the same for these two S. Both the highest dispersion solubility parameter and the lowest hydrogen-bonding solubility parameter for CN phases compared with NH_2 phases explain why

the hydrogen acceptor or donor lipid classes are more retrained than other lipid classes on NH_2 phases. This assertion is confirmed by results of I. Rizov and A. Doulis [38] who were using three different S for separating a great number of lipid classes:

1) NH_2-bonded silica on which chlorophylls and carotenoids (PR) were separated using a mixture of $(CH_3)_2CO/CHCl_3$ 80/20 at 4°C.
2) NH_2-bonded silica in series with pure silica. MGDG, PE, PG, and DGDG were successively eluted using four mobile phases composed of hexane/ $CHCl_3$/THF/CH_3CN/IPA/CH_3OH in different qualitative and quantitative compositions such as the % of hexane decreases when the % of IPA increases.
3) Silica coupled with a weak anion exchanger phase such as protonated NH_2-bonded silica where the more polar lipid classes such as SQDG, PI, and finally PC were successively eluted by using mobile phases composed of $CHCl_3$/CH_3CN/IPA/CH_3OH/aqueous solution of ammonium acetate (pH 8.4) in different quantitative compositions such as the % of IPA decreases when the % of aqueous solution of ammonium acetate increases. With this system MGDG is more rapidly eluted than PE and the same for both PE and PG compared with DGDG, which was the opposite on pure silica. On NH_2-bonded phases, MGDG, DGDG, PE, and PC interact less than on pure silica due to the number of hydrogen bonds.

Similarly, using NH_2-bonded silica percolated by isocratic mode, that is, 2% of CH_3OH in carbon dioxide in supercritical M, Sandra *et al.* reported the separation of less polar lipid classes [39]. In less than 10 min, TAG, SE, DAG, St, MAG, and FFA were perfectly separated with modified order of retention compared with pure silica, PVA-coated silica, or CN-bonded silica. Using 10% CH_3OH in supercritical carbon dioxide, the same authors isolated the free sterol fraction of different oils for characterizing them in a second chromatographic analysis. In this last case TAG, DAG, and SE were co-eluted in the first collected fraction.

In the highest range of polarity, the same reasoning leads to the rationalization of the separation of phospholipid classes whose solubility parameters fall in a range of 16–19 $cal^{1/2}cm^{-3/2}$, keeping in mind that lysophospholipids (LPL) are more polar than the corresponding phospholipids and that PC are cationic molecular species. For this last subclass of lipids, the solubility parameter concept does not apply, but it is sure that their polarity is higher than other phospholipids, which are all nonelectrolyte polar lipid classes.

Most of separations on pure silica used a ternary step gradient [32] with different mixtures of hexane, IPA, and water. In general, the starting M is a mixture of an alkane with either IPA or THF, sometimes mixed with small amount of water followed by the simultaneous small increase of % IPA and water. The range of solubility parameters of such M is very difficult to determine for precisely understanding the range of lipid classes that could be eluted. Considering the reported results, it seems to cover a range from 16 to >23 $cal^{1/2} cm^{-3/2}$. As previously described for all lipid classes, an alternative was given by using PVA-coated silica. Thus Godoy Ramos *et al.* using a complex ternary gradient of heptane/IPA 98/2, $CHCl_3$/IPA 65/35, and CH_3OH/water 95/5 continuously supplemented from the beginning to the end by 1% of acetic acid and 0.08% of triethylamine separated phosphatidic acids PA, PG, CL, PI, PE, and PS, lysophosphatidylethanolamine (LPE), PC, SPH, and LPC in less than 40 min [40].

All reported separations of phospholipid classes used polar S (pure or PVA-coated silica or diol) on which the final eluent always contained a little amount of water. Thus beginning of separation is governed by normal LC mode and ended by HILIC mode. In these conditions it is difficult to compare the gained separations. The different observed selectivities from works of one author to another coming from used M compositions as well as from difference of total δ_T and partial dispersion δ_d and hydrogen-bonding δ_h solubility parameters of three used S.

For example, we could cite the separation gained on glyceropropyl-bonded silica with an elution gradient mixing pure $CHCl_3$ with CH_3OH supplemented by 1% of a mixture of formic acid with ammonia fixed at pH 5.3, as well as 0.05% of triethylamine [41]. The observed order of retention was PG < PC < plasmogenic PE < PE < LPC < PI < LPS in opposite of separation described on PVA-coated silica or pure silica where PI were less retained than PE on which the polarity of beginning M was higher than on diol-bonded silica.

In general, due to the small influence of trace levels of water in organic solvents on their selectivity, using diol-bonded or PVA-coated silicas offers three advantages compared with separation on pure silicas:

1) No activation–deactivation phenomenon
2) Faster re-equilibration time
3) No irreversible adsorptions

This leads to recommending their use instead of pure silica.

5.3.2 Separation of Subclasses of Lipids

Using size exclusion chromatography or argentation chromatography, it is possible to obtain distinct molecular fractions still called subclasses, separated on the basis of a single well-defined property, that is, molecular size [42, 43] or degree of unsaturation [29, 44].

5.3.2.1 Size Exclusion Chromatography

For this type of chromatography, migration of molecules between S and M is only driven by diffusion. The elution order of solutes depends exclusively on molecular size, which is closely related to molecular weight, provided that molecules have the same shape. Applications were mainly dedicated to analysis of polar lipids in frying oils [43].

5.3.2.2 Argentation Chromatography

Despite its potential for providing structural information, argentation chromatography has been used much less often than RP-HPLC, chiefly as a result of the technical difficulties encountered in applying it with HPLC systems. Development of a stable, reproducible chromatographic system has been the main technical stumbling block. Two main types of column have been used: those containing silica gel impregnated with silver nitrate and those based on silver ions attached to cation exchangers (silica based). S using cation-exchange supports achieve good silver-ion retention levels [45–47]. Ag-LC systems separate TAGs in order of unsaturation: that is, SSS, SMS, SMM, SSD, MMM, SMD, MMD, SDD = SST, MDD = SMT, MMT, DDD = SDT, MDT, STT = DDT, MTT, DTT, and finally TTT (where S = saturated FA, M = monounsaturated FA, D = di-unsaturated FA, and T = tri-unsaturated without indication of the position of double bond on alkyl chain as well as of the position of the FAs on the glycerol moiety…) [45, 46, 48–50]. This elution order may vary a little if M is changed substantially. A ternary gradient system and an evaporative detector are required.

Although S was earlier held to be the most important factor affecting argentation chromatography, composition of M exerts a decisive influence on TAG separations. The nature of the interaction between the silver ions, unsaturated solutes, and solvents in M on silver-ion cation-exchange columns has not been fully elucidated. Several researchers have ascribed solute retention on such columns to a mechanism of mixed interaction with the more highly polar TAG groups involving the formation of π complexes between silver ions and double bonds with the participation of the carbonyl oxygen and adsorption by the unbounded polar groups in the support [51, 52]. Considering the different molecular interactions, it could be considered that the silver cations create induced dipoles when interacting with the double bonds. The polarity of the molecular species increases with the number of double bonds as does the induced dipole strength. This explains the trend in values of the total solubility parameter δ_T for FFAs and TAGs. Higher δ_T values are encountered when increasing the double bond numbers in FFA or TAG structures. This increase is due to the higher partial dispersion solubility parameter δ_d, which is only due to the contribution of Debye interactions. From this point of view, argentation S must be considered as of intermediate polarity ($\delta \sim 9\,cal^{1/2}\,cm^{-3/2}$). Thus by varying M composition in a range between 7.4 and $12\,cal^{1/2}\,cm^{-3/2}$, all the subclasses of FFA or TAG could be separated. Chlorinated solvents together with gradients of $(CH_3)_2CO$ or CH_3CN were the first practical systems to be described, but heptane/CH_3CN is often preferred nowadays, even if it shows slightly less resolution.

Elsewhere solvents such as benzene, toluene, and CH_3CN appear to competitively interact with the silver ions, whereas CH_3OH, IPA, and $(CH_3)_2CO$ may block interactions with the unbonded polar groups of the support [53]. Selectivities were greatly influenced by using CH_3OH instead of CH_3CN. It was demonstrated that TAG separation using Ag^+ in M is effective only when using CH_3OH as organic modifier and not CH_3CN [54].

Evaporative detectors have been the most successful to analyze TAGs by argentation HPLC because they do not limit the choice of solvents for M. They also afford good stability and sensitivity when using complex elution gradients with mixtures of three or more solvents [47, 55, 56].

Silver-ion chromatography can also separate FAME and FFA with an increasing degree of unsaturation. As the total solubility parameters of FFA vary from 8.1 (saturated FFA) to 8.5 (three unsaturated FFA), used M is mainly apolar. Generally it is composed of 0.5–1% of CH_3CN in 99.5% or 99% hexane.

This chromatography could be also very efficient for characterizing geometric FAME congeners or positional TAG isomers. Thus the separation of *trans–trans*, *cis–trans*, and *cis–cis* isomeric conjugated octadeca 8–10, 9–11, 10–12, and 11–13 dienoic acids was performed using two columns in series [57–59]. At last silica S systems coated with $AgNO_3$ offer also a separation of regioisomeric TAGs (*sn-1/3, sn-2*), for instance, the separation of SMS and SSM [60, 61].

5.3.3 Separation of Congeners Belonging to Specific Classes of Lipids

Each congener of any lipid class is formed almost invariably by one or more long-chain FA linked to different polar heads. This is true for SE, TAG, DAG, MAG, Cer, WE, CG, MGDG, DGDG, and PL. The only exceptions are HC and St. The first technique used for their analysis involved their transesterification mainly to methyl esters followed by capillary gas–LC. This leads to the knowledge of primary lipid composition, which is necessary but not sufficient for characterizing lipids. The knowledge of secondary composition of lipids, which could be deduced from primary composition, is of fundamental importance. Each congener is defined by the following attributes:

1) Total carbon number (CN) equal to the sum of the carbon atoms contained in the fatty alkyl chains.
2) FA alkyl chain length (odd and even numbered with two to almost one hundred carbon atoms).
3) FA position on the polar head backbone.
4) The total number of double bonds (DBN).
5) Double bond position and configuration in each FA.
6) In addition there can be a host of further structural features including branch points, 3-, 5-, 6-, or 7- membered rings, acetylenic and allenic bonds, oxygenated functions, and many more [1].

As is well known, the common FAs of animals and plants mainly consist of even linear chains of 16–22 carbon atoms with zero to six double bonds of the *cis* (Z) configuration. Polyunsaturated FAs have methylene interrupted double bond system in general. More rarely these double bonds are conjugated [29].

The separation of congeners of different lipid classes is quite challenging because of the presence of numerous involved species with similar physicochemical properties. The first step for solving this problem is evaluation of the number of congeners that have to be separated.

From the primary composition, the possible compositions of them could be obtained by calculation and combinatorial analysis:

- For a lipid class composed of n FAs, without considering the position of FA residue on polar head group, there are N_1 possible different TAGs, M_1 different 1-2 DAG, 1-3 DAG, PL, MGDG, and DGDG, and L_1 different Cer, CG, and WE molecular species. This number is defined by

$$N_1 = \frac{n^3 + 3n^2 + 2n}{6}$$

$$M_1 = \frac{2n^2 + 2n}{4}$$

  $L_1 = n \times n_2$, where n_2 is either the number of fatty alcohols (for waxes: WE) or the number of bases (for SP, Cer, CG).
- Now, if we also take into account the positional isomers of mixed TAGs (XXY and XYX) and (XYZ, XZY, YXZ) or the mixed DAG, PL, MGDG, and DGDG (XY and YX), the possible total number of isomers is N_2 and M_2:

$$N_2 = \frac{n^3 + n^2}{2}$$

$$M_2 = n^2 + n$$

- Lastly, if we take into account all the isomers, including optical isomers, the possible number of TAGs is $N_3 = n^3$.

The more difficult challenge, which is the separation of possible isomers of TAGs, is summarized in Table 5.4.

The problem can be partially solved if interested only in qualitative analysis of the more abundant products, considering the fact that the amount of a given TAG is a mathematical product of the percentage of each of its corresponding

Table 5.4 Number of TAG isomers according to the number (N) of FA residues.

N	N_1 (without position isomer)	N_2 (with sn1–3/sn2)	N_3 (with optical isomers)
1	1	1	1
2	4	6	8
3	10	18	27
4	20	40	64
5	35	75	125
⋮	⋮	⋮	⋮
10	220	550	1000
13	455	1183	2197
15	680	1800	3375
20	1540	4200	8000

FA residue. Such a mixture contains completely randomly distributed FA, so the amount of each TAG is given by

$$\%[\mathrm{TAG_{XXX}}]=(\%\mathrm{X}\cdot\%\mathrm{X}\cdot\%\mathrm{X})\left(10^{-4}\right)$$

$$\%[\mathrm{TAG_{XXY}}]=(\%\mathrm{X}\cdot\%\mathrm{X}\cdot\%\mathrm{Y})\left(3\times10^{-4}\right)$$

$$\%[\mathrm{TAG_{XYZ}}]=(\%\mathrm{X}\cdot\%\mathrm{Y}\cdot\%\mathrm{Z})\left(6\times10^{-4}\right)$$

For the other classes of lipids, the amount of each isomer is given by

$$\%[\mathrm{lipid_{XX}}]=(\%\mathrm{X}\cdot\%\mathrm{X})\left(10^{-2}\right)$$

$$\%[\mathrm{lipid_{XY}}]=(\%\mathrm{X}\cdot\%\mathrm{Y})\left(2\times10^{-2}\right)$$

Even though the calculated percentage of lipid congener is not equal to the experimental lipid congener amount [62], some basic conclusions could be drawn from these theoretical data. If 2 or 3 FA residues present in very low amounts (<1%) form part of a TAG structure, such a TAG will be in very low amount. If the TAG is biosynthesized, it will be present in the oil as traces under the limit of detection of standard apparatus. The same conclusions, as previously stated, could be deduced for 1-2 DAG, 1-3 DAG, PL, MGDG, DGDG, Cer, and CG (or by analogy WE), which contain 2 FA residues in their structure, when 1 or 2 FA will be present in very low amounts in their primary composition.

From a practical standpoint, the number of congeners to be separated and identified is, in fact, much lower than the total number N_1 (or M_1, or L_1), which is calculated from the knowledge of the primary constitution of FA residues n, whatever their relative amount.

The estimation of maximum number of congeners, which could easily be detected experimentally, separated, and discriminated, could be deduced using a criterion of discrimination between all the FA residues. The total number n of FA must be arbitrarily separated into two different groups:

- The first one corresponding to the more abundant FA, that is, the FA (n') whose percentage is higher than 1%
- The second one corresponding to less abundant FA, that is, the FA (n'') whose percentage falls between 1 and 0.1%

The compilation of results reported for vegetable oils in two monographs [60, 61] gives the following ranges of these values: $4 < n' < 9$ and $2 < n'' < 8$. For animal fats $8 < n' < 14$ and $10 < n'' < 20$ [63].

For example, applying these calculations to a classical vegetable oil with $n = 13$ FA residues gave the results reported in following Table 5.5.

It can be seen that statistically, the number of TAGs whose relative amount is ≥0.1 is 67 (13 + 54), of which only 13 are ≥1% (see first line of Table 5.5).

Thus, it could be estimated that the analytical challenge to solve is the separation of TAGs between 70 and 80 and not N_1 TAGs obtained by simple statistical calculation.

The same calculations made from the FA composition of waxes (WE) [60, 62, 64] gave $M_1 = 108$, but considering only both more abundant FAs and alcohols, the challenge is reduced to the separation of 16 congeners.

In the case of Cer, maximum value of n is 10 and n_2 is equal to 16 [63, 65–69]. Considering their primary composition, mean values are around $n = 4$ and $n_2 = 4$ [70]. As for waxes the challenge is reduced to the separation of around 16 congeners. Very recently independent of their relative amount, the variety in Cer structures revealed in skin extracts was investigated. About 236 Cer peaks were already identified, each representing 2–8 unique Cer compounds due to

Table 5.5 Practical example of the number of TAG to be separated in a vegetable oil.

		Statistical TAG number		
TAG abundance (*x*)	***N* total**	***N* with FA from first group**	***N* with 1 FA from second group**	***N* with 2 FA from second group**
$x \geq 1\%$	13	13	0	0
$0.1 \leq x < 1\%$	54	38	16	0

chain isomerism. More than 1000 unique Cer have been elucidated in human skin with high repeatability using LC-QTOF-MS [71].

Each animal or plant tissue can have a characteristic FA composition. Considering a non-exhaustive bibliography, it seems that the typical primary composition of MGDG, DGDG, and PL consists of a set of seven saturated, unsaturated, and polyunsaturated C_{16} and C_{18} FAs whose two or three of them are in great amount compared with the other [64]. The maximum number of congeners to separate is around 50 having ~10 as major components.

In conclusion the challenge of the analysis of congeners of different classes of lipids (excepted TAGs of animal fats) is to find a chromatographic system able to separate a maximum of 90 peaks.

This is entirely consistent with the different chromatographic systems used today. The concept of peak capacity (p.c.), a concept due to Giddings [72, 73], allows one to grasp the opportunity.

The total number of resolvable components is given, in isocratic mode, by the following equation:

$$\text{p.c.} = 1 + \frac{\sqrt{N_{\text{last peak}}}}{4} . \ln\left(1 + k_{\text{last peak}}\right) \tag{5.14}$$

where $k_{\text{last peak}}$ is the retention factor and $N_{\text{last peak}}$ is the number of theoretical plates of last eluted peak.

In gradient mode it is given by p.c. = $\Delta tr/0.93\ \omega_{0.1}$, where $\omega_{0.1}$ is the mean peak width at 10% peak height and Δtr is the range of t_r change in retention time (retention time difference between the first and last eluted peak).

For instance, in isocratic RP-LC using a 25 cm × 4.6 mm × 5 μm ($L_c \times d_c \times d_p$), the p.c. is 88. Thus classical geometries of HPLC columns are suitable to solve the problem of the separation of the congeners of majority of all classes of lipids. The only condition is to find very selective S–M couple for separating closest structural isomers.

5.3.4 Behavior of Lipid Separation in Reversed-Phase Chromatography

Considering, in first approximation, that constitutive FAs of congeners belong mainly to a homologous series, the understanding of their chromatographic separation is very easy.

The solubility parameter theory applied to any homologous series shows that the total solubility parameter δ of different homologues trend to a constant limit value when the number of carbon atoms n_C increases. Considering previously described Equation 5.8, this explains why in RP-LC the retention of homologues is such that log k_A is experimentally proportional to $\tilde{V}$ and consequently proportional to n_C, when n_C is higher than 6.

RP-LC may be considered as the method of choice for the usual separation of congeners. There are two main RP-LC methods for the separation of congener species as a function of the polarity of lipid class to be separated as well as their solubility in polar solvents: NARP-LC for nonpolar lipids ($\delta < 10\,\text{cal}^{1/2}\text{cm}^{-3/2}$) and PARP-LC for polar lipids ($\delta > 13\,\text{cal}^{1/2}\text{cm}^{-3/2}$).

The solvophobic theory leads us to a best understanding on the behavior of lipid congeners in RP-LC in function of their structure [4]:

$$\ln k = \ln\frac{V}{V_M} + \frac{1}{RT}\left[w_S + \Delta A(a_S + N\gamma) + NA_S\gamma\left(\chi^e - 1\right) - \frac{\Delta Z}{\varepsilon}\right] + \ln\frac{RT}{p_0 V_S} \quad (5.15)$$

Considering this equation, it appears that retention of a solute is anticorrelated with dielectric constant ε of M and both correlated with the surface tension γ of M and the increase in the solute size, particularly its hydrocarbon volume. Elsewhere, this leads to a general chromatographic problem when the solutes possess a large hydrocarbon volume. Indeed, when the hydrocarbon volume of the compounds is large, their retentions are higher. Simultaneously they are less soluble in hydro-organic mixtures. More precisely, the important factor is either related to the contact area ΔA between solute and S (solvophobic theory) or to the cavity volume created in the M and S, which is related to $\tilde{V}$ (partition theory). This was experimentally verified on a lot of homologues by many authors but in particular for TAGs [74, 75]. Experimental conditions must be found that allow smaller contact area or smaller partition ratio to get reasonable chromatographic conditions, that is, retention factor not too high.

The first solution is to work with S with more or less short alkyl graft lengths for decreasing ΔA value keeping higher γ values. This is effective if the analyzed solutes are soluble in partially aqueous mixtures. It was not valid for very hydrophobic solutes (with a low δ value and a high $\tilde{V}$ value).

A second solution is to work with keeping long alkyl chain bonded phases (C_{18}, C_{22}, or C_{30}) where ΔA is highest but using M with smaller values of γ and ε that dissolve the compounds much more. The compounds then have more affinity with M than before and the retention factors decrease. However this is to the detriment of selectivity and consequently leads to a loss of resolution. To compensate this effect, columns with high phase ratio V_S/V_M (i.e., using very highly and densely bonded silica as S) must be employed, which permit the recovery of resolution by an increase of the value of retention factors. Such a possibility is offered by the use of organic solvent mixtures or subcritical fluid mixtures such as M. NARP-LC and reversed-phase subcritical fluid chromatography (RP-SbFC) generally uses at least two solvents, the "weak solvent" (with a total solubility parameter of moderately high value, as well as dielectric constant ($\varepsilon > 30$) and surface tension medium ($\gamma > 30$) values) in which the solutes are not very soluble, and the "strong solvent," which is very solubilizing for the solutes (with a total solubility parameter, dielectric constant, and surface tension of low values).

CH_3CN, CH_3OH, and propionitrile are the most widely used "weak" solvents [76, 77]. CH_3CN and propionitrile participate in the interactions with the π electrons of double bonds and affect the unsaturated species more strongly. The "strong" solvents (or modifiers) used in TAG separation are in order of strength: $CHCl_3$, THF, CH_2Cl_2, IPA, and $(CH_3)_2CO$ [76]. Carbon dioxide is the strong solvent in the case of RP-SbFC and reversed-phase supercritical fluid chromatography (RP-SFC) [78]. It is mixed with either a weak solvent or a mixture of weak solvents or a mixture of a weak and a strong solvent [79].

Heron and Tchapla [75] established, for each saturated solute, a correlation between log *k* versus mobile phase composition curves and the surface tension versus mobile phase composition curves. They demonstrated that the effect of the addition of CH_3CN is increased retention due to the surface tension of the solvent [74, 76]. Studies of log *k* versus the percentage of the strong solvent showed a decreasing retention as the percentage of the strong solvent is increased [54, 74, 75]. The major effects of the strong solvent are to modify the geometrical structure surface of S and to increase the solubility of the solutes. When TAGs contain double bonds, the retention time is shorter in RP-LC mode than expected independent of the organic and hydro-organic nature of M [54, 74, 80]. These authors demonstrated the specific π–π interaction between the π electron systems of TAGs and those of CH_3CN in M. Moreover, the chromatographic selectivity between saturated and unsaturated TAGs is modified according to the amount of CH_3CN in M and is different for alkyl-bonded silicas or S possessing π electrons in their structure [54, 74].

A scale of eluent strength based on methylene selectivity was established, allowing the selection of the nonaqueous M composition [81, 82]. However later, for a given binary or ternary mobile phase, more precise studies showed that the eluent strength is dependent on both the alkyl chain length and the nature of bonded graft on silica [83], the nature of S: PGC instead of alkyl-bonded silica [84–86] as well as the temperature at which the experiments were led in NARP-LC [87, 88] and in RP-SbFC [89, 90]. Thus the eluent strength of mixtures of CH_3CN with THF, CH_2Cl_2, $(CH_3)_2CO$, $CHCl_3$, ethyl acetate [83, p. 97] and MTBE, IPA, butanol, and TTE was determined on C_{18}-bonded silicas.

Most RP-HPLC analyses of TAGs have been done with the column at ambient temperature. However, various workers have shown that changes in column temperature resulted in changes in chromatographic resolution [91–95]. The molecular mechanism of retention was not the same at high and low temperature [74]. The two mechanisms are not independent. One can pass to the other by a change of temperature. The results revealed that higher column temperatures decreased retention times and selectivity for the TAGs, especially for critical pairs with the same partition number (PN) [96]. In addition, various linear relationships between the logarithm of the retention factor and the selectivity factor for TAGs and various column temperature functions could be used to establish such variations [97–99]. This demonstrated that the optimized temperature condition is 21°C with a Brownlee C18 column with CH_3CN/CH_2Cl_2 (68/32), 25°C with a Waters ACQUITY UPLC BEH C18 column with CH_3CN/BuOH (74/26), and 12.6°C with a Chromegabond aC22 column (CH_3CN/CH_2Cl_2 60/40) to separate TAGs from soya bean oil. Although lower temperatures result in better separation of the TAGs, the most highly saturated TAGs may precipitate out of M, and consequently many workers have preferred working temperatures of at least 30°C [100]. Furthermore, Singleton and Pattee [94] noted that the number of theoretical plates increased with temperature, enhancing the elution of highly saturated TAGs immensely and yielding narrow, well-defined peaks [93, 96]. The choice of column temperature represents a compromise designed to ensure good solubility of highly saturated TAGs concomitantly with good selectivity of critical pairs with the same PN.

Elsewhere it was demonstrated that working at 150°C instead of 50°C, the substitution of a weak chlorinated solvents by a weak hydroxylated ones ($CHCl_3$ by butan-1-ol or CH_2Cl_2 by propan-1-ol) keeps the eluent strength constant when changing the temperature [87], permitting the development of green analysis of lipids using high-temperature LC (HT-LC). The overall results of these studies led to the optimization of the experimental conditions to be used for the lipid analysis taking into account the nature of the lipid class function of the type of used chromatography (NARP-LC or RP-SbFC) [79].

The systematic study of retention of a lot of saturated and unsaturated long-chain TAGs ($n_C > 14$) has been made on different S (C_{18}, C_{22}, 1-phenylpropyl, mixed C_{18}–CN-bonded silicas, PSDVB) with various nonaqueous binary and ternary mobile phases [76]. It allowed to determine optimal and simple conditions of analysis of TAGs of fats and oils and cerides of waxes (WE) [101]. The use of combined optimization criteria to judge the quality of an entire chromatogram led these authors to propose the use of one column with a polymeric behavior with a binary and isocratic M (CH_2Cl_2–CH_3CN 32–68). This was confirmed later during a computerized optimization in NARP-LC conditions using four gradient runs for providing the optimum composition of M when comparing the powerful mixtures of CH_3CN with either CH_2Cl_2, acetone, or ethyl acetate [102]. A very recent alternative of these optimal conditions is provided when using a binary and isocratic M (BuOH–CH_3CN 26–74) with a UHPLC C18-bonded hybrid silica [88].

Using propionitrile instead of CH_3CN, Podlaha and Töregård [103] demonstrated better resolutions with a shortest analysis time. For fats such as the main FA in the range 6–14, the isocratic M composition must be adapted. In the same study the authors found a 27–73 CH_2Cl_2–CH_3CN mixture as optimal conditions [98].

Elsewhere Hierro *et al.* [104] showed that using an elution gradient with a single monomeric C18-bonded silica and M of CH_3CN–CH_3COCH_3 is better to resolve the short and medium-chain TAGs. At last a flow rate gradient [105] and a temperature-programmed gradient [92] were proven very efficient for the separation of highest saturated TAGs in NARP-LC. As for different homologous series, the comparison of retention mechanisms of homologous TAGs in NARP-LC [82, 106] showed a discontinuity in log k versus the solute carbon atom number for a given critical number of carbon atom, which is characteristic of the bonded chain length of S [15] and dependent on the temperature at which the experiments were led [107]. The selectivity is highest for shorter TAGs than for longer TAGs in NARP-LC [108] as well as in RP-SbFC [109]. A molecular model of interaction involving a close contact between the hydrocarbonaceous parts of both the solute and S when the chain length of solute is smaller than the one of S could be evocated for rationalizing why the lipids with unsaturated *trans*-FA residues are experimentally more retained than their corresponding *cis* isomers [59]. The conformationally more elongated hydrocarbonaceous chain of *trans* isomers leads to a possible tightest contact between the FA alkenyl chain and alkyl chain of S than in the case of *cis* isomers, which is conformationally less linear. For the same chain length, this involves more retention for a *trans* isomer than for a *cis* isomer. This also allows to understand why the influence of two or three *cis* double bonds on retention was experimentally found not equal to twice or thrice the influence of a single *cis* double bond [54, 82].

The mechanisms for separation of TAGs in oils by RP-LC were summarized in two chapters of this book [98].

The mechanisms for separation of Cer and SP were also reported on C_{18}-bonded silica and PGC using NARP-LC [70, 110, 111]. The inversion of elution order of phytosphingosine and sphingosine on PGC, when changing M composition, was justified by the linear solvation energy relationships. It appears that it is due not only to the weak solvent nature but also to the stronger one [111].

By extension it seems that all the conclusions, deduced from the analysis of different structural TAGs and Cer, were also valid for the other classes of lipids.

5.3.5 Behavior of Lipid Separation in Reversed-Phase Sub- and Supercritical Fluid Chromatography

An alternative method for separating TAGs is RP-SFC or RP-SbFC. In comparison with NARP-LC, by combination of temperature and pressure effects, selectivities were modified, permitting best separation of some TAG pairs that were difficult to separate in NARP-LC [78, 90, 112]. However, there was a loss of separation for other pairs of TAGs. The use of modifiers avoids the use of density and temperature elution gradients and enhances the ability to modify specific separation. Very interesting studies were developed for optimizing the separation of lipids in RP-SbFC: when adding organic modifiers with high dielectric constant ($\varepsilon > 30$) to supercritical carbon dioxide in a range of 0–40%

v/v, the retention of high molecular weight compounds, in particular the lipids, is beginning by a decrease, and then after passing by a minimum, it is following by an increase. Curvatures of such curves are not strong. Thus for a modifier content between 0 and 40%, it could be considered that the retention of solutes does not change greatly. Consequently, an isoeluotropic area was defined in which the retention factor of a solute is nearly constant. As the chromatographic efficiency does not vary greatly with addition of modifier in RP-SbFC, this is the variation of selectivity that mainly governs the separation in this isoeluent area. By drawing a window diagram of selectivity, it was easy to determine the percentages of organic modifier that lead to the best separation in the shortest time of analysis. The same concept was also demonstrated as valid for the low level of organic modifier (≤10%) when studying the retention of lipids between 5 and 72°C. The effect of pressure on the separation was almost negligible. Therefore, it was possible to tune the selectivity of separation between unsatisfactorily resolved compounds while maintaining the analysis time when changing the % of organic modifier as well as the temperature of analysis. This methodology was successfully applied to optimize the separation of TAGs on C_{18}-bonded silica. These effects could also be used to fractionate complex mixtures such as fats. Different fractions could be collected for identification purpose or for reanalyzing under more selective chromatographic conditions.

Thus the particular retention behavior of the C_{16} chains in comparison with that of the C_{18} chains enables this fractionation. Finally because the super- or subcritical fluids have low viscosities, a great number of columns could be coupled to increase the total number of theoretical plates and therefore to improve the separation quality. The maximum number (equal to 7) was selected to keep the inlet pressure below 35 MPa. Compared with NARP-LC, one of the advantages of RP-SbFC is its possibility to analyze more easily the lipids with long saturated chain. In the opposite, NARP-LC was more efficient for the separation of isomeric lipids with the same total number of carbon and the same total number of double bonds as well as isomeric lipids differing only in the location of the double bonds in their constitutive FA residue. From this point of view, NARP-LC and RP-SFC or RP-SbFC are complementary techniques for analyzing TAGs [79, 112].

The same type of study was devoted to the development of an RP-SbFC analytical method for Cer analysis. Five C_{18}-bonded silica columns in series allowed to separate Cer with the same total number of carbon containing unsaturated FA residues when the distribution of CN of the two chains is very different [35].

At last, the behavior of waxes (WE) was also studied in RP-SbFC. The authors used only one C_{18}-bonded silica column. The separation of ester when the distribution of CN of the two chains on ester function is inverted was not demonstrated [113].

5.3.6 Multimodal Chromatographic Systems

Although the use of monodimensional chromatographic techniques can often provide useful information in lipidic matrices, a fully comprehensive analytical view of these samples may be attained by combining two independent separation steps with different selectivities. Examples are Ag-LC/NARP-LC [46]. Several multimodal chromatographic tandems for the analysis have been reported [114–120]. Sandra *et al.* [120] developed an automated online comprehensive RP-SFC × Ag-SFC system using octadecyl-bonded silica and silver-loaded S in the first and second dimensions, respectively. At last comprehensive NP-SFC × RP-LC, which is a valuable alternative for TAG separations by the implementation of normal-phase SFC and RP-LC in a comprehensive configuration, has been developed [121]. Recently separation of TAGs in fish oil was performed by comprehensive and offline supercritical fluid chromatography combined with RP-LC [122]. The most common use of multidimensional chromatography is for the pretreatment of a complex matrix in an offline mode.

As reported previously, two-dimensional separation is based on efficient separation systems (S/M) displaying strongly different selectivity in each dimension. The more selectivity difference observed, the more orthogonal the system under consideration. Sometimes it could be interesting to determine the best couple of RP stationary phases displaying the same type of selectivity. For example, such a couple is constituted by Synergy Max and COSMOSIL Cholester stationary phases. This was deduced from the study of the results of isocratic TAG selectivity tests of RP stationary phases (led on all the TAG from soya and black currant seed oils) [123].

As waited, the overall retentions on these two S were correlated, but if considering only TAG with the same equivalent carbon number (ECN), an efficient orthogonal separation appears in the ECN range 36–54, which permits a very easy characterization of TAG isomers (see Figure 5.3).

5.3.7 Identification of the Molecular Species

One of the most challenging research areas in RP-HPLC, RP-SbFC, and RP-SFC of lipidic samples is the identification of the molecular species (congeners of a lipid class) in the peaks on the chromatogram.

Even with MS coupling, the usual method for identification is the comparison with standards analyzed in the same chromatographic conditions. Unfortunately, the number of available standards of pure congeners from each class of lipids is very small compared with all possible congeners in natural samples, making the comparison difficult. To overcome this limitation, the supplementation of unknown fats by another fat of well-known composition such as standard mixture could be used. This was successfully applied for the identification of TAGs of oils [124].

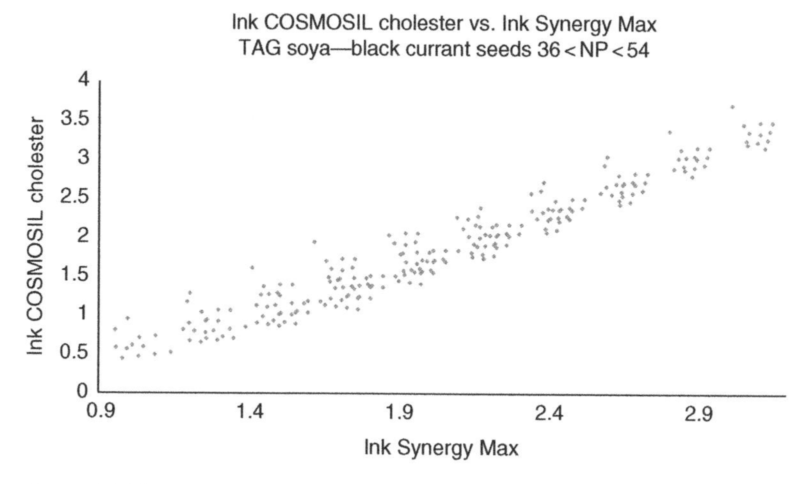

Figure 5.3 TAG retention on two complementary RP stationary phases.

The second mode of identification is the use of retention laws governing RP-LC, RP-SbFC, and RP-SFC. Historically this was used in first for TAGs, but, in fact, the proposed methodology is more general and it was later used for other classes of lipids such as Cer, PL, and cerides (WE). Thus, in the text, most of the examples will concern TAGs.

In NARP-LC several authors have suggested the use of equations to predict the composition of mixtures of TAGs. Wada *et al.* [48] were the first to establish a parameter termed the PN to characterize TAG molecules. The PN is determined by PN = CN-2 DBN where CN is the total number of carbons and DBN is the number of double bonds in the FAs constituting the TAG molecule [49, 76, 125, 126]. They found that on reversed-phase columns, TAGs eluted in increasing order of PN, permitting the prediction of the elution order. This was also valid in subcritical fluid chromatography (SbFC) [127] as well as for RP-LC analysis of both Cer [70] and PL [128, 129]. Moreover specific behaviors were pointed out in RP-SbFC about the TAGs and Cer having the same PN, even if the retention is ordered following PN. In RP-SbFC the retention increases following the DBN, when the opposite effect is obtained in NARP-LC [127, 130].

A very elegant method of TAGs pre-identification in RP-SbFC was achieved, which uses the differences in retention behavior related to TAG structure and to the nature of subcritical mobile phases. These retention differences expressed in terms of ratio of retention factors were produced by the small variation of either outlet pressure (8–12 Mpa) or modifier percentage (7–8%) or of temperature (16–20°C). Discriminating variations of retention were obtained without producing retention order inversions. The ratio of retention factors versus the logarithm of retention factor of TAGs including standards and unknown was drawn. In such maps, all TAGs with the same CN fall on a same line and all TAGs with the same PN fall on another line. The intercept of the two lines led to the identification of unknown TAGs. Whatever the column aging, this method allows the determination of the CN and DBN. Among numerous structures these two criteria restrict the structural hypothesis to three or four possibilities and sometimes only to one [90].

In a general way, the first difficulty is the separation of mixtures of congeners composed of FAs differing only minimally in chain length and degree of unsaturation, that is, compounds with the same PN value. Despite the improvements in efficiency achieved using C_{18} columns with a particle size of 3 μm or less connected in series and elution gradients with appropriate organic modifier, poorly resolved or even unresolved congener pairs and groups still exist, especially in highly complex mixtures of TAGs of natural fats such as milk fat and fish oils [92, 131].

The second difficulty involves the simultaneous analysis of congeners differing markedly in molecular structure, that is, short-, medium-, and long-chain congeners with differing degrees of unsaturation spanning a broad range of PN values. The problem lies in achieving simultaneously both good resolution of the least strongly retained congeners and reasonable analysis times and elution in narrow chromatographic bands for the most strongly retained congeners. In other words, the problem lies in achieving efficient chromatographic systems [91].

The third difficulty involves the detection of congeners at the column outlet. This situation has made necessary to develop new systems such as charged aerosol and different available light scattering detectors and combined HPLC-MS. Even peaks that appear sharp and distinct using an evaporative detector can actually contain multiple overlapped components.

With the development of high-resolution columns online, coupled or not, PN only defines a zone in the chromatogram in which several critical pairs may be distinguished. In fact, it was found later that the coefficient is close but not equal 2 for all acid residues depending on their number and position of double bonds and their chain length. Precisely, it is not constant but decreases with the increasing number of double bonds and also when CN decreases [59, 132–135]. Consequently with the efficient S available nowadays, separation of great number of TAGs with same PN is easily obtained. For example, there is easy separation of groups such as OOO, OOP, OPP, and PPP (with same PN). Also separation has been achieved for pairs such as LLL–OLLn (with same PN, CN, and DBN) [37, 101]. Finally, the most difficult and closest structural pairs with same PN, CN, and DBN but differently located along the same hydrocarbon chain have been resolved after chromatographic optimization, leading to the identification of TAGs with Δ6 or Δ5 FA instead of Δ9 FA [80, 133, 136] as well as FA for which two double bonds are in *trans* configuration instead of in *cis* configuration [101]. Combining RP-LC with evaporative detector and MS detector may furnish indispensable information on the structure of congeners of complex mixture such as TAGs, for instance [137, 138]. Lisa *et al.* reported also the influence of different positions of double bonds in Δ5-UPIFAs on the retention behavior of TAGs [139]. In recent studies, the identification of very rare TAG congeners was reported. They concern TAGs possessing an odd FA residue in their structure (heptadecanoic and nonadecanoic acids) in the particular case of TAGs analysis for characterizing the organic residues that could be found on archeological objects [140] as well as TAGs with branched odd FA residue or very rare polyunsaturated FA residues: 20:1 (Z11), 20:2 (Z5,Z11), 20:3 (Z5,Z11,Z14) [138]. In other areas Holcapek *et al.* reported the identification of various TAGs including ones containing margaric (C17:0) and heptadecenoic (C17:1) acids [141].

Separation of congeners of PLs with FA possessing double bond located in different position in the chain was also reported [142].

Goiffon *et al.* [143, 144] described a very useful method based on the selectivity (α) of TAGs relative to triolein. In isocratic mode a linear relationship exists not only between the number of double bonds (DBN) of TAGs and log α but also between the total CN of TAGs and log α. In the gradient elution mode, this was also valid when drawing α versus DBN and α versus CN plots. Using this method Goiffon *et al.* identified short and common TAGs, Perrin and Naudet [145] established the elution order of 120 TAGs, then Stolyhwo *et al.* [106] identified high molecular weight TAGs, and finally, Acheampong *et al.* identified TAGs isomers whose constitutive FA residues are double bond position isomers (Δ5, Δ7, Δ11) and configuration isomers (anteiso vs. linear saturated FA). This leads to the unambiguous identification of 58 new TAGs [138]. MS is inefficient for precise characterization of isobaric TAG; thus, in this case, Tamba-Sompila *et al.* [137] proposed a method of identification of TAG, which mainly contains polyunsaturated positional isomers and configuration FAs C18:3.

Podlaha and Töregård [146] developed another procedure to identify TAGs based on the ECN. The ECN of each TAG in the sample is the ECN of the hypothetical saturated TAG that has the same retention time. When CN are plotted against ECN, straight parallel lines are found for different unsaturated TAGs. Thus, a theoretical prediction can be made, which has become a useful tool for TAG identification. Herslöf *et al.* [147] estimated the theoretical ECN for unsaturated TAGs on the basis of their relative retention times from an experimental linear relationship between relative retention times and the CN. The linear relationships of both log α versus DBN plots and CN versus ECN plots were used to predict the structure of different TAGs present in complex fractions of oils [148]. For instance, using this method, 84 TAGs were identified in peanuts. Later the same methodology was used for the identification of molecular species of phospholipids (PL) in RP-LC taken in reference to the ECN of FA in position 1 on PL skeleton [142] and for the identification of congeners of Cer in NARP-LC [70]. Elsewhere, the straight parallel lines CN versus DBN were also giving the order of retention of subclasses of congeners. The very recent results gave the following behavior for TAGs: 000 > 001 > 002 > 011 > 003$_{(\Delta_{5})}$ > 003 > 012 > 0b12 > 111 > 013$_{(\Delta_{5})}$ > 013$_{(\Delta_{9})}$ > 0b13 > 112$_{(\Delta_{5})}$ > 022 > 112 > 023$_{(\Delta_{5})}$ > 023 > 113$_{(\Delta_{5})}$ > 122$_{(\Delta_{5})}$ > 122 > 03$_{(\Delta_{5})}$3$_{(\Delta_{5})}$ > 033$_{(\Delta_{5})}$ > 033 > 123$_{(\Delta_{11})}$ > 123$_{(\Delta_{5})}$ > 222$_{(\Delta_{5})}$ > 123 > 222 > 22$_{(\Delta_{5})}$3$_{(\Delta_{5})}$ > 223$_{(\Delta_{5})}$ > 13 3 > 223 > 23$_{(\Delta_{5})}$3$_{(\Delta_{5})}$ > 233$_{(\Delta_{5})}$ > 3$_{(\Delta_{5})}$3$_{(\Delta_{5})}$3$_{(\Delta_{5})}$ > 333, which is more general than the previously published ones that did not include the double bond position isomers of FA [146, 148], where the number indicates the number of double bonds of each FA in the TAG molecule. This confirms previous studies on TAGs and FAs that reported that the influence on the retention of three double bonds is not thrice the influence of a single double bond [54, 149]. Moreover it depends on the position of double bond on the alkyl chain. The same study gave the following behavior for Cer structures and for phospholipids: D20 > S20 > P20 > D18 > S18 > P18 > D16 > S16 > P16 [74] and 18:1n-9 > 20:3n-9 > 20:3n-6 > 22:5n-6 > 22:5n-3 > 18:2n-6 > 20:4 > 22:6 > 16:1 > 18:3n-3 > 20:5 [150].

The congener prediction process becomes tremendously complicated when the fat contains a large number of different FAs, since the number of possible congeners can be extremely high. Therefore, as a second part of the prediction process from the ECN, some authors have proposed application of the equations developed by Takahashi *et al.* [151]. These workers developed a matrix model in which the variables were the CN and DBN of each FA esterified to the glycerol molecule. In spite of all this work, TAG identification is still quite problematic when the ECN is the same or similar.

5.3.7.1 Methodology for Identification of Congeners

Whatever the chromatographic techniques used, the methodology for identification of congeners of different classes of lipids follows in general the three steps:

1) The determination of (FA) composition [152].
2) The identification by comparison with standard analyzed in the same chromatographic conditions. Unfortunately, comparatively to all possible congeners, there are far fewer numbers of available pure congeners for doing that. To overcome these limitations, supplementation of unknown fats by another fat of well-known composition such as standard mixture could be used [80]. Unfortunately, even using these two methods, a lot of congeners remain unidentified.
3) Use of predicting diagrams log *k* versus CN, log *k* versus PN, and log *k* versus DBN.

None of the three methods cited when used alone permit unambiguous identification of all congeners because they are in great number in oils and fats and not always well separated, especially mainly those with similar structures (same CN, same PN, same DBN) because their retention factor (or gradient retention times) could be close together. These methods must be used to complement each other in order to ensure complete and unambiguous characterization.

5.4 Literature Review: Early Use of Corona CAD in Lipid Analysis

As already stated, lipids are essential constituents of cells where they play diverse important roles. They are present in both vegetable and animal kingdoms, and this literature review covers both biosciences and food chemistry. The specific physicochemical properties of lipids and their biological importance explain their use in the pharmaceutical industry that is also covered as a specific item.

For many years, most of the lipids are analyzed in HPLC with detectors adapted for nonvolatile analytes, such as ELSD or MS. Since the development of Corona CAD, numerous papers deal with the comparison of these three aerosol detectors.

5.4.1 Biosciences

C. Merle *et al.* deal with the analysis of stratum corneum lipids [153]. The skin owns a barrier function, which can be altered by a modification of its lipid composition. They developed a new simple quantitative method in NP-LC coupled with a Corona CAD or an ELSD to separate the three major lipid classes of the stratum corneum: FAs, Cer, and cholesterol St. The study points out that Corona CAD has several advantages compared with ELSD, such as repeatability, accuracy, and precision, and allows a study of small concentrations. This proposed method was successfully applied to human stratum corneum lipid extracts obtained by a scraping method.

Other studies dealing with Cer have been reported, especially neurolipids [154]. It has been shown that Cer can be easily profiled with Corona CAD using gradient elution HPLC. In the same way, the ability of Corona CAD to detect and assess purity of nonvolatile FAs has been demonstrated, which can be of value in studying the effect of some FAs on brain function and development.

PLs have been studied with Corona CAD [40, 155–157]. R. Godoy Ramos *et al.* have compared the performance of Corona CAD with ELSD for the analysis of *Leishmania* membrane PL classes in NP-LC with a gradient method [40]. Although the response of both detectors can be fitted to a power function, Corona CAD response can be described by a linear model when a restricted range was used (30 ng to 20 µg). It was also found to be more sensitive at lowest mass range than ELSD. With Corona CAD, the limits of detection ranged from 15 to 249 ng, and the limits of quantification ranged from 45 to 707 ng, whereas with ELSD the values were three times higher. At last, by providing improved detection of PL at low levels, it has been shown that the chromatographic profiles of *Leishmania* cultures obtained with Corona CAD were more informative than with ELSD.

Moreau has developed a method to quantitatively analyze GL and PL in extracts of plant material in HPLC with Corona CAD [155]. With this system, the minimum limits of detection were 25 ng, and the mass-to-peak area relationship was evaluated in the range from about 25 ng to about 10 μg per injection.

At last, concerning PL, the analysis of positional isomers of PI (kind of natural PL found in eukaryotic organisms) has been reported with an HPLC-Corona CAD method [156, 157]. St, which also are membrane components of all eukaryotic cells, were quantified by HPLC-Corona CAD in studies of activity of several antifungals [158, 159].

A study of Corona CAD in the determination of bile acid levels in the upper gastrointestinal lumen has been published [160]. Intraluminal bile acids promote lipid absorption and are essential for adequate absorption of fat-soluble vitamins. A simple, fast isocratic HPLC-Corona CAD method was developed, validated, and applied for the determination of individual bile acids in human gastric and duodenal aspirates. Compared with previously published HPLC-UV or HPLC-ELSD methods, HPLC-Corona CAD method has a number of advantages: low intraday precision (<6%), high recovery (>98.2%), and low minimum limits of detection (<0.6 μM), important especially when aspirates are collected from the fasted stomach or from the colon.

A Japanese team has shown that Corona CAD was a good alternative detection for the determination of impurities in 17β-estradiol reagent [161]. There was correspondence between the results of quantification using Corona CAD and UV detection.

5.4.2 Food Chemistry

Numerous HPLC methods have been developed for the quantitative analysis of plant or animal lipids. Beginning in the mid-1980s, the first HPLC methods were developed for the quantitative analysis of lipids using "universal" detectors such as flame ionization detector (FID), refractive index detector (RID), and ELSD. FID and RID technologies were quickly surpassed by ELSD technology, and the manufacturing of FID ceased in the mid-1990s [155]. In the last 10–20 years, numerous methods have been developed for the analysis of lipids by HPLC-ELSD, and new applications using Corona CAD are now reported.

The first HPLC method to be evaluated with Corona CAD was a normal-phase gradient elution method developed to quantitatively analyze the major nonpolar lipid class components (TAGs, FFAs, phytosterol esters SE, and free phytosterols St) in vegetable oils [155].

Similarly with the ELSD, methyl oleate was not detected; methyl esters were partially evaporated at a detector temperature of 40°C and completely evaporated at higher temperatures. With this HPLC method, the minimum limits of detection with Corona CAD were about 1 ng, and the mass-to-peak area ratio was nearly linear from the range of about 1 to 20 ng per injection. In general, the graph of analyte mass to peak area is nonlinear for both ELSD and Corona CAD, although it sometimes can be quite linear if one examines small ranges of masses [155].

Another method was a reversed-phase method developed to separate the molecular species of TAGs and other nonpolar lipids [155]. The baseline was noisier with reversed-phase system than with the normal-phase systems. But no explanation was given. The minimum limits of detection of Corona CAD with lipids varied with different M. Using solvent systems that were predominantly hexane, the minimum limits of detection of TAGs, CE, and free St were about 1 ng per injection. This compares with 50–100 ng minimum limits of detection with an ELSD. Using hexane predominantly, the mass-to-peak area ratio was nearly linear from the range of about 1 ng to about 10 μg per injection. The mass-to-peak ratio was substantially more linear with Corona CAD than with ELSD, especially around the lower limits of detection. Three other solvents commonly used for HPLC analysis (CH_3OH, IPA, and CH_3CN) caused higher levels of background noise and higher minimum limits of detection.

Lisa *et al.* [162] have used the Corona CAD in NARP-LC in the gradient mode to develop a simple quantitative method, without the need for response factors, for the analysis of complex natural TAG mixtures from plant oils. M compensation was applied, by mixing of the column effluent with the inversed gradient delivered by a second HPLC pump, for the suppression of the response dependency of the analytes on M composition. The developed method was applied for the quantitation of TAGs in oils from sunflower, soybean, grape seed, sesame, linseed, olive and palm, and oils containing mainly saturated and unsaturated FAs with 16 and 18 carbon atoms. The FA content calculated from this method is in good agreement with the results obtained using a quantitative APCI-MS method with knowledge of the response factors. Good reproducibility and excellent limits of detection (about 1.2 ng) were achieved. Compared with a previously published quantitative method, based on the knowledge of response factors, the proposed method does not need TAG standards and is faster because it does not need calibration curves for the determination of relative concentrations.

Recently, a RP-LC-Corona CAD method in the gradient mode has been proposed for the quantification of TAGs in olive oils and in food products containing olive oil using tristearin as an internal quantification standard [163]. The proposed method presents two advantages in comparison with the official method adopted by European Commission for the identification of TAGs in olive oil: lower analysis time and better sensitivity. Moreover, no M compensation was necessary because response factors of the TAGs selected show a variation less than 10%. The same authors used this method for an efficient differentiation of olive oil and several types of vegetable oils using chemometric tools [164].

In our laboratory, we work on lipids in HPLC-ELSD, especially TAGs, for several years [101]. A method development has been proposed for quantitative analysis performed without any standard using ELSD [165, 166]. This has been developed in another part of this chapter.

We have shown that the experimental property (Equation 5.17) could be generalized to the other type of aerosol detector, the Corona CAD [167]. We have also shown that Corona CAD exhibits a linear response over a narrow concentration range, a decade larger than that exhibited by ELSD. The sensitivity of CAD is better at low concentrations, while that of ELSD is better at high concentrations.

Saberi *et al.* have worked on diacylglycerols (DAG), which are found as natural component of various fats and oils. They determined the physicochemical properties of various palm-based DAG oils in comparison with their corresponding palm-based oils [168] and specifically in the preparation of shelf-stable margarine [169]. The same authors studied the crystallization behavior and kinetics of these palm oils in blends with palm DAG [170].

In all these three papers, MAG and DAG composition analyses have been determined by HPLC-Corona CAD method.

Lipid oxidation is a process that is one of the main causes of deterioration in the quality of meat. So, Cascone *et al.* have developed a model system to monitor the changes in the composition of meat PL, considered to be highly susceptible to oxidation. Corona CAD was proved to be a good complementary technique in this investigation of lipids [171].

A comparison of the detection of vitamins (tocopherols and tocotrienols) with fluorescence detector and the CAD has been led [155]. Under the employed conditions, in spite of the low minimum limits of detection of Corona CAD, the fluorescence detection is still the HPLC detection method of choice for the quantitative analysis of these compounds.

Lastly, an HPLC method using Corona CAD has been developed for the analysis of terpene lactones in a ginkgo leaf extract. Linearity and limit of detection were found excellent [172].

5.4.3 Pharmaceutical Sciences

5.4.3.1 Emulsions

Emulsions are liquids containing suspended droplets of FAs. They are of interest in pharmaceutical and vaccine applications for several reasons. For instance, a common challenge in drug discovery is overcoming drug insolubility or instability in order to increase the bioavailability of the active compound. Emulsions can help solubilize lipophilic drugs and decrease aqueous instability by associating them with a hydrophobic oil phase; in addition, emulsions offer a slower release of drug from the formulation. Moreover, since emulsions are particulate in nature, they have longer biological residence times and are more effectively phagocytosed by scavenging cells than aqueous formulations. Thus they can increase drug or vaccine uptake into cells [173].

Squalene is a linear triterpene that is extensively utilized as a principal component of parenteral emulsions for drug and vaccine delivery. Squalene and its hydrogenated form, squalane, have unique properties that are ideally suited for making stable and nontoxic nanoemulsions. Emulsions containing squalene facilitate solubilization, modified release, and cell uptake of drugs, adjuvants, and vaccines. Because of these characteristics, numerous squalene-based emulsions have been effectively developed for drug and vaccine applications. Emulsion stability can be optimized by appropriate selection of oil, surfactants, and aqueous components as well as processing conditions. In a review, Fox evaluated the physicochemical and biological properties of squalene-containing emulsions in the context of parenteral formulations and examined analytical techniques for the characterization of squalene emulsions [173].

It has been pointed out that appropriate quantification of squalene in vaccine or pharmaceutical formulations is essential for manufacturing quality control and regulatory considerations. Detection or characterization of squalene is possible using various analytical methods. For example, squalene can be quantified by RP-LC with UV, light scattering, or refractive index detection. Moreover, Corona CAD with HPLC has been used to effectively detect squalene. A method has been developed by Fox *et al.* for the detection of squalene and impurities from shark or olive sources [174]. In fact, this technique demonstrated a lower limit of detection (>0.2 ng) than evaporative light scattering detection or atmospheric pressure chemical ionization MS.

5.4.3.2 Liposomes

Liposomes are vesicles in the nanometer size range composed of lipid bilayers, primarily PL, which resemble natural membranes. They are able to entrap water-soluble substances inside their aqueous core and to incorporate water-insoluble substances into the lipid bilayer. Due to this universality and the ability of liposomes to improve the pharmacokinetics and pharmacodynamics of associated drugs, pharmaceutical companies are increasingly developing liposomes as drug carrier systems for various drug substances, especially those that are highly toxic and/or highly insoluble. The liposomal formulations can be used for anticancer drugs. Therefore, thorough characterization and quantification of the lipids that form liposomes is wished from both investigators and regulatory authorities when the application in humans is considered. The authorities demand full quantitative and qualitative characterization, not only of the active substance itself but also of the excipients within a drug formulation. In this context HPLC has been well established as a fast and precise method to analyze both drug substances and excipients. However, for lipids that are the predominant component in liposomal formulations, conventional UV detection is often not adequate and limited to chromophores, while other methods have significant limitations in precision, sensitivity, and dynamic range. C. Schönherr *et al.* [175] have showed in a study that for the detection of lipids in liposomal formulations, Corona CAD has the advantage to be independent of the chemical properties of the analytes. This method was used for the quantification of five lipids: cholesterol, α-tocopherol, PC, and mPEG-2000-DSPE. The superiority of this method over UV detection was demonstrated. No absorption effects of the organic solvent in M interfering with the lipid signals were observed with Corona CAD. Corona CAD showed good linearity ($r^2 > 0.90$) for all liposomal compounds. The acceptance criteria for precision including repeatability were met. The average recovery for each of the excipients of the liposomal formulation was in the range of 90.0–110%.

For the purpose of efficient hemostasis, Tokutomi *et al.* [176] have previously developed ADP-encapsulated liposomes modified with a dodecapeptide (HHLGGAKQAGDV), H12-(ADP)Lipo. This liposome actually enhanced platelet aggregation *in vitro* and showed significant hemostatic effect *in vivo*. Since fibrinogen (Fbg) is abundant in the bloodstream, it is unclear why this liposome binds platelets so efficiently, overcoming the competition with Fbg. Therefore, Tokutomi *et al.* investigated the relationship between H12 density on the liposome and the binding ability to platelets and evaluated the inhibitory effect of Fbg on the binding of H12-(ADP)Lipo to platelets. The density of each component of liposomal lipids was measured by HPLC-Corona CAD.

5.4.3.3 Surfactants

Nonionic surfactants are widely used in pharmaceutical formulation as a result of its properties of solubilization, reduction of surface, and interfacial tension or wetting. Nonionic surfactants are not charged, nor do they have a chromophore. Lobback *et al.* [177] have developed a quantitative method of two surfactants, Tween 80 and Span 85, in HPLC-ELSD and HPLC-Corona CAD. Corona CAD offers tenfold advantage in sensitivity, as well as greater linearity. Quantitation is also possible because, although the output of the Corona CAD follows a polynomial second order throughout its entire operating range, there is a linear relationship in the range of 5 μg/mL through 0.1 mg/mL. Corona CAD/MS was developed to characterize the nonionic surfactants D-α-tocopheryl polyethylene glycol (PEG) (1000) and sucrose laurate by A. Christiansen *et al.* [178]. The molecular structure and heterogeneous composition resulting from different isomers and various lengths of PEG chains make it difficult to develop sensitive and specific analytical methods for both surfactants. Sucrose laurate did not possess any chromophore; thus UV detection was not applicable. Therefore, Corona CAD and MS have been used for determination. The aim of the study was to characterize these nonionic surfactants and to examine chemical stability at pH 1.0 and 37°C, simulating harsh gastric conditions. It was shown that both compounds are liable to degradation under these conditions. Sucrose monolaurate exhibited a massive degradation within 8 h incubation due to cleavage of the glycosidic bondage. About 50% of sucrose monolaurate broke down, whereas a marginal amount of 3.4% of TPGs degraded into D-α-tocopheryl succinate and the associated PEG chain.

5.4.3.4 Contrast Agents

Recent ultrasound contrast agents (UCAs) generally consist of polymer microcapsules encapsulating gaseous perfluorocarbons. These agents are injected intravenously in a strategy to better visualize specific tissues. Due to their rather hydrophobic surface, these microcapsules are quickly eliminated by the reticuloendothelial system and end up in the liver. To avoid a rapid clearance from the systemic circulation, it has been shown that covering particle's surface with PEG is very efficient. PEG chains covalently linked to PL are often used in the preparation of lipid or even polymer colloidal particles to avoid recognition and clearance by the reticuloendothelial system and to increase their plasmatic half-life. R. Diaz-Lopez *et al.* [179] have developed the first direct quantitative method of a typical pegylated phospholipid, 1,2-distearoyl-sn-glycero-3-phosphoethalonamine-*N*-[methoxy(polyethylene)-2000] (DSPE-PEG2000), associated with polymeric microcapsules of perfluorooctyl bromide (PFOB) using RP-LC-Corona CAD. This method proved to be selective and sensitive; pegylated PL associated with microcapsules, as well as the PL and total PL in the suspensions, were successfully quantified in three different preparations of microcapsules. Calibration standards consisted of plain microcapsules and pegylated PL (DSPE-PEG2000) in the concentration range of 2.23–21.36 μg/mL (0.22–2.14 μg injected). Calibration curves were evaluated with two different models, linear and power model. The power model described experimental values better than the linear model for pegylated phospholipids with Corona CAD. The correlation coefficient for the power model was 0.996, and limits of detection and quantification obtained were 33 and 100 ng, respectively. This method proved to be selective and sensitive; the accuracy of the method ranged from 90 to 115% and the relative standard deviation was <5.3%. Pegylated PLs in the suspensions were successfully quantified in three different preparations of microcapsules [179–181].

5.4.3.5 Determination of Degradation Product and Impurities

Because Corona CAD presents a good sensitivity, this detector is used for the determination of impurities in formulations in many studies.

For example, Nair and Werling [182] have developed a method to quantify FFA in a pharmaceutical suspension formulated with PL stabilized with Lipoid

E80 that was used as a model system to study the physicochemical stability of an aqueous PL-based suspension for injection. The hydrolysis of the PL during the storage at elevated temperatures necessitated the development of suitable HPLC method for the determination of FFA content in the suspension samples. HPLC methods using two of aerosol detectors were investigated for the aforementioned purpose. Reversed-phase separation coupled with either ELSD or Corona CAD was used. A comparison of the methods indicated that the Corona CAD method provided better sensitivity, precision, recovery, and linearity for the parameters evaluated. The percent RSD ranged from 0.4 to 3.0% for the Corona CAD method and from 0.2 to 11.2% for the ELSD method. As a result, this method was chosen for the stability study of itraconazole suspension and has been incorporated in subsequent formulation studies.

Mengesha and Bummer [183] proposed a simple chromatographic method for the simultaneous analyses of PC and its hydrolytic degradation products: lysophosphatidylcholine (LPC) and FFAs. Quantitative determination of PC, LPC, and FFA is essential in order to assure safety and to accurately assess the shelf life of PL-containing products. The quantitative analysis of PC and LPC has been achieved with external standard method. The FFAs were analyzed as a group using linoleic acid as representative standard. Linear calibration curves were obtained for PC (1.64–16.3 μg, $r^2 = 0.9991$) and LPC (0.6–5.0 μg, $r^2 = 0.9966$), while a logarithmic calibration curve was obtained for linoleic acid (1.1–5.8 μg, $r^2 = 0.9967$). Quantitative HPLC analysis showed that 97% of the total mass balance for PC could be accounted for in liposome formulation.

At last, in the case of Corona CAD used in SFC for pharmaceutical analysis, C. Brunelli *et al.* showed that to obtain uniform response in gradient analysis, it is necessary to perform an M flow compensation by placing a T-piece before a back pressure regulator. In this way, the differences in response were significantly reduced from a factor of 2–3 to a factor of 1.2–1.7 [184, 185].

All these results indicate that Corona CAD is a valuable tool for the quantitative HPLC analysis of lipids. The major advantages are its low minimum limits of detection and its nearly linear mass-to-peak area relationship for many types of lipids.

5.5 Calibration Strategies

5.5.1 Calibration Strategies in Quantitative Analysis of Lipids

As in the case of ELSD [186], the Corona CAD [187] response is a function of the increase in the size and number of solute particles obtained after M nebulization and evaporation.

The nebulization process is known to be influenced by M characteristics such as solvent density and surface tension and also by the gas and liquid flow rate entering the nebulizer [188]. The Nukiyama and Tanasawa [189] empirical equation can be used to predict the Sauter mean diameter of the lognormal distribution of the droplets, but, in studies of ELSD response, this equation is much more usable to predict the effect of liquid or gas flow rates [190] than the complex effect of solvent or solvent mixture on the droplet size distribution and the subsequent detector response.

The size of a particular droplet changes during solvent evaporation in the transfer tube and the diameter (d) of the resulting solute particle is

$$d = D_0 \sqrt[3]{\frac{C}{\rho}} \tag{5.16}$$

where D_0 is the initial droplet diameter at the nebulizer exhaust, C the solute concentration, and ρ the solute density.

The two detectors differ at the stage of the dry aerosol detection. For ELSD, depending on the ratio between the wavelength of the incident light and the diameter (d) of a given particle, different light scattering mechanisms may take place. Rayleigh, Mie, and reflection–refraction mechanisms are involved. The smaller the particle, the lesser the amount of light scattered and the higher the predominance of the Rayleigh scattering. When particles grow, for example, during peak rise or for increasing amounts of solute injected, Mie and even reflection–refraction may take place. These two mechanisms lead to an increased intensity of the light scattered. The response shape of an ELSD detector is thus complex and barely predictable. For this reason, the general purpose "power function" is used to model the response

$$y = Am^b \tag{5.17}$$

where m is the injected amount of solute and A and b are the numerical coefficients. The b term of the equation has a special importance because it is connected to the predominant light scattering phenomenon. b values are from 2/3 when reflection–refraction prevails to 2 when Rayleigh scattering predominates. A is associated with the amount of light scattered per unit of solute injected but is not truly sensitivity since sensitivity varies along the calibration curve.

Corona CAD is based on aerosol charging and measurement of aggregate charge: a stream of nitrogen (or air) is ionized by corona discharge and directed into a chamber at the exhaust of the drift tube. In this chamber, impacting the positively charged nitrogen ionizes the particles. After excess ions and smaller high mobility charged particles are eliminated by an ion trap, larger lower mobility charged particles are detected by an electrometer. As with ELSD, the response curve of Corona CAD is often described in terms of a power function (Equation 5.17). An updated description of theory underlying CAD response is provided in Chapter 1.

5.5.2 Classical Calibration (External Calibration, Normalization)

Although the response coefficients of solutes analyzed under the same M conditions are fairly comparable [191], as with other evaporative detectors, Corona CAD shows an important influence of solvent nature on solute response [192]. Gradient elution employing several organic solvents is commonly used in chromatographic analysis as lipid samples usually cover a wide range of polarity and/or present an important complexity.

Figure 5.4 shows calibration curves obtained with either Corona CAD or ELSD for different lipid classes separated using a gradient elution in normal-phase chromatography. Ten microliters of lipid mixture was injected on a 2.1 × 150 mm Inertsil SIL-100 silica column using a solvent program inspired from [32]. The elution order of lipid classes is as indicated in Table 5.6 that summarizes the values of coefficients A and b from Equation 5.17 together with the determination coefficient R^2. As expected, from their principle of operation and advisable from the graphs, the two detectors exhibit nonlinear calibration curves. Corona CAD shows convex curves with a b coefficient of less than unity, whereas ELSD provides either concave or convex calibration curves with b values in the range 0.75–1.36 as a consequence of the predominant light scattering phenomenon.

As stated previously, and especially when using gradient elution, differences in solute response may come from the differences in solvent composition when they leave the column. The solute peak shape also influences the detector response. For identical amounts injected, compounds for which the highest efficiency is measured show an increased concentration at their peak apex. As both ELSD and Corona CAD are concentration dependent through Equation 5.16, both provide increased sensitivity with narrow peaks. The case is encountered here with CB, the most intense peak, for which the retention (and then the solvent composition at the column outlet) is intermediate between Cer and PE.

The mass or concentration dependence of ELSD or Corona CAD is somewhat controversial. In a study comparing ELSD, Corona CAD, and ESI or APCI-MS coupled with HT-LC [193], we demonstrated that under conditions of high chromatographic efficiency, those nebulizing detectors are neither true mass nor concentration detectors. Accelerating the peak elution (by means of increasing the column temperature) increased peak height that is a characteristic of mass detectors but decrease peak area as with concentration detectors. This topic is further addressed in Chapter 1.

The possibility to fit the detector response to a straight line is also reported in several publications. As already studied in detail in a previous publication from our group [40] comparing ELSD and Corona CAD for PL detection by NP-LC, a better fit was obtained with nonlinear regression using Equation 5.17 over straight-line regression. This study uses the classical regression statistics where F ratios are calculated for regression and lack of fit (the comparison of

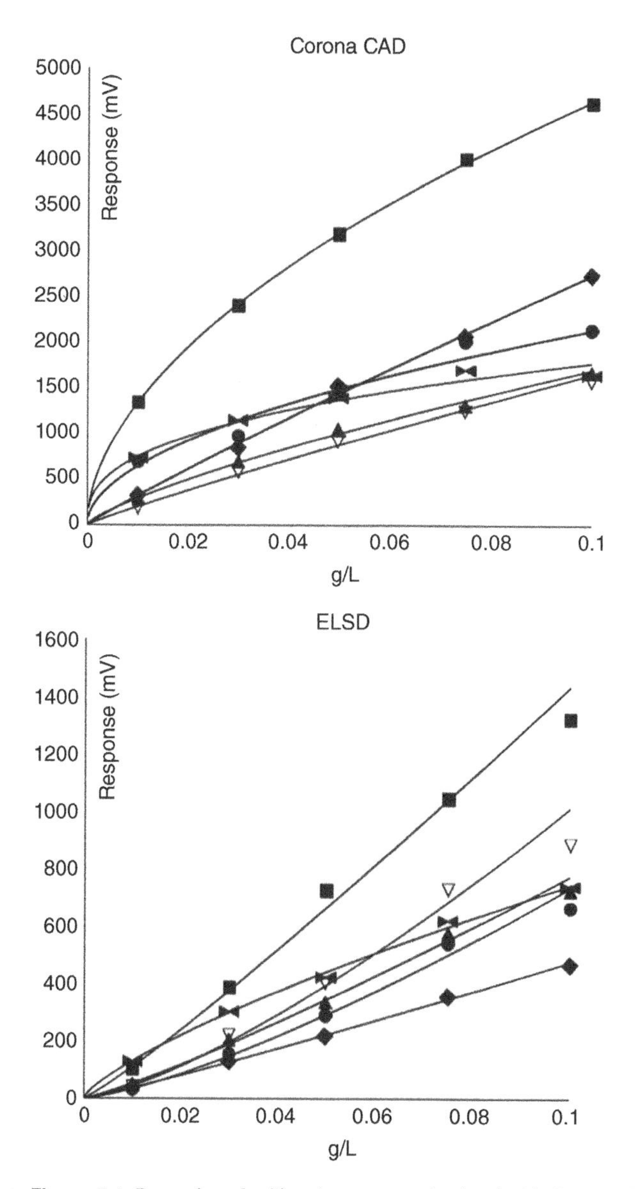

Figure 5.4 Examples of calibration curves obtained with Corona CAD and ELSD.

R^2 is too simplistic for this task). Using ELSD, the nonlinear calibration using Equation 5.17 could be fitted for all of the 10 PLs, whereas the linear model failed to describe the response except in two cases with a probability threshold $p = 0.05$ and 0.09 for lack of fit. Using Corona CAD, the power model could be fitted with acceptable lack of fit for 9 PLs over 10. In one case, the probability

Table 5.6 Coefficients of Equation 5.17 for the calibrations shown in Figure 5.4.

Corona CAD	*A*	*b*	R^2	ELSD	*A*	*b*	R^2
SQL (▽)	13,554	0.91	0.996	SQL	23,295	1.36	0.993
TG (▲)	9,813	0.76	0.997	TG	11,419	1.16	0.996
CO (●)	7,030	0.51	0.958	CO	15,825	1.33	0.997
CER (▶◀)	4,195	0.37	0.977	CER	4,228	0.75	0.998
CB (■)	16,039	0.54	0.997	CB	18,657	1.11	0.995
PE (◆)	23,373	0.93	0.997	PE	5,970	1.1	0.999

threshold $p = 0.03$ was below the accepted risk of 0.05. The b value from Equation 5.17 was found statistically different from unity for 6 PLs, demonstrating a significant curvature of the calibration curve. However, when using the linear model, the goodness of fit was acceptable for 9 PLs over 10 (the remaining one at $p = 0.02$), but the intercept of the calibration line was significantly different from zero for 5 PLs. The linear model could then be used, but the response was not directly proportional to the concentration.

This and similar findings from others [31, 182, 194] suggest that although Corona CAD is intrinsically nonlinear, linear calibration can be used after selection of appropriate domain of linearity. This also suggests that single standard calibration or area (or height) normalization should not be used without prior testing of the linearity statistics.

5.5.3 Calibration in Absence of Standards

In the case of ELSD, several authors have reported that when the coefficient A of Equation 5.17 increases, the value of the exponent b decreases [190, 195, 196]. Recently, it has been demonstrated that the values of these two parameters are correlated [165, 190] and that there is a proportionality relationship between $\log A$ and b (Equation 5.18), which is characteristic of the ELS detector in use:

$$b = \alpha \log A + \beta \tag{5.18}$$

α and β must be determined experimentally and are deduced from the A and b values obtained from the calibration curves of several reference solutes belonging to a given class of chemicals.

Using ELSD, this relationship was verified for different classes of solutes (TAGs, waxes, PEG, antibiotics, parabens or sugars under different nebulization conditions), and several commercial detectors from different manufacturers [165, 197]. Then, a quantitative analysis can be carried out without using pure reference samples for each compound of the involved class. The proposed methodology is composed of three steps.

The **first step** is the determination of α and β parameters defining the general response of a given ELSD, setting of the linear curve: b versus log A.

This step requires to determine experimentally several couples of values (b, log A) from the plot obtained with equation log (Area) = log $A + b$ log m using available standard compounds. These standard compounds can be present or not in the unknown mixture to analyze.

Then, the couples of values lead to the linear correlation of the general response of the ELSD ($b = \alpha \log A + \beta$). In order to get a good precision on the plot of this linear curve, it is necessary to determine a minimum of six different couples (b, log A), covering a wide range of experimental values of b.

The slope α and the x-intercept β are intrinsic characteristics of the ELSD, which is used. The more the selected standard compounds possess chemical structures close to those of the solutes to be analyzed, the higher the quantitation accuracy.

The **second step** is the determination of b value of each solute in the unknown mixture.

This step requires the chromatographic analysis of at least three different dilutions from the unknown mixture. For each solute of the mixture, whatever its initial concentration, parameter b can be deduced by determining the slope of the curve log area versus dilution coefficient.

The **third step** consists in the determination of the quantities of all solutes in the mixture.

Knowing the b value previously determined for each solute of the mixture, log A value can be deduced from the equation established in the first step. Finally from the peak area, it is possible to deduct the mass of each solute present in the unknown sample thanks to Equation 5.19 deduced from previous equations:

$$m = \sqrt[b]{\frac{\text{Area}}{A}} \quad \text{with: } A = e^{(b-\beta)/\alpha} \tag{5.19}$$

Compared with all the quantitative analysis methods described so far, this technique presents the outstanding advantage of allowing the quantitative analysis of all the solutes of an unknown sample even if some of them are not available as standards.

In order to evaluate the results obtained with this method, we have compared the theoretical percentages (with their confidence intervals) of FAME resulting from the calculation of the amounts of TAGs with the experimental percentage of FAME obtained by saponification of four different cucurbitaceous oils. For all FAMEs, calculated and experimental values overlapped. Thus it has been shown that the proposed quantification method provides fairly accurate results [166].

Although the correlation between the A and b coefficients can be understood in term of light scattering phenomenon, the relationship appeared to be experimentally valid using Corona CAD (Figure 5.5).

Figure 5.6 illustrates the Corona CAD response as a function of the amounts of TAGs injected in NARP-LC gradient mode. Over a 2.5–250 ppm concentration range, the detector response follows a power law, with $b < 1$ as predicted by

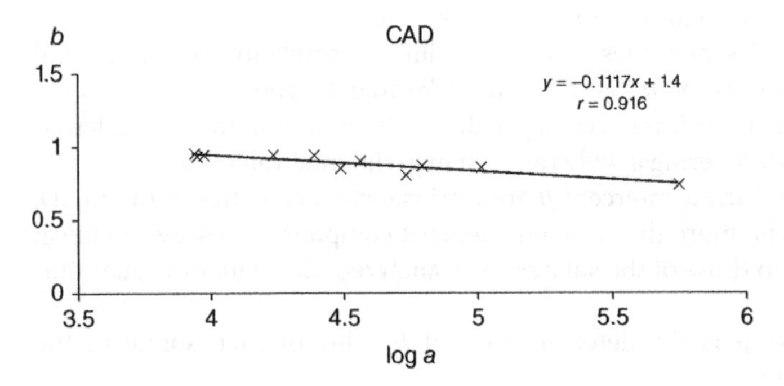

Figure 5.5 Experimental correlation between log A and b.

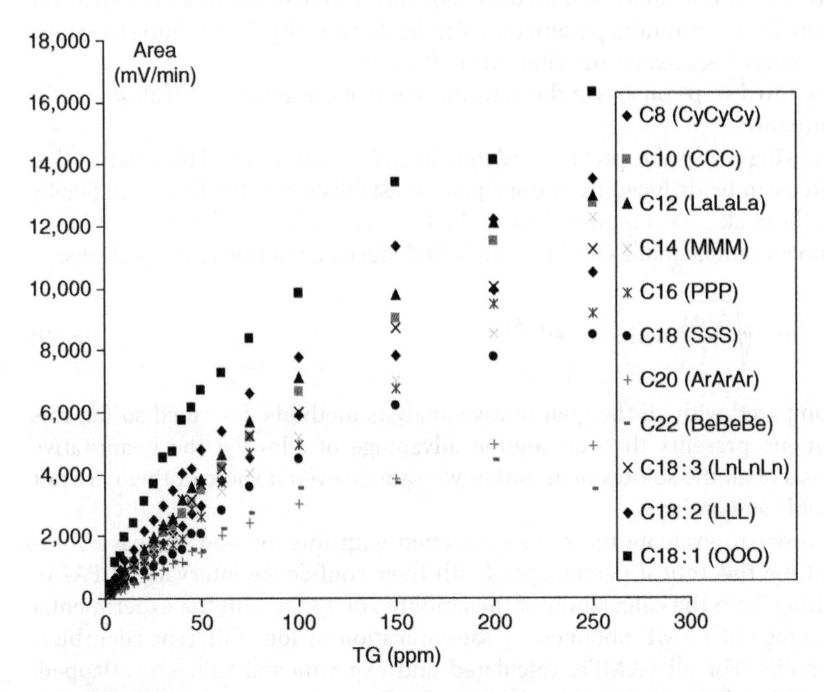

Figure 5.6 TAG calibration curve (peak area vs. sample size) with Corona CAD (range 2.5–250 ppm).

the theory, irrespective of the TAG analyzed. The linear response domain stretches over a 2.5–50 ppm concentration range, with correlation coefficients (r) ranging from 0.993 to 0.999 (4–20 ppm for ELSD).

As for ELSD a linear correlation between log A and b was obtained for Corona CAD (Figure 5.5) under the same chromatographic conditions.

The larger the amplitude of the b and log A values, the more accurately the coefficients of Equation 5.18 can be determined. As the amplitude of the variations of b (and therefore of log A) is greater for ELSD ($0.6 < b < 2.7$)[1] than for CAD ($b < 1$), the proposed calibration approach may be more effective in obtaining an accurate calibration curve for ELSD than for Corona CAD.

In order to validate this quantitative method for Corona CAD, we used TAG mixtures (five unsaturated and six saturated) of known compositions and analyzed them as mixtures of unknown compositions. The mass percentages thus obtained were compared with the known mass percentages. Relative errors were then calculated. In parallel, the quantitative analysis was performed using internal normalization without considering any response coefficient. Assuming that the area percentages are equal to the mass percentages, as usually assumed in the internal normalization method, the relative errors were calculated. As expected, the errors obtained this way can be fairly high because response coefficients depend on the compounds. Errors were between 3.5 and 41.5% for ELSD and between 1 and 36% for Corona CAD. Using the proposed method, the errors, from 0.6 to 10.4% for ELSD and from 1 to 20% for Corona CAD, are definitively smaller.

The improvement is better for ELSD than for Corona CAD because of the corresponding b values with respect to 1. Furthermore the proposed method leads to better quantitative analyses than the internal normalization method does when no standards are available. It even allows leading quantitative analysis in spite of the small modifications of A and b values, which occur with this type of detectors.

As a consequence of the linear relationship between log A and b, ELSD and Corona CAD are the only ones for the time being that currently allow quantitative analyses of all the constituents belonging to a class of chemicals present in complex natural mixtures, even for congeners for which no pure references are available to establish their individual response curves (identified or not/UV absorbent or not/using isocratic or gradient elution mode). Due to the nebulization–evaporation conditions, the only restriction to their use is that the solutes must be much less volatile than M.

The two detectors are complementary with respect to their scope of application: when mixtures are composed of solutes with large ranges of

1 Theoretically, b is bounded between 0.67 for the reflection–refraction phenomenon and 2.0 when Rayleigh diffusion is involved. Slightly lower or higher values than predicted from theory can be experimentally calculated from calibration curves.

concentrations, Corona CAD has better signal-to-noise ratios at very low concentration and has a lower minimum detectable quantity (0.9–2 ppm for Corona CAD and 3–4 ppm for ELSD). On the contrary, ELSD is more sensitive for solutes present at high concentrations.

References

1 E. Fahy, S. Subramaniam, H. A. Brown, C. K. Glass, A. H. Merrill, R. C. Murphy, C. R. H. Raetz, D. W. Russell, Y. Seyama, W. Shaw, T. Shimizu, F. Spener, G. van Meer, M. S. VanNieuwenhze, S. H. White, J. L. Witztum, and E. A. Dennis, "A comprehensive classification system for lipids," J. Lipid Res., vol. 46, no. 5, pp. 839–862, 2005.
2 "A comprehensive classification system for lipids, ERRATA," J. Lipid Res., vol. 51, no. 6, p. 1618, 2010.
3 K. A. Dill, "The mechanism of solute retention in reversed-phase liquid chromatography," J. Phys. Chem., vol. 91, no. 7, pp. 1980–1988, 1987.
4 W. R. Melander and C. Horvath, "Reversed-Phase Chromatography," in High Performance—Liquid Chromatography, Advances and Perspectives, vol. 2, C. Horváth, Ed. New York: Academic Press, 1980, p. 114.
5 C. Tanford, The Hydrophobic Effect: Formation of Micelles and Biological Membranes, 2nd Ed. New York: John Wiley & Sons, Inc., 1980.
6 K. Nakanishi and K. Kinugawa, "Hydration in Dilute Aqueous Solution of Nonelectrolytes," in Structure and Dynamics of Solutions, vol. 79, H. Ohtaki and H. Yamatera, Eds. Amsterdam/New York: Elsevier, 1992.
7 S. Heron and A. Tchapla, "Les théories prévisionnelles en chromatographie liquide à polarité de phases inversée," Analusis, vol. 22, no. 4, pp. 161–177, 1994.
8 J. G. Dorsey and K. A. Dill, "The molecular mechanism of retention in reversed-phase liquid chromatography," Chem. Rev., vol. 89, no. 2, pp. 331–346, 1989.
9 K. Heinzinger, "P.L. Huyskens, W.A.P. Luck and T. Zeegers-Huyskens (Eds.): Intermolecular Forces – An Introduction to Modern Methods and Results. Springer-Verlag, Berlin Heidelberg New York 1991, ISBN 3-540-53410-5, 490 Seiten, Preis: DM 198.00," Ber. Bunse. Phys. Chem., vol. 97, p. 1069, 1993.
10 J. H. Hildebrand and R. L. Scott, The Solubility of Nonelectrolytes, 3rd Ed. New York: Dover Publications, 1964.
11 P. Jandera, H. Colin, and G. Guiochon, "Interaction indexes for prediction of retention in reversed-phase liquid chromatography," Anal. Chem., vol. 54, no. 3, pp. 435–441, 1982.
12 H. Colin, G. Guiochon, and P. Jandera, "Interaction indexes and solvent effects in reversed-phase liquid chromatography," Anal. Chem., vol. 55, no. 3, pp. 442–446, 1983.

13 L. R. Snyder, "Role of the Mobile Phase in Liquid Chromatography," in Modern Practices of Liquid Chromatography," J. J. Kirkland, Ed. New York: John Wiley & Sons, Inc., 1971.

14 B. L. Karger, L. R. Snyder, and C. Eon, "Expanded solubility parameter treatment for classification and use of chromatographic solvents and adsorbents," Anal. Chem., vol. 50, no. 14, pp. 2126–2136, 1978.

15 A. Tchapla, S. Héron, E. Lesellier, and H. Colin, "General view of molecular interaction mechanisms in reversed-phase liquid chromatography," J. Chromatogr. A, vol. 656, no. 1–2, pp. 81–112, 1993.

16 R. Lumry and S. Rajender, "Enthalpy–entropy compensation phenomena in water solutions of proteins and small molecules: a ubiquitous properly of water," Biopolymers, vol. 9, no. 10, pp. 1125–1227, 1970.

17 H. Colin, J. C. Diez-Masa, G. Guiochon, T. Czjkowska, and I. Miedziak, "The role of the temperature in reversed-phase high-performance liquid chromatography using pyrocarbon-containing adsorbents," J. Chromatogr. A, vol. 167, pp. 41–65, 1978.

18 P. Schoenmakers, Optimization of Chromatographic Selectivity—A Guide to Method Development, vol. 35. Amsterdam, the Netherlands: Elsevier, 1986.

19 B. L. Karger, L. R. Snyder, and C. Horváth, Introduction to Separation Science. New York: John Wiley & Sons, Inc., 1973.

20 J. Roire, Les Solvants. Puteaux: EREC, 1989.

21 P. A. Small, "Some factors affecting the solubility of polymers," J. Appl. Chem., vol. 3, no. 2, pp. 71–80, 1953.

22 D. H. Kaelble, Physical Chemistry of Adhesion. Toronto, ON: John Wiley & Sons Canada Ltd, 1971.

23 T. C. Patton, Paint Flow and Pigment Dispersion. New York: John Wiley & Sons, Inc., 1979.

24 R. Tijssen, H. A. H. Billiet, and P. J. Schoenmakers, "Use of the solubility parameter for predicting selectivity and retention in chromatography," J. Chromatogr. A, vol. 122, pp. 185–203, 1976.

25 C. M. Hansen and K. Skaarup, "The three dimensional solubility parameter—key to paint component affinities. III. Independent calculation of the parameter components," J. Paint Technol., vol. 39, no. 511, pp. 511–514, 1967.

26 J. P. Teas, "Graphic analysis of resin solubilities," J. Paint Technol., vol. 40, no. 516, pp. 19–25, 1968.

27 L. R. Snyder, "Classification off the solvent properties of common liquids," J. Chromatogr. Sci., vol. 16, no. 6, pp. 223–234, 1978.

28 S. Rutan, P. Carr, W. Cheong, J. Park, and L. Snyder, "Re-evaluation of the solvent triangle and comparison to solvatochromic based scales of solvent strength and selectivity," J. Chromatogr., vol. 463, no. 1, pp. 21–37, 1989.

29 W. W. Christie, Lipid Analysis: Isolation, Separation, Identification and Structural Analysis of Lipids. Bridgwater, UK: Oily Press, 2003.

30 F. S. Deschamps, A. Baillet, and P. Chaminade, "Mechanism of response enhancement in evaporative light scattering detection with the addition of triethylamine and formic acid," Analyst, vol. 127, no. 1, pp. 35–41, 2002.

31 N. Vervoort, D. Daemen, and G. Török, "Performance evaluation of evaporative light scattering detection and charged aerosol detection in reversed phase liquid chromatography," J. Chromatogr. A, vol. 1189, no. 1–2, pp. 92–100, 2008.

32 W. Christie, "Separation of lipid classes by high-performance liquid-chromatography with the mass detector," J. Chromatogr., vol. 361, pp. 396–399, 1986.

33 R. A. Moreau, "Plant Lipid Class Analysis by HPLC," in Plant Lipid Biochemistry, Structure and Utilization: The Proceedings of the Ninth International Symposium on Plant Lipids, Held at Wye College, Kent, July 1990, P. J. Quinn and J. L. Harwood, Eds. London: Portland Press, 1990, pp. 20–22.

34 E. D. Hudson, R. J. Helleur, and C. C. Parrish, "Thin-layer chromatography-pyrolysis-gas chromatography-mass spectrometry: a multidimensional approach to marine lipid class and molecular species analysis," J. Chromatogr. Sci., vol. 39, no. 4, pp. 146–152, 2001.

35 E. Lesellier, K. Gaudin, P. Chaminade, A. Tchapla, and A. Baillet, "Isolation of ceramide fractions from skin sample by subcritical chromatography with packed silica and evaporative light scattering detection," J. Chromatogr. A, vol. 1016, no. 1, pp. 111–121, 2003.

36 W. Christie and R. Urwin, "Separation of lipid classes from plant-tissues by high-performance liquid-chromatography on chemically bonded stationary phases," HRC J. High Resolut. Chromatogr., vol. 18, no. 2, pp. 97–100, 1995.

37 A. H. El-Hamdy and W. W. Christie, "Separation of non-polar lipids by high performance liquid chromatography on a cyanopropyl column," J. High Resolut. Chromatogr., vol. 16, no. 1, pp. 55–57, 1993.

38 I. Rizov and A. Doulis, "Separation of plant membrane lipids by multiple solid-phase extraction," J. Chromatogr. A, vol. 922, no. 1–2, pp. 347–354, 2001.

39 A. Medvedovici, F. David, and P. Sandra, "Analysis of sterols in vegetable oils using off-line SFC/capillary GC-MS," Chromatographia, vol. 44, no. 1–2, pp. 37–42, 1997.

40 R. Godoy Ramos, D. Libong, M. Rakotomanga, K. Gaudin, P. M. Loiseau, and P. Chaminade, "Comparison between charged aerosol detection and light scattering detection for the analysis of *Leishmania* membrane phospholipids," J. Chromatogr. A, vol. 1209, no. 1–2, pp. 88–94, 2008.

41 S. Uran, Å. Larsen, P. B. Jacobsen, and T. Skotland, "Analysis of phospholipid species in human blood using normal-phase liquid chromatography coupled with electrospray ionization ion-trap tandem mass spectrometry," J. Chromatogr. B Biomed. Sci. Appl., vol. 758, no. 2, pp. 265–275, 2001.

42 M. C. Dobarganes and G. Márquez-Ruiz, "Analytical evaluation of fats and oils by size-exclusion chromatography," Analusis, vol. 26, no. 3, pp. 61–65, 1998.
43 M. C. Dobarganes and G. Marquez-Ruiz, "High-Performance Size-Exclusion Chromatography Applied to the Analysis of Edible Fats," in New Trends in Lipid and Lipoprotein Analyses, J.-L. Sebedio and E. G. Perkins, Eds. Champaign, IL: AOCS Press, 1995, pp. 81–92.
44 P. Juanéda, "Utilization of Silver Ion HPLC for Separation of the Geometrical Isomers of Alpha-Linolenic Acid," in New Trends in Lipid and Lipoprotein Analyses, J.-L. Sebedio and E. G. Perkins, Eds. Champaign, IL: AOCS Press, 1995, pp. 75–80.
45 W. Christie, "Separation of molecular-species of triacylglycerols by high-performance liquid-chromatography with a silver ion column," J. Chromatogr., vol. 454, pp. 273–284, 1988.
46 P. Laakso and W. Christie, "Combination of silver ion and reversed-phase high-performance liquid-chromatography in the fractionation of herring oil triacylglycerols," J. Am. Oil Chem. Soc., vol. 68, no. 4, pp. 213–223, 1991.
47 P. Laakso, W. Christie, and J. Pettersen, "Analysis of North-Atlantic and Baltic fish oil triacylglycerols by high-performance liquid-chromatography with a silver ion column," Lipids, vol. 25, no. 5, pp. 284–291, 1990.
48 S. Wada, C. Koizumi, and J. Nonaka, "Analysis of triglycerides of soybean oil by high-performance liquid chromatography in combination with gas liquid chromatography," J. Jpn. Oil Chem. Soc., vol. 26, no. 2, pp. 95–99, 1977.
49 V. Ruiz-Gutiérrez and L. J. R. Barron, "Methods for the analysis of triacylglycerols," J. Chromatogr. B Biomed. Sci. Appl., vol. 671, no. 1–2, pp. 133–168, 1995.
50 G. Dobson, W. W. Christie, and B. Nikolova-Damyanova, "Silver ion chromatography of lipids and fatty acids," J. Chromatogr. B Biomed. Sci. Appl., vol. 671, no. 1–2, pp. 197–222, 1995.
51 S. Lam and E. Grushka, "Silver loaded aluminosilicate as a stationary phase for the liquid chromatographic separation of unsaturated compounds," J. Chromatogr. Sci., vol. 15, no. 7, pp. 234–238, 1977.
52 E. C. Smith, A. D. Jones, and E. W. Hammond, "Investigation of the use of argentation high-performance liquid chromatography for the analysis of triglycerides," J. Chromatogr. A, vol. 188, no. 1, pp. 205–212, 1980.
53 R. R. Heath, J. H. Tumlinson, R. E. Doolittle, and A. T. Proveaux, "Silver nitrate-high pressure liquid chromatography of geometrical isomers," J. Chromatogr. Sci., vol. 13, no. 8, pp. 380–382, 1975.
54 G. Thevenonemeric, A. Tchapla, and M. Martin, "Role of Pi-Pi interactions in reversed-phase liquid-chromatography," J. Chromatogr., vol. 550, no. 1–2, pp. 267–283, 1991.
55 A. McGill and C. Moffat, "A study of the composition of fish liver and body oil triglycerides," Lipids, vol. 27, no. 5, pp. 360–370, 1992.

56 B. Jeffrey, "Silver-complexation liquid-chromatography for fast, high-resolution separations of triacylglycerols," J. Am. Oil Chem. Soc., vol. 68, no. 5, pp. 289–293, 1991.
57 V. Fournier, P. Juanéda, F. Destaillats, F. Dionisi, P. Lambelet, J.-L. Sébédio, and O. Berdeaux, "Analysis of eicosapentaenoic and docosahexaenoic acid geometrical isomers formed during fish oil deodorization," J. Chromatogr. A, vol. 1129, no. 1, pp. 21–28, 2006.
58 S. Banni and J.-C. Martin, "Conjugated Linoleic Acid and Metabolites," in Trans Fatty Acids in Human Nutrition, J.-L. Sebedio and W. W. Christie, Eds. Dundee: Oily Press, 1998, pp. 261–302.
59 B. Nikolova-Damyanova, "Lipid Analysis by Silver Ion Chromatography," in Advances in Lipid Methodology, R. O. Adlof, Ed. Elsevier, 2003, pp. 43–124.
60 A. Merrien, J. Morice, A. Pouzet, O. Morin, C. Sultana, J. P. Helme, A. Bockelee-Morvan, M. Cognée, F. Rognon, W. Wuidart, J. Pontillon, B. Monteuuis, E. Ucciani, A. Uzzan, C. Foures, J. L. Sedebio, M. Chambon, J. Graille, and C. Demanze, "Sources et Monographie des principaux corps gras," in Manuel des corps gras, A. Karleskind, Ed. Lavoisier, Cachon, France: Technique et Documentation—Lavoisier, 1992, pp. 115–316.
61 E. Ucciani, Nouveau dictionnaire des huiles végétales: compositions en acides gras. Paris/Londres/New York: Technique et Documentation—Lavoisier, 1995.
62 G. Bianchi, "Plant Waxes," in Waxes: Chemistry, Molecular Biology and Functions, R. J. Hamilton, Ed. Dundee: Oily Press, 1995, pp. 175–222.
63 P. Bowser, D. Nugteren, R. White, U. Houtsmuller, and C. Prottey, "Identification, isolation and characterization of epidermal lipids containing linoleic-acid," Biochim. Biophys. Acta, vol. 834, no. 3, pp. 419–428, 1985.
64 W. W. Christie, "Structural Analysis of Lipids," in Lipid Analysis: Isolation, Separation, Identification and Structural Analysis of Lipids, 3rd Ed. Bridgwater, UK: Oily Press, 2005, p. 127.
65 W. Abraham, P. Wertz, and D. Downing, "Linoleate-rich acylglucosylceramides of pig epidermis—structure determination by proton magnetic-resonance," J. Lipid Res., vol. 26, no. 6, pp. 761–766, 1985.
66 S. Hamanaka, C. Asagami, M. Suzuki, F. Inagaki, and A. Suzuki, "Structure determination of glucosyl beta-1-N-(omega-O-linoleoyl)-acylsphingosines of human-epidermis," J. Biochem. (Tokyo), vol. 105, no. 5, pp. 684–690, 1989.
67 P. Wertz and D. Downing, "Glucosylceramides of pig epidermis—structure determination," J. Lipid Res., vol. 24, no. 9, pp. 1135–1139, 1983.
68 P. Wertz and D. Downing, "Ceramides of pig epidermis—structure determination," J. Lipid Res., vol. 24, no. 6, pp. 759–765, 1983.
69 G. Rawlings, "Skin Waxes," in Waxes: Chemistry, Molecular Biology and Functions, R. J. Hamilton, Ed. Dundee: Oily Press, 1995, pp. 223–256.

70 K. Gaudin, P. Chaminade, A. Baillet, D. Ferrier, J. Bleton, S. Goursaud, and A. Tchapla, "Contribution to liquid chromatographic analysis of cutaneous ceramides," J. Liq. Chromatogr. Relat. Technol., vol. 22, no. 3, pp. 379–400, 1999.
71 R. t'Kindt, L. Jorge, E. Dumont, P. Couturon, F. David, P. Sandra, and K. Sandra, "Profiling and characterizing skin ceramides using reversed-phase liquid chromatography-quadrupole time-of-flight mass spectrometry," Anal. Chem., vol. 84, no. 1, pp. 403–411, 2012.
72 J. C. Giddings, "Maximum number of components resolvable by gel filtration and other elution chromatographic methods," Anal. Chem., vol. 39, no. 8, pp. 1027–1028, 1967.
73 J. M. Davis and J. C. Giddings, "Statistical theory of component overlap in multicomponent chromatograms," Anal. Chem., vol. 55, no. 3, pp. 418–424, 1983.
74 S. Heron and A. Tchapla, "Properties and characterizations of stationary and mobile phases used in reversed-phase liquid-chromatography," Analusis, vol. 21, no. 8, pp. 327–347, 1993.
75 S. Heron and A. Tchapla, "Description of retention mechanism by solvophobic theory: influence of organic modifiers on the retention behaviour of homologous series in reversed-phase liquid chromatography," J. Chromatogr. A, vol. 556, no. 1–2, pp. 219–234, 1991.
76 S. Heron and A. Tchapla, "Choice of stationary and mobile phases for separation of mixed triglycerides by liquid-phase chromatography," Analusis, vol. 22, no. 3, pp. 114–126, 1994.
77 J. Myher, A. Kuksis, L. Marai, and F. Manganaro, "Quantitation of natural triacylglycerols by reversed-phase liquid-chromatography with direct liquid inlet mass-spectrometry," J. Chromatogr., vol. 283, pp. 289–301, 1984.
78 J. W. King, "Supercritical Fluid Chromatography (SFC) Global Perspective and Applications in Lipid Technology," in Advances in Lipid Methodology, 3rd Ed., R. Adlof, Ed. Bridgwater, UK: Oily Press, 2003, p. 301.
79 E. Lesellier and A. Tchapla, "Subcritical Fluid Chromatography with Organic Modifiers on Octadecyl Packed Columns: Recent Developments for the Analysis of High Molecular Organic Compounds," in Supercritical Fluid Chromatography with Packed Columns: Techniques and Applications, K. Anton and C. Berger, Eds. New York, Basel, Hong Kong: CRC Press, 1997, pp. 195–222.
80 S. Heron, E. Lesellier, and A. Tchapla, "Analysis of triacylglycerols of borage oil by RPLC identification by coinjection," J. Liq. Chromatogr., vol. 18, no. 3, pp. 599–611, 1995.
81 H. Colin, G. Guiochon, Z. Yun, J. Diezmasa, and J. Jandera, "Selectivity for homologous series in reversed-phase LC: 3. Investigation," J. Chromatogr. Sci., vol. 21, no. 4, pp. 179–184, 1983.

82 M. Martin, G. Thevenon, and A. Tchapla, "Comparison of retention mechanisms of homologous series and triglycerides in non-aqueous reversed-phase liquid-chromatography," J. Chromatogr., vol. 452, pp. 157–173, 1988.
83 S. Heron and A. Tchapla, "Validity of a notion of eluent force taking into account the nature of the graft on silica in aqueous and non-aqueous liquid chromatography," Analusis, vol. 25, no. 8, pp. 257–262, 1997.
84 K. Gaudin, P. Chaminade, D. Ferrier, and A. Baillet, "Use of principal component analysis for investigation of factors affecting retention behaviour of ceramides on porous graphitized carbon column," Chromatographia, vol. 50, no. 7–8, pp. 470–478, 1999.
85 H. Mockel, A. Braedikow, H. Melzer, and G. Aced, "A comparison of the retention of homologous series and other test solutes on an ODS column and a hypercarb carbon column," J. Liq. Chromatogr., vol. 14, no. 13, pp. 2477–2498, 1991.
86 J. Kriz, E. Adamcova, J. Knox, and J. Hora, "Characterization of adsorbents by high-performance liquid-chromatography using aromatic-hydrocarbons—porous graphite and its comparison with silica-gel, alumina, octadecylsilica and phenylsilica," J. Chromatogr. A, vol. 663, no. 2, pp. 151–161, 1994.
87 A. Hazotte, D. Libong, and P. Chaminade, "High-temperature micro liquid chromatography for lipid molecular species analysis with evaporative light scattering detection," J. Chromatogr. A, vol. 1140, no. 1–2, pp. 131–139, 2007.
88 D. Hmida, M. Abderrabba, A. Tchapla, S. Héron, and F. Moussa, "Comparison of iso-eluotropic mobile phases at different temperatures for the separation of triacylglycerols in Non-Aqueous Reversed Phase Liquid Chromatography," J. Chromatogr. B, vol. 990, pp. 45–51, 2015.
89 K. Gurdale, E. Lesellier, and A. Tchapla, "Methylene selectivity and eluotropic strength variations in subcritical fluid chromatography with packed columns and CO_2-modifier mobile phases," Anal. Chem., vol. 71, no. 11, pp. 2164–2170, 1999.
90 E. Lesellier, J. Bleton, and A. Tchapla, "Use of relationships between retention behaviors and chemical structures in subcritical fluid chromatography with CO_2/modifier mixtures for the identification of triglycerides," Anal. Chem., vol. 72, no. 11, pp. 2573–2580, 2000.
91 E. Frede, "Improved HPLC of triglycerides by special tempering procedures," Chromatographia, vol. 21, no. 1, pp. 29–36, 1986.
92 C. Maniongui, J. Gresti, M. Bugaut, S. Gauthier, and J. Bezard, "Determination of bovine butterfat triacylglycerols by reversed-phase liquid-chromatography and gas-chromatography," J. Chromatogr., vol. 543, no. 1, pp. 81–103, 1991.
93 L. Barron and G. Santamaria, "Nonaqueous reverse-phase HPLC analysis of triglycerides," Chromatographia, vol. 23, no. 3, pp. 209–214, 1987.
94 J. Singleton and H. Pattee, "Optimization of parameters for the analysis of triglyceride by reverse phase HPLC using a UV detector at 210 nm," J. Am. Oil Chem. Soc., vol. 61, no. 4, pp. 761–766, 1984.

95 G. W. Jensen, "Improved separation of triglycerides at low temperatures by reversed-phase liquid chromatography," J. Chromatogr., vol. 204, pp. 407–411, 1981.
96 L. Barron, G. Santamaria, and J. Masa, "Influence of bonded-phase column type, mobile phase-composition, temperature and flow-rate in the analysis of triglycerides by reverse-phase high-performance liquid-chromatography," J. Liq. Chromatogr., vol. 10, no. 14, pp. 3193–3212, 1987.
97 T. Sawada, K. Takahashi, and M. Hatano, "Molecular-species analysis of fish oil triglyceride by light-scattering mass detector equipped liquid chromatograph. 1. Effect of column temperature on improvement of resolution in separating triglyceride molecular-species containing highly unsaturated fatty-acids by reverse phase high-performance liquid-chromatography," Nippon Suisan Gakkaishi, vol. 58, no. 7, pp. 1313–1317, 1992.
98 S. Héron, J. Bleton, and A. Tchapla, "Mechanism for Separation of Triacylglycerols in Oils by Liquid Chromatography: Identification by Mass Spectrometry," in New Trends in Lipid and Lipoprotein Analyses, J.-L. Sebedio, Ed. Urbana, IL: The American Oil Chemists Society, 1995, pp. 205–231.
99 S. Heron and A. Tchapla, "Using a molecular interaction-model to optimize the separation of fatty triglycerides in CLPI—fingerprinting the different types of fats," Ocl Ol. Corps Gras Lipides, vol. 1, no. 3, pp. 219–228, 1994.
100 K. Aitzetmuller and M. Gronheim, "Gradient elution HPLC of fats and oils with laser-light scattering detection," Fett Wiss. Technol.-Fat Sci. Technol., vol. 95, no. 5, pp. 164–168, 1993.
101 S. Heron and A. Tchapla, Fingerprints of Triacylglycerols from Oils and Fats. Alfortville: Sedere, 1993.
102 S. Heinisch, E. Lesellier, C. Podevin, J. L. Rocca, and A. Tchapla, "Computerized optimization of RP-HPLC separation with nonaqueous or partially aqueous mobile phases," Chromatographia, vol. 44, no. 9–10, pp. 529–537, 1997.
103 O. Podlaha and B. Töregård, "Some new observations on the equivalent carbon numbers of triglycerides and relationship between changes in equivalent carbon number and molecular-structure," J. Chromatogr., vol. 482, no. 1, pp. 215–226, 1989.
104 M. Hierro, A. Najera, and G. Santamaria, "Analysis of triglycerides by reversed-phase HPLC with gradient elution using a light-scattering detector," Rev. Esp. Cienc. Tecnol. Aliment., vol. 32, no. 6, pp. 635–651, 1992.
105 K. Aitzetmuller, "Flow gradients in the HPLC analysis of triacylglycerols in fats and oils," HRC J. High Resolut. Chromatogr., vol. 13, no. 5, pp. 375–377, 1990.
106 A. Stolyhwo, H. Colin, and G. Guiochon, "Analysis of triglycerides in oils and fats by liquid-chromatography with the laser-light scattering detector," Anal. Chem., vol. 57, no. 7, pp. 1342–1354, 1985.

107 S. Héron and A. Tchapla, "Reversible solvent and temperature induced 'monomeric-like'—'Polymeric-like' transitions in alkyl bonded silica," Chromatographia, vol. 36, no. 1, pp. 11–18, 1993.

108 A. Tchapla and S. Héron, "Property-structure relationship of solute-stationary phase complexes occurring in a molecular mechanism by penetration of eluite in bonded alkyl chains in reversed-phase liquid chromatography," J. Chromatogr. A, vol. 684, no. 2, pp. 175–188, 1994.

109 E. Lesellier, K. Gurdale, and A. Tchapla, "Interaction mechanisms on octadecyl packed columns in subcritical fluid chromatography with CO_2-modifier mobile phases," J. Chromatogr. A, vol. 975, no. 2, pp. 335–347, 2002.

110 K. Gaudin, P. Chaminade, D. Ferrier, A. Baillet, and A. Tchapla, "Analysis of commercial ceramides by non-aqueous reversed-phase liquid chromatography with evaporative light-scattering detection," Chromatographia, vol. 49, no. 5–6, pp. 241–248, 1999.

111 C. West, G. Cilpa, K. Gaudin, P. Chaminade, and E. Lesellier, "Modelling of ceramide interactions with porous graphite carbon in non-aqueous liquid chromatography," J. Chromatogr. A, vol. 1087, no. 1–2, pp. 77–85, 2005.

112 E. Lesellier and A. Tchapla, "Separation of vegetable oil triglycerides by subcritical fluid chromatography with octadecyl packed columns and CO_2/modifier mobile phase," Chromatographia, vol. 51, no. 11–12, pp. 688–694, 2000.

113 S. Brossard, M. Lafosse, M. Dreux, and J. Becart, "Abnormal composition of commercial waxes revealed by supercritical fluid chromatography," Chromatographia, vol. 36, pp. 268–274, 1993.

114 H. R. Mottram and R. P. Evershed, "Structure analysis of triacylglycerol positional isomers using atmospheric pressure chemical ionisation mass spectrometry," Tetrahedron Lett., vol. 37, no. 47, pp. 8593–8596, 1996.

115 C. Borchjensen, A. Staby, and J. Mollerup, "Supercritical-fluid chromatographic analysis of a fish-oil of the sand eel (*Ammodytes* sp.)," HRC J. High Resolut. Chromatogr., vol. 16, no. 10, pp. 621–623, 1993.

116 J. Gresti, M. Bugaut, C. Maniongui, and J. Bezard, "Composition of molecular-species of triacylglycerols in bovine-milk fat," J. Dairy Sci., vol. 76, no. 7, pp. 1850–1869, 1993.

117 H. Nakashima and Y. Hirata, "Proceedings of the 22nd International Symposium on Capillary Chromatography, Gifu, Japan, Naxos Software Solutions, M. Schaefer." November 1999.

118 A. Dermaux, A. Medvedovici, M. Ksir, E. Van Hove, M. Talbi, and P. Sandra, "Elucidation of the triglycerides in fish oil by packed-column supercritical fluid chromatography fractionation followed by capillary electrochromatography and electrospray mass spectrometry," J. Microcolumn Sep., vol. 11, no. 6, pp. 451–459, 1999.

119 C. BorchJensen and J. Mollerup, "Determination of vernolic acid content in the oil of *Euphorbia lagascae* by gas and supercritical fluid chromatography," J. Am. Oil Chem. Soc., vol. 73, no. 9, pp. 1161–1164, 1996.

120 P. Sandra, A. Medvedovici, and F. David, "Comprehensive pSFCxpSFC-MS for the characterization of triglycerides in vegetable oils," LC GC Europe, vol. 16, no. 12A, pp. 32–34, 2003.

121 I. Francois and P. Sandra, "Comprehensive supercritical fluid chromatography × reversed phase liquid chromatography for the analysis of the fatty acids in fish oil," J. Chromatogr. A, vol. 1216, no. 18, pp. 4005–4012, 2009.

122 I. Francois, A. dos S. Pereira, and P. Sandra, "Considerations on comprehensive and off-line supercritical fluid chromatography × reversed-phase liquid chromatography for the analysis of triacylglycerols in fish oil," J. Sep. Sci., vol. 33, no. 10, pp. 1504–1512, 2010.

123 A. Tchapla, C. Bordes, D. Charbonneau, B. Chabot, and S. Héron, "How to choose the appropriate class of stationary phases in relation to the nature of the solutes to be analyzed," in *Poster CMTR25, HPLC2013*, Amsterdam, 2013.

124 P. Dugo, O. Favoino, P. Q. Tranchida, G. Dugo, and L. Mondello, "Off-line coupling of non-aqueous reversed-phase and silver ion high-performance liquid chromatography-mass spectrometry for the characterization of rice oil triacylglycerol positional isomers," J. Chromatogr. A, vol. 1041, no. 1–2, pp. 135–142, 2004.

125 N. K. Andrikopoulos, "Chromatographic and spectroscopic methods in the analysis of triacylglycerol species and regiospecific isomers of oils and fats," Crit. Rev. Food Sci. Nutr., vol. 42, no. 5, pp. 473–505, 2002.

126 M. Buchgraber, F. Ullberth, H. Emons, and E. Anklam, "Triacylglycerol profiling by using chromatographic techniques," Eur. J. Lipid Sci. Technol., vol. 106, no. 9, pp. 621–648, 2004.

127 E. Lesellier and A. Tchapla, "Retention behavior of triglycerides in octadecyl packed subcritical fluid chromatography with CO_2/modified mobile phases," Anal. Chem., vol. 71, no. 23, pp. 5372–5378, 1999.

128 A. Larsen, E. Mokastet, E. Lundanes, and E. Hvattum, "Separation and identification of phosphatidylserine molecular species using reversed-phase high-performance liquid chromatography with evaporative light scattering and mass spectrometric detection," J. Chromatogr. B Analyt. Technol. Biomed. Life Sci., vol. 774, no. 1, pp. 115–120, 2002.

129 N. U. Olsson and N. Salem, "Molecular species analysis of phospholipids," J. Chromatogr. B Biomed. Sci. Appl., vol. 692, no. 2, pp. 245–256, 1997.

130 K. Gaudin, E. Lesellier, P. Chaminade, D. Ferrier, A. Baillet, and A. Tchapla, "Retention behaviour of ceramides in sub-critical fluid chromatography in comparison with non-aqueous reversed-phase liquid chromatography," J. Chromatogr. A, vol. 883, no. 1–2, pp. 211–222, 2000.

131 L. Barron, M. Hierro, and G. Santamaria, "HPLC and GLC analysis of the triglyceride composition of bovine, ovine and caprine milk-fat," J. Dairy Res., vol. 57, no. 4, pp. 517–526, 1990.

132 M. Narce, J. Gresti, and J. Bezard, "Method for evaluating the bioconversion of radioactive poly-unsaturated fatty-acids by use of reversed-phase liquid-chromatography," J. Chromatogr., vol. 448, no. 2, pp. 249–264, 1988.

133 M. Ozcimder and W. E. Hammers, "Fractionation of fish oil fatty acid methyl esters by means of argentation and reversed-phase high-performance liquid chromatography, and its utility in total fatty acid analysis," J. Chromatogr., vol. 187, no. 2, pp. 307–317, 1980.

134 P. T. Pei, R. S. Henly, and S. Ramachandran, "New application of high pressure reversed-phase liquid chromatography in lipids," Lipids, vol. 10, no. 3, pp. 152–156, 1975.

135 J. Yoo and V. McGuffin, "Determination of fatty-acids in fish oil dietary-supplements by capillary liquid-chromatography with laser-induced fluorescence detection," J. Chromatogr., vol. 627, no. 1–2, pp. 87–96, 1992.

136 S. Héron, S. Therrey, R. L. Wolff, and A. Tchapla, Identification and Fingerprint Maps of Triacylglycerols of Oils Containing Delta 5 Olefinic Acids in NARP Chromatography. *1st Int Symp of Eur Section of AOCS*, Dijon, France, September 20, 1996.

137 A. W. G. Tamba-Sompila, M. G. Maloumbi, J. Bleton, A. Tchapla, and S. Héron, "Identification des triacylglycérols en HPLC. Comment se passer du couplage HPLC-SM ? Dans quel cas la chromatographie est-elle encore indispensable?," OCL, vol. 21, no. 6, p. A601, 2014.

138 A. Acheampong, N. Leveque, A. Tchapla, and S. Heron, "Simple complementary liquid chromatography and mass spectrometry approaches for the characterization of triacylglycerols in *Pinus koraiensis* seed oil," J. Chromatogr. A, vol. 1218, no. 31, pp. 5087–5100, 2011.

139 M. Lisa, M. Holcapek, T. Rezanka, and N. Kabatova, "High-performance liquid chromatography-atmospheric pressure chemical ionization mass spectrometry and gas chromatography-flame ionization detection characterization of Delta 5-polyenoic fatty acids in triacylglycerols from conifer seed oils," J. Chromatogr. A, vol. 1146, no. 1, pp. 67–77, 2007.

140 A. Charrie-Duhaut, J. Connan, N. Rouquette, P. Adam, C. Barbotin, M. de Rozieres, A. Tchapla, and P. Albrecht, "The canopic jars of Rameses II: real use revealed by molecular study of organic residues," J. Archaeol. Sci., vol. 34, no. 6, pp. 957–967, 2007.

141 M. Holcapek, P. Jandera, P. Zderadicka, and L. Hruba, "Characterization of triacylglycerol and diacylglycerol composition of plant oils using high-performance liquid chromatography-atmospheric pressure chemical ionization mass spectrometry," J. Chromatogr. A, vol. 1010, no. 2, pp. 195–215, 2003.

142 G. M. Patton, J. M. Fasulo, and S. J. Robins, "Separation of phospholipids and individual molecular species of phospholipids by high-performance liquid chromatography," J. Lipid Res., vol. 23, no. 1, pp. 190–196, 1982.
143 J. Goiffon, C. Reminiac, and M. Olle, "High performance liquid chromatography for fat triglyceride analysis. I. Search for the best operating conditions for soya-bean oil," Rev. Fr. Corps Gras, vol. 28, pp. 167–170, 1981.
144 J. Goiffon, C. Reminiac, and D. Furon, "High performance liquid chromatography for fat triglyceride analysis. II. Retention indices of triglycerides," Rev. Fr. Corps Gras, vol. 28, pp. 199–207, 1981.
145 J.-L. Perrin and M. Naudet, "Identification et dosage des triglycérides des corps gras naturels par CLHP," Rev. Fr. Corps Gras, vol. 30, no. 7–8, pp. 279–285, 1983.
146 O. Podlaha and B. Töregård, "A system for identification of triglycerides in reversed phase HPLC chromatograms based on equivalent carbon numbers," J. High Resolut. Chromatogr., vol. 5, no. 10, pp. 553–558, 1982.
147 B. Herslöf, O. Podlaha, and B. Töregård, "HPLC of triglycerides," J. Am. Oil Chem. Soc., vol. 56, no. 9, pp. 864–866, 1979.
148 G. Sempore and J. Bezard, "Qualitative and quantitative-analysis of peanut oil triacylglycerols by reversed-phase liquid-chromatography," J. Chromatogr., vol. 366, pp. 261–282, 1986.
149 D. Firestone, "Liquid-chromatographic method for determination of triglycerides in vegetable-oils in terms of their partition numbers—summary of collaborative study," J. AOAC Int., vol. 77, no. 4, pp. 954–957, 1994.
150 W. W. Christie, Advances in Lipid Methodology. Dundee: Oily Press, 1997.
151 K. Takahashi, T. Hirano, M. Egi, and K. Zama, "A mathematical-model for the prediction of triglyceride molecular-species by high-performance liquid-chromatography," J. Am. Oil Chem. Soc., vol. 62, no. 10, pp. 1489–1492, 1985.
152 S. de Koning, B. van der Meer, G. Alkema, H.-G. Janssen, and U. A. T. Brinkman, "Automated determination of fatty acid methyl ester and cis/trans methyl ester composition of fats and oils," J. Chromatogr. A, vol. 922, no. 1–2, pp. 391–397, 2001.
153 C. Merle, C. Laugel, P. Chaminade, and A. Baillet-Guffroy, "Quantitative study of the stratum corneum lipid classes by normal phase liquid chromatography: comparison between two universal detectors," J. Liq. Chromatogr. Relat. Technol., vol. 33, no. 5, pp. 629–644, 2010.
154 J. Waraska and I. N. Acworth, "Neurolipids and the use of a charged aerosol detector," Am. Biotechnol. Lab., vol. 26, no. 1, pp. 12–13, 2008.
155 R. A. Moreau, "The analysis of lipids via HPLC with a charged aerosol detector," Lipids, vol. 41, no. 7, pp. 727–734, 2006.

156 Y. Iwasaki, A. Masayama, A. Mori, C. Ikeda, and H. Nakano, "Composition analysis of positional isomers of phosphatidylinositol by high-performance liquid chromatography," J. Chromatogr. A, vol. 1216, no. 32, pp. 6077–6080, 2009.

157 A. Masayama, K. Tsukada, C. Ikeda, H. Nakano, and Y. Iwasaki, "Isolation of phospholipase D mutants having phosphatidylinositol-synthesizing activity with positional specificity on *myo*-inositol," ChemBioChem, vol. 10, no. 3, pp. 559–564, 2009.

158 J. Cernicka, Z. Kozovska, M. Hnatova, M. Valachovic, I. Hapala, Z. Riedl, G. Hajós, and J. Subik, "Chemosensitisation of drug-resistant and drug-sensitive yeast cells to antifungals," Int. J. Antimicrob. Agents, vol. 29, no. 2, pp. 170–178, 2007.

159 P. Kohut, D. Wüstner, L. Hronska, K. Kuchler, I. Hapala, and M. Valachovic, "The role of ABC proteins Aus1p and Pdr11p in the uptake of external sterols in yeast: dehydroergosterol fluorescence study," Biochem. Biophys. Res. Commun., vol. 404, no. 1, pp. 233–238, 2011.

160 M. Vertzoni, H. Archontaki, and C. Reppas, "Determination of intralumenal individual bile acids by HPLC with charged aerosol detection," J. Lipid Res., vol. 49, pp. 2690–2695, 2008.

161 T. Yamazaki, T. Ihara, S. Nakamura, and K. Kato, "Determination of impurities in 17-b-estradiol reagent by HPLC with charged aerosol detector," Bunseki Kagaku, vol. 59, no. 3, pp. 219–224, 2010.

162 M. Lisa, F. Lynen, M. Holcapek, and P. Sandra, "Quantitation of triacylglycerols from plant oils using charged aerosol detection with gradient compensation," J. Chromatogr. A, vol. 1176, no. 1–2, pp. 135–142, 2007.

163 P. Mata-Espinosa, J. M. Bosque-Sendra, and L. Cuadros-Rodríguez, "Quantification of triacylglycerols in olive oils using HPLC-CAD," Food Anal. Methods, vol. 4, no. 4, pp 574–581, 2011.

164 P. Mata-Espinosa, J. M. Bosque-Sendra, R. Bro, and L. Cuadros-Rodríguez, "Discriminating olive and non-olive oils using HPLC-CAD and chemometrics," Anal. Bioanal. Chem., vol. 399, no. 6, pp. 2083–2092, 2010.

165 S. Heron, M. Maloumbi, M. Dreux, E. Verette, and A. Tchapla, "Method development for a quantitative analysis performed without any standard using an evaporative light-scattering detector," J. Chromatogr. A, vol. 1161, no. 1–2, pp. 152–156, 2007.

166 S. Heron, M. G. Maloumbi, T. Silou, E. Verette, M. Dreux, and A. Tchapla, "Calibration of an evaporative light-scattering detector for the universal quantitative analyses in liquid chromatography-application to the determination of triacylglycerols in cucurbitaceous oils," Food Anal. Methods, vol. 3, no. 2, pp. 67–74, 2010.

167 A. Tchapla, A. Ait Adoubel, and S. Heron, "Analyse de triglycérides à l'aide de détecteurs d'HPLC basés sur la formation d'aérosols," LC-GC En Fr., vol. 2008, pp. 5–10, 2008.

168 A. H. Saberi, B. B. Kee, L. Oi-Ming, and M. S. Miskandar, "Physico-chemical properties of various palm-based diacylglycerol oils in comparison with their corresponding palm-based oils," Food Chem., vol. 127, no. 3, pp. 1031–1038, 2011.

169 A. H. Saberi, L. O. Ming, and M. S. Miskandar, "Physical properties of palm-based diacylglycerol and palm-based oils in the preparation of shelf-stable margarine," Eur. J. Lipid Sci. Technol., vol. 113, pp. 627–636, 2011.

170 A. H. Saberi, O.-M. Lai, and J. F. Toro-Vázquez, "Crystallization kinetics of palm oil in blends with palm-based diacylglycerol," Food Res. Int., vol. 44, no. 1, pp. 425–435, 2011.

171 A. Cascone, S. Eerola, A. Ritieni, and A. Rizzo, "Development of analytical procedures to study changes in the composition of meat phospholipids caused by induced oxidation," J. Chromatogr. A, vol. 1120, no. 1–2, pp. 211–220, 2006.

172 Y. Kakigi, N. Mochizuki, T. Icho, T. Hakamatsuka, and Y. Goda, "Analysis of terpene lactones in a Ginkgo leaf extract by high-performance liquid chromatography using charged aerosol detection," Biosci. Biotechnol. Biochem., vol. 74, no. 3, pp. 590–594, 2010.

173 C. B. Fox, "Squalene emulsions for parenteral vaccine and drug delivery," Molecules, vol. 14, pp. 3286–3312, 2009.

174 C. B. Fox, R. C. Anderson, T. S. Dutill, Y. Goto, S. G. Reed, and T. S. Vedvick, "Monitoring the effects of component structure and source on formulation stability and adjuvant activity of oil-in-water emulsions," Colloids Surf. B Biointerfaces, vol. 65, no. 1, pp. 98–105, 2008.

175 C. Schönherr, S. Touchene, G. Wilser, R. Peschka-Süss, and G. Francese, "Simple and precise detection of lipid compounds present within liposomal formulations using a charged aerosol detector," J. Chromatogr. A, vol. 1216, no. 5, pp. 781–786, 2009.

176 K. Tokutomi, T. Tagawa, M. Korenaga, M. Chiba, T. Asai, N. Watanabe, S. Takeoka, M. Handa, Y. Ikeda, and N. Oku, "Decoration of fibrinogen [gamma]-chain peptide on adenosine diphosphate-encapsulated liposomes enhances binding of the liposomes to activated platelets," Int. J. Pharm., vol. 407, no. 1–2, pp. 151–157, 2011.

177 C. Lobback, T. Backensfeld, A. Funke, and W. Weitschies, "Quantitative determination of nonionic surfactants with CAD," Chromatogr. Tech., no. November, pp. 18–20, 2007.

178 A. Christiansen, T. Backensfeld, S. Kühn, and W. Weitschies, "Investigating the stability of the nonionic surfactants tocopheryl polyethylene glycol succinate and sucrose laurate by HPLC-MS, DAD, and CAD," J. Pharm. Sci., vol. 100, no. 5, pp. 1773–1782, 2011.

179 R. Diaz-Lopez, D. Libong, N. Tsapis, E. Fattal, and P. Chaminade, "Quantification of pegylated phospholipids decorating polymeric microcapsules of perfluorooctyl bromide by reverse phase HPLC with a charged aerosol detector," J. Pharm. Biomed. Anal., vol. 48, no. 3, pp. 702–707, 2008.

180 R. Diaz-Lopez, N. Tsapis, D. Libong, P. Chaminade, C. Connan, M. M. Chehimi, R. Berti, N. Taulier, W. Urbach, V. Nicolas, and E. Fattal, "Phospholipid decoration of microcapsules containing perfluorooctyl bromide used as ultrasound contrast agents," Biomaterials, vol. 30, no. 8, pp. 1462–1472, 2009.

181 R. Diaz-Lopez, N. Tsapis, M. Santin, S. L. Bridal, V. Nicolas, D. Jaillard, D. Libong, P. Chaminade, V. Marsaud, C. Vauthier, and E. Fattal, "The performance of PEGylated nanocapsules of perfluorooctyl bromide as an ultrasound contrast agent," Biomaterials, vol. 31, no. 7, pp. 1723–1731, 2010.

182 L. M. Nair and J. O. Werling, "Aerosol based detectors for the investigation of phospholipid hydrolysis in a pharmaceutical suspension formulation," J. Pharm. Biomed. Anal., vol. 49, no. 1, pp. 95–99, 2009.

183 A. E. Mengesha and P. M. Bummer, "Simple chromatographic method for simultaneous analyses of phosphatidylcholine, lysophosphatidylcholine, and free fatty acids," AAPS PharmSciTech, vol. 11, no. 3, pp. 1084–1091, 2010.

184 C. Brunelli, T. Górecki, Y. Zhao, and P. Sandra, "Corona-charged aerosol detection in supercritical fluid chromatography for pharmaceutical analysis," Anal. Chem., vol. 79, no. 6, pp. 2472–2482, 2007.

185 L. T. Taylor, "Supercritical fluid chromatography," Anal. Chem., vol. 80, no. 12, pp. 4285–4294, 2008.

186 J. M. Charlesworth, "Evaporative analyzer as a mass detector for liquid chromatography," Anal. Chem., vol. 50, no. 11, pp. 1414–1420, 1978.

187 R. W. Dixon and D. S. Peterson, "Development and testing of a detection method for liquid chromatography based on aerosol charging," Anal. Chem., vol. 74, no. 13, pp. 2930–2937, 2002.

188 G. Guiochon, A. Moysan, and C. Holley, "Influence of various parameters on the response factors of the evaporative light-scattering detector for a number of non-volatile compounds," J. Liq. Chromatogr., vol. 11, no. 12, pp. 2547–2570, 1988.

189 S. Nukiyama and Y. Tanasawa, "An experiment on the atomization of liquid by means of air stream," Trans. Jpn. Soc. Mech. Eng., vol. 4, pp. 86–93, 1938.

190 K. Gaudin, A. Baillet, and P. Chaminade, "Adaptation of an evaporative light-scattering detector to micro and capillary liquid chromatography and response assessment," J. Chromatogr. A, vol. 1051, no. 1–2, pp. 43–51, 2004.

191 P. H. Gamache, R. S. McCarthy, S. M. Freeto, D. J. Asa, M. J. Woodcock, K. Laws, and R. O. Cole, "HPLC analysis of non-volatile analytes using charged aerosol detection," LC-GC Eur., vol. 18, no. 6, pp. 345–354, 2005.

192 T. Górecki, F. Lynen, R. Szucs, and P. Sandra, "Universal response in liquid chromatography using charged aerosol detection," Anal. Chem., vol. 78, no. 9, pp. 3186–3192, 2006.

193 A. Hazotte, D. Libong, M. Matoga, and P. Chaminade, "Comparison of universal detectors for high-temperature micro liquid chromatography," J. Chromatogr. A, vol. 1170, no. 1–2, pp. 52–61, 2007.

194 L. Nováková, S. A. Lopéz, D. Solichová, D. Satínský, B. Kulichová, A. Horna, and P. Solich, "Comparison of UV and charged aerosol detection approach in pharmaceutical analysis of statins," Talanta, vol. 78, no. 3, pp. 834–839, 2009.

195 M. Righezza and G. Guiochon, "Effects of the nature of the solvent and solutes on the response of a light-scattering detector," J. Liq. Chromatogr., vol. 11, no. 9–10, pp. 1967–2004, 1988.

196 F. S. Deschamps, A. Baillet, and P. Chaminade, "Mechanism of response enhancement in evaporative light scattering detection with the addition of triethylamine and formic acid," The Analyst, vol. 127, no. 1, pp. 35–41, 2002.

197 S. Heron, M.-G. Maloumbi, M. Dreux, E. Verette, and A. Tchapla, "Experimental proofs of a correlation between the coefficients for the slope of the response line and the response factor of an ELSD," LC-GC Eur., vol. 19, no. 12, pp. 664–672, 2006.

6

Inorganic and Organic Ions

Xiaodong Liu[1], *Christopher A. Pohl*[1], *and Ke Zhang*[2]

[1] *Thermo Fisher Scientific, Sunnyvale, CA, USA*
[2] *Genentech Inc., South San Francisco, CA, USA*

CHAPTER MENU

6.1 Introduction

Ions are ubiquitous and important in everyday life. Sodium, potassium, calcium, chloride, and other ions play an important role in the cells of living organisms, particularly in regard to transport through cell membranes. Ion analysis is important for determination of ionic contaminants in water sources, monitoring biological activity, development of pharmaceuticals, and ensuring proper operation of nuclear power plants.

Ions can be measured by potentiometric titration, high-performance liquid chromatography (HPLC), capillary electrophoresis (CE), ion chromatography (IC), and inductively coupled plasma-atomic emission spectrometry (ICP-AES). Among these, IC is the technique most commonly used for quantification of a variety of ions, such as inorganic ions, small organic ions, and ionic surfactants. IC can be accomplished using a variety of different stationary phases, mobile phases, and detectors. In this chapter we will focus on the use of charged aerosol detection (CAD) for the analysis of ions, but it is worthwhile to start with a bit of background on other ion analysis techniques before delving into the topic.

Charged Aerosol Detection for Liquid Chromatography and Related Separation Techniques, First Edition. Edited by Paul H. Gamache.

First and foremost it must be said that IC is the primary tool used by analytical chemists for the analysis of ions. The high sensitivity of the technique, coupled with the wide dynamic operating range made possible with modern high-capacity stationary phases, makes it ideal for the analysis of ions in a wide range of applications. The combination of gradients and suppressed conductivity detection provides a powerful screening tool for the analysis of ions in drug substances and in pharmaceutical formulations, providing the basis for analysis of counter ions, additives, and manufacturing by-products. It is also widely used in a number of other areas including environmental science and food science. IC as we know it today began with the pioneering work of Small, Stevens, and Bauman [1]. The topic has been well covered in a number of books [2–5] and reviews [6, 7].

Ion-exchange chromatography was widely used in HPLC well before the advent of IC as a vehicle for ion analysis. However, IC evolved into a subdiscipline of HPLC with specialized instrumentation, consumables, and detectors optimized for separation and detection of ions. Furthermore, IC has come to mean specifically the HPLC separation of inorganic anions, small organic acids, inorganic cations, and simple amines when the ultimate in sensitivity is required.

With regard to the detection of ions, ultraviolet (UV) detection, while widely used in liquid chromatography (LC) because of its high sensitivity, broad linear range, low cost, ease of use, and compatibility with most mobile phase systems, is unsuitable for detection of most common ionic analytes since most such analytes absorb UV light weakly, if at all. Suppressed conductivity provides high sensitivity for most ionic analytes but is limited to the detection of either anionic or cationic species since the suppressor is designed to remove analyte co-ions (i.e., simultaneous detection of anions and cations is not generally possible). Mass spectrometry (MS) provides a specific and universal detection method for HPLC, but quantitative analysis using MS coupled with LC is sometimes limited in robustness. Furthermore, the high price of such instruments limits their use for routine analysis. Chemiluminescent nitrogen detection (CLND) may have poor precision, may require high maintenance, and is not compatible with nitrogen-containing mobile phases such as those containing acetonitrile and triethylamine. Refractive index detectors (RID) have significant limitations in sensitivity and are incompatible with gradient elution. Evaporative light scattering detectors (ELSD) exhibit significant limitations in precision, sensitivity, dynamic range, and the nature of calibration curves. The relatively new condensation nucleation light scattering detector (CNLSD) is appropriate for detecting semi-volatile and nonvolatile compounds with better sensitivity than ELSD.

CAD represents a significant technological improvement over ELSD as a new universal detection option. While CAD exhibits substantially lower sensitivity than suppressed conductivity detection, CAD delivers outstanding performance with regard to all important universal detection criteria.

In contrast, each of the other detection technologies has significant weaknesses in its performance, limiting its usefulness as a universal HPLC detector. CAD overcomes the shortcomings of the other detection methods in a single platform. The main benefits of CAD include universal detection of nonvolatile analytes, response largely independent of chemical properties, broad dynamic response range with detection limits spanning from low ng to high μg, good precision for a diverse range of analytes, and simple and reliable operation. However, like other aerosol-based detectors, such as ELSD and CNLSD, CAD has several limitations. First, the response of the detector varies with the mobile phase composition. Second, such detectors provide no analyte-specific information, so it is not possible to identify unknown analytes or perform peak purity analysis as in UV diode array detectors or MS detectors. Nevertheless, when properly applied, CAD can be successfully used in a variety of applications. This chapter focuses on the technical aspects of ion analysis by CAD as well as a collection of applications.

6.2 Technical Considerations

6.2.1 Instrumentation Platform

Ions can be analyzed using ion-exchange columns on standard stainless steel LC systems equipped with a pump, a sample injector, and a detector. Detection of ions can be achieved by UV, indirect UV, MS, CLND, RID, conductivity, ELSD, CNLSD, and CAD. Among all, conductivity detection is most routinely used for ion analysis. To maximize sensitivity, minimize interferences, eliminate metal contamination, and improve reproducibility, specially designed nonmetallic IC system with dedicated IC columns, automobile phase generation, suppressors, and conductivity detection is highly recommended. Because of its excellent sensitivity, high selectivity, linear calibration curve, wide dynamic range, and ease of use, a wealth collection of IC applications have been developed [1–3, 5].

In the recent years, HPLC-CAD has been considered and attempted as an alternative to IC for ion analysis. It has been proved that CAD can detect ions in pharmaceutical samples with the limit of quantification (LOQ) of single digit parts per million (ppm) [3, 5]. The operation of CAD requires a volatile mobile phase while the analyte of interest must be nonvolatile or semi-volatile. The CAD-compatible mobile phases usually contain deionized (DI) water, volatile acids (e.g., acetic acid, formic acid, trifluoroacetic acid), volatile buffer salts (e.g., ammonium acetate, ammonium formate), and/or common HPLC organic solvents (e.g., acetonitrile, acetone, methanol, ethanol, isopropanol, hexane, heptanes, etc.). It should be noted that when using reagent-free ion chromatography (RFIC) instrument, the effluent after the ion suppressor contains only high-purity water and the analytes [8], which is suitable for CAD operation. Therefore, both HPLC and IC instruments can be used with CAD for ion analysis.

6.2.2 Separation Column

Column selectivity and column bleed are two factors for selecting a column for ion analysis with CAD. Liu *et al.* reported a novel anion-exchange/cation-exchange/reversed-phase trimodal column that separates active pharmaceutical ingredients (APIs) and corresponding counterions (including anions and cations) using CAD [9, 10]. Its column chemistry is shown in Figure 6.1 and an exemplary application is shown in Figure 6.2. Zhang *et al.* used this trimodal column (Acclaim Trinity P1, Thermo Fisher Scientific) to develop a generic screening LC-CAD method for simultaneous determination of positive and negative pharmaceutical counterions [11]. In this study, several mixed-mode columns including Primesep AB (SIELC), ZIC-pHILIC (Merck), Obelisc N (SIELC), and Acclaim Trinity P1 (Thermo Fisher Scientific) were compared, and Acclaim Trinity P1 was found to provide most desired selectivity for separating a total of 25 pharmaceutical-related counterions.

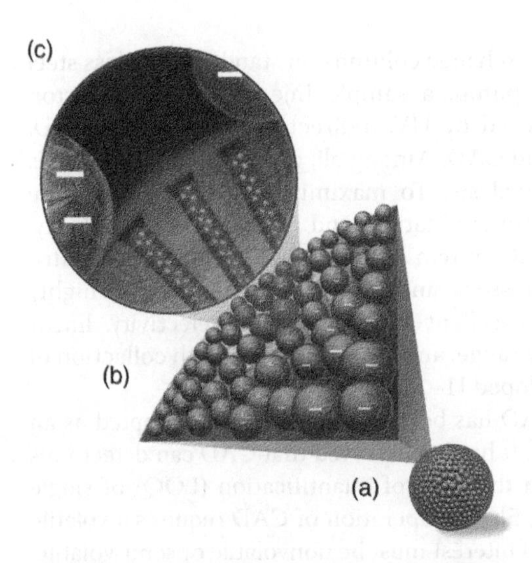

Figure 6.1 Column chemistry of Acclaim Trinity P1—a reversed-phase/anion-exchange/cation-exchange trimodal mixed-mode column based on Nanopolymer Silica Hybrid (NSH) technology. (a) Overview of a silica particle coated with nanopolymer beads; (b) enlarged view of the silica surface coated with negatively charged nanopolymer beads; (c) inner-pore area consists of reversed-phase and weak anion-exchange functionalities and outer surface provides strong cation-exchange interaction. *Source*: Reproduced with permission of Thermo Fisher Scientific Inc.

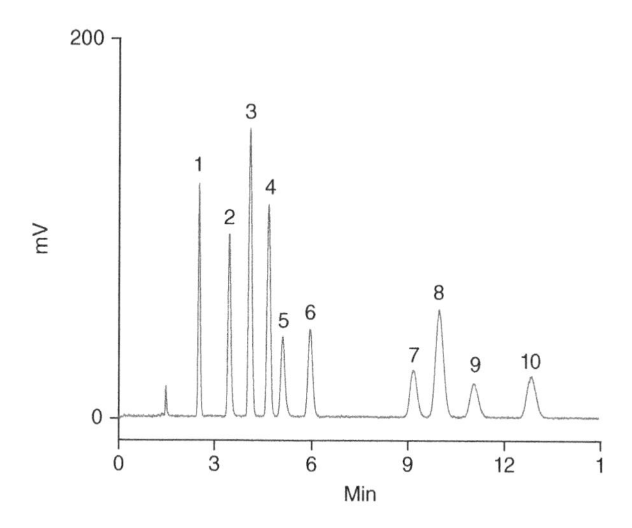

Figure 6.2 Separation of pharmaceutical counterions using a reversed-phase/anion-exchange/cation-exchange trimodal mixed-mode column. Column, Acclaim Trinity P1, 3.0 × 100-mm format; mobile phase, acetonitrile/ammonium acetate buffer, pH 5 (20 mM total concentration) (60 : 40, v/v); flow rate, 0.5 mL/min; injection volume, 2 µL; temperature, 30°C; and detection, Corona ultra (gain = 100 pA; filter = med; nebulizer temperature = 30°C). Sample: 0.05–0.1 mg/mL. Peaks: (1) Choline or *N,N,N*-trimethylethanolammonium; (2) tromethamine or tris(hydroxymethyl)aminomethane; (3) sodium; (4) potassium; (5) meglumine or *N*-methyl glucamine; (6) mesylate or methanesulfonate; (7) nitrate; (8) chloride; (9) bromide; and (10) iodide. *Source*: Xiu *et al.* [9]. Reproduced with permission of Elsevier.

A zwitterionic column (ZIC-pHILIC, Merck) combined with CAD was also used in a generic method for the determination of inorganic pharmaceutical counterions in drug substances in HILIC mode [12]. In this study, three silica-based HILIC columns were evaluated. The experimental data suggested that the presence of ion-exchange functionality on the stationary phase was necessary to obtain desirable selectivity and decent peak shape for ions. A later study on comparison between a silica-based zwitterionic HILIC column (ZIC-HILIC, Merck) and an anion-exchange/cation-exchange/reversed-phase trimodal column (Acclaim Trinity P1, Thermo Fisher Scientific) for determination of APIs and counterions indicated that the trimodal column provided several advantages over the zwitterionic column (Figure 6.3), such as desirable selectivity for pharmaceutical counterions, better versatility for simultaneous separation of APIs, and corresponding counterions, greater flexibility in method development, and less usage of organic solvent [13]. Recently, CAD and suppressed conductivity detection were compared for inorganic cation and anion analysis by connecting CAD in series after the ion suppressor and the conductivity detector (Liu X, Tracy M, Pohl C, unpublished result). Since

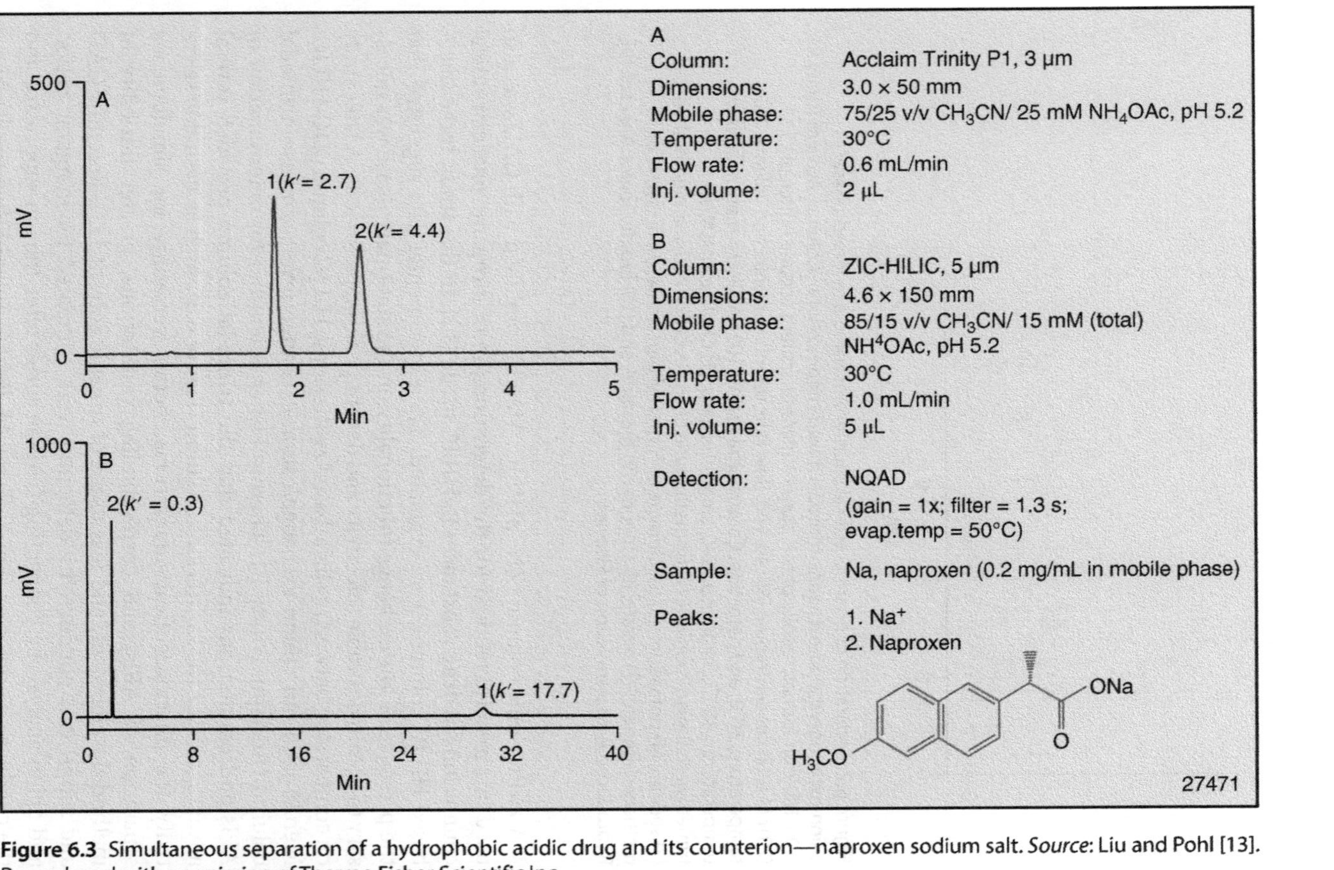

Figure 6.3 Simultaneous separation of a hydrophobic acidic drug and its counterion—naproxen sodium salt. *Source*: Liu and Pohl [13]. Reproduced with permission of Thermo Fisher Scientific Inc.

the effluent from the conductivity detector contains only high-purity water or carbonic acid with trace amount of analyte ions, a variety of IC columns can be used with CAD [14].

Since CAD is a mass-dependent "universal" detector with high sensitivity, the dissolved silica or column bleed can be easily detected, causing an increase in detector baseline response. As a result, when using CAD, separation columns should have minimal bleed, which requires chemically stable linkage between the functional group and the substrate surface, and thorough cleanup after synthesis. It has been reported that some silica-based columns exhibited high background levels with CAD. In fact, CAD was used for detecting column bleed as an indicator of induced degradation of the stationary phase [15]. Column bleed can be less problematic for polymer-based separation columns because of high chemical stability compared with their silica counterparts.

Huang *et al.* evaluated the column bleed of several HILIC columns with CAD [12]. They studied three silica-based columns (including ZIC-HILIC, Waters Atlantis HILIC, and Phenomenex Luna HILIC) and one polymer-based column (ZIC-*p*HILIC) and found that baseline noise varied significantly among the columns. When operating on silica-based columns (ZIC-HILIC and Atlantis HILIC), significant baseline noise (>50 pA) of the Corona® CAD was observed, which was believed to be a result of column bleeding. Under the same condition, both ZIC-*p*HILIC and Luna HILIC exhibited negligible baseline noise, suggesting minimal column bleed. In a separate study, the baseline noise of different mixed-mode columns including Primesep AB, ZIC-pHILIC, Obelisc N, and Acclaim Trinity P1 were compared [11]. Among these columns, the Primesep AB column exhibited significant baseline noise (above 50 pA out of 200 pA output scale). The remaining three columns gave a baseline lower than 5 pA, thus compatible with CAD. A recent study on comparison of conductivity detection and CAD in IC indicated that both IonPac AS18 (Thermo Fisher Scientific) and IonPac CS12A (Thermo Fisher Scientific) column showed negligible baseline noise and were fully compatible with CAD when suppressors were used (Liu X, Tracy M, Pohl C, unpublished result).

6.2.3 Mobile Phase

CAD requires volatile mobile phases. Mobile phases usually contain DI water, volatile acids (e.g., acetic acid, formic acid, trifluoroacetic acid), volatile buffer salts (e.g., ammonium acetate, ammonium formate), and/or common HPLC organic solvents (e.g., acetonitrile, acetone, methanol, ethanol, isopropanol, hexane, heptanes, etc.). The response of CAD is sensitive to contaminants in the mobile phase. To achieve lowest baseline noise and best sensitivity, mobile phase must be free of nonvolatile impurities, which requires the use of high-purity buffer salts, DI water, acid or base additives, and organic solvents.

Mobile phase composition, such as organic solvent, buffer concentration, and pH, are critical for the retention of ionic analytes. Risley and Pack [16] reported a comprehensive investigation on how organic composition, pH, and buffer concentration of the mobile phase impacted retention of ionic analytes using a zwitterionic stationary phase in HILIC mode. Liu and Pohl evaluated an anion-exchange/cation-exchange/reversed-phase trimodal column and suggested that the retention behavior of ionic analytes was heavily affected by organic solvent, buffer concentration, salt type, and pH levels. In other words, selectivity can be optimized by adjusting these variables according to specific requirement [9]. This finding was substantiated by Zhang *et al.* for pharmaceutical counterion ion analysis [11]. Therefore, the mobile phase needs to be accurately prepared for reproducible result. For inorganic ions using the Acclaim Trinity P1 (an anion-exchange/cation-exchange/reversed-phase trimodal column) and the Corona ultra detector, it was found that the best sensitivity for Na^+ and Cl^- was obtained when the mobile phase contained 20% and 60% acetonitrile, respectively (Liu X, Tracy M, Pohl C, unpublished result).

The response of CAD is influenced by the size of the mist aerosols generated in the nebulizer. Since the aerosol size is dependent on the mobile phase density, viscosity, and interfacial tension, CAD is sensitive to the types and concentration of buffers and organic solvent added to the mobile phase. Mitchell *et al.* [17] found that CAD was approximately 5–10 times more sensitive in HILIC mode (90% acetonitrile/10% buffer) than in RP mode (5% acetonitrile/95% buffer) for a group of 12 hydrophilic compounds. The response of CAD is also sensitive to the additives in the mobile phase. Vervoort *et al.* studied response between ELSD and CAD at varied concentrations of ammonium acetate (5, 10, and 20 mM) in the mobile phase (60% water/40% acetonitrile) and concluded that CAD performed significantly better than ELSD at low buffer concentrations but, at higher buffer concentrations, the S/N ratio for CAD dropped markedly. Using volatile acids such as formic or acetic acid did not pose any problem for CAD [18]. It is generally recommended that the total buffer concentration in the mobile phase should be below 100 mM for best sensitivity.

CAD is a mass-dependent detector, and the generated response does not depend on the spectral or physicochemical properties of the analyte as in a specific UV detector. Theoretically, CAD can generate a similar response for identical amounts of different analytes. For example, only a slight variation of the response for equal amount of compounds analyzed was observed by Gamache *et al.* over a test set of 17 chemically different compounds under isocratic elution conditions [19]. However this variation was about 7% relative standard deviation (RSD) between all responses of all 17 chemically different compounds, which indicates that CAD response depends upon analyte volatility [19]. The response of CAD is influenced by the diameter of generated particles, thus dependent on the mobile phase composition. In gradient elution chromatography, the response factor will vary significantly with the mobile phase composition, which is a main drawback of evaporative aerosol detectors [20]. Higher organic content in the mobile phase leads to greater transport efficiency of the nebulizer, which results in a larger number of particles reaching the detector chamber and a higher signal [21]. Gorecki *et al.* [20] proposed an elegant approach based on mobile phase compensation to solve this problem. The principle is to provide the detector at all times with a constant composition of the mobile phase. In this method, a secondary stream of the mobile phase of exactly reverse composition is provided by a second pump and was added to the column effluent to ensure a constant mobile phase composition at the detector inlet. This resulted in constant response, independent of the mobile phase composition in the column. In modern IC, mobile phases are usually hydroxide, carbonate, or methanesulfonic acid aqueous solutions. With the use of automobile phase generator and suppressor, the effluent after the ion suppressor contains only high-purity water (volatile) and the analytes (nonvolatile), which is a perfect condition for CAD operation. In addition, IC manufacturers often prescribe the "standard" separation conditions for most ion analysis. Especially for RFIC in which only water is needed to facilitate ion analysis, CAD can be used after the suppressor, and no mobile phase compensation is necessary for gradient methods [22].

6.2.4 CAD Parameter Setting

The operating process of CAD is described as follows. First, the mobile phase of a chromatographic system is nebulized using a flow of gas. Then the resulting aerosol is transported through a drift tube where the volatile components and solvents are evaporated. In the last step, the dried particle stream is charged with a secondary stream of nitrogen that has passed a high-voltage platinum needle, and the resulting charged particle flux is measured by an electrometer. The operation of CAD is simple, requiring only setting of few parameters, such as gas (nitrogen or air) input pressure, nebulizer temperature or temperature range, and signal output range. Typically, the gas pressure is set at 35 psi, nebulizer temperature in the range from 10 to 50°C, and signal output in the range from 50 to 200 pA. Eom *et al.* reported that in most cases, the purity of nitrogen gas did not have a significant effect on the sensitivity of CAD [23]. Generally, while it is not necessary to use highly pure nitrogen for CAD operation, it is critical that the gas should be free of particulates. For inorganic ion analysis, it was found that a nebulizer temperature higher than 30°C was necessary to achieve good sensitivity (Liu X, Tracy M, Pohl C, unpublished result).

6.2.5 Sensitivity

Compared with ELSD, CAD provides higher sensitivity with low limits of detection (LOD) for a wide range of analytes, irrespective of their chemical structure. In general, CAD can readily detect compounds present in single digit ng quantities, approximately 2–10 times better than ELSD. This sensitivity improvement is very important for determination of pharmaceutical impurities that have no chromophore. Detections of impurities of 0.05% (or single digit ng) along with the parent compound are routinely achievable. A recent study (Liu X, Tracy M, Pohl C, unpublished result) using CAD, ELSD, and suppressed conductivity detection for ion analysis showed that CAD offered 5 and 12 times better LOD for Na^+ and Cl^-, respectively, than ELSD (Figure 6.4). In the same study, it was also shown that conductivity detection was approximately two orders of magnitude more sensitive than CAD.

6.2.6 Calibration Curve, Dynamic Range, Accuracy, and Precision

CAD is a mass-dependent detector, and the response does not depend on the spectral or physicochemical properties of the analyte as in a specific UV detector, which is a concentration-dependent detector. The relationship between signal and amount of analyte is nonlinear in CAD, in which the relationship between area response and analyte mass can be described by Vehovec and Obreza [24]:

$$A = aM^b \tag{6.1}$$

where A is the area response of the detector, M is the mass of the analyte, and a and b are values that depend on the analyte and chromatographic conditions.

Equation 6.1 can be transformed into

$$\log A = b \log M + \log a \tag{6.2}$$

which can be used for calibration as a linear log–log plot of peak area versus quantity of analyte. This allows accurate quantification when using a two- or three-point calibration curve with CAD.

CAD gives parabolic calibration curve. The study by Vervoort *et al.* [18] showed that recovery for a high concentration sample was always in the 98–102% interval when using a two- or three-point calibration curve. The same was true for low concentrations. However, when the concentration level is very low or when the concentration range is small, the calibration curve is close to linear. Another study on etidronate disodium and related impurities (phosphate and phosphite salts) using CAD showed good correlation coefficients (0.9981, 0.9953, and 0.9956 for etidronate, phosphate, and phosphite, respectively) and excellent recoveries (95.9–102.6%, 94.5–103.6%, and 96.5–107.9% for etidronate, phosphate, and phosphite, respectively) [25]. Huang *et al.* evaluated linearity and recoveries of various pharmaceutical counterions using Corona CAD [12]. Although the relationship of response as a function of ion concentration was nonlinear over ranges of approximately two orders of magnitude, linear regressions with $r^2 > 0.995$ were achieved at a

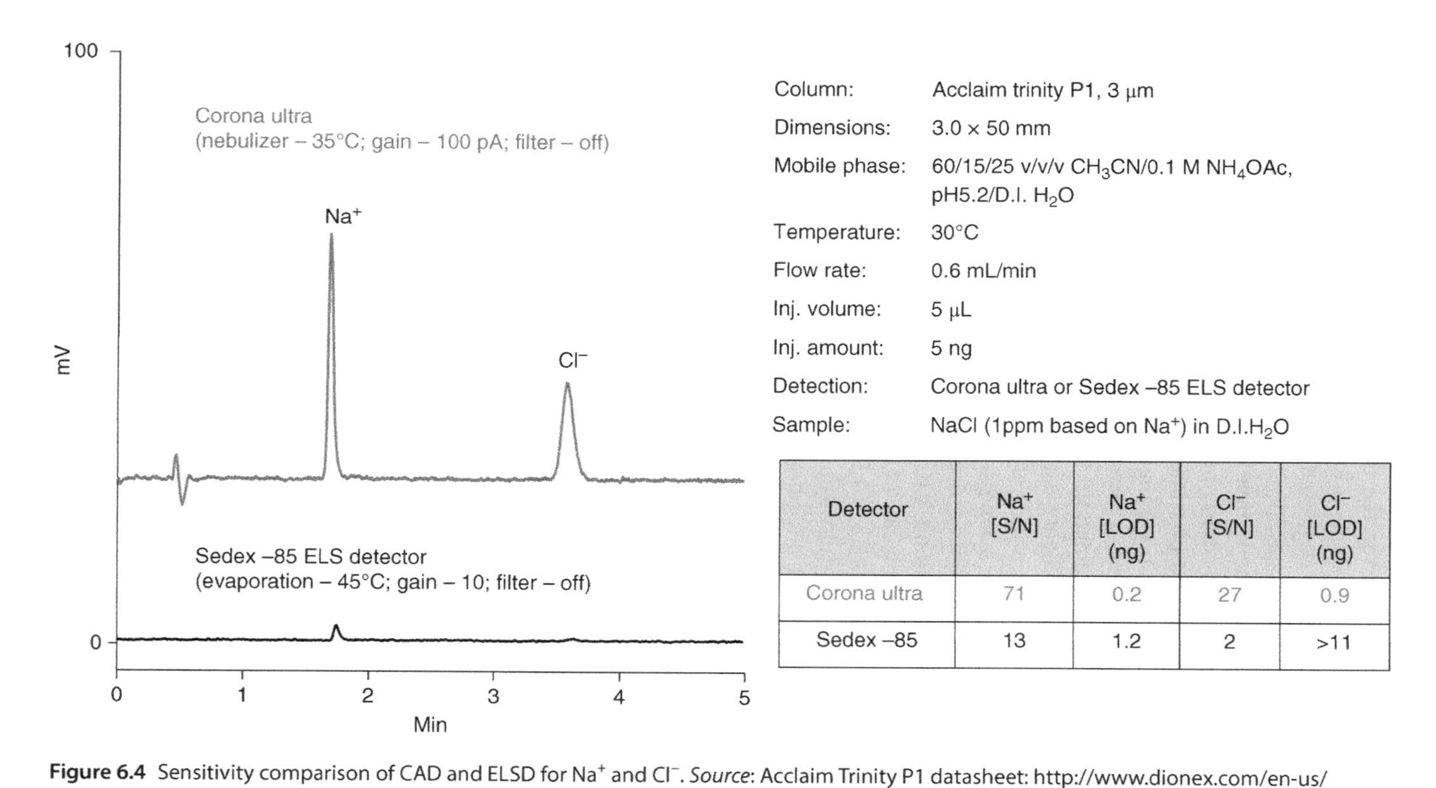

Detector	Na^+ [S/N]	Na^+ [LOD] (ng)	Cl^- [S/N]	Cl^- [LOD] (ng)
Corona ultra	71	0.2	27	0.9
Sedex –85	13	1.2	2	>11

Figure 6.4 Sensitivity comparison of CAD and ELSD for Na^+ and Cl^-. *Source*: Acclaim Trinity P1 datasheet: http://www.dionex.com/en-us/webdocs/70761-DS-Acclaim-Trinity-12Feb2010-LPN2239-02.pdf. Reproduced with permission of Thermo Fisher Scientific Inc.

smaller range (75–125%) of the nominal concentration (typically 50 mg/L) for the inorganic ions. However, the percent bias between the theoretical peak areas determined from a single point line that is forced through zero and the raw peak areas cannot be neglected.

The precision, or more specifically the system repeatability, is inferred from repetitive injections of the sample mixture at different concentration levels. CAD provides significantly better precision than ELSD. The study by Vervoort *et al.* on ten hydrophilic pharmaceutical molecules indicated an RSD of 2.0–3.0% in the concentration range of 1–0.05 mg/mL. At concentrations less than 0.05 mg/mL, imprecision was higher with RSD values in the range of 10–20% [18]. For determination of inorganic pharmaceutical counterions, Huang *et al.* reported excellent system precision for Cl^-, K^+, Ca^{2+}, and SO_4^{2-} in six drug substances with an RSD of less than 2.0% for each set of six standard injections [12]. In a separate study on determination of pharmaceutical counterions by Zhang *et al.*, three drug substances naproxen sodium, adenine hydrochloride, and compound X in hemifumarate salt were analyzed for their counterions; the calibration curves were linear in the defined range; the accuracy was in the range of 99.0–101.0% of theoretical values; and the RSD was less than 2.0% [11]. In the study using CAD to detect etidronate salt and related impurities (e.g., phosphate and phosphite) [25], Liu *et al.* evaluated the RSD at LOQ (S/N = 10 : 1) and LOD (S/N = 3 : 1) levels during method validation. The LOQ and LOD for etidronate were 144 and 50 ng, with %RSD of 3.1 and 14.7, respectively. The LOQ for phosphate and phosphite was 75 ng with RSD of 2.0%. When the injection volume was 5 µL, the LOQ level for phosphate and phosphite was equivalent to 0.015 mg/mL or 0.3% of the etidronate assay concentration. The LOD for phosphate and phosphite is 50 ng, with RSD of 4.4% for phosphate and 10.7% for phosphite.

The dynamic range of CAD allows quantification across a range that exceeds four orders of magnitude, from low ng to high µg scale. Combined with the excellent sensitivity, accuracy, and precision, CAD provides significant advantages over ELSD for measurements of an analyte and low level impurities in a single run [24].

6.3 Applications

As many ions either have a weak or no chromophore, UV detection only has limited applications for ion analysis. Conductivity detectors, RID, and ELSD are commonly utilized to detect ions. CAD is a universal detector for nonvolatile compounds. CAD has good sensitivity, accuracy and repeatability and adequate linearity. Since the introduction of commercial CAD detector in 2005, the interest of using CAD as an alternative technology for ion analysis has been growing rapidly, especially in pharmaceutical industry [26].

6.3.1 Pharmaceutical Counterions and Salts

Salts have been routinely used as counterions in pharmaceutical drug substances to improve drug solubility, stability, and processability. About half of the active drug substances approved by the US FDA are in salt forms [27]. The final concentrations of counterions affect drug potency, stability, and bioavailability; thus the control of counterions is a key element in drug substance specifications. Because most counterions have no UV chromophores, CAD has gained great popularity in counterion analysis [11, 12, 28].

One attraction of using CAD is that it does not require a dedicated IC system to perform ion analysis, provided that the sensitivity requirement is larger than ng level, such as pharmaceutical counterion analysis. In this case, CAD can be connected to a conventional HPLC system that is typically equipped with a UV detector, so the same instrument can be used not only for ion analysis but also for other uses, like assay, impurity, and content uniformity tests. This also makes method transfer easier between labs because conventional HPLC systems are readily available in different analytical labs. Another attraction of using CAD is that when using an appropriate separation column, there is no need to have separate methods for cations and anions, which calls for different separation columns and instrument configuration by IC. This way, both negative and positive ions can be analyzed simultaneously. In this aspect, it is greatly attributed to the successful mixed-mode column technologies in the recent years and the HILIC separation mode [9–12, 16, 28]. The coupling of mixed-mode columns and CAD detection is extremely helpful for the analysis of ionic, polar, and non-UV-absorbing compounds. A wide variety of ions, whether cations or anions, inorganic or organic, monovalent or multivalent, can be separated simultaneously by using conventional HPLC and CAD. The separation mechanisms include ion-exchange, reversed-phase, and HILIC interactions.

As shown in Figure 6.5, using mixed-mode column Dionex Acclaim Trinity P1 separation and CAD [11], Zhang *et al.* developed a generic method for simultaneous separation of 25 commonly used anionic and cationic pharmaceutical counterions within 20 min. The reported ions are:

- Anions (order according to the pharmaceutical occurrence): chloride, sulfate, bromide, maleate, mesylate, tartrate, citrate, phosphate, fumarate, nitrate, lactate, succinate, besylate, malate, gluconate, and tosylate
- Cations (order according to the pharmaceutical occurrence): sodium, calcium, potassium, meglumine, tromethamine (Tris), zinc, magnesium, procaine, and choline

Because of the significant physicochemical property differences of counterions and API, they are generally well separated, and there is no interference from API matrix. Figure 6.6 shows the separation of counterions and their APIs.

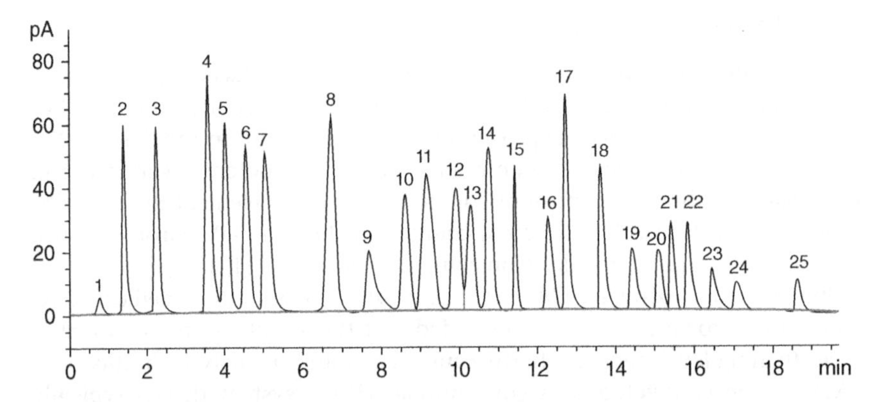

Figure 6.5 Chromatogram of the separation of 25 common pharmaceutical ions. Column, Acclaim Trinity P1, 3.0×50-mm format; mobile phase, A, acetonitrile; B, 200 mM ammonium format (pH 4.0); C, DI water; gradient elution; flow rate, 0.5 mL/min; injection volume, 5 µL; temperature, 35°C; and detection, Corona CAD. Peaks: (1) lactate, (2) procaine, (3) choline, (4) tromethamine, (5) sodium, (6) potassium, (7) meglumine, (8) mesylate, (9) gluconate, (10) maleate, (11) nitrate, (12) chloride, (13) bromide,(14) besylate, (15) succinate, (16) tosylate, (17) phosphate, (18) malate, (19) zinc, (20) magnesium, (21) fumarate, (22) tartrate, (23) citrate, (24) calcium, and (25) sulfate. *Source*: Zhang *et al.* [11]. Reproduced with permission of Elsevier.

The method showed good linearity in the defined range, and the accuracy was in the range of 99.0–101.0% with %RSD less than 2.0%.

Huang *et al.* [12] reported the separation of inorganic counterions using Sequant ZIC-*p*HILIC columns by two different methods based on the valence of the ions: (i) a 150-mm column with pH 7.0 ammonium acetate buffer: acetonitrile (25:75) as mobile phase for monovalent ions such as NO_3^-, Cl^-, Br^-, Na^+, and K^+ (ii) a 50-mm column with pH 3.5 ammonium formate: acetonitrile (30:70) as mobile phase for multivalent ions such as Ca^{2+}, Mg^{2+}, SO_4^{2-}, and PO_4^{3-}. Excellent system precision was obtained for all tested ions with RSD less than 2.0%. In the same study, a better accuracy was observed for Cl^- in several drug substances as compared to IC. The authors used a three-point standard calibration curve for counterion quantification to overcome the percent bias from a single point line.

Since counterions may constitute 5–30 wt% of drug substances, the sensitivity by most of the detection techniques should be adequate for assay and stoichiometric calculation purpose. However, salts can also be present in the drug product as undesirable impurities carried from raw materials, upstream synthesis, or contaminations; thus the quantification and control of trace ions is sometimes critical, especially when potential genotoxic impurities can be formed, such as the genotoxic alkyl mesylates. Good sensitivity can be achieved for most of the ions by CAD at ng level or ppm and sub-ppm level depending on the injection volume [11, 28].

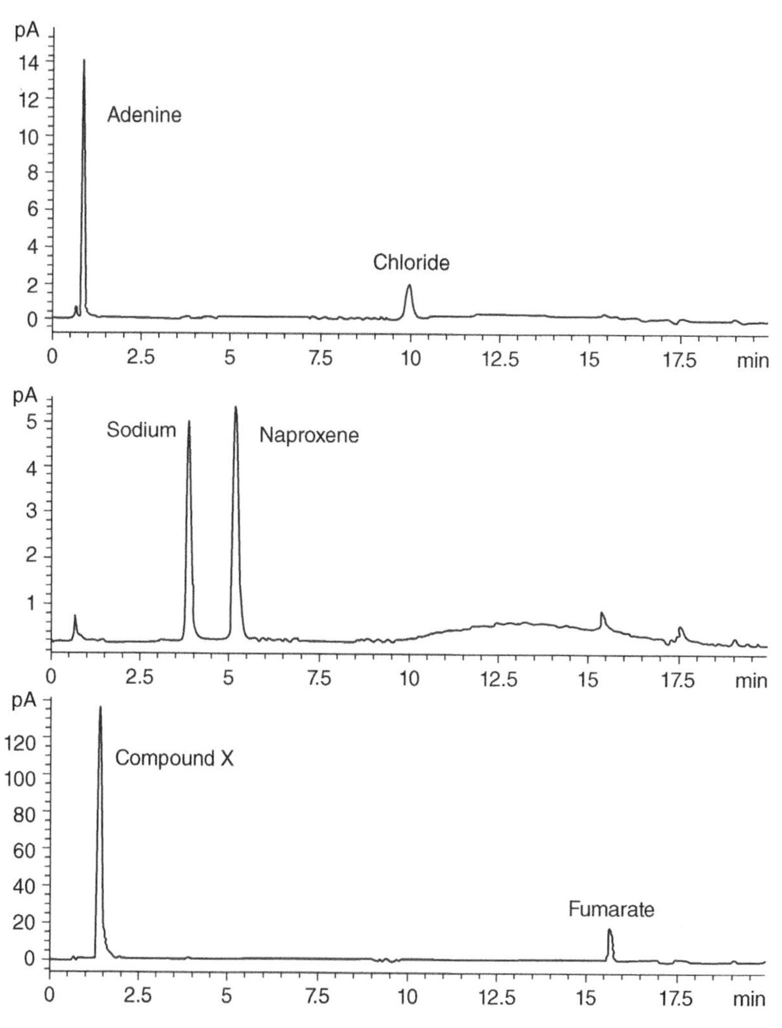

Figure 6.6 Chromatograms of the separation of active pharmaceutical ingredients and their counterions. Column, Acclaim Trinity P1, 3.0×50-mm format; mobile phase, A, acetonitrile; B, 200 mM ammonium format (pH 4.0); C, DI water; gradient elution; flow rate, 0.5 mL/min; injection volume, 5 μL; temperature, 35°C; and detection, Corona CAD. Samples: adenine hydrochloride (top), naproxen sodium (middle), and compound X fumarate salt (bottom). *Source*: Zhang *et al*. [11]. Reproduced with permission of Elsevier.

6.3.2 Bisphosphonate

Bisphosphonate is a class of drugs that prevent the loss of bone mass. The analysis of bisphosphonate has been challenging because these molecules are polar and ionic and most of them have no UV chromophore. Pre-column derivatization, indirect UV analysis, ion pair, and CE methods were reported to separate and detect these molecules. Etidronate, (1-hydroxyethylidene) bisphosphonate, and its impurities phosphate and phosphite have no UV chromophore and are highly ionic, so it is difficult to separate these compounds by conventional HPLC columns and to detect by ordinary spectrophotometric methods. Liu *et al.* [25] developed a stability-indicating method for the analysis of etidronate disodium and its related substances by using a mixed-mode column (Primesep SB) and CAD. The method was also validated in terms of specificity, linearity, accuracy, precision, sensitivity, and stability. The correlation of peak area versus concentration of etidronate was not linear. The authors used a logarithmic calibration equation and found that the recoveries of etidronate at different levels were in the range of 95.9–102.6%. Similarly, the recoveries of phosphate and phosphite were found in the range of 94.5–103.6% and 96.5–107.9%, respectively. Good accuracy and precision was demonstrated for etidronate and the impurities. The LOQ and LOD for etidronate were 144 and 50 ng, with %RSD of 3.1 and 14.7, respectively. The LOQ for phosphate and phosphite was 75 ng with %RSD of 2.0. The LOD for phosphate and phosphite is 50 ng, with %RSD of 4.4% for phosphate and 10.7% for phosphite. It was noted that although CAD is reportedly a universal detector with response magnitude independent of the analyte chemical properties, Liu *et al.* observed different recovery responses of phosphate and phosphite by using etidronate standard at impurity levels.

6.3.3 Phosphorylated Carbohydrates

Phosphorylated carbohydrates are important metabolites in various central metabolic pathways. Chromatographic separation of individual isomeric forms of phosphorylated carbohydrates is critical as these isomers give similar fragmentation patterns and cannot be distinguished by MS/MS detection. However, the chromatographic separation and detection is challenging because of the hydrophilicity of phosphorylated carbohydrates and lack of strong UV-absorbing groups of most carbohydrates. Hinterwirth *et al.* [29] reported the separation of isomeric sugar phosphates by reversed-phase/weak anion-exchanger mixed-mode column and CAD. The employed stationary phases were 3-aminoquinuclidine-derived and 3-α-aminotropane-derived reversed-phase/weak anion-exchanger mixed-mode columns. The best results were obtained when the column was operated under HILIC mode. Acidic conditions led to the complete separation of α- and β-anomers of glucose 6-phosphate at low temperature.

6.3.4 Ionic Liquids

Ionic liquid, also called liquid salt, is a class of organic salts with melting point below the boiling point of water. The interest of ionic liquid is rapidly growing because of the variety of industry applications. Ionic liquids with UV-absorbing groups like imidazolium- and pyridinium-based molecules can be analyzed by UV detector, but for the ionic liquids such as ammonium- and phosphonium-based molecules that lack chromophoric groups, CAD is very attractive. Stojanovic *et al.* applied the coupling of CAD for the non-chromophoric aliphatic cations with diode array detection for the aromatic anions to simultaneous analysis of a set of new ionic liquids derived from either tricaprylmethylammonium chloride (Aliquat 336) or trihexyltetradecylphosphonium chloride [30]. In this study, a gradient method was developed using Gemini C18 column and 0.1% TFA in water as mobile phase A and 0.1% TFA in ACN as mobile phase B. The authors established a unified calibration function for the quantitative analysis of the quaternary ammonium cations of the ionic liquids, due to lack of single component standards. The method sensitivity LOD is at the ng level for all tested ions, which is in agreement with the reported figures for other compounds.

6.3.5 Pesticides

Some bacteria can release halide ions from halogenated hydrocarbons that are widely used as organic pesticides. The level of halogenated hydrocarbons in the environment can be analyzed by quantification of the halide anions. Mikelova *et al.* [31] compared amperometric, coulometric, and CAD for estimation of contamination of the environment by pesticides and found the chloride ion detection limits were 30 μM for CAD, 100 mM for coulometric, and 1 nM for amperometric detection. An anion-exchange column (Hamilton PRP-X100) was employed in this study.

6.3.6 Other Applications

CAD was also applied for bile salts analysis. CAD has been reported to have better sensitivity than UV, RI, and ELSD for bile acid determinations. Vertzoni *et al.* [32] developed and validated an isocratic HPLC method with CAD to determine individual bile salts and applied the method to determine bile acids in human gastric and duodenal aspirates. CAD with gradient HPLC method was also applied to characterize bile salt content of ascending colon [33].

Phospholipids are major constituents of cell membranes and are widely used in foods, pharmaceutical, and cosmetic products. CAD was also used for the analysis of amphoteric lipids like phosphatidylcholine (PC), phosphatidylethanolamine (PE), and their hydrolyzed forms Lyso-PC and Lyso-PE [34, 35]. Chojnacha *et al.* [36] reported using CAD and normal phase separation to analyze phospholipids to study the enzymatic enrichment of egg-yolk PC with α-linolenic acid.

Besides the applications described earlier, CAD is also used in the analysis of polar ionic species that have no or weak UV chromophores, like amino acids [37, 38], antibiotics (e.g., gentamicin sulfate) [39], and pharmaceutical starting materials [40].

6.4 Concluding Remarks

CAD represents the latest advancement in aerosol-based detector technology. Its main benefits include universal detection of nonvolatile or semi-volatile analytes, response largely independent of chemical properties, broad dynamic response range with high sensitivity from low ng to high μg amounts of analytes, good precision for a diverse range of analytes, and simple and reliable operation. While IC using suppressed conductivity detection has been routinely used for ion analysis because of its excellent sensitivity and selectivity for ionic species, CAD is very attractive for those applications with simple sample matrix and less demanding sensitivity requirement (>1 ppm), such as pharmaceutical counterion analysis.

For ion analysis using CAD, several factors should be considered to achieve satisfactory results, such as instrument platform, CAD parameter setting, column selection, and mobile phases. CAD can be coupled with either LC or IC instruments. Its operation is simple, requiring only setting of few controllable parameters, such as gas input pressure, nebulization temperature, and signal output range. With newer CAD instruments, evaporation temperature is an additional parameter that can be used for detector optimization. The separation column is critical for ion analysis with CAD. Not only should the column provide desired selectivity but also be of least column bleed for good sensitivity. Equally important, the CAD-compatible mobile phase needs to be volatile and free of nonvolatile impurities, requiring the use of highest quality of organic solvent, DI water, and other additives. In the event of a gradient elution, mobile phase compensation (inverse gradient) can be used to achieve constant response, independent of the mobile phase composition in the column.

Since its debut in 2005, the first commercial CAD detector (Corona CAD) has been successful for pharmaceutical analysis, such as determination of counterions, and sometimes drug substances as well. Although CAD is not as sensitive or as selective as conductivity detection for inorganic ions and small organic ions, it is compatible with both LC and IC instrument, has superior sensitivity and precision to ELSD, has close to linear response for a narrow calibration range and with mixed-mode columns can provide simultaneous analysis of anions and, cations and of ions and APIs. It is expected that CAD will gain even more acceptance in pharmaceutical applications as well as other areas where its benefits can be fully utilized.

References

1 Small H, Stevens TS, Bauman WC. Novel ion exchange chromatographic method using conductimetric detection. Analytical Chemistry 1975, 47(11), 1801–1809.

2 Small H, Ion Chromatography. Plenum: New York, 1989.
3 Haddad PR, Jackson PE. Ion chromatography principles and applications. Journal of Chromatography Library 1990, 46, 1.
4 Fritz JS, Gjerde DT, Ion Chromatography. Wiley-VCH: Weinheim, 2000.
5 Weiss J, Ionenchromatographie. 3rd ed.; Wiley-VCH: Weinheim, 2001.
6 Pohl C, Stillian JR, Jackson PE. Factors controlling ion-exchange selectivity in suppressed ion chromatography. Journal of Chromatography A 1997, 789, 29–41.
7 Jackson PE, Pohl CA. Advances in stationary phase development in suppressed ion chromatography. Trends in Analytical Chemistry 1997, 16(7), 393–400.
8 Liu Y, Srinivasan K, Pohl C, Avdalovic N. Recent developments in electrolytic devices for ion chromatography. Journal of Biochemical and Biophysical Methods 2004, 60, 205–232.
9 Liu X, Pohl C, Woodruff A, Chen J. Chromatographic evaluation of reversed-phase/anion-exchange/cation-exchange trimodal stationary phases prepared by electrostatically driven self-assembly process. Journal of Chromatography A 2011, 1218(22), 3407–3412.
10 Liu X, Pohl C. HILIC behavior of a reversed-phase/cation-exchange/anion-exchange trimode column. Journal of Separation Science 2010, 33(6–7), 779–786.
11 Zhang K, Dai L, Chetwyn NP. Simultaneous determination of positive and negative pharmaceutical counterions using mixed-mode chromatography coupled with charged aerosol detector. Journal of Chromatography A 2010, 1217(37), 5776–5784.
12 Huang Z, Richards MA, Zha Y, Francis R, Lozano R, Ruan J. Determination of inorganic pharmaceutical counterions using hydrophilic interaction chromatography coupled with a Corona® CAD detector. Journal of Pharmaceutical and Biomedical Analysis 2009, 50(5), 809–814.
13 Liu X, Pohl C, Is HILIC the Best Way for Determination of Active Pharmaceutical Ingredients and Counterions? Poster, presented at HPLC 2010, June 21, 2010, Boston.
14 Pohl C. Recent developments in ion-exchange columns for inorganic ions and low molecular weight ionizable molecules. LC-GC North America 2010, 28, 24–31.
15 Teutenberg T, Tuerk J, Holzhauser M, Kiffmeyer TK. Evaluation of column bleed by using an ultraviolet and a charged aerosol detector coupled to a high-temperature liquid chromatographic system. Journal of Chromatography A 2006, 1119(1–2), 197–201.
16 Risley DS, Pack BW. Simultaneous determination of positive and negative counterions using a hydrophilic interaction chromatography method. LC-GC North America 2006, 24, 82–90.
17 Mitchell CR, Bao Y, Benz NJ, Zhang S. Comparison of the sensitivity of evaporative universal detectors and LC/MS in the HILIC and the reversed-phase HPLC modes. Journal of Chromatography B 2009, 877(32), 4133–4139.

18 Vervoort N, Daemen D, Torok G. Performance evaluation of evaporative light scattering detection and charged aerosol detection in reversed phase liquid chromatography. Journal of Chromatography A 2008, 1189(1–2), 92–100.

19 Gamache PH, Mccarthy RS, Freeto SM, Asa DJ, Woodcock MJ, Laws K, Cole RO. HPLC analysis of nonvolatile analytes using charged aerosol detection. LCGC North America 2005, 23(2), 155–161.

20 Gorecki T, Lynen F, Szucs R, Sandra P. Universal response in liquid chromatography using charged aerosol detection. Analytical Chemistry 2006, 78, 3186–3192.

21 Cobb Z, Shaw PN, Lloyd LL, Wrench N, Barrett DA. Evaporative light-scattering detection coupled to microcolumn liquid chromatography for the analysis of underivatized amino acids: Sensitivity, linearity of response and comparisons with UV absorbance detection. Journal of Microcolumn Separations 2001, 13(4), 169–175.

22 Thermo Fisher Scientific. Website: Datasheet of Eluent Suppressors for Ion Chromatography, https://tools.thermofisher.com/content/sfs/brochures/PS-70690-IC-Eluent-Suppressors-PS70690-EN.pdf (accessed February 1, 2017).

23 Eom HY, Park SY, Kim MK, Suh JH, Yeom H, Min JW, Kim U, Lee J, Youm JR, Han SB. Comparison between evaporative light scattering detection and charged aerosol detection for the analysis of saikosaponins. Journal of Chromatography A 2010, 1217(26), 4347–4354.

24 Vehovec T, Obreza A. Review of operating principle and applications of the charged aerosol detector. Journal of Chromatography A 2010, 1217(10), 1549–1556.

25 Liu X-K, Fang JB, Cauchon N, Zhou P. Direct stability-indicating method development and validation for analysis of etidronate disodium using a mixed-mode column and charged aerosol detector. Journal of Pharmaceutical and Biomedical Analysis 2008, 46(4), 639–644.

26 Swartz M, Emanuele M, Awad A, Hartley D. Charged aerosol detection in pharmaceutical analysis: An overview. LC-GC Chromatography Online 2009, 27, 40–48.

27 Paulekuhn GS, Dressman JB, Saal C. Trends in active pharmaceutical ingredient salt selection based on analysis of the orange book database. Journal of Medicinal Chemistry 2007, 50(26), 6665–6672.

28 Crafts C, Bailey B, Plante M, Acworth I. Evaluation of methods for the simultaneous analysis of cations and anions using HPLC with charged aerosol detection and a zwitterionic stationary phase. Journal of Chromatographic Science 2009, 47, 534–539.

29 Hinterwirth H, Lammerhofer M, Preinerstorfer B, Gargano A, Reischl R, Bicker W, Trapp O, Brecker L, Lindner W. Selectivity issues in targeted metabolomics: Separation of phosphorylated carbohydrate isomers by mixed-mode hydrophilic interaction/weak anion exchange chromatography. Journal of Separation Science 2010, 33, 3273–3282.

30 Stojanovic A, Lämmerhofer M, Kogelnig D, Schiesel S, Sturm M, Galanski M, Krachler R, Keppler BK, Lindner W. Analysis of quaternary ammonium and phosphonium ionic liquids by reversed-phase high-performance liquid chromatography with charged aerosol detection and unified calibration. Journal of Chromatography A 2008, 1209(1–2), 179–187.

31 Mikelova R, Prokop Z, Stejskal K, Adam V, Beklova M, Trnkova L, Kulichova B, Horna A, Chaloupkova R, Damborsky J, Kizek R. Enzymatic reaction coupled with flow-injection analysis with charged aerosol, coulometric, or amperometric detection for estimation of contamination of the environment by pesticides. Chromatographia 2008, 67, 47–53.

32 Vertzoni M, Archontaki H, Reppas C. Determination of intralumenal individual bile acids by HPLC with charged aerosol detection. Journal of Lipid Research 2008, 49, 2690–2695.

33 Diakidou A, Vertzoni M, Goumas K, Söderlind E, Abrahamsson B, Dressman J, Reppas C. Characterization of the contents of ascending colon to which drugs are exposed after oral administration to healthy adults. Pharmaceutical Research 2009, 26(9), 2141–2151.

34 Nair LM, Werling JO. Aerosol based detectors for the investigation of phospholipid hydrolysis in a pharmaceutical suspension formulation. Journal of Pharmaceutical and Biomedical Analysis 2009, 49(1), 95–99.

35 Moreau RA. The analysis of lipids via HPLC with a charged aerosol detector. Lipids 2006, 41, 727–734.

36 Chojnacka A, Gładkowski W, Kiełbowicz G, Wawrzeńczyk C. Enzymatic enrichment of egg-yolk phosphatidylcholine with α-linolenic acid. Biotechnology Letters 2009, 31(5), 705–709.

37 Holzgrabe U, Nap C-J, Beyer T, Almeling S. Alternatives to amino acid analysis for the purity control of pharmaceutical grade L-alanine. Journal of Separation Science 2010, 33(16), 2402–2410.

38 Holzgrabe U, Nap C-J, Almeling S. Control of impurities in L-aspartic acid and L-alanine by high-performance liquid chromatography coupled with a corona charged aerosol detector. Journal of Chromatography A 2010, 1217(3), 294–301.

39 Joseph A, Rustum A. Development and validation of a RP-HPLC method for the determination of gentamicin sulfate and its related substances in a pharmaceutical cream using a short pentafluorophenyl column and a charged aerosol detector. Journal of Pharmaceutical and Biomedical Analysis 2010, 51(3), 521–531.

40 Soman A, Jerfy M, Swanek F. Validated HPLC method for the quantitative analysis of a 4-methanesulfonyl-piperidine hydrochloride salt. Journal of Liquid Chromatography and Related Technologies 2009, 32(7), 1000–1009.

7

Determination of Carbohydrates Using Liquid Chromatography with Charged Aerosol Detection

Jeffrey S. Rohrer[1] *and Shinichi Kitamura*[2]

[1] *Thermo Fisher Scientific, Sunnyvale, CA, USA*
[2] *Graduate School of Life and Environmental Sciences, Osaka Prefecture University, Osaka, Japan*

CHAPTER MENU

7.1 Summary

Charged aerosol detection (CAD) has been commercially available for less than 10 years at this writing, yet we already have witnessed its rapid adoption in the analytical laboratory due to its ability to detect analytes lacking a good UV chromophore with better sensitivity and dynamic range than similar detectors. In the field of carbohydrate analysis, CAD had found a place between those applications requiring high sensitivity and those where sensitivity is not critical. This chapter showed that CAD has been substituted for other detectors in applications that already had a volatile mobile phase, and either more sensitivity or a greater dynamic response range was needed. The applications ranged from small sugars to large oligosaccharides and small polysaccharides. We expect that we will continue to see CAD substituted for RI, UV, and ELSD in carbohydrate applications due to the aforementioned benefits of CAD relative to those detectors. For more updated applications of CAD to carbohydrate analysis, please see Chapter 2.

Charged Aerosol Detection for Liquid Chromatography and Related Separation Techniques,
First Edition. Edited by Paul H. Gamache.

7.2 Liquid Chromatography of Carbohydrates

There are numerous liquid chromatography (LC) methods for the determination of carbohydrates. In part, the number of methods reflects the wide variety of carbohydrates, the large number of carbohydrate-containing samples, and the broad carbohydrate concentration range those samples represent. The variety of methods also reflects the challenges in designing an LC method for carbohydrate analysis that can be applied to all samples. While some of the methods enjoy more popularity than others, there is no single dominant method of analysis as found, for example, with peptide mapping, where nearly all separations use a Widepore C18 column with an acetonitrile/trifluoroacetic acid eluent.

Most carbohydrates cannot be determined using typical HPLC conditions, that is, separation on a C18 column and detection by UV absorbance. The two major reasons for this situation are that most carbohydrates are polar compounds and lack a good chromophore. This too led to the development of a number of LC solutions for the determination of carbohydrates. The choice of technique is typically dictated by the type of carbohydrate(s) being determined and the concentration(s) of the carbohydrate(s) in the sample. For example, simple carbohydrates (i.e., mono- and disaccharides) when present in samples at high concentrations, with the presence of few other compounds at high concentration, are usually separated with one of two types of stationary phases, a metal-loaded cation-exchange phase or an amine- or carbamoyl-bonded reversed phase. The second of the latter two phases is usually referred to as an amido-bonded stationary phase. On the metal-loaded cation-exchange phase, carbohydrates are separated by differences in their affinity for the metal ion (e.g., Pb^{2+}) that has been previously bound to the cation-exchange stationary phase. Water is the mobile phase and it is heated to 80°C so that α and β anomers elute as one peak. Carbohydrates are separated on the amine or amido stationary phase using a mobile phase with a high organic solvent content (usually between 50 and 80%) either isocratically or with a gradient of decreasing organic solvent. Refractive index (RI) detection is used with the metal-loaded cation-exchange separations and when the amine or amido stationary phase separation uses isocratic conditions. Absorbance at a low UV wavelength (e.g., 202 nm) is sometimes used with an amine or amido stationary phase column. The detection technique is neither sensitive (low to mid nmol detection limits) nor specific, but because the carbohydrate concentrations are high and the separation delivers some specificity, both techniques are effective for most samples with high carbohydrate concentrations and low concentrations of other compounds. While there are many samples with high concentrations of larger carbohydrates (i.e., oligo- and polysaccharides), gradient separations are required for most samples, and neither RI nor low UV absorbance is suitable for use with mobile phase gradients. The exception is size-exclusion chromatography (SEC) where a gradient is not required and water or dilute salt solution is used as the mobile phase. Unfortunately SEC can only effectively resolve straight chain oligosaccharides up to about a degree of polymerization (DP) between 10 and 12, and the separations can require hours to complete. Unless the determination of larger oligosaccharides (>100 DP) and polysaccharides is required, SEC-RI methods for oligo- and polysaccharides have been replaced with techniques that resolve much larger oligosaccharides, which will be discussed in the forthcoming pages.

Samples containing low concentrations of carbohydrates and/or larger oligosaccharides than can be resolved by SEC spurred the development of chromatographic techniques for determining carbohydrates. One such example was the development of a new detector, the evaporative light scattering detector (ELSD). Although newer than RI or UV, ELSD has been available for well over 20 years. An ELSD requires a volatile mobile phase that is sprayed and evaporated, leaving only the separated nonvolatile carbohydrate to scatter light and thus be detected. The ELSD can be substituted for the RI detector in the metal-loaded cation-exchange separations and in the amine- or amido-bonded separations when the mobile phase is volatile. It can also be used when gradients are used with the amine- or amido-bonded columns. Unfortunately ELSD offers only a three- to fivefold improvement in sensitivity over RI and low UV absorbance detections.

When high sensitivity is required, the available techniques can be roughly divided into two groups, methods that require sample derivatization for detection and those that do not. One way to achieve greater sensitivity is to derivatize the carbohydrate with a fluorophore. While some early work derivatized carbohydrates with chromophores [1], today nearly all derivatization is with a fluorophore using a reductive amination reaction via a Schiff base in the presence of an excess of the fluorophore, because it provides approximately 10-fold more sensitivity than UV detection and is less subject to inference from other compounds in the sample. After fluorophore attachment the derivatized carbohydrate is often separated on an amine- or amido-bonded silica phase with mobile phases having a high organic solvent content. Over the years these separations have been referred to as reversed-phase chromatography, normal-phase chromatography, a mix of normal-phase and anion-exchange chromatography, partition chromatography, and hydrophilic interaction chromatography (HILIC) [2, 3]. The latter mechanism is most commonly referred to today, although a 2006 review of fluorescent labeling techniques and subsequent chromatography for carbohydrates does not use the term HILIC to describe the amine- or amido-bonded stationary phase separations [3]. Other high-performance separation techniques used are ion-pairing reversed-phase in a C8 or C18 stationary phase [4], reversed-phase on a porous graphitized carbon (PGC) column [5], and weak anion-exchange chromatography (nearly always in tandem with one of the other techniques such as HILIC) [6].

In addition to fluorescence detection, separations that use volatile mobile phases can also be coupled to mass spectrometry detection with electrospray ionization. The two most commonly used fluorophores for LC of carbohydrates are 2-aminopyridine and 2-aminobenzamide (2-AB). These labels allow detection of carbohydrates at mid femtomolar concentrations, which is at least 10,000-fold more sensitive than RI detection.

High-performance anion-exchange chromatography with pulsed amperometric detection (HPAE-PAD) delivers high femtomole to low pmol sensitivity without sample labeling. Carbohydrates are separated at high pH where even a carbohydrate that is uncharged at pH 7 is an anion. Mono- and disaccharides are separated on a high-performance anion-exchange column, which is compatible with the strong alkaline mobile phase, using a sodium or potassium hydroxide mobile phase. Charged and large uncharged (at pH 7) carbohydrates are separated on the same column with a gradient of sodium acetate in a strong sodium hydroxide (≥100 mM) mobile phase. From both classes of HPAE-PAD separations, carbohydrates are detected at high pH by oxidation on a gold working electrode set at one potential. After analyte detection, a series of voltage potentials are applied to clean and restore the working electrode surface for subsequent analyte detection. This sequence of voltage potentials, including the detection potential, is executed twice a second and referred to as a waveform. Unlike the carbohydrate analysis techniques previously discussed, HPAE-PAD is applicable to a broad range of carbohydrates including monosaccharides, disaccharides, sugar alcohols, oligosaccharides (charged and uncharged, branched, and unbranched), oligosaccharide alditols, small polysaccharides, sugar acids, sialic acids, chemically modified carbohydrates, aminoglycosides, and aminosugars [7–9].

7.3 Charged Aerosol Detection

Charged aerosol detection (CAD) detects carbohydrates and other compounds without derivatization. As CAD is the subject of this book, the basic principles of CAD will only be briefly discussed here. CAD is a relatively new technique as it was first described in a scientific article in 2002 by Dixon and Peterson [10]. Like MS and ELSD, CAD requires that the mobile phase be volatile and, like ELSD, that the analyte not be volatile. As carbohydrates are not volatile without a derivatization step, this is not a limitation for analyzing carbohydrates by CAD. As the mobile phase exits the chromatography column, it is mixed with flowing gas (nitrogen or air) and nebulized to form an aerosol. While that aerosol transverses a heated drift tube, the mobile phase and other volatile compounds are evaporated to produce dried aerosol particles consisting of nonvolatile components including impurities and any analyte present. This aerosol flow is then mixed with a second stream of gas that has been ionized by passing it by a platinum needle set at high voltage. This ion jet in turn charges the aerosol particles, and the resulting charged particles are measured by an electrometer. More details on the principles of CAD operation can be found in Chapter 1 and in a recent review by Vehovec and Obreza [11].

7.4 Why LC-CAD for Carbohydrate Analysis?

With all the LC techniques for determining carbohydrates described earlier in this chapter, where does CAD fit within the field of carbohydrate determinations? First, like RI, ELSD, and PAD, CAD does not require a chromophore for detection, and as discussed earlier, nearly all carbohydrates lack a good chromophore. Additionally, like the three detection techniques mentioned in the previous sentence, CAD is a direct detection technique and therefore requires no sample derivatization. This saves time and money and eliminates the possibility of selective or incomplete analyte derivatization. Because CAD, like ELSD, is a mass-sensitive detector, its sensitivity is a bit difficult to compare with concentration-sensitive detectors (PAD, RI, UV, fluorescence), but CAD has low ng on column sensitivity and for many samples this will translate to mid pmol sensitivity. CAD has been reported to be as much as 10× more sensitive than low UV and RI for carbohydrates and has also been reported to be more sensitive than ELSD [12]. Because CAD response is independent of an analyte's structure and its chromophoric properties (not true for ELSD), carbohydrates are not needed to compare CAD sensitivity with ELSD. Vervoort *et al.* compared the two detection techniques and concluded that CAD was more sensitive than ELSD and under certain conditions up to six times more sensitive [13]. This result was confirmed by Eom *et al.* who evaluated CAD and ELSD for a set of ten saikosaponins and concluded that CAD was two to six times more sensitive than ELSD [14]. Another more recent study that separated antidiabetic drugs found that CAD was as much as two times more sensitive than ELSD [15].

In addition to better sensitivity, why has CAD been chosen instead of RI and ELSD? First, CAD has a larger linear range than RI and ELSD. While CAD's linearity is limited to just over an order of magnitude, that range can start at the detection limit and measure low concentrations, unlike ELSD that is inherently nonlinear at low concentrations. Second, using a log–log calibration plot, CAD has a large dynamic range. This is a property not shared by RI and ELSD. Although CAD response has sensitivity to eluent composition (as does ELSD), CAD can be used with gradient separations unlike RI detection and many separations that use low UV absorbance detection. Although CAD does have better sensitivity than many other techniques used for carbohydrate analysis, it does not have enough sensitivity to replace PAD and fluorescence detection in applications requiring high sensitivity.

Before discussing the application of CAD to carbohydrate analysis, it is prudent to discuss the limitations of CAD. First, CAD is limited to separations that use volatile eluents. The carbohydrate must also be soluble in the eluent. This usually is not a limitation but does apply to large oligosaccharides and small polysaccharides in highly organic mobile phases. As noted earlier, CAD response does vary with mobile phase composition as well as flow rate and temperature. The latter two variables are not a problem with a modern LC system as flow rate and temperature are easily and precisely controlled. The change in response with mobile phase conditions is only an issue when one wishes to compare the responses of unknowns in a separation with the response of a known when a gradient separation is used. More specifically, this applies to applications where one expresses the peak area of an unknown peak as a percentage of the peak area of a known peak, or a percentage of the area of all peaks in a separation. For applications where those calculations are required, there is a solution [16]. A second pump can deliver a solution that is mixed with the mobile phase after the column before the combined solution enters CAD. The added solution has the same components as the mobile phase but compensates for the gradient change so the solution that enters the detector always has the same composition, which is the mobile phase at the midpoint of the separation. Therefore there is no response variation due to mobile phase.

7.5 Early Applications of CAD to Carbohydrate Analysis

We have arbitrarily chosen a publication year of 2007 and earlier to define early applications of CAD to carbohydrate analysis. This choice was mainly a result of the numerous publications that have appeared since then. Three publications fit our criteria and each pairs CAD with one of the separation techniques discussed earlier in this chapter. A 2006 publication paired CAD with a metal-loaded cation-exchange separation to determine levoglucosan in an air sample for detecting wood burning [17]. In this application the CAD replaces an RI detector. CAD was undoubtedly chosen because it delivers better sensitivity. The second early paper paired CAD with a polymeric amine-bonded column to monitor the four enzyme synthesis of lacto-*N*-biose I, a building block from human milk oligosaccharides [18]. It is important to note that silica-based amine (and diol) columns are often observed to produce high background signal and noise with both ELSD and CAD. This is attributed to column bleed. Polymer-based amino columns are therefore used in most applications employing these types of stationary phases. CAD detected *N*-acetylglucosamine, fructose, sucrose, and glucose in the reaction mixture. The third paper paired CAD with a HILIC separation to determine a sialylglycopeptide from chicken egg yolk [19]. The title of the paper states that the oligosaccharides are detected, but it is important to note that those oligosaccharides remain attached to the peptide. Though a column often used for HILIC mode separations is used in this paper and the initial conditions could be considered HILIC conditions (85% acetonitrile), a gradient is run down to 20% acetonitrile, certainly not HILIC. This publication does state that CAD is 10× more sensitive than ELSD and notes CAD's compatibility with gradient elution.

7.6 Additional Applications of CAD to Carbohydrate Analysis

A search of the scientific literature shows that one of the more popular applications of CAD to carbohydrate analysis is the substitution of an RI or UV detector with a CAD detector in an application where simple carbohydrates are determined and a volatile mobile phase is used. This substitution is generally made for the improved sensitivity and dynamic range of CAD relative to those other detectors. Like Ref. [18], three additional publications have paired CAD with an amine-bonded column for determination of a monosaccharide or other small carbohydrate [20–22]. One publication followed cellulose hydrolysis [20], a second followed the enzymatic conversion of galactose to tagatose [21], and the third followed the synthesis of a fucose–galactose disaccharide that can be used to synthesize fucosyloligosaccharides [22]. In the first and third publications, no reason was given for the selection of CAD, but in the second publication, the authors noted that they switched from RI detection (different column, a metal-loaded cation-exchange column) when the product concentration was less than 10 mM. In the synthesis of the fucose–galactose disaccharide, the authors measured the substrates, fucose and lactose, and product. A publication from the same authors as Ref. [18] where they followed lacto-*N*-biose I synthesis used the same aminopropyl column-based chromatography to measure the lacto-*N*-biose I [23] as they investigated its utilization by bifidobacteria, bacteria that colonize the human intestine and are believed to be beneficial to one's health. Figure 7.1 shows the separation of fructose, glucose, and sucrose with amine-bonded column and detection by CAD. The experiment follows the sugar content of rice leaves during the diurnal cycle. The sensitivity of CAD allowed the accurate measurement of the low amounts of glucose and fructose present at some time points, which would have been difficult with UV or RI detection. In addition, the wide dynamic range of a log–log calibration plot allowed for determination of very different concentrations at the same time.

Others have paired CAD with a reversed-phase column run in either a traditional reversed-phase mode or HILIC mode for the determination of small

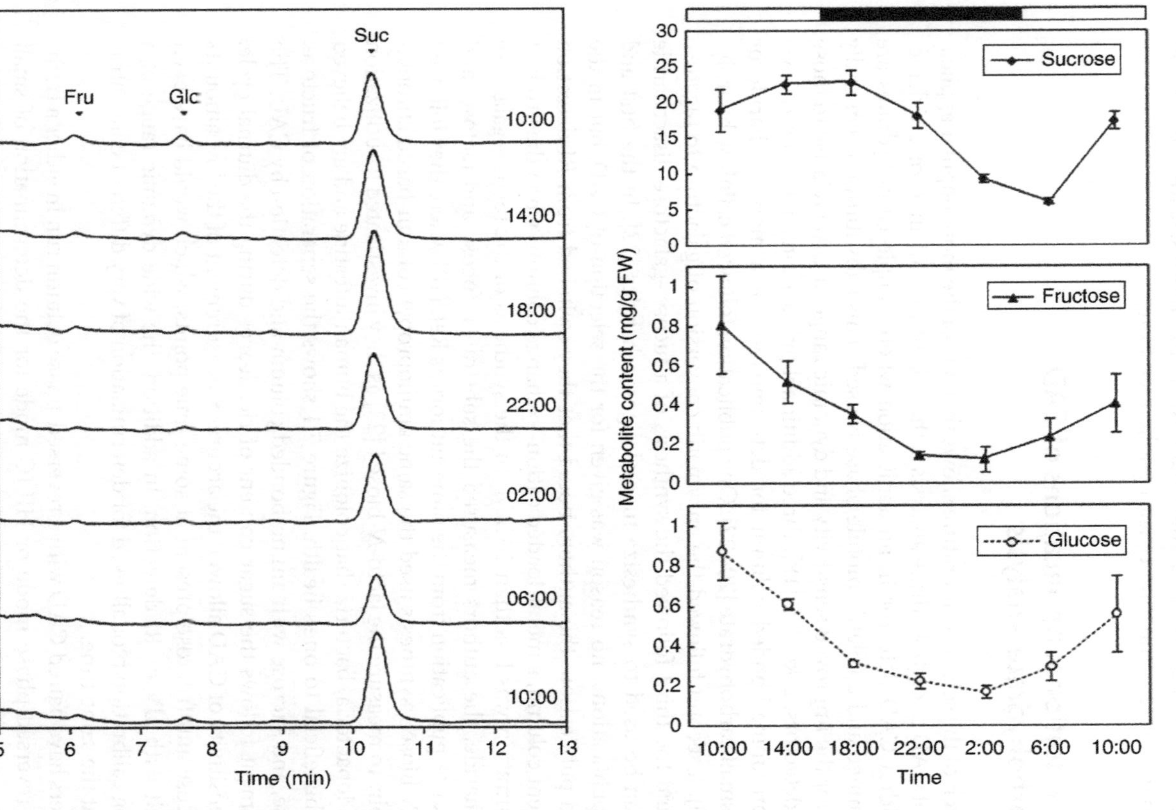

Figure 7.1 Sugar content of rice leaves over diurnal cycle. The top chromatogram corresponds to the time point on the left side of the graphs on the right. Sugars were separated on an AsahiPak NH2P-50G 4E (6 × 250 mm column) at 35°C and 1.0 mL/min. The mobile phase was 85% acetonitrile. The horizontal bars at the top of the graphs on the right side of the figure indicate transitions between light (white) and darkness (black). The sugar content is expressed in milligrams per fresh weight grams of leaves. Each point and vertical bar indicates the average and standard deviation of three measurements.

carbohydrates. As previously noted in the discussion of CAD sensitivity, CAD has been paired with a polar-embedded reversed-phase separation for the determination of glycosides from stevia [12]. These glycosides are used as low calorie sugar substitutes with one rebaudioside A, having been granted GRAS (generally recognized as safe) status by the US Food and Drug Administration. The authors noted that for this application, CAD was three to five times more sensitive than UV. This was despite their observation that the CAD detector caused slight peak broadening. In the design of their separation they made sure that the analytes of interest eluted in a part of the chromatogram where the mobile was not changing so that there would not be a response change with the mobile phase change. CAD was used to follow the reversed-phase purification of aza-*C*-glycosides, inhibitors of glycosylceramide metabolism [24]. No reason was given for choosing CAD as the method of detection. Figure 7.2 shows the reversed-phase separation with CAD detection of an α-glucosidase inhibitor 6-*O*-desulfated-kotalanol. This compound was recently isolated from the plant *Salacia reticulata* [25, 26]. The sensitivity of CAD allows the detection of impurities of this inhibitor, which, while perhaps not a

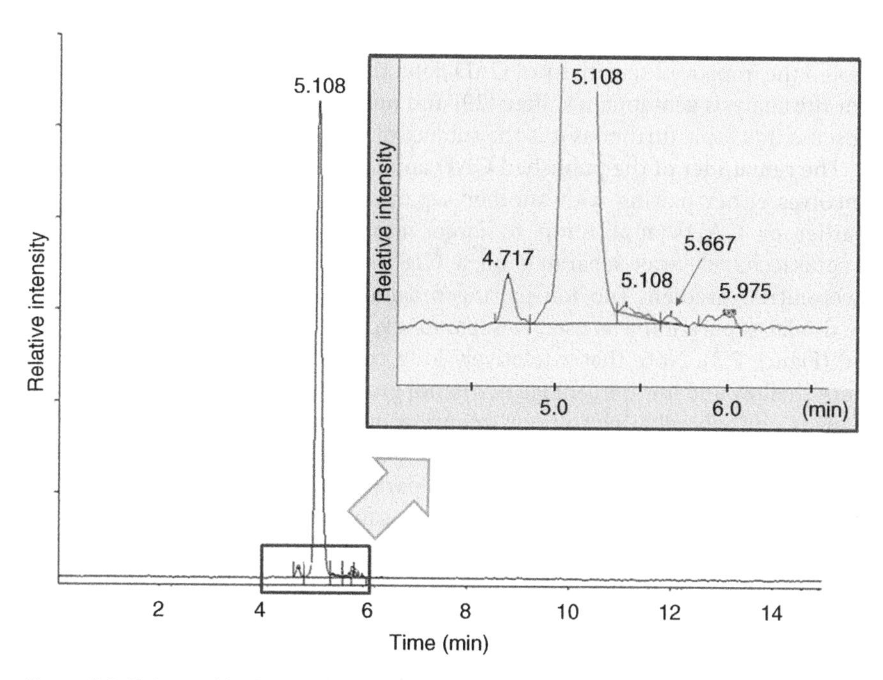

Figure 7.2 Estimated by CAD purity of a glucosidase inhibitor (6-*O*-desulfated kotalanol). The separation used a Unison UK-C18 (4 × 250 mm) with formic acid/water/acetonitrile mobile phase (0.1/100/0.2) flowing at 0.5 mL/min at a column temperature of 30°C. The inhibitor and the four impurity peaks were detected by CAD.

carbohydrate, is a polyol like most carbohydrates and lacks a good chromophore. Peak area quantification of the main peak and four impurities reveals that the inhibitor is 97.7% pure. A mixed-mode reversed-phase/weak anion-exchange column run in the HILIC mode was paired with CAD to determine sugar phosphates [27]. Carbohydrates were among the many pharmaceutical compounds evaluated for CAD response in an experimental setup that had a reversed-phase column preceding a column operated in the HILIC mode [28]. Because the HILIC mode requires high organic solvent content in the mobile phase, the reversed-phase column was a 2 mm diameter column, acetonitrile was added after the reversed-phase column, and the HILIC mode column was a 4 mm diameter column to accommodate the additional flow. Dilution does not significantly affect detection because, as discussed before, CAD is a mass-sensitive detector. In this experiment a UV detector preceded the CAD detector so that they could be compared, and in some experiments the flow path bypassed the HILIC separation.

CAD has also been applied to LC of aminoglycoside antibiotics [29, 30]. These antibiotics and their related compounds lack good UV chromophores, and this has led to the development of methods for assay and impurity determinations of these compounds that use pre- or post-column derivatization or electrochemical detection. In the referenced publications the authors noted the improved sensitivity of CAD detection compared with UV detection for the analysis gentamicin sulfate [29] and netilmicin sulfate [30]. We will not discuss this topic further as it is the subject of Chapter 12.

The remainder of the published CAD applications to carbohydrate analysis involves either pairing with another separation mechanism than described earlier or CAD's application to larger carbohydrates. A cyclic β-glucan heptasaccharide was separated on a C18 reversed-phase column with an acetonitrile gradient (no ion-pairing reagent) and detected by CAD [31]. A similar separation was used to separate cycloamyloses up to a DP of about 80 (Figure 7.3). Note that a relatively long column was used (50 cm) with a very shallow and low percentage methanol gradient (maximum 6% methanol). This is undoubtedly due to poor retention of these hydrophilic oligosaccharides on the C18 stationary phase and possibly poor solubility in higher percentages of acetonitrile. It is a fine separation but requires over 13 h. The average molecular weight of cycloamylose samples can be determined from the chromatograms [32].

One application with a different separation mechanism used the CarboPac PA1, the first HPAE-PAD column, with an ammonium acetate mobile phase rather than the typical sodium hydroxide/sodium acetate mobile phase, to monitor the purification of a natural sulfur-containing sugar from *Amycolatopsis orientalis* by CAD [33]. There may be other applications where the selectivity and high-resolution of a CarboPac column might be

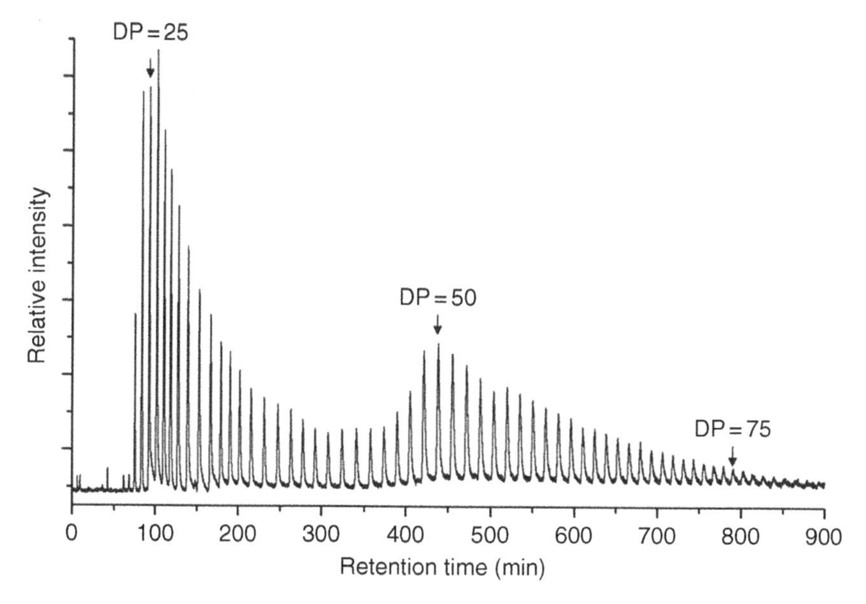

Figure 7.3 Chromatography of a cycloamylose sample. The sample was separated on a Cadenza C18 column (4.6 × 500 mm) with a gradient of 3.5–6% methanol over 900 min. The flow rate was 0.5 mL/min and the column temperature was 35°C. The cycloamyloses with DP from 21 to 80 were detected by CAD.

paired with CAD when the higher sensitivity of PAD is not required. Another interesting paper pairs CAD with a PGC column to detect malto-oligosaccharides [34]. This paper was a study in the mechanism of separation of 2-AB-labeled malto-oligosaccharides on PGC and used unlabeled, but reduced, malto-oligosaccharides detected by CAD as a control to ensure that the observed retention effects were not caused by the fluorescent label. CAD has also been paired with a PGC column to detect asparagine-linked oligosaccharides after release from a glycoprotein by PNGase or serine/threonine-linked oligosaccharides after release from a glycoprotein by reductive beta-elimination [35]. Figure 7.4 shows the *O*-linked glycans released from bovine submaxillary mucin separated on a PGC column and detected by CAD. The peaks were identified by mass spectrometry. Release of *O*-linked glycans is usually conducted under reductive conditions to prevent oligosaccharide degradation. Because the reduced oligosaccharides are not amenable to reductive amination, the typically used fluorophore labeling reactions are not possible. This makes CAD ideal for their detection, when there is enough sample, and the mobile phase is compatible for mass spectrometry detection.

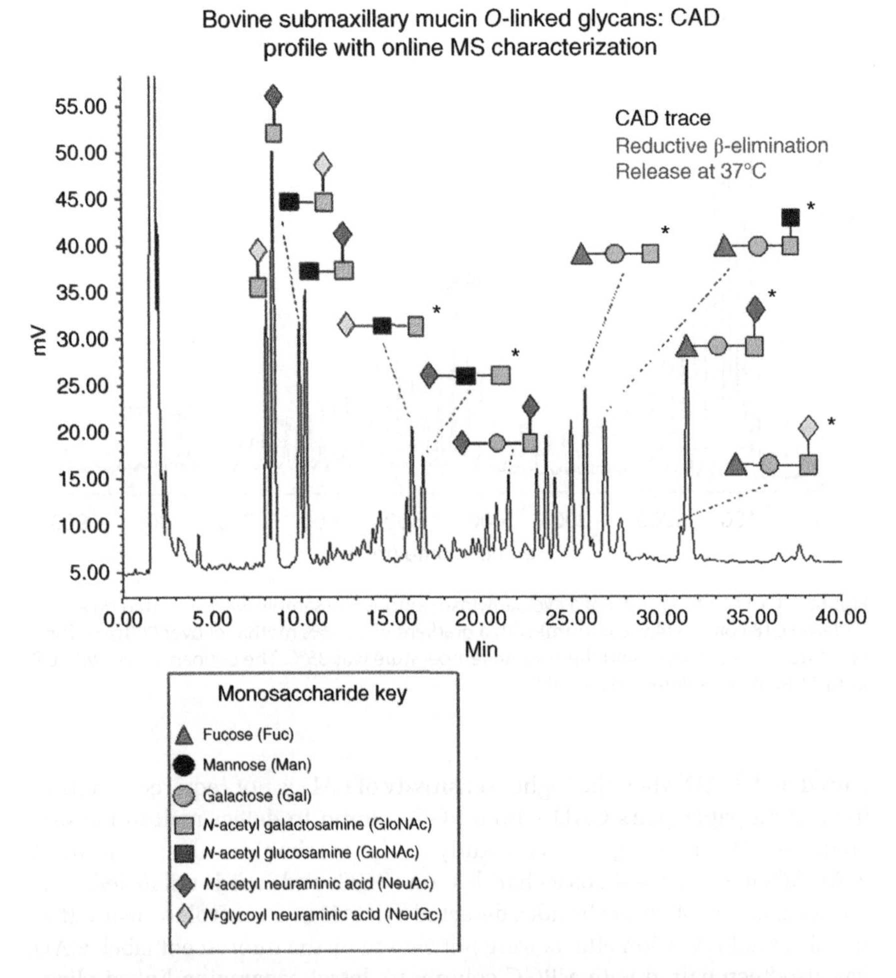

Figure 7.4 Chromatography of *O*-linked oligosaccharides from bovine submaxillary mucin. The oligosaccharides were separated on a Hypercarb PGC column (4.6×150 mm) with a water/acetonitrile/trifluoroacetic acid mobile phase with 90% of the flow going to the CAD and 10% going to a mass spectrometer for oligosaccharide identification. Identified structures are shown on the figure. Peaks marked with * depict proposed structures based on online accurate mass determination and literature [36].

References

1 Jentoft N. Analysis of sugars in glycoproteins by high-pressure liquid chromatography. Anal. Biochem. 1985;148:424–433.

2 Verhaar LAY, Kuster BFM. Contribution to the elucidation mechanism if sugar retention on amine-modified silica in liquid chromatography. J. Chromatogr. 1982;234:57–64.

3 Anumula KR. Advances in fluorescence derivatization methods for high-performance liquid chromatographic analysis of glycoprotein carbohydrates. Anal. Biochem. 2006;350:1–23.

4 Tomiya N, Awaya J, Kurono M, Endo S, Arata Y, Takahashi N. Analyses of N-linked oligosaccharides using a two-dimensional mapping technique. Anal. Biochem. 1988;174:73–90.

5 Davies M, Smith KD, Harbin A, Hounsell EF. High-performance liquid chromatography of oligosaccharide alditols and glycopeptides on a graphitized carbon column. J. Chromatogr. 1992;609:125–131.

6 Deguchi K, Keira T, Yamada K, Ito H, Takegawa Y, Nakagawa H, Nishimura S-I. Two-dimensional hydrophilic interaction chromatography coupling anion-exchange and hydrophilic interaction columns for separation of 2-pyridylamino derivatives of neutral and sialylated N-glycans. J. Chromatogr. A 2008;1189:169–174.

7 Lee YC. Carbohydrate analyses with high-performance anion-exchange chromatography. J. Chromatogr. A 1996;720:137–149.

8 Cataldi TRI, Campa C, De Benedetto GE. Carbohydrate analysis by high-performance anion-exchange chromatography with pulsed amperometric detection: The potential is still growing. Frensenius J. Anal. Chem. 2000;368:739–758.

9 Rohrer JS. Analyzing sialic acids using high-performance anion-exchange chromatography with pulsed amperometric detection. Anal. Biochem. 2000;283:3–9.

10 Dixon RW, Peterson DS. Development and testing of a detection method for liquid chromatography based on aerosol charging. Anal. Chem. 2002;74:2930–2937.

11 Vehovec T, Obreza A. Review of operating principles and applications of charged aerosol detection. J. Chromatogr. A 2010;1217:1549–1556.

12 Clos JF, DuBois GE, Prakash I. Photostability of rebaudioside A and stevioside in beverages. J. Agric. Food Chem. 2008;56:8507–8513.

13 Vervoort N, Daemen D, Torok G. Performance evaluation of evaporative light scattering detection and charged aerosol detection in reversed phase liquid chromatography. J. Chromatogr. A 2008;1189:92–100.

14 Eom HY, Park S-Y, Kim MK, Suh JH, Yeom H, Min JW, Kim U, Lee J, Youm J-R, Han SB. Comparison between evaporative light scattering detection and charged aerosol detection for the analysis of saikosaponins. J. Chromatogr. A 2010;1217:4347–4354.

15 Shaodong J, Lee WJ, Ee JW, Park JH, Kwon SW, Lee J. Comparison of ultraviolet detection, evaporative light scattering detection and charged aerosol detection methods for liquid-chromatographic determination of anti-diabetic drugs. J. Pharm. Biomed. Anal. 2010;51:973–978.

16 Gorecki T, Lynen F, Szucs R, Sandra P. Universal response in liquid chromatography using charged aerosol detection. Anal. Chem. 2006;78:3186–3192.

17 Dixon RW, Baltzell G. Determination of levoglucosan in atmospheric aerosols using high performance liquid chromatography with aerosol charge detection. J. Chromatogr. A 2006;1109:214–221.
18 Nishimoto M, Kitaoka M. Practical preparation of lacto-N-biose I, a candidate for the bifidus factor in human milk. Biosci. Biotechnol. Biochem. 2007;71:2101–2104.
19 Inagaki S, Min JZ, Toyo'oka T. Direct detection method of oligosaccharides by high-performance liquid chromatography with charged aerosol detection. Biomed. Chromatogr. 2007;21:338–342.
20 Igarashi K, Ishida T, Hori C, Samejima M. Characterization of an endoglucanase belonging to a new subfamily of glycoside hydrolase family 45 of the basidiomycete *Phanerochaete chrysosporium*. Appl. Environ. Microbiol. 2008;74:5628–5634.
21 Kim JH, Lim BC, Yeom SJ, Kim YS, Kim HJ, Lee JK, Lee SH, Kim SW, Oh DK. Differential selectivity of the *Escherichia coli* cell membrane shifts the equilibrium for the enzyme-catalyzed isomerization of galactose to tagatose. Appl. Environ. Microbiol. 2008;74:2307–2313.
22 Wada J, Honda Y, Nagae M, Kato R, Wakatsuki S, Katayama T, Taniguchi H, Kumagai H, Kitaoka M, Yamamoto K. 1,2-α-L-Fucosynthase: A glycosynthase derived from an inverting α-glycosidase with an unusual reaction mechanism. FEBS Lett. 2008;582:3739–3743.
23 Xiao J, Takahashi S, Nishimoto M, Odamaki T, Yaeshima T, Iwatsuki K, Kitaoka M. Distribution of in vitro fermentation ability of lacto-N-biose I, a major building block of human milk oligosaccharides, in Bifidobacterial strains. Appl. Environ. Microbiol. 2010;76;54–59.
24 Wennekes T, van den Berg RJBHN, Boltje TJ, Donker-Koopman WE, Kuijper B, van der Marel GA, Strijland A, Verhagen CP, Aerts JMFG, Overkleeft HS. Synthesis and evaluation of lipophilic aza-C-glycosides as inhibitors of glucosylceramide metabolism. Eur. J. Org. Chem. 2010;2010:1258–1283.
25 Ozaki S, Oe H, Kitamura S. α-Glucosidase inhibitor from Kothala-himbutu (*Salacia reticulata* WIGHT). J. Nat. Prod. 2008;71:981–984.
26 Muraoka O, Xie W, Tanabe G, Amer MFA, Minematsu T, Yoshikawa M. On the structure of the bioactive constituent from ayurvedic medicine *Salacia reticulata*: Revision of the literature. Tetrahedron Lett. 2008;49:7315–7317.
27 Hinterwirth H, Lämmerhofer M, Preinerstorfer B, Gargano A, Reischl R, Bicker W, Trapp O, Brecker L, Lindner W. Selectivity issues in targeted metabolomics: Separation of phosphorylated carbohydrate isomers by mixed-mode hydrophilic interaction/weak anion exchange chromatography. J. Sep. Sci. 2010;33:3273–3282.
28 Louw S, Pereira AS, Lynen F, Hanna-Brown M, Sandra P. Serial coupling of reversed-phase and hydrophilic interaction liquid chromatography to broaden the elution window for the analysis of pharmaceutical compounds. J. Chromatogr. A 2008;1208:90–94.

29 Joseph A, Rustum A. Development and validation of a RP-HPLC method for the determination of gentamicin sulfate and its related substances in a pharmaceutical cream using a short pentafluorophenyl column and a charged aerosol detector. J. Pharm. Biomed. Anal. 2010;51:521–531.
30 Joseph A, Patel S, Rustum A. Development and validation of a RP-HPLC method for the estimation of netilmicin sulfate and its related substances using charged aerosol detection. J. Chromatogr. Sci. 2010;48:607–612.
31 Vasur J, Kawai R, Jonsson KHM, Widmalm G, Engström Å, Frank M, Andersson E, Hansson H, Forsberg Z, Igarashi K, Samejima M, Sandgren M, Ståhlberg J. Synthesis of cyclic β-glucan using laminarinase 16A glycosynthase mutant from the basidiomycete *Phanerochaete chrysosporium*. J. Am. Chem. Soc. 2010;132:1724–1730.
32 Suzuki S, Kitamura S. Unfrozen water in amylosic molecules is dependent on the molecular structures: A differential scanning calorimetric study. Food Hydrocoll. 2008;22:862–867.
33 Sasaki E, Ogasawara Y, Liu H-W. A biosynthetic pathway for BE-7585A, a 2-thiosugar-containing angucycline-type natural product. J. Am. Chem. Soc. 2010;132:7405–7417.
34 Melmer M, Stangler T, Premstaller A, Linder W. Solvent effects of the retention of oligosaccharides in porous graphitic carbon liquid chromatography. J. Chromatogr. A 2010;1217:6092–6096.
35 Bailey B, Acworth I, Hanneman A, Rouse J. Glycan analysis using HPLC with charged aerosol detection and MS detectors. LCGC N. Am. June 2012; Supplement: p 13.
36 Chai WG, Hounsell EF, Cashmore GC, Rosankiewicz JR, Bauer CJ, Feeney J, Feizi T, Lawson AM. Neutral oligosaccharides of bovine submaxillary mucin. A combined mass spectrometry and 1H-NMR study. Eur. J. Biochem. 1992;203:257–268.

8

Polymers and Surfactants

Dawen Kou[1], Gerald Manius[2], Hung Tian[3], and Hitesh P. Chokshi[4]

[1] *Genentech Inc., South San Francisco, CA, USA*
[2] *Retired from Hoffmann-La Roche Inc., Nutley, NJ, USA*
[3] *Novartis, East Hanover, NJ, USA*
[4] *Roche Innovation Center, New York, NY, USA*

CHAPTER MENU

8.1 Summary

In this chapter, the use of CAD with various chromatographic techniques such as HPLC, SEC, and SFC for the analysis of polymers and surfactants has been discussed. Following a brief introduction, general discussions on polymer analysis including assay and impurity analysis and polydispersity determination were given, and then specific application examples such as PEGs and PEGylated molecules and nonionic surfactants (e.g., Tween 80 and Span 85) were described.

The analytical performances of CAD have also been evaluated in comparison with RI and ELSD. CAD was found to offer superior performance, especially in terms of sensitivity, precision, and linearity (compared with ELSD). It has also been observed that different types of universal detectors can generate considerably different polydispersity profiles. ELSD consistently gives artificially lower polydispersity and often fails to differentiate the polydispersity of

Charged Aerosol Detection for Liquid Chromatography and Related Separation Techniques, First Edition. Edited by Paul H. Gamache.

different lots of polymers. CAD provides more accurate polydispersity profiles, which better differentiate various batches of materials. In conclusion, CAD has proven to be well suited for the analysis of polymers and surfactants. More importantly, it is a better choice for impurity analysis and a valuable tool for polydispersity profiling for polymeric compounds.

8.2 Introduction

This chapter describes a number of different approaches to the analysis of polymers and surfactants, primarily in the context of their use as pharmaceutical excipients and reagents, by charged aerosol detection (CAD). A commonality of the compounds discussed here is that they are polymeric in nature and generally lack chromophores required for UV detection. Therefore a universal detector is necessary for the direct detection of such compounds without cumbersome procedures of derivatization.

CAD has emerged as a new type of universal detection technique, with reported analytical performances better than other commonly used universal detectors such as refractive index (RI) and evaporative light scattering detection (ELSD) [1, 2]. The operational principles and mechanisms of CAD have been described in detail in Chapter 1 and are not repeated here. CAD has been applied with high performance liquid chromatography (HPLC), supercritical fluid chromatography (SFC), or size exclusion chromatography (SEC) for the analysis of polymers and surfactants. This includes the use of HPLC-CAD for analysis of various surfactants, SFC-CAD for lower MW polyethylene glycol (PEG), SEC-CAD for the analysis of PEG reagents, and HPLC-CAD for PEGylated molecules. This chapter provides a comparison of CAD to other techniques for determining quantity, purity, and impurities and for profiling polydispersity.

8.3 Polymer Analysis

Although the specific examples of polymers used in this chapter are mainly PEG and PEGylated molecules, the general discussions on polymer analysis can be applied to all types of polymers. Interested readers can also find the analysis of industrial polymers in Chapter 15.

Polymer analysis generally involves two main aspects. The first is determination of the polymer and its related impurities, including higher molecular weight aggregation products. HPLC and SEC provide different mechanisms of separation and thus different selectivity and are often used to complement each other in polymer analysis. SEC is especially useful for the separation of aggregation products.

Polydispersity profiling is the second aspect of polymer analysis. A polymer sample generally consists of a mixture of compounds with a range of molecular weights. The distribution of molecular weights, referred to as polydispersity, is defined as the average molecular weight divided by the number average molecular weight (M_w/M_n):

$$M_w = \frac{\sum_i N_i(M_i M_i)}{\sum_i N_i M_i}$$

$$M_n = \frac{\sum_i N_i M_i}{\sum_i N_i}$$

The higher the polydispersity number, the wider the molecular weight distribution, and vice versa.

A common approach to polydispersity measurement is using SEC with a universal detector. Peak identification and system calibration is carried out by using reference standards of known molecular weights. There are other techniques to determine molecular weight and polydispersity without the use of reference standards. One such technique is matrix-assisted laser desorption ionization time-of-flight (MALDI-TOF) mass spectrometry (MS), which can directly measure the molecular weights of the different fractions of the polymer. Another technique is to use a multi-angle light scattering (MALS) detector in connection with a differential RI detector. In its "classical" or "static" operation mode, the MALS signal is proportional to both molecular weight and concentration, while the RI signal is only proportional to concentration. The ratio of MALS/RI signals can then be used to determine molecular weight. The MALS technique should not be confused with ELSD. The former measures scattered light signals in solution in a flow cell, while the latter responds to light scattered by solid particles after the evaporation of eluent. MALDI-TOF-MS and MALS are powerful detection techniques but are not as widely available as simpler and less expensive universal detectors. Polydispersity can be used to differentiate the purity or grade of the polymer. The cost of different grades of polymers can vary a great deal. The ability to accurately measure and control the polydispersity distribution of a polymer is of high importance.

8.4 Polyethylene Glycol

The molecular weight of PEG can range from a few hundred to a few million daltons. Lower molecular weight PEGs are widely used as emulsifiers and surfactants in pharmaceuticals, cosmetics, and other products. PEG can also be used as a reagent or modifier to the drug molecule. The process of attaching the PEG to the drug molecule is referred to as PEGylation. PEGylated molecules generally take longer to be cleared from the body, resulting in a longer half-life. This can help to achieve improved efficacy at lower doses or longer dosing intervals. PEG with molecular weight lower than 30 kDa can be readily discharged from the body and is generally nontoxic [3]. Examples of PEGylated medicines on the market include PEGASYS® (peg-interferon alfa-2a) and PEG-Intron® (peg-interferon alfa-2b) for the treatment of hepatitis C and Neulasta® for neutropenia. The physicochemical properties of the PEGylated molecule are largely dependent upon the molecular weight ratio of the PEG and the drug molecule, with the larger one being the more dominant in the PEGylated molecule.

8.4.1 PEG Reagents

Kou *et al.* studied the use of SEC with RI, ELSD, or CAD for the analysis of PEG reagents used in PEGylation [4]. The PEG reagent had an average molecular weight of 32 kDa. The main impurity was the aggregation product, or "dimer," of the PEG. An isocratic method was used with DMF as the mobile phase. Five polymer reference standards of molecular weights of 10.0, 18.3, 32.6, 50.1, and 73.4 kDa were used for SEC calibration.

Figure 8.1 shows the SEC chromatograms from the three detectors. In the early stages of drug development, reference standards of individual impurities are generally unavailable. Quantification of impurities is usually done by peak area normalization as the best estimation, assuming that the impurities and the main compound have the same relative response factor. Using peak area normalization, the dimer was found to be 21.5, 9.1, and 25.6% by RI, ELSD, and CAD. RI is known to give linear response and can be used as a reference value. Its main drawbacks are its incompatibility with gradient elution and its low sensitivity. A larger sample injection volume is needed for impurity analysis (100 μL for RI vs. 50 μL for ELSD and CAD), leading to sample overload and peak broadening. Compared with RI, ELSD significantly underestimated the amount of dimer by 58%, while CAD slightly overestimated the dimer amount by 19%.

The detector response of ELSD and CAD were evaluated over 50–150% of the nominal concentration, a typical range used for assay validation in a pharmaceutical analysis setting. The CAD detector was found to be more linear than ELSD ($r = 0.9957$ for CAD vs. $r = 0.9856$ for ELSD). The signal-to-concentration ratio of CAD was slightly higher at lower concentrations than at higher concentrations, whereas the relative response ratio of ELSD was dramatically lower at lower concentrations indicating severe nonlinearity.

The same HPLC method with the three detectors was applied to the analysis of seven different lots of PEG reagents. Table 8.1 shows the percentage of dimer found in these lots by each detector. The trend is consistent with what was observed in Figure 8.1. RI was not sensitive enough to quantify the dimer at

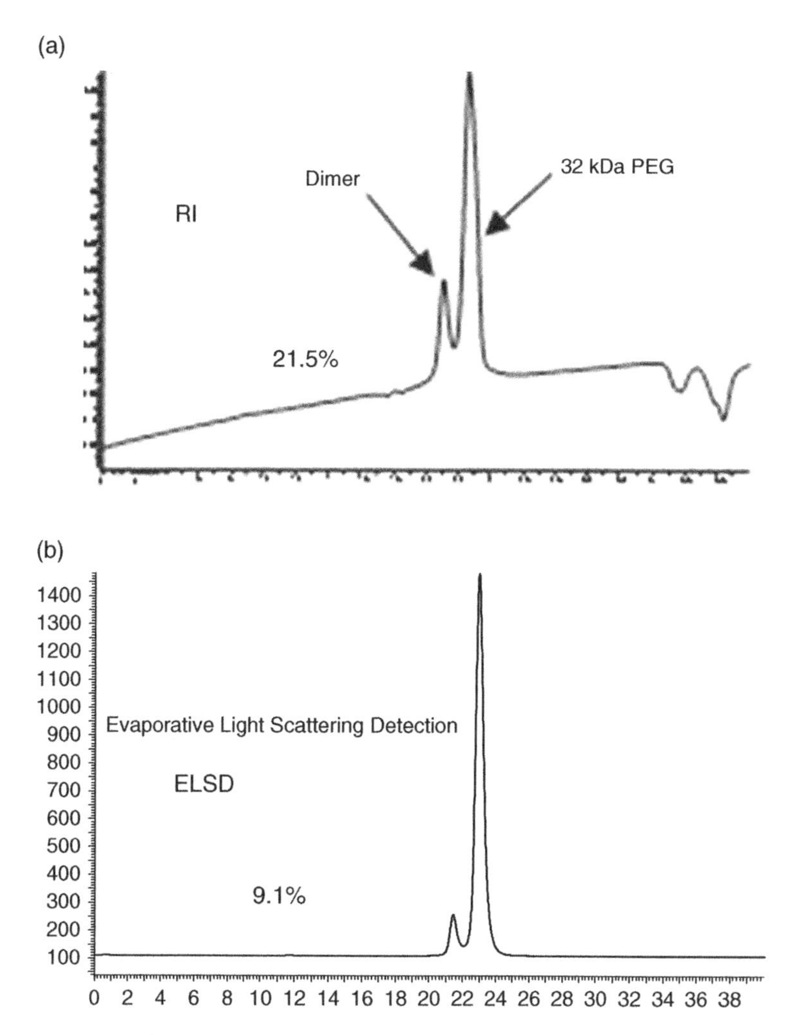

Figure 8.1 Chromatograms of the same lot of PEG 32 kDa reagent by three universal detectors: (a) RI, (b) ELSD, (c) CAD. *Source*: Kou *et al.* [4]. Reproduced with permission of Elsevier.

concentrations lower than 14.4%. The lower the dimer concentration, the lower the relative response of ELSD compared with CAD, resulting in significant underestimation of the impurity. From a quality control point of view, slight overestimation of impurities by CAD is more acceptable than severe underestimation by ELSD.

A commercially available 32.6 kDa polymer standard was analyzed with the HPLC-CAD method, with polydispersity calculated via TurboSEC software.

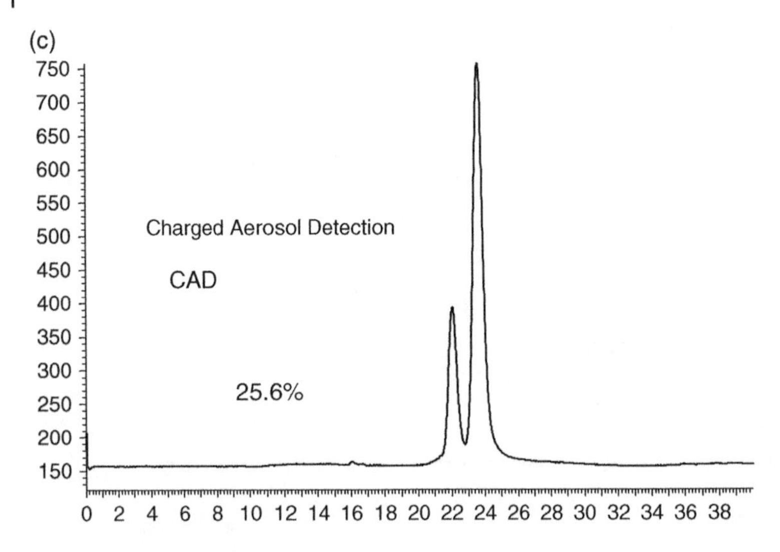

Figure 8.1 (Continued)

Table 8.1 Percentage of the dimer in seven lots of PEG reagents determined by RI, ELSD, and CAD.

PEG reagent	% Dimer (by peak area)			
Lot number	By RI	By ELSD	By CAD	ELSD result/CAD result (%)
5	Not detected	0.4	7.8	5.1
6	Not detected	2.1	11.7	18
7	Not detected	3.2	12.0	27
1	Not detected	2.5	13.5	19
4	14.4	5.2	19.7	26
2	15.6	5.4	19.9	27
3	21.5	9.1	25.6	36

Source: Kou *et al.* [4]. Reproduced with permission of Elsevier.

The determined polydispersity was 1.032 with an RSD of 0.16% for a triplicate analysis. The result was in good agreement with the value 1.03 (1.026 to be exact) reported in the certificate of analysis from the manufacturer.

The polydispersity of the seven lots of PEG reagents determined by CAD and ELSD are shown in Table 8.2. It is apparent that the polydispersity of the PEG reagents obtained from ELSD were consistently lower than those from CAD. This again was due to severe underestimation of lower concentration

Table 8.2 Polydispersity data of seven lots of PEG reagents determined by CAD and ELSD with TurboSEC software.

PEG reagent	By CAD			By ELSD		
Lot number	Polydispersity	M_w	M_n	Polydispersity	M_w	M_n
1	1.025	32,264	31,474	1.015	34,270	33,748
2	1.033	32,171	31,147	1.015	33,332	32,854
3	1.034	31,904	30,867	1.023	32,880	32,136
4	1.038	32,065	30,887	1.016	34,252	33,706
5	1.192	28,924	24,265	1.020	36,493	35,790
6	1.195	30,357	25,398	1.018	33,170	32,574
7	1.375	25,039	18,210	1.033	31,666	30,651

Source: Kou *et al.* [4]. Reproduced with permission of Elsevier.

fragments in the polymer by ELSD, giving falsely narrower polydispersity profiles. Moreover, ELSD was not able to meaningfully differentiate the polydispersity of six out of seven lots. CAD gave more accurate polydispersity profiles and better differentiated the polydispersity and quality of the PEG reagents.

8.4.2 Low Molecular Weight PEGs

Takahashi *et al.* compared the quantitative performance of a CAD detector with an ELSD detector for the analysis of low molecular weight PEG by SFC [5]. The PEGs studied included a certified reference material of PEG 1000 and a well-defined equimass mixture of nine uniform PEG oligomers with degree of polymerization $n = 6, 8, 10, 12, 18, 21, 25, 30$, and 42. A silica gel column was used for the separation. Both CAD and ELSD were found to be capable of operating at the high pressure of SFC.

Compared with ELSD, CAD provided a more uniform response to PEGs of different molecular weights over a range of concentrations from 0.4 to 10 mg/mL, as shown in Figure 8.2. While ELSD signals decreased with the increase in molecular weight (degree of polymerization), CAD responses were more consistent. Also, the signal response differences between the two detectors widened with the decrease in concentration. CAD was also found to be nearly linear in the range studied. It was also more sensitive than ELSD, being able to detect the PEGs at the concentration of 10 μg/mL, 10 times lower than the lowest ELSD could detect at 100 μg/mL.

Figure 8.3 shows the mass fraction or polydispersity profiles of the PEG 1000 certified reference material, determined by CAD and ELSD and compared with the certified values. The profile from CAD matched the certified values

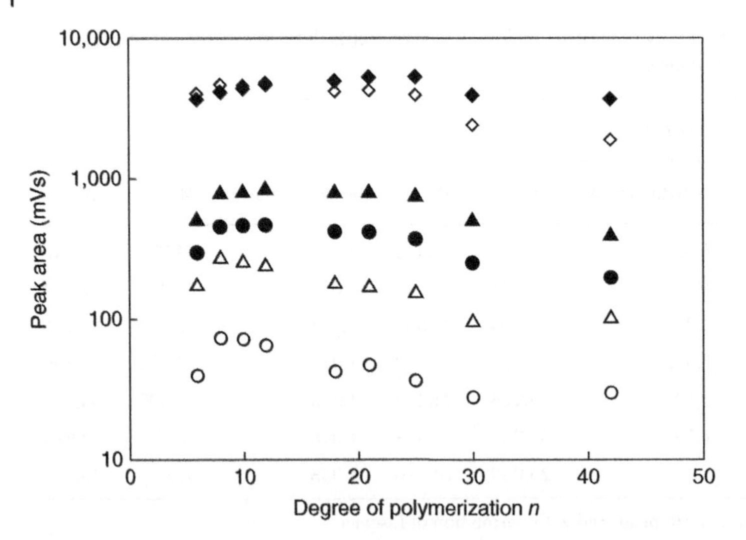

Figure 8.2 Changes in CAD and ELSD peak area as a function of the degree of polymerization *n* for the equimass mixture of uniform PEGs ($n = 6$, 8, 10, 12, 18, 21, 25, 30, and 42) at the concentration of 10 mg/mL (◆), 1 mg/mL (▲), and 0.4 mg/mL (●) for CAD and 10 mg/mL (◇), 1 mg/mL (△), and 0.4 mg/mL (○) for ELSD. *Source*: Takahashi *et al.* [5]. Reproduced with permission of Elsevier.

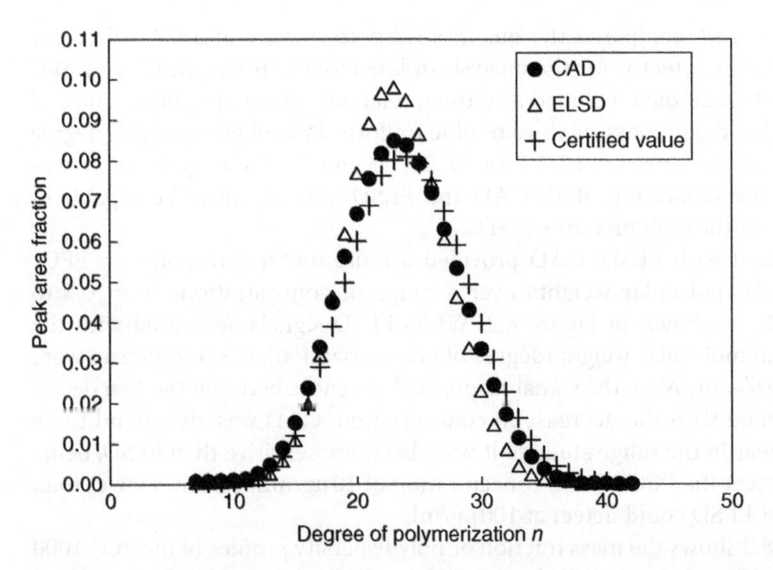

Figure 8.3 Comparison between the mass fractions of CRM PEG 1000 detected by CAD and ELSD. *Source*: Takahashi *et al.* [5]. Reproduced with permission of Elsevier.

Table 8.3 M_w, M_n, and molecular mass distributions (M_w/M_n) of CRM PEG 1000 determined by CAD and ELSD compared to the certified values.

	M_n	M_w	M_w/M_n
CAD	1012	1054	1.042
ELSD	1001	1035	1.034
Certified value	1041	1085	1.042

Source: Takahashi *et al.* [5]. Reproduced with permission of Elsevier.

very well, while ELSD gave artificially narrower molecular weight distribution. The M_w, M_n, and polydispersity (M_w/M_n) are listed in Table 8.3. This observation is consistent with what was reported in the paper by Kou *et al.*, as discussed in the previous section.

8.4.3 PEGylated Molecules

Polymeric microcapsules of perfluorocarbons have been used as ultrasound contrast agents administered intravenously. Due to the hydrophobic nature of the polymeric microcapsules, they can be quickly cleared by the mononuclear phagocyte system. It has been reported that adding hydrophilic PEGylated phospholipids to the polymeric shells could slow its clearance from the body and thus enhance its contrasting ability [6]. It is important to measure the amount of PEGylated phospholipids actually associated to the polymeric surfaces and shells, which affects the surface properties of the resulted colloidal particles. Diaz-Lopez *et al.* developed a method using HPLC-CAD for the analysis of PEGylated phospholipids in polymeric microcapsules of perfluorooctyl bromide (PFBO), which was the first reported quantitative method for such an analysis [7].

The PEGylated phospholipid was 1,2-distearoyl-sn-glycero-3-phosphoethanolamine-*N*-[methoxy(PEG)-2000], referred to as DSPE-PEG2000. The polymeric microcapsule material was poly(lactide-co-glycolide) acid (PLGA). The HPLC method used nonaqueous reverse phase separation, with acetonitrile and methanol as the mobile phase. Using chloroform and methanol as cosolvents to partition the PEGylated phospholipids and other components of the suspensions, the method was able to quantify the total PEGylated phospholipids in the suspension, the free form of PEGylated phospholipids, and the PEGylated phospholipid attached to polymeric microcapsules.

The calibration curve of CAD response over a concentration range of 2.23–21.36 μg/mL was evaluated with a linear model and a power (nonlinear) model. The regression analysis results are shown in Table 8.4. The power model was found to be a better fit over the entire calibration range, based on the larger *F*-value and smaller *p*-value for the lack of fit.

Table 8.4 Regression analysis with the two models used.

Parameter	Linear model $y = a_0 + a_1x$
Range (μg/mL)	2.23–21.36 μg/mL (0.22–2.14 μg injected)
Calibration curve equation	$y = 1.10 + 1.87x$
Correlation coefficient, R^2	0.9942
Intercept, $a_0 \pm$ SD	1.10 ± 0.47
Slope, $a_1 \pm$ SD	1.87 ± 0.04
F-value for the regression	2749
F-value for lack of fit	7.25
p-Value of the lack of fit	0.003
Parameter	**Power model $y = Ax^b$**
Range (μg/mL)	2.23–21.36 μg/mL (0.22–2.14 μg injected)
Calibration curve equation	$y = 2.47x^{0.91}$
Correlation coefficient, R^2	0.9961
Coefficient $A \pm$ SD	2.47 ± 0.14
Coefficient $b \pm$ SD	0.91 ± 0.02
F-value for the regression	4374
F-value for lack of fit	3.2
p-Value of the lack of fit	0.052

Source: Diaz-Lopez *et al.* [7]. Reproduced with permission of Elsevier.

The analytical performance of the HPLC-CAD method was further evaluated through method validation. The detection limit and quantification limit were determined to be 33 and 100 ng, respectively, corresponding to 0.33 and 1.0 μg/mL with a 100 μL of injection volume. Injection repeatability ($n = 6$) was 1.2% RSD. The accuracy (recovery) in the concentration range of 4.44–21.36 μg/mL was 90–115%, with an intraday precision of no more than 5.3% RSD for triplicate analyses. The method was then applied to three different amounts of DSPE-PEG added to PFOB. It was found that, regardless of the initial amount of DSPE-PEG added, approximately 10% of the DSPE-PEG was associated with PFOB, with the remaining 90% in its free unassociated form.

8.5 Surfactants

Surfactants are sometimes used in pharmaceutical products as excipients to improve wetting and/or increase the solubility of drug molecules. Protein formulations usually contain surfactants to minimize adsorption to contact surfaces of containers and syringes and reduce protein denaturing and resulted aggregation by decreasing the surface tensions [8, 9]. Certain surfactants are routinely used in pharmaceutical manufacturing equipment cleaning and should be subsequently removed at the end of the cleaning process.

Surfactants can be cationic, anionic, zwitterionic, or nonionic. Without charges, nonionic surfactants cannot be analyzed with ion chromatography and a conductivity detector, and the ones without chromophore require a universal detector. Examples of commonly used nonionic surfactants are polyoxyethylene sorbitans, polyoxyethylene ethers, and polyethylene–polypropylene glycols. Many of the nonionic surfactants used in the pharmaceutical setting are polymeric and heterogeneous, consisting of a mixture of compounds of similar structures and molecular weights.

Polysorbate 20 (polyoxyethylene sorbitan monolaurate, brand name Tween 20) and Polysorbate 80 (polyoxyethylene sorbitan monooleate, Tween 80) are the most commonly used nonionic surfactants in protein formulations. An HPLC-CAD method for fast and sensitive determination of Polysorbate 80 in solutions containing proteins was reported by Fekete *et al.* [10]. The polysorbates (chemical structure of Tween 80 shown in the following text) share a common backbone with slightly different fatty acid side chains.

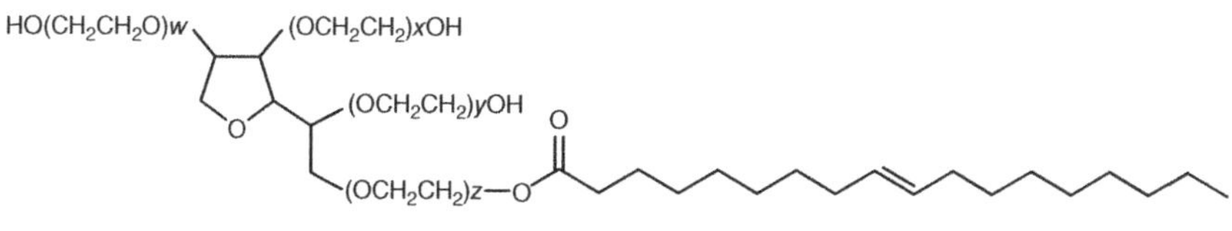

$HO(CH_2CH_2O)w$ $(OCH_2CH_2)xOH$

$(OCH_2CH_2)yOH$ O

O

$(OCH_2CH_2)z$—O

where $W+X+Y+Z \approx 20$.

The method used a HPLC column with fused core technology connected to a CAD detector. The method was found to be specific and capable of separating Tween 80 from the native protein or its oxidized and reduced forms. The chromatograms in Figure 8.4 show two major bands of Tween 80, consistent with its heterogeneous composition. The first band was used for quantification. The method was further validated for its linearity, accuracy, precision, limit of quantitation (LOQ), and limit of detection (LOD), as well as stability of sample and stock solutions. The LOQ and LOD were determined to be 10 and 5 μg/mL, respectively. The linearity and accuracy was assessed at 10–60 μg/mL, corresponding to the LOQ to 150% of the nominal sample concentration. The linearity correlation coefficient r^2 was 0.9998, and the equation curve was $y = 3008x - 249$. The method precision for six repeated analyses was 2.5 and 2.1% of RSD, performed on two separate days. The standard and sample solutions were stable for 48 h.

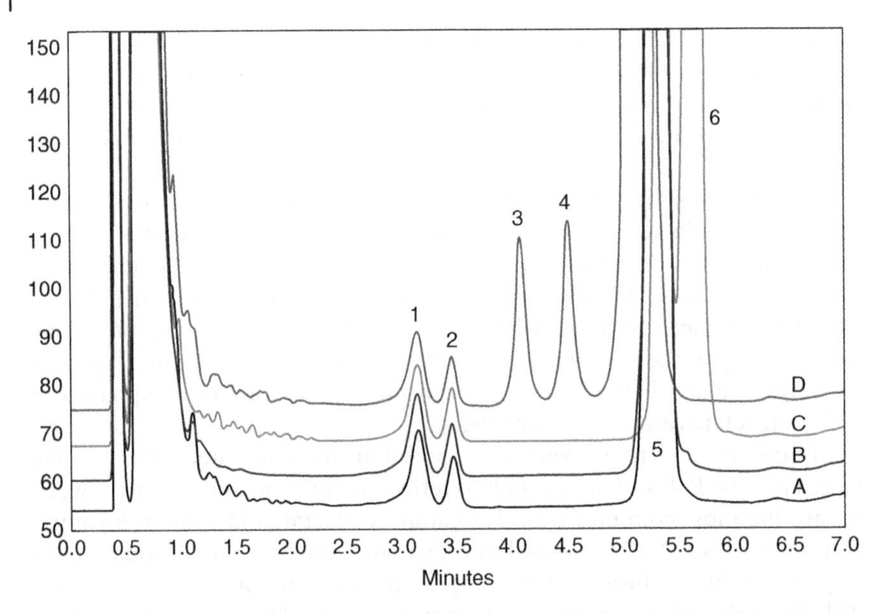

Figure 8.4 Chromatograms of stressed protein solution containing 1 mg/mL protein and 40 μg/mL Polysorbate 80. Chromatograms: protein solution (A), protein solution containing deamidated protein (B), protein solution containing reduced form (C), and protein solution containing two major oxidized protein form (D). Peaks: 1, 2: major peaks of Polysorbate 80 origin; 3, 4: oxidized protein; 5: protein (native); and 6: reduced protein form. Chromatographic conditions: Poroshell 300 SB-C18 column packed with 5 μm shell particles (75 mm × 2.1 mm), mobile phase: acetonitrile–methanol–water–trifluoroacetic acid gradient elution (50–100% B, in 6 min), flow: 0.65/min, column temperature: 20°C, injection volume: 5 μL, sample temperature 4°C, detection: charged aerosol detection, range: 50 pA, nebulizing temperature: 30°C, gas (nitrogen) pressure: 37–39 psi. *Source*: Fekete *et al.* [10]. Reproduced with permission of Elsevier.

Lobback *et al.* reported on the quantitative determination of nonionic surfactants Tween 80 and Span 85 with CAD in comparison with ELSD [11, 12]. CAD had a linear response in the range of 5–100 μg/mL and was found to be 10 times more sensitive than ELSD. Fukushima *et al.* also reported HPLC with CAD detection for the measurement of different nonionic surfactants, including Tween 80, polysorbate 20, and Triton X100 [13]. Two HPLC methods, one isocratic and one gradient, were developed to measure the residual surfactants in cleaning validation. Both methods provided excellent sensitivity with 25 ng of analytes on column, wide dynamic range of four orders of magnitude, and excellent reproducibility with RSDs of typically 4% or less even at trace levels. Christiansen *et al.* conducted a study on the stability of nonionic surfactants tocopheryl PEG succinate and sucrose laurate in simulated conditions of the GI tract by HPLC-MS, DAD, and CAD [14].

References

1 Dixon RW, Peterson DS. Development and testing of a detection method for liquid chromatography based on aerosol charging. Anal Chem, 2002; 74; 2930–2937.

2 Gamache P, McCarthy R, Freeto SM, Asa D, Woodcock M, Laws K, Cole R. HPLC analysis of nonvolatile analytes using charged aerosol detection. LCGC N Am, 2005; 23; 150–161.

3 Yamaoka T, Tabata Y, Ikada Y. Distribution and tissue uptake of poly (ethylene glycol) with different molecular weights after intravenous administration to mice. J Pharm Sci, 1994; 83(4); 601–606.

4 Kou D, Manius G, Zhan S, Chokshi HP. Size exclusion chromatography with Corona charged aerosol detector for the analysis of polyethylene glycol polymer. J Chromatogr A, 2009; 1216; 5424–5428.

5 Takahashi K, Kinugasa S, Senda M, Kimizuka K, Fukushima K, Matsumoto T, Shibata Y, Christensen J. Quantitative comparison of a corona-charged aerosol detector and an evaporative light-scattering detector for the analysis of a synthetic polymer by supercritical fluid chromatography. J Chromatogr A, 2008; 1193; 151–155.

6 Lindner JR. Microbubbles in medical imaging: current applications and future directions. Nat Rev Drug Discov, 2004; 3(6); 527–533.

7 Diaz-Lopez R, Libong D, Tsapis N, Fattal E, Chaminade P. Quantification of pegylated phospholipids decorating polymeric microcapsules of perfluorooctyl bromide by reverse phase HPLC with a charged aerosol detector. J Pharm Biomed Anal, 2008; 48; 702–707.

8 Chawla A, Hinberg I, Blais E, Johnson D. Aggregation of insulin, containing surfactants, in contact with different materials. Diabetes, 1985; 34; 420–424.

9 Lougheed WD, Albisser AM, Martindale HM, Chow JC, Clement JR. Physical stability of insulin formulations. Diabetes, 1983; 32; 424–432.

10 Fekete S, Ganzler K, Fekete J. Fast and sensitive determination of Polysorbate 80 in solutions containing proteins. J Pharm Biomed Anal, 2010; 52; 672–679.

11 Lobback C, Backensfeld T, Funke A, Weitschies W. Quantitative determination of nonionic surfactants with CAD. Chromatogr. Techniques, November 2007; 18–20.

12 Lobback C, Backensfeld T, Funk A, Weitschies W. Analysis of polysorbate 80 using fast HPLC and charged aerosol detection. Pharm Technol, 2010; 34(5); 48–50, 52, 54.

13 Fukushima K, Matsumoto T, HashiguchiI K, Senda M, Carreiro D, Asa D, Christensen J, Acworth I. HPLC with charged aerosol detection for the measurement of different non-ionic surfactants. Chromatography, 2006; 27; 139–142.

14 Christiansen A, Backensfeld T, Kühn S, Weitschies W. Investigating the stability of the nonionic surfactants tocopheryl polyethylene glycol succinate and sucrose laurate by HPLC-MS, DAD, and CAD. J Pharm Sci, 2011; 100(5); 1773–1782.

9

Application of Charged Aerosol Detection in Traditional Herbal Medicines

Lijuan Liang[1,2], *Yong Jiang*[3], *and Pengfei Tu*[3]

[1] *Beijing Friendship Hospital, Capital Medical University, Beijing, China*
[2] *State Key Laboratory of Natural and Biomimetic Drugs, School of Pharmaceutical Sciences, Peking University Health Science Center, Beijing, China*
[3] *State Key Laboratory of Natural and Biomimetic Drugs, Peking University, Beijing, China*

CHAPTER MENU

9.1 Summary

Traditional herbal medicines generally contain many compounds that may be relevant to their putative activities. Therefore, a more rational approach that can simultaneously analyze a suite of compounds for the authentication and quality control of traditional medicines is of great importance. The need for universal HPLC detection is widespread, especially for the analysis of compounds lacking strong UV chromophores. As a complement to UV/Vis detection, mass spectrometry (MS), evaporative light scattering detection (ELSD), and refractive index detection (RID) could be used. In the recent years, a new mass-flow-sensitive technique, charged aerosol detection (CAD), has also been used for the analysis of traditional medicines. In this paper, CAD operating principle, the factors affecting its sensitivity, and its application in analyzing traditional herbal medicines, such as Notoginseng, Ginseng, and Radix Astragali have been summarized. The positive characteristics and drawbacks

Charged Aerosol Detection for Liquid Chromatography and Related Separation Techniques, First Edition. Edited by Paul H. Gamache.

of CAD compared with other detectors, such as ELSD and UV, were also specified through these concrete examples.

9.2 Introduction

Traditional herbal medicines (THMs) have been used for thousands of years and have provided a unique theoretical and practical approach for the treatment of diseases. Recently, THMs have gained increasing popularity in many countries due to their few side effects and integrated adjustment of the whole body. Therefore, a more rational and effective approach for the authentication and quality evaluation of THMs has become more and more urgent and important.

As each herbal medicine contains many compounds that may be relevant to the medicine's putative activity, analytical techniques that look at a suite of compounds are of great importance. Ultraviolet (UV) detection is often used as the preferred detection technique due to its high sensitivity [1–4], broad linear range, and ease of operation. However, in case the active ingredient does not have a UV-absorbing chromophore, it will not be detected in routine HPLC analysis, thus the need for universal HPLC detection is widespread. Several universal detectors could be used in conjunction with HPLC, such as MS, nuclear magnetic resonance (NMR) detector, refractive index detection (RID), chemiluminescence nitrogen detector (CLND), evaporative light scattering detection (ELSD), and condensation nucleation light scattering detector (CNLSD).

RID, based on changes in the refractive index (RI), is widely used since it is nonspecific [5]. But this detector also has its intrinsic weaknesses, such as lack of sensitivity and the use of gradient elution conditions usually results in a variable baseline. The temperature and gradient elution must be both kept constant as RID is very sensitive to small changes in temperature and pressure. MS has been the universal detection system of choice due to its high sensitivity and additional mass spectral information [6, 7]. However, MS is known to suffer from variable response factors, which can be attributed to the fact that a universal ionization interface is still being sought. Besides, it is too expensive to use in a routine manner and requires highly trained people to operate. CLND is extremely sensitive and is regarded as a universal detector [8]. However, CLND responds only to analytes containing nitrogen, and the mobile phase must be kept free of nitrogen-containing components, which excludes the use of acetonitrile as the organic modifier in the HPLC eluent [9]. Another downside is that the analyte must either be fluorescent or be made fluorescent by tagging with a suitable fluorophore. ELSD has made an important progress toward successfully detecting analytes that do not contain UV-absorbing chromophore [10]. Up to now, it has been successfully applied for quantitative determination of herbal medicines [11–16]. However, it also suffers from some disadvantages; in some cases, unsatisfactory quantitation, reproducibility, sensitivity, and dynamic range have been reported, and its response varied with the solvent composition [10].

Recently, a new type of detector, the ESA Biosciences Corona® CAD® charged aerosol detector (CAD), which is designed to measure nonvolatile and some semivolatile analytes, has been introduced for LC applications [17]. As a universal detector, CAD is considered to be more sensitive than ELSD [18, 19] and has a wide dynamic response range of approximately four orders of magnitude, high sensitivity, and good precision. The response of the detector does not rely on the optical properties of analytes nor on the ability of analytes be ionized in the gas phase. These characteristics, along with reliability and simple operation, make it a superior detector for a wide range of HPLC analyses [20]. Furthermore, this detector can be used in combination with a variety of different separation modes (isocratic and gradient reversed-phase [21], mixed-mode chromatography [22], hydrophilic interaction liquid chromatography [23, 24], supercritical fluid chromatography [25], and size exclusion chromatography [26]) in normal and narrow-bore column formats, for a wide range of different analytes. Applications of CAD in literature are quite diverse, for example, the analyses of synthetic polymers, inorganic ions, and lipids, the determination of enantiomeric ratios, and the analyses of pharmaceuticals and their purity. Though CAD is known to exhibit uniform relative response factors under isocratic conditions for a wide range of nonvolatile analytes, it is not perfect enough. The use of this detector has been restricted due to the variable response observed under gradient elution conditions. Besides, CAD exhibits reduced response if the analyte is volatile or if particle formation is incomplete. In the following, the factors affecting the sensitivity of CAD will be summarized.

9.3 Factors that Affect the Sensitivity of CAD

Parameters that could affect the response of CAD include the ratio of water/organic solvent of mobile phase; the kind of additives, such as formic acid, acetic acid, and ammonium acetate buffers; the flow rate changes; and nitrogen gas purity.

9.3.1 Mobile Phase Composition

The response factor under gradient conditions for CAD is not as universal as initially inferred, for the response of CAD system is affected by the diameter of the generated particles, which is in turn related to the size of the aerosol droplets, given by the equation

$$d_{\mathrm{p}} = d_{\mathrm{d}}\left(\frac{C}{\rho_{\mathrm{p}}}\right)^{1/3}$$

where ρ_p is the density of the particle (given by the density of the analyte), C is the analyte concentration, d_p is the particle diameter, and d_d is the droplet diameter [17]. Since the droplet diameter is related to several other factors, including density and viscosity of the mobile phase, which depend on the mobile phase composition, so in gradient elution chromatography, the response factor will vary significantly. Higher organic content in the mobile phase leads to greater transport efficiency of the nebulizer, which results in a larger number of particles reaching the detector chamber and a higher signal [27].

For aerosol detectors to gain wider acceptance, it is also necessary to overcome the "gradient effect" of these detectors. Efforts to mitigate this gradient effect have involved a gradient compensation approach, whereby a second pump has been used to deliver a post-column inverse gradient prior to the aerosol detector [28, 29]. This process ensured that the composition of the mobile phase entering the detector was constant and thus resulted in constant response, independent of the mobile phase composition in the column. In the report of Miroslav Lísa, the chromatograms of triacylglycerols (TGs) from plant oils showed an increasing baseline signal and TG response during gradient elution. The increased response is caused by an improved transport efficiency of the nebulizer, leading to a higher number of charged particles reaching the detector because of the increasing content of apolar solvent in the gradient [30]. These investigators therefore used gradient compensation and reported significant improvement in response factors (i.e., only 5% variation) [30]. Another approach for overcoming the gradient effect is to construct a three-dimensional calibration plot, such as that performed by Mathews *et al.* [31].

9.3.2 Effects of Nitrogen Gas Purity on the Sensitivity of CAD

To assess the effect of nitrogen gas purity on the sensitivities of CAD, Han Young Eom *et al.* compared two types of nitrogen gas, one with ultrahigh purity (99.999%) and the other with normal purity (99.9%) [19]. They found that in most cases, the purity of the nitrogen gas did not have a significant effect on the sensitivity of CAD. However, they proposed that more studies were needed to fully investigate the effect of gas purity on the response of the ELSD and CAD systems.

9.3.3 The Effect of Mobile Phase Modifiers

Modifier gradients would cause the composition of the mobile phase to differ for analytes, which are eluted at different parts of the gradient. This in turn affects the nebulization and droplet evaporation process in the detector and can lead to a 5–10-fold change in the response of an individual analyte due to variations in the transport efficiency of droplets/particles within the detector [27].

To evaluate the influence of various concentrations of mobile phase modifiers on the sensitivity of CAD, Han Young Eom *et al.* tested ammonium acetate buffer, ammonium formate buffer, acetic acid, and formic acid at different concentrations. They found that the sensitivity of the detector was improved by decreasing the salt concentrations to a certain extent in the mobile phase [19].

9.3.4 Comparison of Flow Rate Effect on the Sensitivity of CAD

Han Young Eom *et al.* evaluated the effect of flow rate on the sensitivity of CAD [19]. The flow rates of 0.8, 1.0, and 1.2 mL/min were tested using 0.1 mM ammonium acetate (pH 4.0)/acetonitrile, and the highest sensitivity was obtained at a 1.0 mL/min flow rate.

9.4 Application of CAD in Quality Analysis of Traditional Herbal Medicines

CAD has been widely used in the analysis of TGs [30], aminoglycosides [21], lipids [32], and polymers [25, 26] and in the analysis of squalene, cholesterol, and ceramide [33], but its application in the quality analysis of THMs is not very widely reported. With the increasing popularity of THMs, more and more attention has been paid to their quality and efficacy. Each herbal medicine contains many compounds that may be relevant to its therapeutic effect, among which saponins are an important group due to their significant anticancer properties and other putative activities, so saponins are generally regarded as an important chemical marker for the quality control of herbal medicines. Here we used the analyses of saponins in Notoginseng and Ginseng as examples to highlight the possible application of CAD in the analysis and identification of THMs.

9.4.1 Determination of Saponins in Radix et Rhizoma Notoginseng by CAD Coupled with HPLC

Chang-Cai Bai [34] examined the feasibility and performance of CAD on the analysis of Radix et Rhizoma Notoginseng and finally established an HPLC-CAD method for the simultaneous determination of seven saponins, namely, notoginsenoside R_1; ginsenosides Rg_1, Re, Rb_1, Rg_2, Rh_1, and Rd in 30 batches of crude drugs (Figure 9.1). The limit of detections (LODs) and LOQs of CAD, ELSD, and UV were compared by injecting various amounts of standard compound solutions.

Results showed that CAD method exhibited a lower LOD (0.01–0.15 μg) and LOQ (0.04–0.41 μg) than UV and ELSD (Table 9.1). Furthermore, the CAD exhibited a steadier baseline in gradient elution compared with UV at 203 nm. By comparison of the peak areas of seven compounds, it can be found that response of CAD is apparently higher than those of the other two detectors, except ginsenosides Rb_1, which exhibited a fairly sensitive response on ELSD

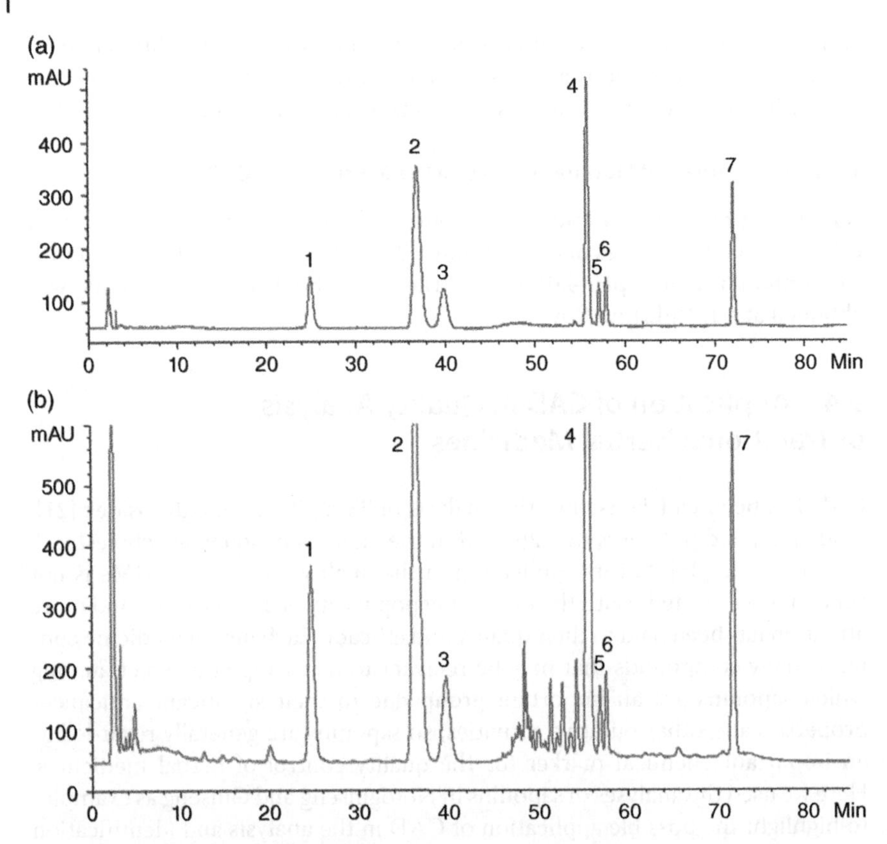

Figure 9.1 Typical HPLC chromatograms of mixed standards with CAD (a) and methanol extracts of Radix et Rhizoma Notoginseng with the detectors of CAD (b), UV (c), and ELSD (d). Compounds 1–7 are notoginsenoside R_1; ginsenosides Rg_1, Re, Rb_1, Rg_2, Rh_1, and Rd, respectively. *Source:* Bai *et al.* [34]. Reproduced with permission of Taylor & Francis.

(Figures 9.1 and 9.2). Mean values of peak area were calculated as 7.66 times of CAD to ELSD while 16.76 times of CAD to UV. Furthermore, the main performance index of detector and the sensitivities of CAD and ELSD were compared by using three representative saponins: notoginsenoside R_1, ginsenosides Rg_1, and Rb_1. The result (Table 9.2) showed that the mean sensitivity of CAD was 2.07 times of ELSD.

9.4.2 Determination of Ginsenosides by LC-CAD

Li Wang *et al.* [35] developed a simple and sensitive method for the quantification of seven major saponins, namely, ginsenosides Rg_1, Re, Rb_1, Rc, Rb_2, Rb_3, and Rd in *Panax ginseng* collected from different locations in Liaoning, Jilin, and Heilongjiang provinces by LC-CAD (Figure 9.3). In her study, CAD was

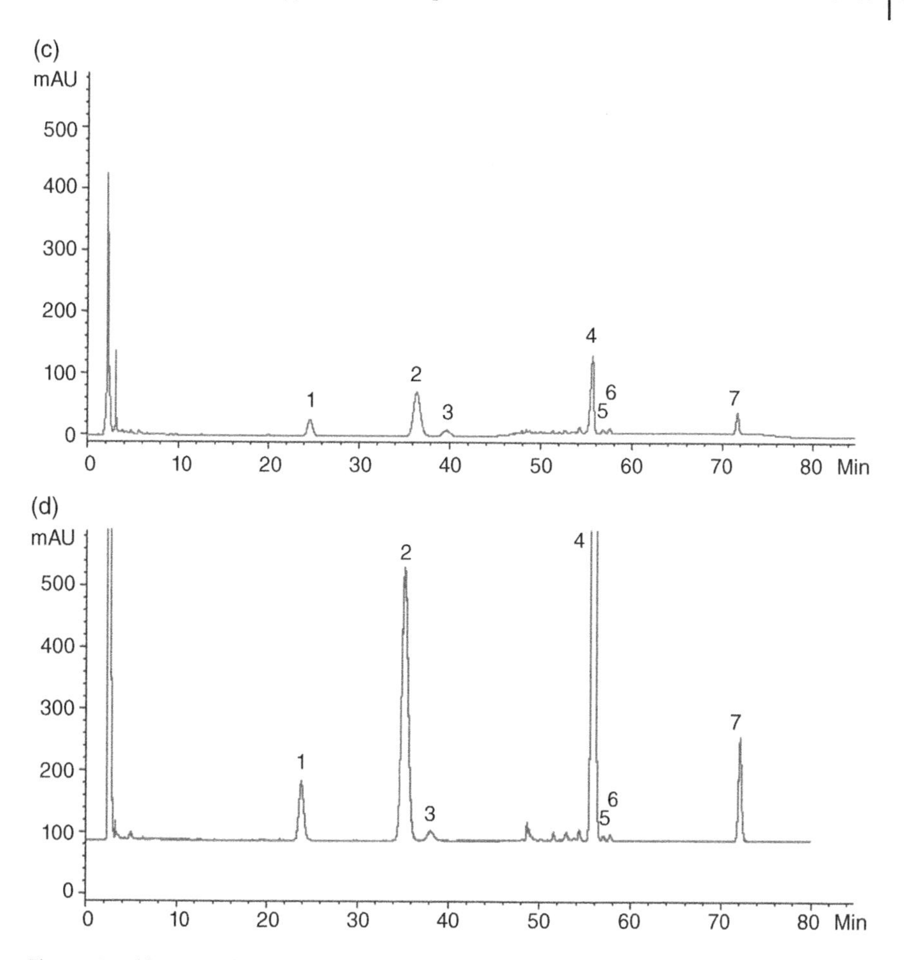

Figure 9.1 (Continued)

compared with ELSD and UV in sensitivity, linearity, and precision for detection of ginsenosides. The regression data, LODs, and LOQs of the three detectors for seven compounds are shown in Table 9.3. It showed that CAD was more sensitive than the other two detectors for it had the lowest LOD and LOQ values for the analyses of ginsenosides. For example, for ginsenoside Rg_1 and Re, CAD was able to detect the concentration of six times lower than that of ELSD and UV detection.

CAD and ELSD are both mass-dependent detectors. The relation between the area response and the corresponding concentration of the analytes is nonlinear and fits $A = aM^b$, where A is the peak area, M is the mass of the sample, and a/b are experimental constants. After log–log transformation, $\log A = b \log M + \log a$ was obtained and thus can be used for calibration.

Table 9.1 LODs and LOQs for seven saponins in Radix et Rhizoma Notoginseng by UV, ELSD, and CAD.

	UV		ELSD		CAD	
Compound	**LOD (μg)**	**LOQ (μg)**	**LOD (μg)**	**LOQ (μg)**	**LOD (μg)**	**LOQ (μg)**
N-R_1	0.03	0.13	0.04	0.13	0.05	0.16
G-Rg_1	0.04	0.28	0.08	0.16	0.03	0.13
G-Re	0.02	0.63	0.19	0.75	0.15	0.41
G-Rb_1	0.05	0.30	0.05	0.17	0.01	0.05
G-Rg_2	0.12	0.45	0.15	0.82	0.02	0.05
G-Rh_1	0.13	0.45	0.18	0.48	0.12	0.04
G-Rd	1.83	4.60	0.08	0.30	0.02	0.15

Source: Bai *et al.* [34]. Reproduced with permission of Taylor & Francis.
G-, ginsenoside; N-, notoginsenoside.

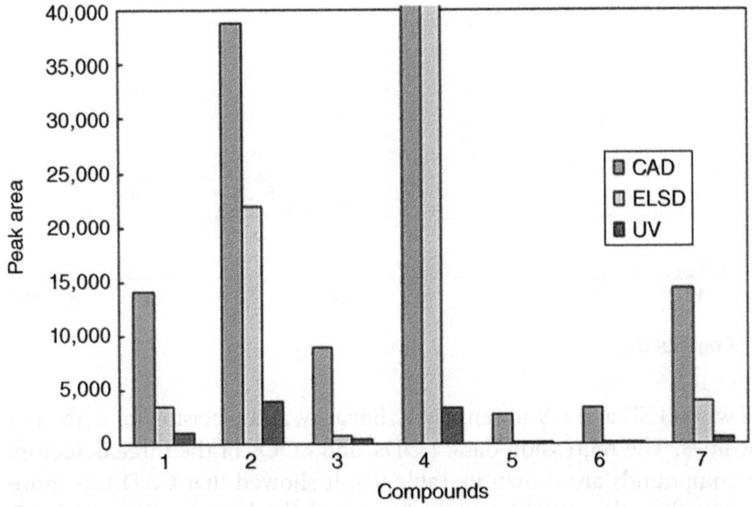

Figure 9.2 CAD, UV, and ELSD response of seven investigated saponins (10 μL of 0.048 g/mL sample solution was injected). *Source*: Bai *et al.* [34]. Reproduced with permission of Taylor & Francis.

In Table 9.3, the linearity coefficients ranged from 0.9975 to 0.9990 for ELSD, from 0.9975 to 0.9995 for CAD, and from 0.9995 to 0.9999 for UV. These data showed that the regression quality of the CAD was somewhat better than that of ELSD. For intraday and interday precision test, the CAD provided a

Table 9.2 Comparison of sensitivity of ELSD and CAD by QI, PA, ratio of Sc, and Se parameters.

	CAD			ELSD			Ratio
Compound	QI	PA	Sc	QI	PA	Se	Sc/Se
N-R_1	1.00	2,990.60	—	2.80	2,313.30	—	—
	2.00	5,629.00	2,638.40	3.60	3,470.10	1,446.00	1.82
	4.00	11,325.00	2,848.00	4.40	4,784.80	1,643.38	1.73
	5.00	14,352.00	3,027.00	5.20	6,516.00	2,164.00	1.40
	6.00	16,792.00	2,440.00	6.00	8,664.20	2,685.25	0.91
	7.50	19,954.50	2,108.33	7.00	10,558.90	1,894.70	1.11
G-Rg_1	0.70	2,174.40	—	2.80	3,615.65	—	—
	3.50	10,727.40	3,054.64	3.60	5,105.05	1,861.75	1.64
	7.00	20,254.00	2,721.89	4.40	6,288.80	1,479.69	1.84
	10.50	29,484.20	2,637.20	5.20	7,650.25	1,701.81	1.55
	14.00	38,301.60	2,519.26	6.00	9,709.00	2,573.44	0.98
	17.50	46,131.90	2,237.23	8.00	17,341.20	3,816.10	0.59
G-Rb_1	5.00	33,449.60	—	1.60	928.00	—	—
	7.50	45,316.60	4,746.80	2.00	1,170.00	605.00	7.85
	10.00	56,400.60	4,433.60	3.00	3,157.40	1,987.40	2.23
	12.50	66,266.80	3,946.48	4.00	6,598.90	3,441.50	1.15
	15.00	74,895.30	3,451.40	6.00	9,850.10	1,625.60	2.12
	17.50	91,006.40	6,444.44	7.00	11,436.00	1,585.90	4.06
Mean							2.07

Source: Bai *et al.* [34]. Reproduced with permission of Taylor & Francis.
PA, peak area; QI, quantity of injecting sample; S (sensitivity) = $\Delta PA/\Delta QI$; Sc, sensitivity of CAD; Se, sensitivity of ELSD.

better reproducibility (RSD ≤ 2.94%) over ELSD (≤3.99%) and close to that of the UV detector (RSD ≤ 2.50%).

9.4.3 Other Applications of CAD

Lijuan Liang developed a method to simultaneously determine 13 pharmacologically active flavonoids and astragalosides in Radix Astragali (RA) by using HPLC-UV-CAD. Through comparison, UV detector exhibited to be more sensitive for the determination of flavonoids, while CAD was suitable for the analysis of astragalosides (data is being submitted). The contents of 13 active compounds in 45 samples collected from different cultivating regions

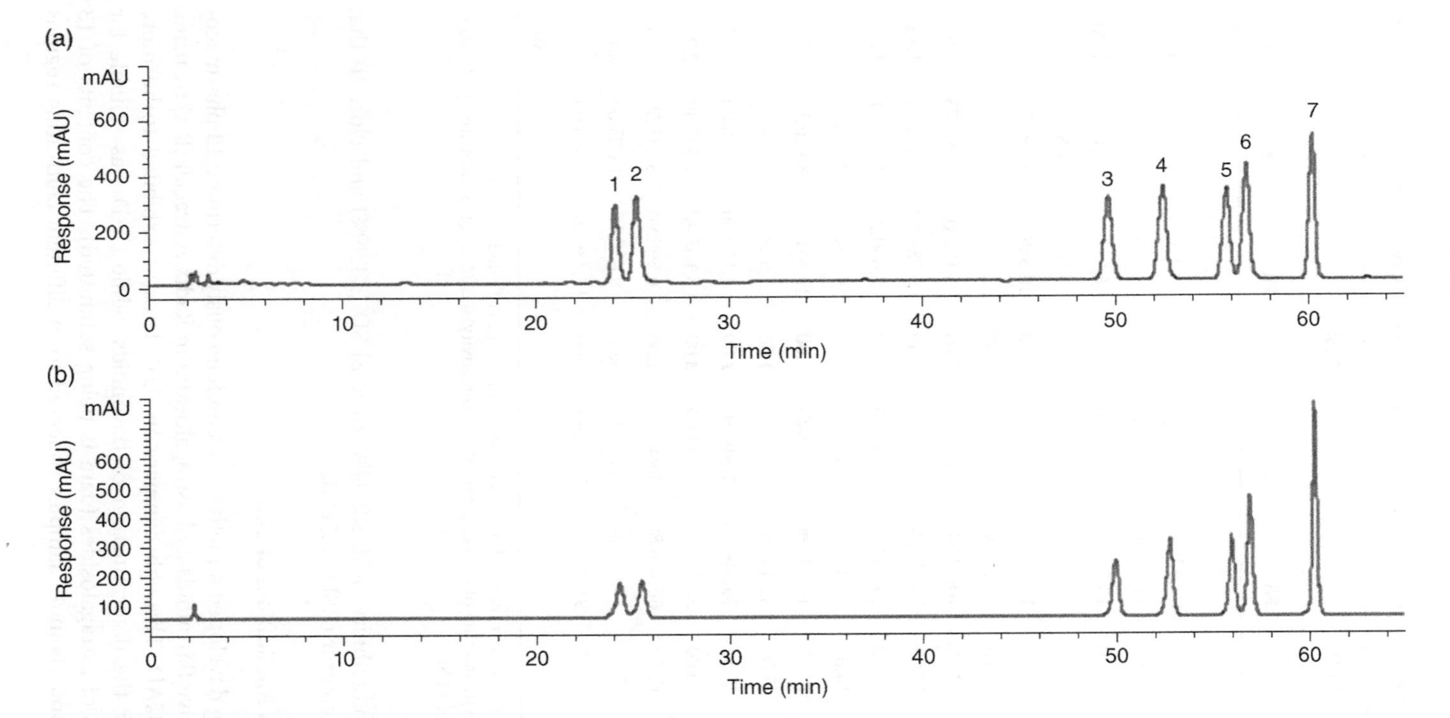

Figure 9.3 Representative LC chromatograms of *P. ginseng*. (a) Representative LC chromatograms of the mixed standards solutions of seven compounds of CAD signal, (b) corresponding ELSD signal, (c) corresponding UV signal, (d) representative LC chromatograms of the samples of *P. ginseng* of CAD signal. Peaks: 1, Ginsenoside Rg_1; 2, Ginsenoside Re; 3, Ginsenoside Rb_1; 4, Ginsenoside Rc; 5, Ginsenoside Rb_2; 6, Ginsenoside Rb_3; 7, Ginsenoside. *Source*: Wang *et al*. 2009 [35]. Reproduced with permission of Springer.

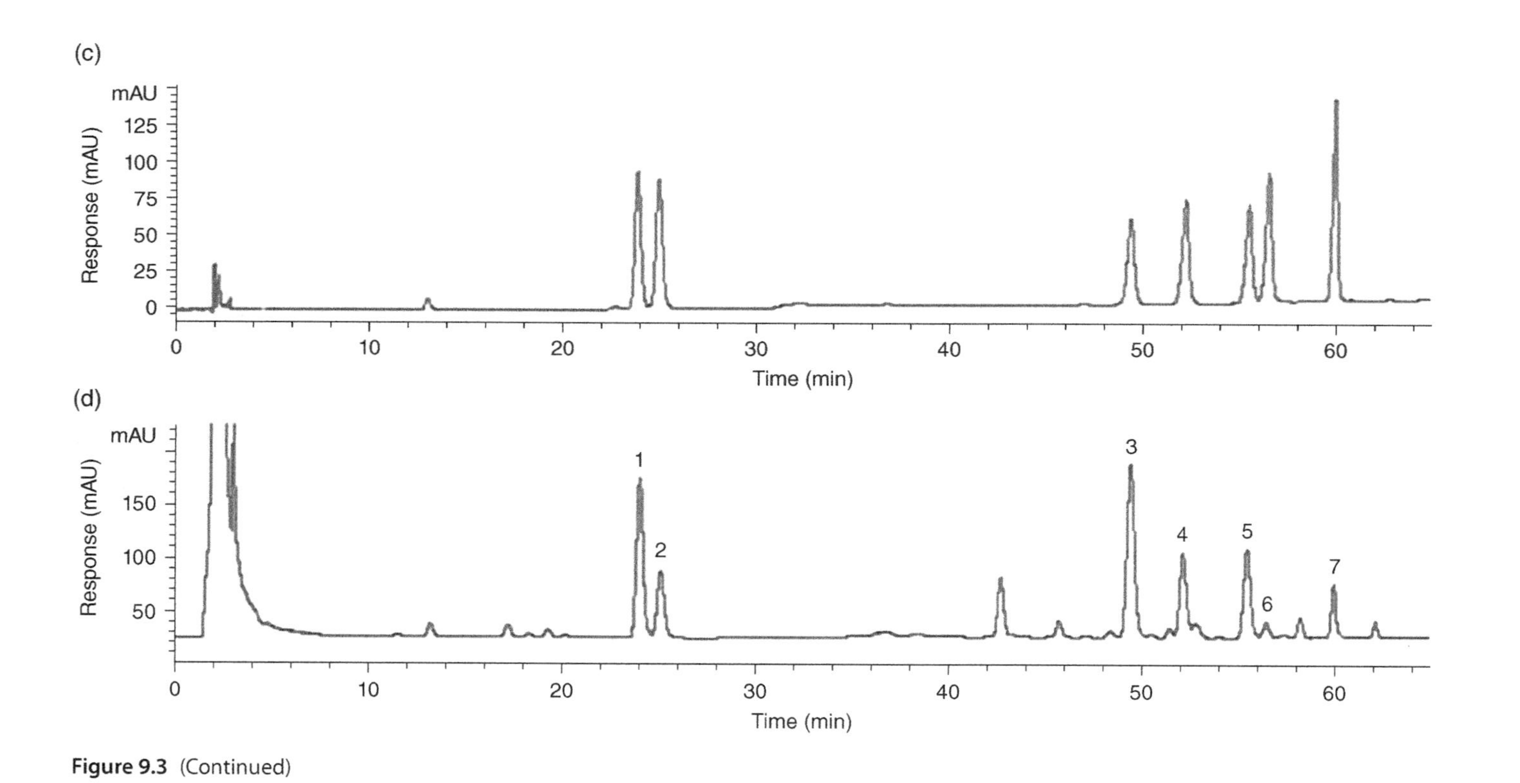

Figure 9.3 (Continued)

Table 9.3 Regression data, LODs, and LOQs of seven ginsenosides for the three different detectors.

Analyte	Rg_1	Re	Rb_1	Rc	Rb_2	Rb_3	Rd
Test range (mg/mL)	0.0158–0.63	0.015–0.06	0.0136–0.546	0.0164–0.654	0.0136–0.546	0.0147–0.588	0.0169–0.675
CAD[a]							
r^2	0.9995	0.9991	0.9985	0.9984	0.9975	0.9991	0.9993
LOD (ng)	9.64	6.57	13.9	15.0	11.3	11.6	10.9
LOQ (ng)	24.1	17.5	34.8	40.1	25.5	20.9	18.2
ELSD							
r^2	0.9983	0.9987	0.9990	0.9975	0.9979	0.9988	0.9988
LOD (ng)	60.2	43.8	46.4	58.4	47.2	38.6	24.2
LOQ (ng)	144.6	102.2	116.0	150.3	113.4	96.5	72.6
UV							
r^2	0.9998	0.9998	0.9999	0.9996	0.9995	0.9996	0.9996
LOD (ng)	62.3	43.8	36.0	67.0	85.0	57.9	54.4
LOQ (ng)	124.6	119.0	108.0	101.0	128.0	133.0	142.0

Source: Bai *et al.* [34]. Reproduced with permission of Taylor & Francis.
[a] The logarithm of the peak areas (y) and the logarithm of the concentration of the analyte (x).

Table 9.4 Comparison of CAD and ELSD for the determination of astragalosides.

	LOD		LOQ	
	CAD	ELSD	CAD	ELSD
Astragaloside IV	48	75	130	156
Astragaloside I	50	110	130	270
Acetylastragaloside I	40	90	120	200

were determined by the developed method. The results indicated that there were significant differences regarding the marker contents among the crude drugs from different regions. Components in samples from Jilin, Liaoning, and Shanxi were higher than those from other resources.

Lianwen Qi also established a HPLC-DAD-ELSD method for the determination of six isoflavonoids and four saponins in RA [36]. A comparison between the LODs and LOQs of ELSD [36] and CAD was made. The results (Table 9.4) showed that LODs and LOQs of CAD were much lower than those of ELSD.

Han Young Eom [19] optimized the analytical conditions for the simultaneous determination of the 10 saikosaponins in the extracts of Bupleuri Radix and compared the detectors of ELSD and CAD. Results indicated that even though both CAD and ELSD were sensitive enough to simultaneously analyze the 10 saikosaponins, the linearity, sensitivity, reproducibility, and peak sharpness were superior with CAD.

Natural plant oils are complex mixtures of various nonpolar compounds of which the TGs make up to 90% of the content. The identification and quantitation of TGs in plant oils that contain several hundred types of compound is complicated and challenging. Common HPLC detectors, such as MS, UV, or ELSD, may lead to much differing response factors. Miroslav Lísa developed a simple approach for the quantitation of TGs in complex natural mixtures from plant oils by using nonaqueous reversed-phase high-performance liquid chromatography (NARP-HPLC) in a gradient mode combined with CAD and mobile phase compensation [30]. Good reproducibility and excellent LOD with CAD was achieved, and the results were in good agreement with the results obtained using a quantitative APCI-MS method.

9.5 Conclusion

CAD is the dawn of a new era in HPLC detection for its high sensitivity and precision, wide dynamic range, and ease of operation for the universal detection of nonvolatile or semivolatile compounds, especially of the compounds with weak or no UV absorption chromophores. In herbal medicines or traditional Chinese medicines, there are lots of such constituents like saponins, saccharides, terpenoids, amino acids, and so on that needed a more sensitive method for the routine analysis, which supplies a broad and brilliant prospect of CAD application in this field. On the other hand, how to eliminate those factors affecting the sensitivity of CAD and to make it more sensitive and more stable is also a noticeable issue for pushing a wider application of CAD in THMs.

References

1 K. Jones, D. J. Malcolme-Lawes. J. Chromatogr. A 441 (1988) 387–393.
2 Y. Shi, C. J. Sun, B. Zheng, Y. Li, Y. Wang. Food Chem. 123 (2010) 1322–1327.
3 L. Nováková, S. A. Lopéz, D. Solichová, D. Šatínský, B. Kulichová, A. Horna, P. Solich. Talanta 78 (2009) 834–839.
4 H. J. Lee, C. Y. Kim. Food Chem. 120 (2010) 1224–1228.
5 S. G. Westerbuhr, K. L. Rowlen. J. Chromatogr. A 886 (2000) 9–18.
6 M. Vogeser, C. Seger. Clin. Biochem. 41 (2008) 649–662.
7 W. A. Korfmacher. Drug Discov. Today 10 (2005) 1357–1367.

8 C. A. Lucy, C. R. Harrison. J. Chromatogr. A 920 (2001) 135–141.
9 M. C. Allgeier, M. A. Nussbaum, D. S. Risley. LCGC N. Am. 78 (2003) 376.
10 S. Almeling, U. Holzgrabe. J. Chromatogr. A 1217 (2010) 2163–2170.
11 H. L. Wang, J. Gao, D. N. Zhu, B. Y. Yu. J. Pharm. Biomed. Anal. 43 (2007) 1552–1556.
12 Y. Cong, Y. B. Zhou, J. Chen, Y. M. Zeng, J. H. Wang. J. Pharm. Biomed. Anal. 48 (2008) 573–578.
13 W. J. Kong, C. Jin, W. Liu, X. H. Xiao, Y. L. Zhao, Z. L. Li, P. Zhang, X. F. Li. Food Chem. 120 (2010) 1193–1200.
14 X. Y. Chai, S. L. Li, P. Li. J. Chromatogr. A 1070 (2005) 43–48.
15 S. N. Kim, Y. W. Ha, H. Shin, S. H. Son, S. J. Wu, Y. S. Kim. J. Pharm. Biomed. Anal. 45 (2007) 164–170.
16 B. S. Sun, L. J. Gu, Z. M. Fang, C. Y. Wang, Z. Wang, M. R. Lee, Z. Li, J. J. Li, C. K. Sung. J. Pharm. Biomed. Anal. 50 (2009) 15–22.
17 R. W. Dixon, D. S. Peterson. Anal. Chem. 74 (2002) 2930.
18 N. Vervoort, D. Daemen, G. Torok. J. Chromatogr. A 189 (2008) 92–100.
19 H. Y. Eom, S. Y. Park, M. K. Kim. J. Chromatogr. A 1217 (2010) 4347.
20 T. Vehovec, A. Obreza. J. Chromatogr. A 1217 (2010) 1549.
21 A. Joseph, A. Rustum. J. Pharm. Biomed. Anal. 51 (2010) 521–531.
22 K. Zhang, L. L. Dai, N. P. Chetwyn. J. Chromatogr. A 1217 (2010) 5776–5784.
23 L. Novakova, D. Solichova, P. Solich. J. Chromatogr. A 1216 (2009) 4574–4581.
24 Z. Huang, M. A. Richards, Y. Zha, R. Francis, R. Lozano, J. Ruan. J. Pharm. Biomed. Anal. 50 (2009) 809–814.
25 K. Takahashi, S. Kinugasa, M. Senda, K. Kimizuka, K. Fukushima, T. Matsumoto, Y. Shibata, J. Christensen. J. Chromatogr. A 1193 (2008) 151–155.
26 D. Kou, G. Manius, S. D. Zhan, H. P. Chokshi. J. Chromatogr. A 1216 (2009) 5424–5428.
27 Z. Cobb, P. Shaw, L. Lloyd, N. Wrench, D. Barrett. J. Microcol. 13 (2001) 169.
28 T. Gorecki, F. Lynen, R. Szucs, P. Sandra. Anal. Chem. 78 (2006) 3186.
29 A. de Villiers, T. Gorecki, F. Lynen, R. Szucs, P. Sandra. J. Chromatogr. A 1161 (2007) 183.
30 M. Lísa, F. Lynen, M. Holčapek, P. Sandra. J. Chromatogr. A 1176 (2007) 135.
31 B. T. Mathews, P. D. Higginson, R. Lyons, J. C. Mitchell, N. W. Sach, M. J. Snowden, M. R. Taylor, A. G. Wright. Chromatographia 60 (2004) 625.
32 R.A. Moreau. Lipids 41 (2006) 727.
33 A. Hazzotte, D. Libong, M. Matoga, P. Chaminade. J. Chromatogr. A 1170 (2007) 52.
34 C. C. Bai, S. Y. Han, X. Y. Chai, Y. Jiang, P. Li, P. F. Tu. J. Liq. Chromatogr. Relat. Technol. 32 (2009) 242.
35 L. Wang, W. S. He, H. X. Yan, Y. Jiang. Chromatographia 70 (2009) 603.
36 L. W. Qi, Q. T. Yu, P. Li, S. L. Li, Y. X. Wang, L. H. Sheng, L. Yi. J. Chromatogr. A 1134 (2006) 162–169.

Section 3

Industrial Applications of Charged Aerosol Detection

10

Charged Aerosol Detection in Pharmaceutical Analysis

An Overview

Michael Swartz[1], *Mark Emanuele*[2], *and Amber Awad*[3]

[1] *Analytical Development, Validation Science, Uxbridge, MA, USA*
[2] *Westford, MA, USA*
[3] *Dominion Diagnostics, North Kingstown, RI, USA*

CHAPTER MENU

10.1 Summary

Over the past several years, charged aerosol detection (CAD) has become a widely used technology in the pharmaceutical laboratory. From formulation to stability and even QC, many analysts are turning to this technology due to its advantages of sensitivity, ease of use, dynamic range, and applicability to a wide range of analyses in the drug development process. This chapter examines the operation and use of CAD in a regulated pharmaceutical environment; addresses method development, validation, and transfer; and highlights a few examples illustrating some advantages of using CAD in the pharmaceutical laboratory.

Charged Aerosol Detection for Liquid Chromatography and Related Separation Techniques, First Edition. Edited by Paul H. Gamache.

10.2 Introduction

CAD was first introduced commercially in 2004 (Corona® CAD®, ESA Biosciences, Chelmsford, MA) and is based upon a combination of HPLC with electrical aerosol technology available since the 1970s [1–6]. CAD is a unique technology in which the HPLC column eluent is first nebulized with a nitrogen (or air) carrier gas to form droplets that are then dried to remove mobile phase, producing analyte particles. The primary stream of analyte particles is met by a secondary stream that is positively charged as a result of having passed a corona discharge formed at the tip of a high-voltage platinum needle. The charge transfers difusionally to the opposing stream of analyte particles and is further transferred to a collector where it is measured by a highly sensitive electrometer, generating a signal in direct proportion to the quantity of analyte present. More detail on the design and operation of CAD is provided in Chapter 1; therefore it will not be repeated here. However, because the entire process involves particles and direct measurement of charge, CAD is highly sensitive, provides a consistent response, and has a broad dynamic range, offering real advantages to researchers and analysts in the pharmaceutical laboratory, particularly when analyzing compounds lacking UV chromophores. Often compared to other universal-type HPLC detectors, like refractive index (RI) and evaporative light scattering detection (ELSD), CAD has been shown to be much easier to use and, unlike RI, can accommodate gradients. In addition, CAD response is not dependent upon the chemical characteristics of the compounds of interest but on the initial mass concentration of analyte in the droplets formed upon nebulization, providing a much more uniform response as opposed to, for example, UV, where responses can vary dramatically according to the wavelength used and the extinction coefficient. It is precisely these advantages that make it an attractive addition to the pharmaceutical laboratory throughout all phases of drug development.

CAD has been used for a wide range of analyses throughout the drug development process, for example, drug discovery [7], formulations research and development (R&D) [8, 9], natural product isolation [10], impurities [11–18], cleaning validation [19], drug substance and drug product characterization [20, 21], and stability [22], among other examples. In most aspects, CAD is simple and easy to use and can be described as a "plug and play" detector requiring little in the way of special attention, unlike ELSD; a comprehensive list of CAD applications by compound type is available [1]. While many of the reported uses of CAD in the literature are for R&D/method development, the use in a regulated environment in support of GMP, has also been reported [22], where method validation and method transfer are important considerations. However, whether implementing CAD in an R&D, or quality control (QC) laboratory, in addition to highlighting its use, this chapter also discusses a few things to keep in mind to ensure success.

10.3 Analytical Method Development

The first question to answer during analytical method development (AMD) is "will my compound respond?" While CAD certainly has advantages for detecting compounds that do not have UV chromophores, it can provide advantages (e.g., equivalent relative responses independent of the extinction coefficient) even for compounds that do have a chromophore because of its near-universal response. The single overriding criterion for determining analyte response is that the volatility of compounds of interest must be nonvolatile. Molecular weight, melting point, or boiling point cannot be used to predict a compound's volatility with any great accuracy since compounds that have similar molecular weights may have very different volatilities due to polarity and hydrogen bonding. For example, glycerol (MW 92; BP 290°C) is easily detected to less than 10 ng on column, but propylglycerol (propanediol) (MW 76; BP 188°C) is not. A better indicator of volatility is vapor pressure. Since substances with higher vapor pressure vaporize more readily than substances with a lower vapor pressure, the latter respond better to CAD. Table 10.1 lists a few compounds and their responsiveness to CAD along with their vapor pressure.

Table 10.1 Physical characteristics vs. CAD response.

Compound	Form	MW	BP (°C)	Vapor pressure (mm Hg)	Detected by corona
2,4-Dimethylaniline	Liquid, covalent	121.19	218	0.16	No
Ammonium acetate	Solid, ionic	77.08	Decomp.	HOAc 11 at 20°C NH_4OH 115 at 20°C	No
Caffeine	Solid, covalent	194.19	178	$<1 \times 10^{-8}$	Yes
Theophylline	Solid, covalent	180.17	454.1	5.1×10^{-9} at 25°C	Yes
Ethyl carbamate	Solid, covalent	89.09	185	0.36 at 25°C	No
Glycerol	Liquid, covalent	92.09	290	0.001	Yes
Menthol	Solid, covalent	156.27	212	0.8 at 20°C	No
Naphthalene	Solid, covalent	128.16	218	0.018 at 25°C	No
Propylglycerol (1,3-propanediol)	Liquid, covalent	76.09	214	0.8 at 20°C	No
Sodium chloride	Solid, ionic	58.44	1465	N/A	Yes

Solvents also play a role in CAD response; purity, volatility, and viscosity are important factors. In general, the background current is lower using higher-purity solvents, leading to less noise and baseline drift due to fewer particles formed from nonvolatile impurities, particularly when using gradients. One major requirement, however, is that since the CAD process involves nebulization to evaporate the mobile phase, volatile mobile phases must be used. That generally means aqueous/organic solvents (water/methanol/acetonitrile mixtures), with volatile buffer additives (when necessary) such as formic acid, acetic or trifluoroacetic acid, and ammonium acetate, similar to mass spectrometry (MS) mobile phase requirements. It should also be mentioned that there is a much wider choice of organic modifiers for CAD compared with UV detection. For example, acetone, with a UV cutoff of 330 nm, is not normally used in UV detection but is completely compatible with CAD. The topic of mobile phase requirements for CAD is further discussed in Chapter 3.

Finally mobile phase viscosity is also important because it can affect both the nebulizer and drying process. Low-viscosity mobile phases (i.e., high organic) produce a greater number of droplets, and particle generation is more efficient than those of high viscosity (i.e., aqueous), increasing detector response/sensitivity. Also, with low viscosity mobile phases, more analyte is available for detection; with aqueous phases, more analyte goes to waste affecting sensitivity.

A good general approach to determine analyte response and solvent affects during AMD is to perform a flow injection analysis (FIA) experiment by injecting the analyte of interest into the mobile phase without the column in line [23]. A typical AMD system might include multiple detectors in addition to CAD, for example, UV/PDA or MS. Detectors in series are preferable to a parallel configuration to avoid flow splitting; however when used in combination with other destructive detectors (e.g., MS), flow splitting is unavoidable. In series configurations, CAD should be placed last in line. A CAD instrument causes about seven bars of backpressure, well within the range of a typical UV/PDA detector flow cell limitations. In multiple detector system configurations, extra care should be taken to make proper connections and to avoid excessive tubing lengths so as to not contribute additional dead volume that can lead to increased band spread.

Of course, in any AMD process, column choice is very important, a fact that naturally does not change with CAD. However, care should be taken to choose a column with minimal "bleed," as bleed in the form of nonvolatile compounds contributed by the column can result in increased background noise [24]. For this reason, method developers sometimes choose polymeric-based columns over silica to reduce background noise due to column bleed when sensitivity is of prime importance.

One additional AMD note, CAD is fully compatible with fast HPLC, or ultrahigh performance liquid chromatography (UHPLC) approaches.

10.4 Analytical Method Validation

Analytical method validation (AMV) is one part of the overall validation process that also includes software qualification, analytical instrumentation qualification (AIQ), and system suitability [25–28]. AIQ is the process of ensuring that an instrument is suitable for its intended application. In general, AIQ and AMV generally ensure the quality of analysis before conducting a test; system suitability and QC checks ensure the quality of analytical results immediately before or during sample analysis.

Method validation establishes through laboratory testing that the performance characteristics of the method meet the requirements of the intended analytical application. Method validation provides an assurance of reliability of laboratory studies during normal use and is sometimes referred to as the process of providing documented evidence that the method does what it is intended to do. In addition to being good science, regulated laboratories must carry out method validation in order to be in compliance with governmental or other regulatory agencies. In addition to providing proof that acceptable scientific practices are used, method validation is therefore a critical part of the overall validation process. A well-defined and documented method validation process not only satisfies regulatory compliance requirements but also provides evidence that the system and method are suitable for their intended use and aids in method transfer.

In the late 1980s, the US Food and Drug Association (FDA) first designated the specifications listed in the current edition of the United States Pharmacopeia (USP) as those legally recognized to determine compliance with the Federal Food, Drug, and Cosmetic Act [29, 30], and every USP since has included guidelines on method validation. More recently, new information has been published, updating the previous guidelines and providing more detail and harmonization with International Conference on Harmonization (ICH) guidelines [31, 32]. The inclusion and/or definition of some terms differs between the FDA, USP, and ICH. But as a process, harmonization on a global basis has provided much more detail than what was available in the past, and it helps to minimize the differences between global regulatory requirements.

Validation is regulated by the FDA and has roots in manufacturing practice guidelines for the laboratory environment. Two of the most common references to these practices are current Good Manufacturing Practice (cGMP) [29, 30] and the International Organization for Standardization (ISO) 9000 family of quality management systems standards and related ISO documents. The two of the most important guidelines for any method validation process are USP Chapter 1225 (Validation of Compendial Methods) [26] and the ICH Guideline (Validation of Analytical Procedures: Text and Methodology Q2 (R1)) [31]. Although the main focus of this chapter is on CAD, both the USP and ICH guidelines are generic, that is, they do not specifically reference individual technology or specific instrumentation. So in that regard, CAD should be treated just like any other detector used for AMV in the regulated laboratory and should be held to the same standards. Indeed, there are many literature reports of validating methods using CAD (e.g., [14–22, 33, 34] and those reviewed in Chapter 2).

During AMV, several analytical performance characteristics are potentially evaluated, including specificity, accuracy, precision, linearity, and range, limits of detection (LOD) and limits of quantitation (LOQ), and robustness, depending upon the requirements of the method. While a detailed discussion of AMV is outside the scope of this chapter, guidance for AMV is available from a number of sources and should be consulted for more detail [25, 26, 31, 32, 35]. However it is worthwhile here to examine some specific considerations, regarding robustness, linearity, intermediate precision, LOD/LOQ, and system suitability that should be made when validating a method that includes CAD.

Robustness is one parameter that, if not investigated during AMD, is usually investigated early in AMV. The robustness of an analytical procedure is defined as a measure of its capacity to obtain comparable and acceptable results when perturbed by small but deliberate variations in procedural parameters listed in the documentation. Robustness provides an indication of the method's suitability and reliability during normal use. During a robustness study method, parameters are intentionally varied to see if the method results are affected. The key word in the definition is *deliberate*. For example, HPLC variations include parameters such as temperature, flow rate, pH, buffer concentration, and so on. Since CAD response can be dependant upon solvent composition, any potential for change in the mobile phase is one parameter that should be investigated in some detail, particularly when using gradients. In addition, CAD response is somewhat sensitive to the nitrogen flow rate to the detector; therefore it is a CAD-specific parameter that should be investigated during any robustness study.

Linearity is the ability of the method to provide test results that are directly proportional to analyte concentration within a given range. While the dynamic range of CAD has been shown to be over four orders of magnitude [23], over wide ranges CAD response is often nonlinear and approximates a quadratic function. However for smaller ranges, CAD response can be treated as linear ([19]; Swartz ME, Emanuele M, and Awad A, Synomics Pharmaceutical Services, LLC, unpublished results). Additional theoretical insight and explanation of the shape of the response curves for CAD can be found in Chapter 1, and Chapter 3 provides some practical guidance for the use of nonlinear calibration curves and related characteristics.

With a CAD dynamic range of four orders of magnitude, determining the LOQ and LOD does not prove to be an issue for most compounds at the 0.1 and 0.05% concentration levels (relative to the target compound), respectively, as specified in the ICH guidelines [36].

Resolution and efficiency are common criteria for system suitability and can be affected by the band-spread in the system. When used by itself, CAD contributes little more than any other type of detector in terms of band-spread. Where band-spread might become important, however, is when CAD is used in combination with other detectors, as mentioned previously. The detector used before CAD may cause band broadening and this effect (if any) should be measured in separations where resolution is critical, for example, closely eluting-related substances.

Intermediate precision refers to the agreement between the results from within-laboratory variations due to random events, which might normally occur during the use of a method, such as different days, analysts, or equipment. From the authors' experience, it can be expected that differences between different CADs are minimal (Swartz ME, Emanuele M, and Awad A, unpublished results). Satisfactory repeatability results have also been reported (e.g., [11, 20, 22]).

10.5 CAD in Analytical Method Transfer

The process that establishes documented evidence that the analytical method works as well in the receiving laboratory as in the originator's laboratory, or the transferring laboratory, is called analytical method transfer (AMT). AMT is required for good manufacturing practices (GMP) "reportable data" from laboratory results [25, 37, 38]. The goal of AMT is to ensure that the receiving laboratory is well trained, qualified to run the method in question, and gets the same results—within experimental error—as the initiating laboratory. The success of AMT depends upon the development and validation of robust methods and strict adherence to well-documented standard operating procedures (SOPs).

The AMT process can proceed by any one of four options: comparative testing, complete or partial method validation or revalidation, co-validation between the two laboratories, or omission of a formal transfer, sometimes called a transfer waiver. The option chosen depends upon many factors, like the complexity of the method, or level of experience. In general, transferring methods that use CAD is really no different than transferring methods using any other type of HPLC detector. The same experimental design and statistical treatment can be used; however it is a good idea to make sure that the receiving laboratory personnel has adequate training in CAD technology in advance of the transfer. In this respect, regardless of the AMT process option chosen, it is a good idea for the originating laboratory to share any experience gained in the development and validation of the method, particularly related to instrument configuration, nitrogen source, purity and flow, solvent source/purity, and temperature considerations, among other parameters.

10.6 CAD in Formulation Development and Ion Analysis

Formulation plays a critical role in the drug development process. The counterion chosen during formulation can have a significant impact on the bioavailability, manufacturability, and stability of the potential drug candidate. A critical requirement of the formulation process is the development of an analytical procedure that separates and quantitates the counterions (anions and/or cations) involved in the drug formulation being developed. Counterion analysis requires the repeatability and robustness that will enable it to be easily transferred to other analytical laboratories where the quality of the active pharmaceutical ingredient (API) will be carefully monitored.

New chemical entities (NCEs) are most commonly either weak acids or weak bases with a low molecular weight. In the majority of cases, the free acid or base does not provide sufficient aqueous solubility and is also often lacking in solid-state stability. These weaknesses can usually be overcome by pairing a basic or acidic drug molecule with a counterion in order to yield the salt form of the NCE. The properties of the counterion that is selected can have a major impact on the performance of the resulting drug, not to mention its safety and manufacturability. Salt selection was previously performed relatively late in the drug development process. However, there have been numerous occasions when the salt that was initially selected later proved to have problems that made it necessary to repeat toxicology, biological, and stability studies. As a result, the salt selection process is now typically initiated early in the development process for ionizable compounds that pass initial toxicology screening [39].

Chemists frequently search for the optimal crystalline form in a linear manner, creating and analyzing one substance after another in the search for a salt with the right physicochemical properties. In some cases, integrated salt selection systems are used to enable crystallization, salt selection, and polymorph studies to be completed in a fraction of the time. The first screen for salt forms that are identified by these methods is to assess their crystallinity. A crystalline salt form is usually easier to handle, transport, and use. The salt forms that are identified as crystalline during these screens are selected for further evaluation and require analysis to confirm their identity and stoichiometry.

The next step is to assess the salt form's hygroscopicity profile to ensure that it will maintain its properties under the sometimes humid conditions experienced in pharmaceutical manufacturing. Other screens that are performed at this stage of the process include solubility, stability, polymorphism, and processability. The ion analysis method selected for the formulation process is also typically used in downstream processes when the API is monitored to ensure its safety, quality, strength, and purity. This helps to explain the considerable attention that has been devoted in recent years to developing ion analysis methods with the accuracy, precision, and robustness to be easily transferred from one analytical laboratory to another.

Several alternatives for ion analysis are in use today. The most common method, ion chromatography (IC) with conductivity detection, uses weak ionic resins for its stationary phase and an additional suppressor column to remove background eluent ions. IC has proven to provide very sensitive detection of both anions and cations. However it uses relatively expensive nonstandard chromatography instruments and consumables and requires two runs with different columns to measure anions and cations. Lengthy changeover times or duplicate systems are normally required to measure both anions and cations. Capillary electrophoresis (CE) also has demonstrated a high level of sensitivity in ion analysis but requires two columns, two mobile phases, and two runs to detect both anions and cations.

A new approach to ion analysis overcomes these difficulties. This approach is based on hydrophilic interaction liquid chromatography (HILIC) that uses very polar stationary phases such as diol (neutral), silica (charged), amino (charged), or zwitterionic (charged) columns. The mobile phase is highly organic but contains a small amount of aqueous/polar solvent. This establishes a stagnant-enriched water layer around the polar stationary phase, allowing analytes to partition between the two phases based on polarity. Water or a polar solvent is the strong eluting solvent. With zwitterionic columns and chromatographic conditions, the partition function between the two phases permits easier access for electrostatic interaction of anionic analytes to the positively charged group, enhancing anion retention. Using a zwitterionic stationary phase operating in the HILIC mode, the separation and quantitation of 33 commonly used pharmaceutical counterions, 12 cations, and 21 anions was reported [40]. This work used ELSD; however CAD can better provide the high sensitivity, dynamic range, precision, and robustness desirable from a workhorse instrument in these types of critical pharmaceutical applications [41].

The HILIC/CAD approach is extremely flexible and can be used to simultaneously measure an API and its counterion (e.g., penicillin and its counter-cation, potassium; metformin and its counter-anion, chloride; Figure 10.1).

In addition to formulation analysis, CAD can also be used for other types of ion analysis, such as API characterization and detection of ionic impurities. Figure 10.2 illustrates how the HILIC/CAD method was used to analyze nonsteroidal anti-inflammatory drug (NSAID) diclofenac, its sodium counterion, and chloride impurity in a single run. Figure 10.3 shows a further advantage of the HILIC/CAD combination—the simultaneous analysis of several anions and cations in a single run, on a single system using one mobile phase, and one column. This method has been completely validated by determining linearity and range, accuracy, precision, specificity, and LOD and LOQ, showing the utility of CAD in a regulated environment [42]. Recently, this approach was extended to 25 counterions by using gradient elution (Figure 10.4). Another example run

(a)

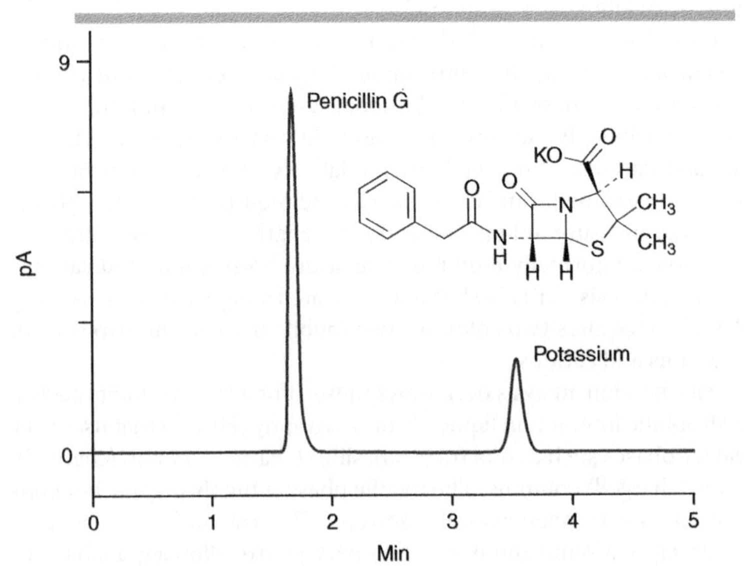

(b) Metformin and its counterion, Cl⁻

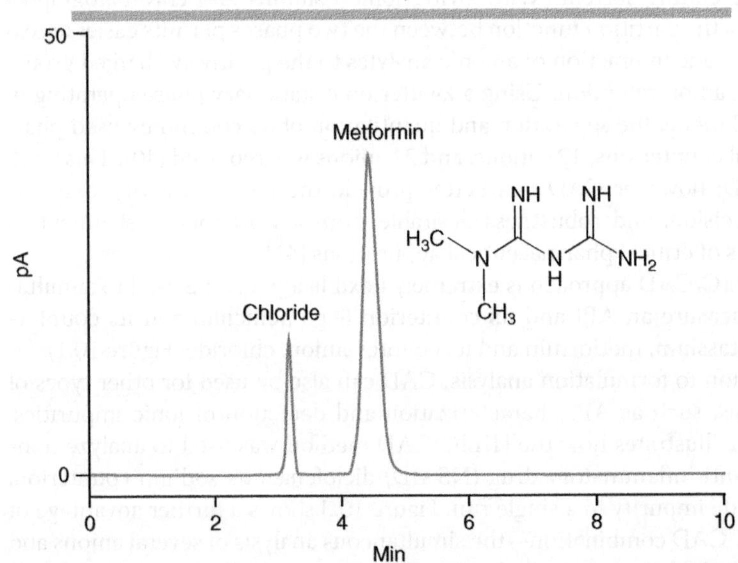

Figure 10.1 HILIC–CAD analysis of API and counterion in a single run. (a) Penicillin G and potassium. Conditions: HPLC: Thermo Scientific™ Dionex™ UltiMate™ 3000 RSLC. Column: Thermo Scientific™ Acclaim™ Trinity P2, 3 μm, 3 × 50 mm. Col. Temp: 30°C; Flow Rate: 0.5 mL/min. Inj. Volume: 1 μL. Eluent A: Acetonitrile, Eluent B: Water, Eluent C: 100 mM ammonium formate, pH 3.65. Isocratic: 25 : 50 : 25 (v/v) A : B : C. Detector: Thermo Scientific™ Dionex™ Corona™ Veo™ RS; Sample: Potassium Penicillin G, 100 ng/μL in DI water.

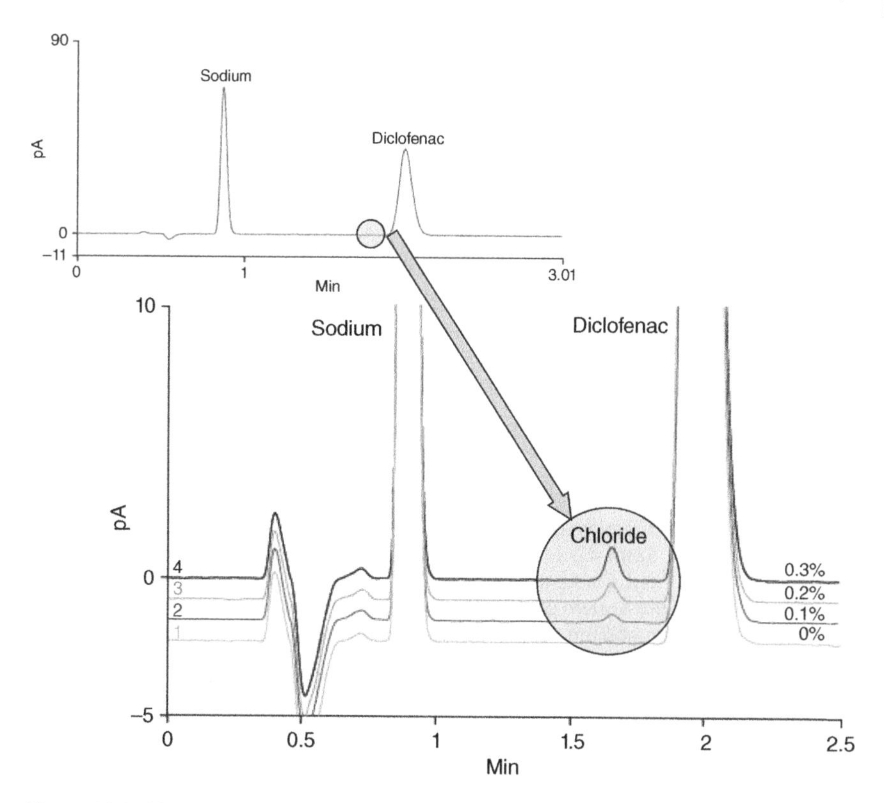

Figure 10.2 API, counterions, and trace impurities in a single analysis. Figure shows the measurement of chloride impurity at 0–0.3% of API. Conditions: HPLC: Thermo Scientific™ Dionex™ UltiMate™ 3000 RSLC. Column: Thermo Scientific™ Acclaim™ Trinity P1, 3 μm, 3 × 50 mm. Col. Temp: 30°C; Flow Rate: 0.8 mL/min. Inj. Volume: 5 μL. Eluent A: Acetonitrile, Eluent B: 200 mM Ammonium acetate pH 4.00. Isocratic: 75 : 25 (v/v) A : B. Detector: Thermo Scientific™ Dionex™ Corona™ Veo™ RS, Evap. Temp: 60°C. Sample: Diclofenac sodium 1 mg/mL in DI water. *Source*: Reproduced with permission of Thermo Scientific Inc.

under roughly the same conditions as those used for Figure 10.3 is presented in Figure 10.5 and shows the separation of some organic acids. Organic bases such as lysine and arginine can also be run under identical conditions. Good linearity and low ng on-column detection limits can be obtained in all cases.

Figure 10.1 (Continued) (b) Metformin and chloride. HPLC: Thermo Scientific™ Dionex™ UltiMate 3000 RSLC. Column: Thermo Scientific™ Acclaim™ Trinity P2, 3 μm, 3 × 50 mm. Col. Temp: 30°C. Flow Rate: 0.5 mL/min. Inj. Volume: 1 μL. Eluent A: Acetonitrile, Eluent B: 100 mM ammonium formate, pH 3.65. Isocratic: 20 : 80 (v/v) A : B. Detector: Thermo Scientific™ Dionex™ Corona™ Veo™ RS, Evap. Temp: 55°C. Sample: Metformin hydrogen chloride, 100 ng/μL in DI water. *Source*: Reproduced with permission of Thermo Scientific Inc.

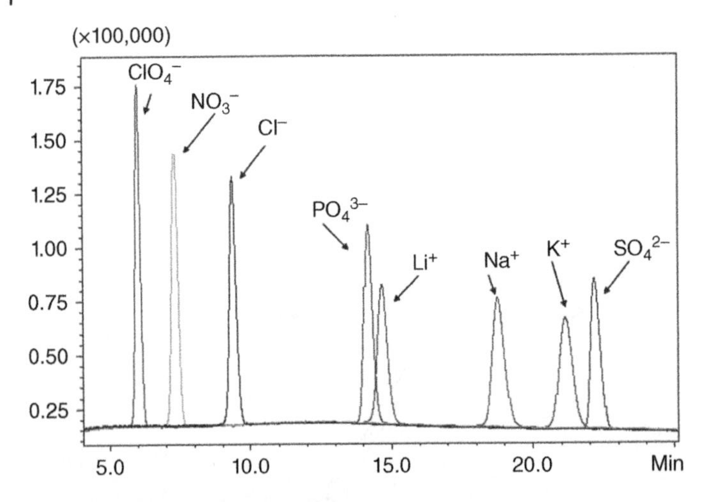

Figure 10.3 Simultaneous analysis of anions and cations using HILIC–CAD. Conditions: A Sequant ZIC®-pHILIC 5 mm, 4.6 × 150 mm column (The Nest Group, Southborough, MA) operated at 30°C was used. Gradient conditions: 20–70% B over 26 min; Mobile Phase A: 15% 100 mM Ammonium Acetate pH 4.68, 5% Methanol, 20% IPA, 60% Acetonitrile; mobile phase B: 50% 30 mM Ammonium Acetate pH 4.68, 5% Methanol, 20% IPA, 25% Acetonitrile, at a flow rate of 0.5 mL/min and a 10 μL injection.

10.7 Carbohydrate Analysis by CAD

CAD response to a wide range of compounds has been explored [7, 23, 43]. In many laboratories, there is increasing interest in analyzing compounds that do not have good UV chromophores, from complex (e.g., polyethylene glycol polymers, oligosaccharides, cyclodextrins, and lipids) to more simple compounds (e.g., sugars). One particular area of focus is the analysis of sugars from fermentation broths.

In many companies, the development and production of proteins, peptides, or antibodies through recombinant DNA technology is becoming increasingly important as new generations of therapeutics are based on these types of biomolecules. However, direct assays to assess both the nutrients supporting the cells in the fermentation broth and measurements of the proteins or peptides themselves can be problematic to perform, especially if real-time measurements are needed. Good, rapid analytical tools and assays are lacking to accurately measure many of the nutritional components of fermentation broths (sugars, amino acids, salts, etc.) as well as the levels of proteins themselves.

As assays for these nutrients and proteins are time consuming, many researchers are forced to default to using simple glucose/lactic acid measurements as a substitute for a real assessment of the metabolic state and protein

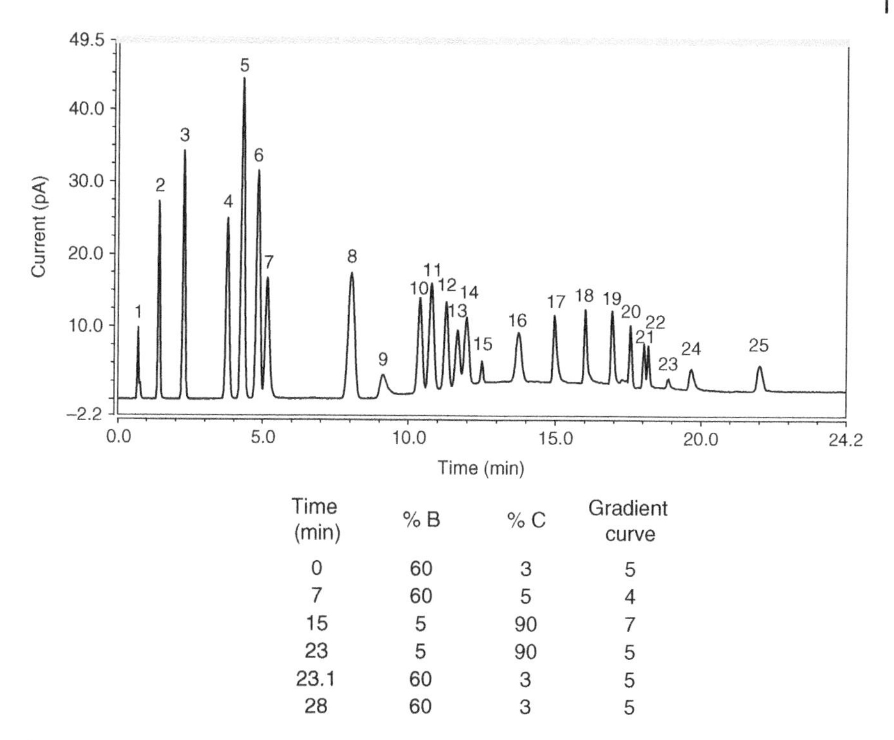

Time (min)	% B	% C	Gradient curve
0	60	3	5
7	60	5	4
15	5	90	7
23	5	90	5
23.1	60	3	5
28	60	3	5

Figure 10.4 Simultaneous analysis of 25 pharmaceutical counterions using HILIC–CAD. HPLC: Thermo Scientific™ Dionex™ UltiMate™ 3000 RSLC. Column: Thermo Scientific™ Acclaim™ Trinity P1, 3 µm, 3 × 50 mm. Col. Temp: 35°C; Flow Rate: 0.5 mL/min. Inj. Volume: 5 µL. Eluent A: water, Eluent B: Acetonitrile, Eluent C: 200 mM Ammonium formate pH 4.00. Gradient—see Figure. Detector: Thermo Scientific™ Dionex™ Corona™ Veo™ RS, Evap. Temp: 50°C. 1, Lactate; 2, procaine; 3, choline; 4, tromethamine; 5, sodium; 6, potassium; 7, meglumine; 8, mesylate; 9, glucoronate; 10, maleate; 11, nitrate; 12, chloride; 13, bromide; 14, besylate; 15, succinate; 16, tosylate; 17, phosphate; 18, malate; 19, zinc; 20, magnesium; 21, fumarate; 22, tartrate; 23, citrate; 24, calcium; 25, sulfate. *Source*: Reproduced with permission of Thermo Scientific Inc.

production levels of their cells. Measurement of typical components of a fermentation broth (sugars, amino acids, etc.) can often involve very lengthy and complex techniques. For example, sugars are often analyzed by pulsed amperometric electrochemical detection coupled to HPLC. CAD has also been used for such analytes.

Figure 10.6 highlights an overlay of five injections of a separation of four simple sugars often found in fermentation broth analyses. Amino acids and peptides from the same sample matrices, as well as sugars from glycopeptides and proteins, can also be analyzed using standard reversed-phase HPLC with CAD [44, 45].

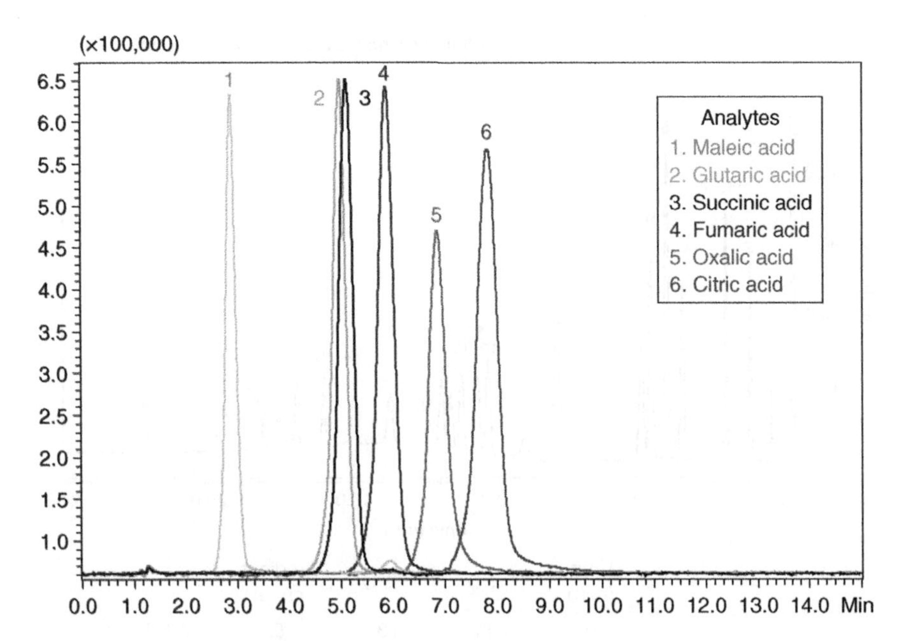

Figure 10.5 CAD–HILIC analyses of organic acids conditions: Column: SeQuant ZIC®-pHILIC, 5 m, 4.6 × 150 mm (The Nest Group, Southborough, MA) operated at 30°C was used with an isocratic mobile phase of 70 : 30 ACN/200 mM ammonium acetate, pH 6.7 operated at a flow rate of 1.0 mL/min and a 10-μL injection.

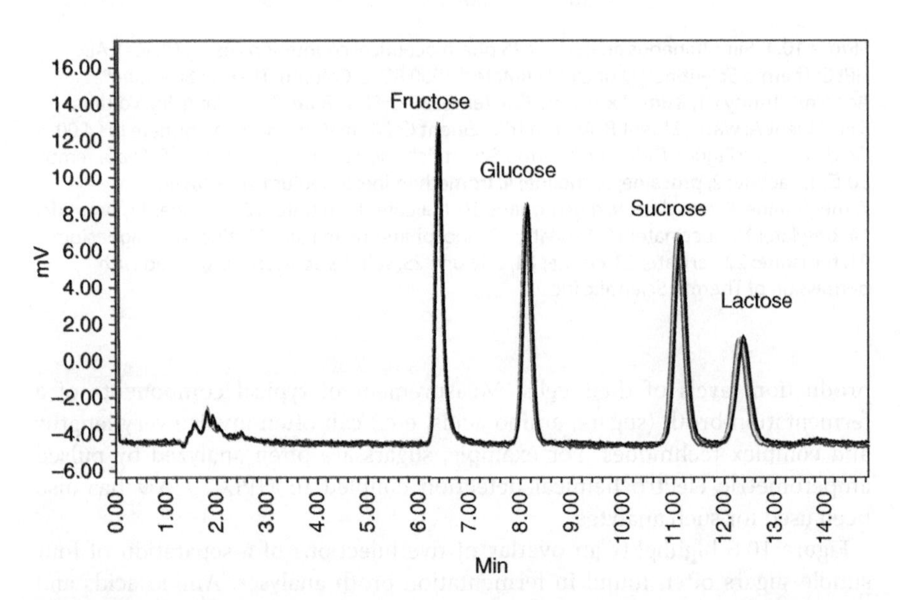

Figure 10.6 Analysis of fermentation broth sugars by HPLC/CAD. Conditions: A Shodex Asahipak NH2P-50 4.6 × 250 mm 5 μm column operated at 35°C was used with an isocratic mobile phase of 25/75 water/ACN and a flow Rate: 1.0 mL/min. Samples are 10 μL injections of 10 μg/mL each in 30/70 water/ACN.

10.8 CAD in Stability Analyses

As mentioned previously, applications of CAD have included monitoring stability [22]. Often a precursor to a formal stability study is a forced degradation or chemical stress study [46, 47]. A forced degradation study is undertaken to understand the reactive chemistry of the drug substance and to help anticipate future stability issues of both drug substance and drug product and can also provide useful information for formulation development. Forced degradation studies are also often required for various regulatory submissions. In addition, they can be used to demonstrate specificity when developing stability-indicating methods (SIMs) and to actually generate a sample that can be used for method development in support of the stability study. Some typical conditions that might be used include extremes of acid and base pH, oxidation, heat, hydrolysis, and light. A SIM is a validated method that can accurately and precisely quantitate the decrease of the API content due to degradation. It is specific for the drug substance; shows a decrease in assay value (correlated to drug substance loss) due to degradation; has no interference from excipients, impurities, or degradation products; and might also detect and quantitate impurities and degradation products of the target compound. For validation, SIMs fall into USP Category 2, that is, methods for the quantitation of impurities or degradation products [26]. For a Category 2 method, a complete list of analytical performance parameters that should be investigated include specificity, linearity and range, accuracy, precision, LOQ, and robustness.

Peptide therapeutics is an interesting case study for the use of CAD in the stability laboratory. Peptides themselves can often be readily detected by UV or MS detection, but their degradation products, or impurities, are potentially smaller peptides or amino acids that in many cases do not possess UV chromophores. (This fact leads one to ask what might be missing in other stability studies.) One example is the analysis of a peptide called adrenocorticotropic hormone fragment 4–10. This peptide, in several respects, is a good representative example of some of the proprietary therapeutic peptide drugs and is readily available commercially and not too expensive. It consists of seven distinct amino acids and it has been studied therapeutically for memory improvement.

The following example looks at this peptide both as an example of how CAD can be used during a stability study, to detect degradants or impurities that do not have UV chromophores, and as a way to look at amino acid composition as a QC tool.

The advantage in using CAD for this application is that the potential amino acid degradants, and/or the composition, can be analyzed without any derivatization in a single simple reversed-phase HPLC system.

Figure 10.7 illustrates the analysis of a pure reference standard of the adrenocorticotropic hormone fragment 4–10 peptide, by both UV (PDA) detection

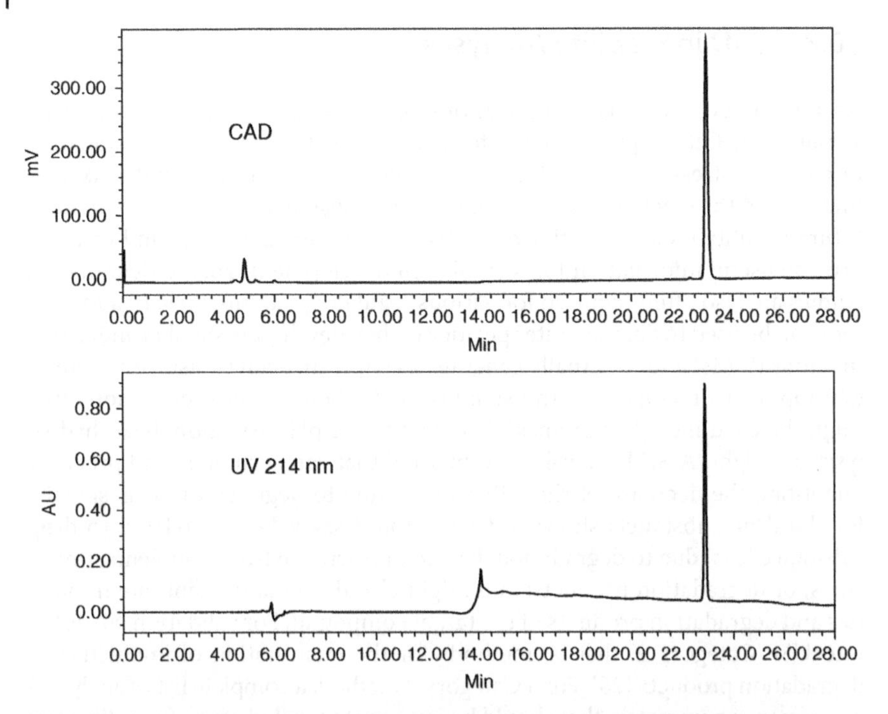

Figure 10.7 Analysis of adrenocorticotropic hormone fragment 4–10 peptide standard by UV/PDA and CAD. Conditions: A C_{18} 4.6 × 250 mm 5 μm column was used at ambient temperature with a mobile phase of 0.1% TFA in water (A) and ACN (B). The gradient consisted of a 5-min hold at 100% A, then a linear gradient from 0 to 40% B over 20 min at a flow rate of 0.6 mL/min; 10 μL of a 1 mg/mL sample concentration in mobile phase A was injected.

and CAD. The PDA detector was run in series before CAD; the S/N is higher on CAD versus the UV detector even at this low UV wavelength where the peptide bond absorbs strongly. We have obtained CAD sensitivity for peptides down to the low ng level, enough sensitivity to pick up potential impurities and degradants, particularly as the response to these analytes will be independent of their physical/chemical characteristics.

To fully test the sensitivity hypothesis, the reversed-phase HPLC separation of all of the amino acids present in the peptide, underivatized, using the same conditions used for the intact peptide analysis, again with both UV/PDA detection and CAD, is shown in Figure 10.8. Although not shown in this chromatogram, the parent peptide is also well resolved from all of the amino acids. As can be seen, some of the amino acids respond well to UV detection, however others without chromophores do not. Also, note that CAD response is roughly equivalent, although some of the later-eluting amino acids benefit from the organic-rich mobile phase environment. Separations like these can be evaluated

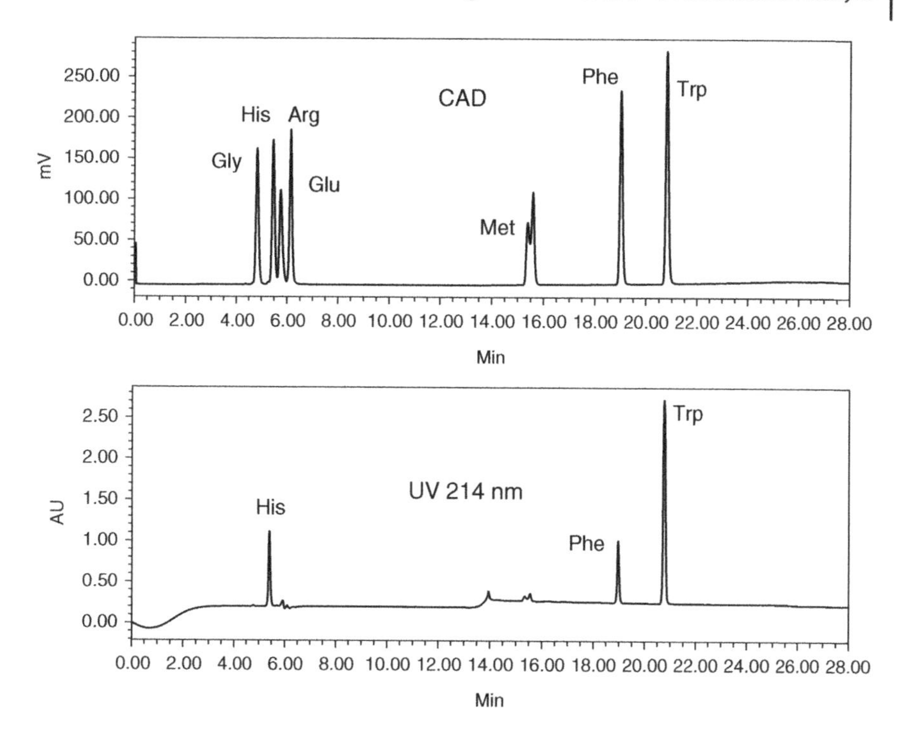

Figure 10.8 Analysis of free, underivatized amino acids that compose the adrenocorticotropic hormone fragment 4–10 peptide by HPLC/UV/PDA detection and CAD (see Figure 10.7 for conditions).

in forced degradation studies, to determine if the method is stability indicating and can be used in support of stability storage studies. This method can also of course be used as a QC tool to analyze amino acid composition following hydrolysis using the same conditions, without the need for a complex and time-consuming derivatization processes.

10.9 Conclusion

The use of CAD can provide significant advantages to analysts in both the research and regulated pharmaceutical laboratory when compared with more traditional technology, such as UV, RI, or ELSD. CAD provides universal detection of nonvolatile analytes with a response independent of chemical properties, a wide dynamic range, high sensitivity, and good precision. These characteristics, along with reliability and its ease of use, make this a superior detector for a wide range of HPLC analyses throughout the drug development process.

References

1 Thermo Scientific. http://www.thermoscientific.com/CAD (accessed February 2, 2017).

2 Dixon RW and Peterson DS. Development and testing of a detection method for liquid chromatography based on aerosol charging. Analytical Chemistry, 2002; 74; 2930–2937.

3 Liu BYH and Pul DYH. On the performance of the electrical aerosol analyzer. Journal of Aerosol Science, 1975; 6; 249–250.

4 Medved A, Dorman F, Kaufman SL, and Pocher A. A new corona-based charger for aerosol particles. Journal of Aerosol Science, 2000; 31(Supplement 1); S616–S617.

5 Flanagan RC. History of electrical aerosol measurements. Aerosol Science and Technology, 1998; 28; 301–380.

6 Kaufman SL. Evaporative electrical detector. US 6,568,245. May 27, 2003.

7 Reilly J, Everett B, and Aldcroft C. Implementation of charged aerosol detection in routine reversed phase liquid chromatography methods. Journal of Liquid Chromatography and Related Technologies, 2008; 31; 3132–3142.

8 Schonherr C, Touchene S, Wilser G, Peshka-Suss R, and Francese G. Simple and precise detection of lipid compounds present within liposomal formulations using a charged aerosol detector. Journal of Chromatography A, 2009; 1216; 781–786.

9 Diaz-Lopez R, Libong D, Tsapis N, Fattal E, and Chaminade P. Quantification of pegylated phospholipids decorating polymeric microcapsules of perfluorooctyl bromide by reverse phase HPLC with a charged aerosol detector. Journal of Pharmaceutical and Biomedical Analysis, 2008; 48; 702–707.

10 Lisa M, Lynen F, Holcapek M, and Sandra P. Quantitation of triacylglycerols from plant oils using charged aerosol detection with gradient compensation. Journal of Chromatography A, 2007; 1176; 135–142.

11 Sun P, Wang X, Alquier L, and Maryanoff CA. Determination of relative response factors of impurities in paclitaxel with high performance liquid chromatography equipped with ultraviolet and charged aerosol detectors. Journal of Chromatography A, 2008; 1177; 87–91.

12 Asa D. Impurity testing with a universal HPLC detector. Genetic Engineering News, 2005; 25(19); 33–34.

13 Rystov L, Chadwick R, Krock K, and Wang T. Simultaneous determination of Maillard reaction impurities in memantine tablets using HPLC with charged aerosol detector. Journal of Pharmaceutical and Biomedical Analysis, 2011; 56; 887–894.

14 Stypulkowska K, Blazewicz A, Brudzikowska A, Warowna–Grzeskiewicz M, Sarna K, and Fijalek Z. Development of high performance liquid chromatography methods with charged aerosol detection for the determination

of lincomycin, spectinomycin and its impurities in pharmaceutical products. Journal of Pharmaceutical and Biomedical Analysis, 2015; 112; 8–14.

15 Blazewicz A, Fijalek Z, Warowna-Grzeskiewicz M, and Jadach M. Determination of atracurium, cisatracurium and mivacurium with their impurities in pharmaceutical preparations by liquid chromatography with charged aerosol detection. Journal of Chromatography A, 2010; 1217; 1266–1272.

16 Blazewicz A, Fijalek Z, Sarna K, and Warowna-Grzeskiewicz M. Determination of gentamicin sulphate composition and related substances in pharmaceutical preparations by LC with charged aerosol detection. Chromatographia, 2010; 72; 183–186.

17 Holzgrabe U, Nap C-J, and Almeling S. Control of impurities in L-aspartic acid and L-alanine by high-performance liquid chromatography coupled with a corona charged aerosol detector. Journal of Chromatography A, 2010; 1217; 294–301.

18 Wahl O and Holzgrabe U. Impurity profiling of carbocisteine by HPLC-CAD, qNMR and UV/vis spectroscopy. Journal of Pharmaceutical and Biomedical Analysis, 2014; 95; 1–10.

19 Forsatz B and Snow NH. HPLC with charged aerosol detection for pharmaceutical cleaning validation. LCGC North America, 2007; 25; 960–968.

20 Novakova L, Lopex SA, Solichova D, Satinsky D, Kulichova B, Horna A, and Solich P. Comparison of UV and charged aerosol detection approach in pharmaceutical analysis of statins. Talanta, 2009; 78; 834–839.

21 Holzgrabe U, Nap CJ, Kunz N, Almeling S. Identification and control of impurities in streptomycin sulfate by high-performance liquid chromatography coupled with mass detection and corona charged-aerosol detection. Journal of Pharmaceutical and Biomedical Analysis, 2011; 56(2); 271–279.

22 Liu XK, Fang JB, Cauchon N, and Zhou P. Direct stability-indicating method development and validation for analysis of etidronate disodium using a mixed-mode column and charged aerosol detector. Journal of Pharmaceutical and Biomedical Analysis, 2008; 46; 639–644.

23 Gamache PH, McCarthy RD, Freeto SM, Asa DJ, Woodcock MJ, Laws K, and Cole RO. HPLC analysis of nonvolatile analytes using charged aerosol detection. LCGC North America, 2005; 23; 150–161.

24 Teutenberg T, Tuerka J, Holzhausera M, and Kiffmeyera TK. Evaluation of column bleed by using an ultraviolet and a charged aerosol detector coupled to a high-temperature liquid chromatographic system. Journal of Chromatography A, 2006; 1119; 197–201.

25 Swartz ME and Krull IS. Handbook of Analytical Validation, CRC Press, New York, 2012.

26 Chapter 1225, United States Pharmacopeia, 2016; No. 39.

27 Chapter 621, United States Pharmacopeia, 2016; No. 39.

28 Chapter 1058, United States Pharmacopeia, 2016; No. 39.

29 US FDA. Current Good Manufacturing Practice in Manufacturing, Processing, Packing, or Holding of Drugs: General; 21 CFR, Part 210. www.accessdata.fda.gov/scripts/cdrh/cfdocs/cfcfr/CFRSearch.cfm?CFRPart=210 (accessed April 06, 2017).
30 US FDA. Current Good Manufacturing Practice for Finished Pharmaceuticals; 21 CFR, Part 211. www.accessdata.fda.gov/scripts/cdrh/cfdocs/cfcfr/cfrsearch.cfm?cfrpart=211 (accessed March 6, 2017).
31 Harmonized Tripartite Guideline, Validation of Analytical Procedures, Text and Methodology, Q2 (R1), International Conference on Harmonization, November 2005. www.ich.org (accessed March 6, 2017).
32 US FDA. Guidance for Industry: Analytical Procedures and Methods Validation for Drugs and Biologics, February 2014.
33 Joseph A, Patel S, and Rustum A. Development and validation of a RP-HPLC method for the estimation of netilmicin sulfate and its related substances using charged aerosol detection. Journal of Chromatographic Science, 2010; 48; 607–612.
34 Joseph A and Rustum A. Development and validation of a RP-HPLC method for the determination of gentamicin sulfate and its related substances in a pharmaceutical cream using a short pentafluorophenyl column and a charged aerosol detector. Journal of Pharmaceutical and Biomedical Analysis, 2010; 51; 521–531.
35 Swartz ME and Krull I. LCGC Validation Viewpoint Columns; see http://chromatographyonline.findanalytichem.com/columns (accessed February 13, 2017).
36 Harmonized Tripartite Guideline, Impurities in New Drug Substances/Products, Q3A (R2)/Q3B (R2), 2006. www.ich.org (accessed March 6, 2017).
37 ISPE Good Practice Guide: Technology Transfer, March 2003. www.ispe.org (accessed March 6, 2017).
38 Chapter 1224, United States Pharmacopeia, 2016; No. 39. www.usp.org (accessed March 6, 2017).
39 Kumar L, Amin A, and Bansal A. Preparation and characterization of salt forms of enalapril. Pharmaceutical Development and Technology, 2008; 13; 345–357.
40 Risley DS and Pack BW. Simultaneous determination of positive and negative counterions using a hydrophilic interaction chromatography method. LCGC, 2006; 24(8); 776–785.
41 Zhang K, Dai L, and Chetwyn NP. Simultaneous determination of positive and negative pharmaceutical counterions using mixed-mode chromatography coupled with charged aerosol detector. Journal of Chromatography A, 2010; 1217; 5776–5784.
42 Crafts C, Bailey B, Plante M, and Acworth IN. Evaluation of methods for the simultaneous analysis of cations and anions using HPLC with charged aerosol detection and a zwitterionic stationary phase. Journal of Chromatographic Science, 2009; 47; 534–539.

43 Almeling S, Ilko D, and Holzgrabe U. Charged aerosol detection in pharmaceutical analysis. Journal of Pharmaceutical and Biomedical Analysis, 2012; 69; 50–63.

44 Thomas D, Acworth I, Bauder R, Plante M, and Kast L. Label-free profiling of O-linked glycans by UHPLC with charged aerosol detection. Thermo Scientific Poster Note 71846. http://www.thermoscientific.com/content/dam/tfs/ATG/CMD/cmd-documents/sci-res/posters/ms/events/eas2015/PN-71846-HPLC-CAD-Glycans-O-linked-EAS2015-PN71846-EN.pdf (accessed February 2, 2017).

45 Thomas D, Acworth I, Bailey B, Plante M, and Zhang Q. Direct determination of native *N*-linked glycans by UHPLC with charged aerosol detection. Thermo Scientific Poster Note 70903. http://www.thermoscientific.com/content/dam/tfs/ATG/CMD/CMD%20Documents/Application%20&%20Technical%20Notes/Chromatography/Liquid%20Chromatography/Liquid%20Chromatography%20Modules/PN-70903-Determination-N-Linked-Glycans-UHPLC-CAD-PN70903-EN.pdf (accessed February 2, 2017).

46 Baertschi SW, editor. Pharmaceutical Stress Testing, Predicting Drug Degradation, Drugs and the Pharmaceutical Sciences, Volume 153. Informa Healthcare Publishers, New York, 2007.

47 Swartz ME and Krull IS. Developing and validating stability-indicating methods. LCGC North America, June 2005; 23(6); 586–593.

11

Impurity Control in Topiramate with High Performance Liquid Chromatography

Validation and Comparison of the Performance of Evaporative Light Scattering Detection and Charged Aerosol Detection

David Ilko[1], Robert C. Neugebauer[2], Sophie Brossard[2], Stefan Almeling[2], Michael Türck[3], and Ulrike Holzgrabe[1]

[1] *Institute of Pharmacy and Food Chemistry, University of Wuerzburg, Wuerzburg, Germany*
[2] *European Directorate for the Quality of Medicines & HealthCare (EDQM), Strasbourg, France*
[3] *Merck KGaA, Darmstadt, Germany*

CHAPTER MENU

11.1 Summary

An HPLC method using a pentafluorophenyl column for the impurity control in topiramate is presented. The performance of an ELSD and a Corona® CAD® for the detection of substances lacking a suitable UV-chromophor was investigated. The method was validated according to ICH guideline Q2(R1) and the "technical guide for the elaboration of monographs" of the Ph. Eur. for both detectors. Although CAD appeared to be superior in terms of repeatability, sensitivity, and linearity, both detectors gave satisfactory results in the accuracy studies. However, the use of ELSD was considered not feasible due to the appearance of ghost peaks when injecting the highly concentrated test solution of topiramate. Due to its relatively high vapor pressure, one of the impurities (bisacetonide of β-D-fructopyranose) gave no or little response in ELSD and

Charged Aerosol Detection for Liquid Chromatography and Related Separation Techniques, First Edition. Edited by Paul H. Gamache.

CAD. We were able to achieve a ninefold increase in sensitivity by means of post-column addition of acetonitrile and a lower nebulizer temperature with a Corona® ultra® RS. As the sensitivity for all other impurities was still about a factor 10^3 higher, simultaneous detection of all impurities was not feasible. Thus, an HPTLC limit test has to be applied for impurity A.

11.2 Introduction

Topiramate was developed as an anticonvulsant drug substance. Moreover, its use for the prophylaxis of migraine [1] and the treatment of obesity [2], trigeminus neuralgia [3], and substance-related diseases [4, 5] are currently discussed.

Numerous studies dealing with the determination of topiramate in plasma for therapeutic drug monitoring and pharmacokinetic and bioequivalence investigations have been reported [6–13]. Concerning the control of impurities, a method controlling topiramate and impurity C in liquid oral solutions was presented [14]. Moreover, six impurities could be determined by Biro *et al.*, but a total of four different high performance liquid chromatographic (HPLC) methods were employed for impurity profiling [15]. To the best of our knowledge, to date, no comprehensive method for the control of topiramate is available.

The aim of this study was to develop and validate a HPLC method for the determination of impurities in topiramate. The possible impurities derive from the given synthetic route [16] or degradation (Figure 11.1) are as follows:

Impurity A: 2,3:4,5-bis-*O*-(1-methylethylidene)-β-D-fructopyranose
ImpurityB:*N*-[(diethylamino)carbonyl]-2,3:4,5-bis-*O*-(1-methylethylidene)-β-D-fructopyranose sulfamic acid
Impurity C: 2,3-*O*-(1-methylethylidene)-β-D-fructopyranose sulfamic acid
Impurity D: *N*-{[2,3:4,5-bis-*O*-(1-methylethylidene)-β–D-fructopyranosyl] oxycarbonyl}-2,3:4,5-bis-*O*-(1-methylethylidene)-β–D-fructopyranose sulfamic acid
Impurity E: D-Fructose

The control of impurities in topiramate is challenging. Due to the lack of a suitable UV-chromophor of the substituted monosaccharides, the analytes show poor or even no UV absorption. To overcome the detection problem, we decided to evaluate the suitability of aerosol-based detectors, that is, an evaporative light scattering detector (ELSD) and a charged aerosol detector (CAD), which are suitable for the detection of all nonvolatile substances regardless of their chemical properties.

Preliminary experiments revealed that, due to its relatively high vapor pressure, one of the impurities (impurity A) gave little or no response in CAD and ELSD.

Enhancement of the sensitivity for impurity A was on the one hand explored by using a Corona ultra RS, where the temperature of the nebulizer can be

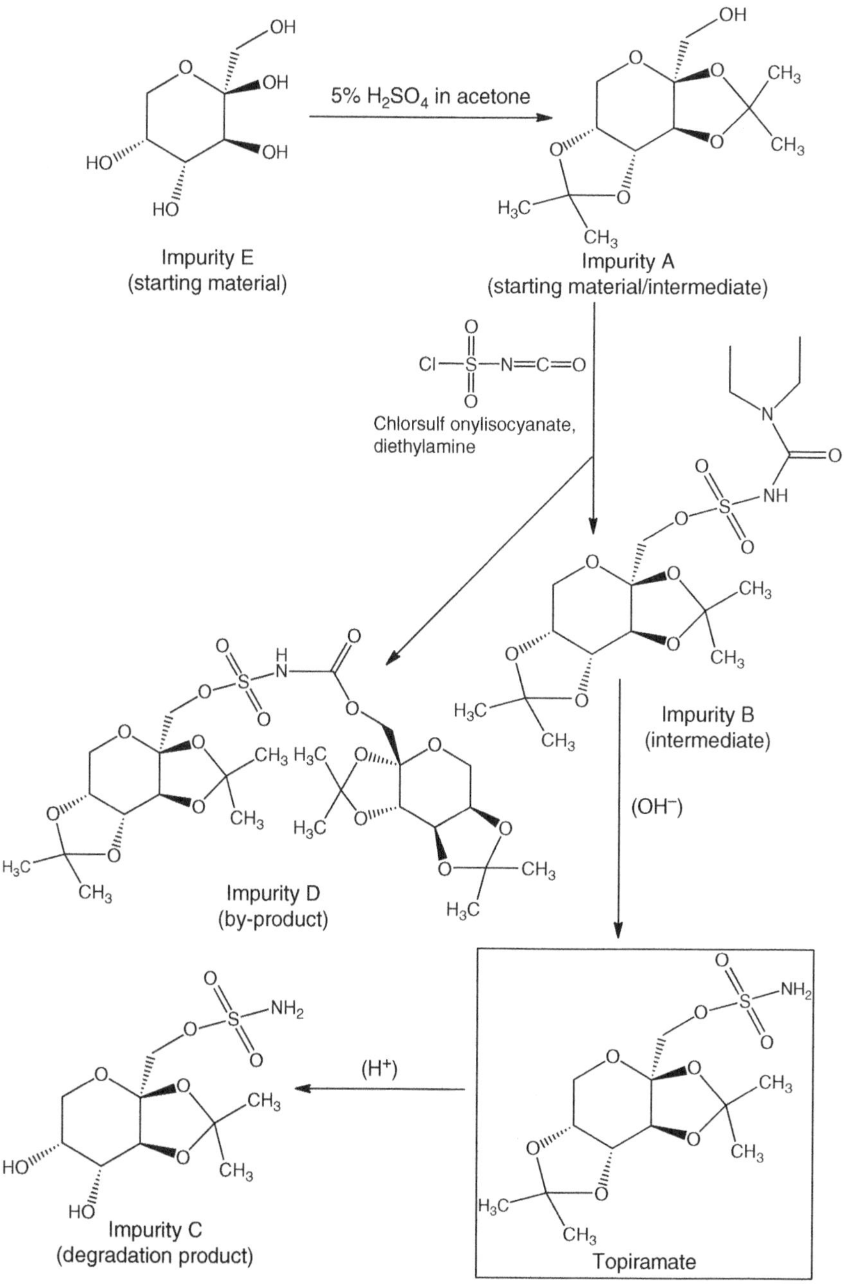

5% H_2SO_4 in acetone

Impurity E
(starting material)

Impurity A
(starting material/intermediate)

Chlorsulf onylisocyanate,
diethylamine

Impurity B
(intermediate)

(OH^-)

Impurity D
(by-product)

(H^+)

Impurity C
(degradation product)

Topiramate

Figure 11.1 Structural formulae of topiramate and its impurities derived from the synthetic route [16]. Impurity A is commercially available and can thus be either starting material or intermediate.

varied. On the other hand, it is known that high amounts of organic modifier in the mobile phase result in an increased response in evaporation-based detectors [17]. Therefore, post-column addition of an organic solvent is a viable option of exerting influence on the composition of the eluent reaching the detector without modifying the HPLC parameters. Chromatographic systems possessing two individual pumps are commercially available and were successfully applied for gradient compensation after the column in order to provide consistent response over the entire chromatographic run [18].

11.3 Material and Methods

11.3.1 Reagents and Material

Topiramate and its impurities A, B, C, and D were provided by the European Directorate for the Quality of Medicines & HealthCare (EDQM). Ammonium acetate for HPLC, fructose (impurity E), glacial acetic acid, and sodium chloride were purchased from Sigma-Aldrich (St-Quentin Fallavier, France).

Ultrapure water (>18.2 MΩ) was delivered by an ELGA PureLab Ultra system (Elga Antony, France) or a Milli-Q Synthesis system (Billerica, MA, USA). Gradient grade acetonitrile was purchased from Sigma-Aldrich (Chromasolv®) (St-Quentin Fallavier, France) and VWR International (HiPerSolv Chromanorm®) (Darmstadt, Germany).

11.3.2 HPLC–ELSD/CAD

The optimized method employs a Kinetex PFP (100 × 4.6 mm, 2.6 μm particle size) analytical column from Phenomenex (Aschaffenburg, Germany). The column compartment was maintained at 40°C. Flow rate was 1.0 mL/min and the injection volume 20 μL. A gradient was applied with 25 mM ammonium acetate (pH 3.5, adjusted with glacial acetic acid) as mobile-phase A and acetonitrile as mobile-phase B. The proportion of mobile-phase B was 20% (v/v) during the initial 5 min and then increased to 50% (v/v) within further 10 min.

A PL-ELS 2100 ELS detector (Polymer Laboratories, Marseille, France) and a Corona ultra RS (Thermo Fisher, Courtaboeuf, France) were operated with a Dionex UltiMate® 3000 ×2 chromatographic system (Dionex, Courtaboeuf, France) equipped with two ternary pumps, an online degasser, a thermostated autosampler, and a thermostated column compartment. The first pump delivered the analytical gradient as described previously. The second pump was used for experiments involving post-column addition of organic modifier. In these experiments, acetonitrile at a flow rate of 1.0 mL/min was mixed with the column eluent via a post-column tee connection. The evaporation temperature of the ELSD was set to 80°C, the nebulizer temperature to 50°C, and the flow rate of the nebulization gas (nitrogen) to 1.0 standard liter per minute (SLM). The gas inlet pressure (nitrogen) for the Corona ultra RS was 35 psi, the range to 100 pA, and the filter to "0," and the nebulizer temperature was varied between 18 and 35°C.

The method validation using the Corona CAD (Thermo Scientific, Idstein, Germany) was conducted on an Agilent 1100 LC system (Waldbronn, Germany) equipped with a binary pump, an online degasser, and a thermostated column compartment. The settings for the Corona CAD were as follows: gas inlet pressure (nitrogen): 35 psi, filter: "none," range: 100 pA.

A pH-meter 780 (Metrohm, Villebon-sur-Yvette, France) and a PHM220 Lab pH Meter (Radiometer Analytical SAS, Lyon, France) were used for pH adjustment. For measurements using ELSD, the test solution was prepared by dissolving 20 mg topiramate in 1.0 mL of a mixture of mobile-phase A and mobile-phase B (80:20, v/v). The concentration of the test solution when using CAD was 5 mg/mL. For the content determination with CAD, the concentration of topiramate was 25 μg/mL.

Impurities A, B, C, and D were stored as 1 mg/mL stock solutions in acetonitrile at −20°C. Impurity E was dissolved and stored in water at 4°C. Spiked solutions were then prepared immediately before use by dilution of the stock solutions.

11.3.3 TLC and HPTLC Limit Test for Impurity A

The TLC and HPTLC test is based on the USP monograph for topiramate [19] with modifications from Cilag AG (personal communication, 2011). The preconditioning, the development of the plates, and the detection were performed according to Ref. [19]. Ethylsilyl TLC plates (Silica Gel RP-2 TLC plates, Merck, Molsheim, France) and HPTLC plates (Silica Gel RP-2 HPTLC plates, Merck, Molsheim, France) were used.

11.4 Results and Discussion

11.4.1 Method Validation: Impurity Control

The HPLC method employed a pentafluorophenyl column. Mobile-phase A consisted of 25 mM aqueous solution of ammonium acetate (pH 3.5), and mobile-phase B was acetonitrile. A gradient was applied starting with 20% (v/v) mobile-phase B for 5 min and increasing the proportion of mobile-phase B from 20 to 50% (v/v) within further 10 min. For detection of substances lacking a suitable UV-chromophore, ELSD and CAD were employed. The proposed specification limits were 0.15% for impurity E; 0.10% for impurities A, B, C, and D; 0.10% for unspecified impurities; and 0.20% for total impurities.

Specificity. A solution of topiramate (20 mg/mL for ELSD and 5 mg/mL for CAD) spiked with each impurity at concentrations of 0.1% was analyzed to demonstrate specificity. The structural formulae of the impurities are given in

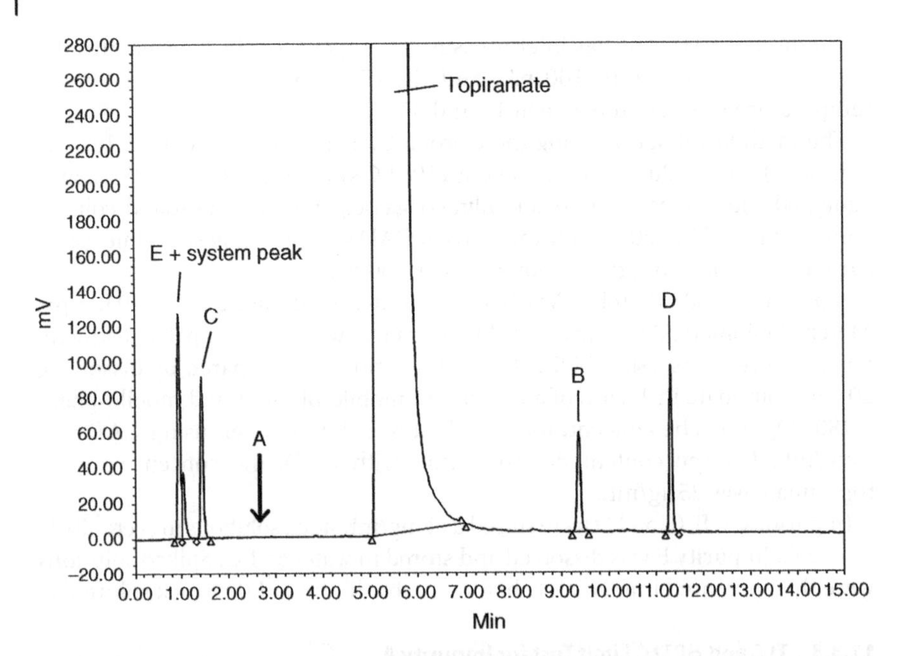

Figure 11.2 Example chromatogram of the test solution of topiramate (c=5 mg/mL) spiked with each impurity at concentrations of 0.1%. Chromatographic conditions are described in Section 11.3.2. Detection: Corona ultra RS (nebulizer temperature: 35°C). Impurity A was not detectable at this concentration. The retention time derived from the experiments for sensitivity enhancement with the Corona ultra RS, cf. Figure 11.3.

Figure 11.1. All impurities were separated from topiramate and each other (Figure 11.2). However, impurity E is eluted before the void volume. Usually, retention times less than the hold-up time indicate the exclusion of an analyte from the pores of the column material, usually by reason of the analyte size. Here, we assume that the intrusion of impurity E (fructose) into the particles is hindered by the repulsion of the highly polar nature of the analyte, which is opposite to the nonpolar surfaces of the column. Due to the partial overlap with the system peak, impurity E was quantified using peak height.

Repeatability was tested on three levels (0.05, 0.10, and 0.15% relative to the concentration of the test solution, n = 3). The relative standard deviations (% RSD) found were between 0.24 and 2.43% for CAD and 1.22 and 18.3% for ELSD, respectively. These exceptionally high % RSD values for ELSD were observed at the lowest concentration only. Regarding the mid and high concentration level, % RSD was less than 6%. Still, CAD was found to provide a more consistent response compared with ELSD.

Sensitivity. The limits of quantification (LOQ) of topiramate and impurities B, C, D, and E were 29, 20, 22, and 48 ng on column, respectively, for ELSD and 5.9, 7.7, 4.5, and 4.7 ng on column, respectively, for CAD. Thus, the sensitivity of CAD was about three to nine times higher compared with ELSD.

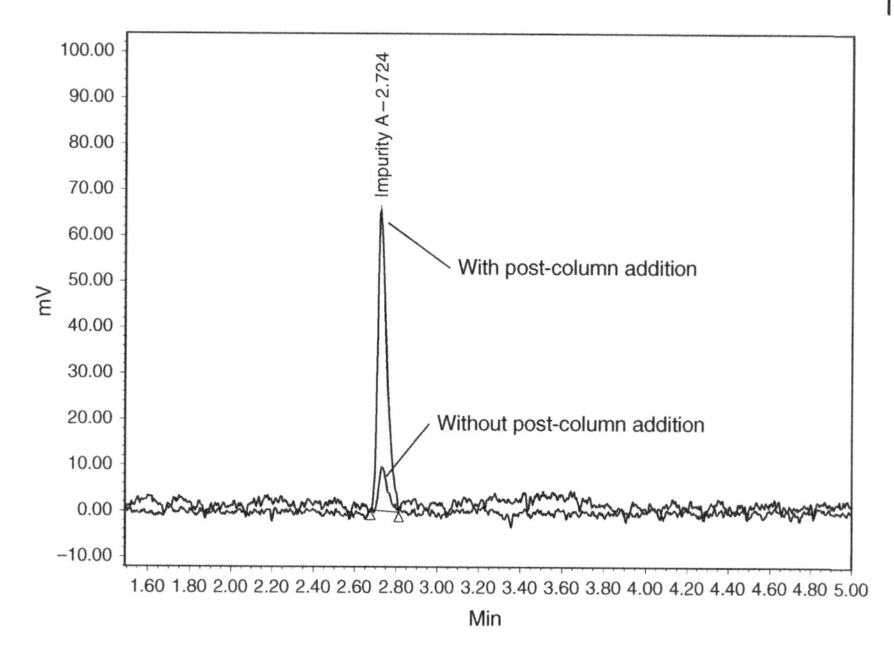

Figure 11.3 Influence of post-column addition of acetonitrile on the response of impurity A (c = 2.6 mg/mL). Detection: Corona ultra RS (nebulizer temperature: 35°C). The chromatographic conditions are described in Section 11.3.2.

Impurity A (bisacetonide) gave no signal in the ELSD due to its relatively high vapor pressure. It is entirely vaporized along with the mobile phase in the evaporation tube at the set temperature of 80°C. It was not possible to detect impurity A by lowering the temperature of the nebulizer (T_n) or the evaporation tube.

Using the Corona ultra RS, impurity A could be detected at a concentration of 12 μg on column. The LOQ is 38 μg on column. By post-column addition of acetonitrile at a flow rate of 1.0 mL/min (Figure 11.3), a gain in sensitivity by a factor of about five was achieved (LOQ = 7.8 μg on column).

Furthermore, the influence of T_n on the response was investigated. It can be varied between 5 and 35°C in the Corona ultra RS. However, according to the instrument manual, the minimum T_n that can actually be reached is depending on several factors such as the temperature of the mobile phase, the mobile-phase flow rate, and the ambient temperature. Thus, it was impossible to set a T_n lower than 18°C in the experimental setup. A decreased T_n correlates with a decreasing LOQ, thus increased sensitivity. A maximum gain in sensitivity of about a factor 2 was observed (data not shown). In combination with the post-column addition of acetonitrile, a ninefold increase in sensitivity could be achieved (LOQ of 4.2 μg on column with post-column addition of acetonitrile at 18°C vs. 38 μg on column without addition of acetonitrile at 35°C).

However, the detector is still about 10^3 times more sensitive for other impurities than for impurity A. Proper control of all impurities within a single

HPLC–CAD run was not feasible. Therefore, impurity A is more appropriately controlled by an HPTLC test (cf. [19] chapter 3.3).

Linearity. We investigated linearity in the range from the LOQ of the respective substance to 0.15% (0.50% for impurity E) relative to the concentration of topiramate in the test solution.

All evaporation-based detectors show a nonlinear response, which is described by following equation:

$$A = a \times m^b \tag{11.1}$$

where A is the peak area, m is the analyte mass, and a and b are numerical coefficients. Log–log transformation gives a linear relation expressed by:

$$\log A = b \times \log m + \log a \tag{11.2}$$

A coefficient of $b = 1$ will result in a linear response curve. Within a small concentration range of about two orders of magnitude, the response curve of CAD is reported to be sufficiently linear without log–log transformation [20].

The results of the linearity studies are summarized in Table 11.1. As can be seen from the coefficients of determination (R^2) of the linear regression and the

Table 11.1 Parameters of the linear regression before and after log–log transformation of analyte concentration and peak area for both ELSD and CAD are shown.

	Linear			Logarithmic		
	R^2	Slope	y-Intercept	R^2	b	Log a
ELSD						
Topiramate	0.9860	34,869,781	−381,799	0.9981	1.5375	7.9551
Impurity B	0.9854	30,520,686	−199,509	0.9992	1.5174	8.0115
Impurity C	0.9981	23,431,043	−91,228	0.9982	1.3822	7.8052
Impurity D	0.9845	41,574,416	−407,210	0.9966	1.4310	7.9952
Impurity E	0.9965	45,985,745	−171,743	0.9974	1.3650	8.0251
CAD						
Topiramate	0.9982	4481	5.4067	0.9998	0.9338	3.5794
Impurity B	0.9991	3984	4.9791	0.9992	0.9275	3.5199
Impurity C	0.9956	4511	5.8785	0.9974	0.9603	3.6316
Impurity D	0.9924	5126	11.6463	0.9972	0.9041	3.6201
Impurity E	0.9890	3276	10.1519	0.9966	0.8620	3.3762
Impurity E*	0.9880	1108	4.0472	0.9966	0.8330	2.8698

Impurity E* represents calibration results using peak height. All other results are based on peak area.

values for the coefficient *b* of the logarithmic calibration curves, the signal provided by CAD showed better linearity than ELSD. After log–log transformation, however, the differences were negligible.

Correction factors and accuracy. As it is common practice in the impurity control of drug substances in pharmacopeias, a dilution of the test solution serves as external standard for the determination of the impurity content [21]. In order to adjust possible differences in response, correction factors were assessed for each impurity. Correction factors were calculated based on the peak areas of each impurity compared with that of topiramate. Correction factors were not constant over the entire concentration range. Decreasing correction factors with higher concentrations for all impurities were observed. The factors applied for the calculations were the mean values of all determinations. A possible reason is that CAD response depends on the amount of organic modifier in the mobile phase. Response factors can differ for peaks eluting at different times of the chromatogram when gradient elution is applied.

Accuracy was expressed as the recovery of each impurity in spiked test solutions. Three concentration levels were investigated covering the range from the reporting threshold to at least 150% of the specification limit. Results are shown in Table 11.2. Impurity contents found with CAD showed a better closeness to the true value. In accordance to the trend in correction factors, we

Table 11.2 Correction factors (CFs) and accuracy expressed as % recovery of the spiked impurities.

	Correction factor	% Recovery			
		0.05%	0.10%	0.15%	0.30%
ELSD					
Impurity B	0.6	93.2	90.3	88.5	ND
Impurity C	0.6	119.7	103.8	99.0	ND
Impurity D	0.5	109.1	98.0	97.8	ND
Impurity E[a]	0.4	101.0	95.0	95.0	ND
CAD					
Impurity B	0.8	99.3	104.9	101.1	ND
Impurity C	0.9	112.6	108.5	104.4	ND
Impurity D	0.6	99.8	101.6	99.1	ND
Impurity E[a]	0.4	104.8	ND	101.5	94.5

[a] For the calculation of the correction factor and the impurity content of impurity E, the peak height was used instead of the peak area. Measurements were carried out in triplicate.
ND, not determined.

observed a decreasing recovery with increasing concentrations of the impurities, except for impurities B and D assessed with CAD. This trend seemed to be more pronounced for ELSD than for CAD. Still, both detectors gave satisfactory recoveries with averaged correction factors.

When ELSD was used, we observed peaks eluting after the principle peak due to topiramate in the test solution (see Figure 11.4). These peaks were not reproducible and added up to a content of more than 0.3% in total, which exceeds the limit for total impurities and would result in the refusal of an actually compliant batch.

Such ghost or "spike peaks" have previously been reported and investigated [22]. Their occurrence can be minimized or even avoided by adjusting ELSD and HPLC parameters but results in a loss in sensitivity. As the method presented here is intended to be part of a European Pharmacopoeia (Ph. Eur.) monograph, it has to be applicable to ELS detectors different from the model used in this study. For that reason, we cannot state generally valid detector settings that ensure the avoidance of ghost peaks and provide a sufficiently sensitive detection at the same time. Therefore, we considered the ELSD not to be feasible in this particular case. In contrast, when using CAD, no ghost peaks occurred (cf. Figure 11.2).

11.4.2 Method Validation: Assay

The HPLC–CAD method for the impurity control was also used for the content determination of topiramate. Some of the validation parameters had to be reassessed for this purpose, that is, linearity and range, repeatability, and accuracy.

Linearity and range. It is reported that CAD delivers a linear signal up to a concentration of approximately 250–500 ng on column [23]. Accordingly, the concentration of the test solution was set to 25 μg/mL (equivalent to 500 ng on column with an injection volume of 20 μL). Linearity was therefore investigated from 20 to 30 μg/mL, which is equivalent to 80–120% of the concentration of the test solution. The detector response was found to be sufficient linear ($R^2 = 0.9983$).

Repeatability was assessed at 80, 100, and 120% of the concentration of the test solution in triplicate. The % RSD values found were 0.22, 0.51, and 0.18%, respectively, which is considered highly satisfactory for aerosol-based detectors [23].

Accuracy. Recovery was determined at three levels (80, 100, and 120% of the concentration of the test solution) against the reference solution at 100% of the concentration of the test solution. All solutions were prepared in triplicate. Recoveries ranged from 98.89 to 102.05% (mean: 100.27, 1.04% RSD).

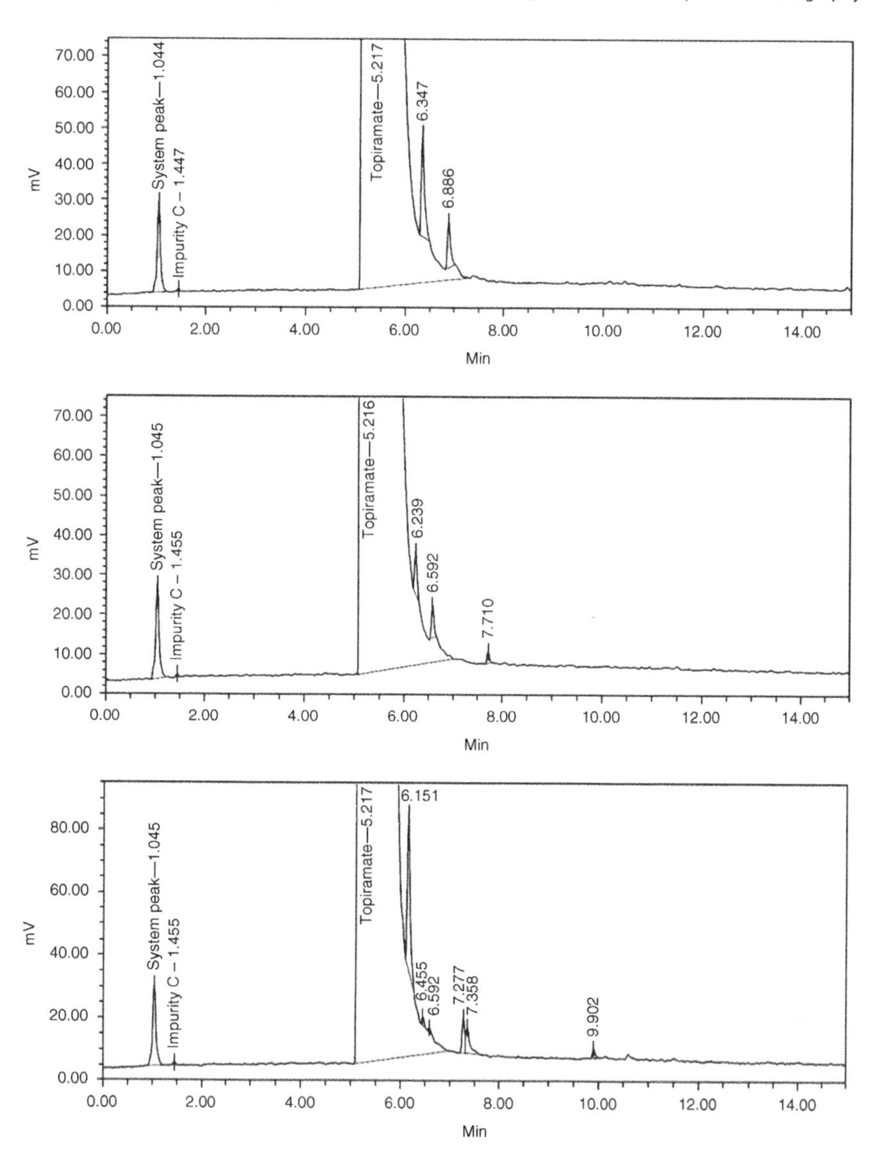

Figure 11.4 Three consecutive injections of a solution of topiramate ($c = 20$ mg/mL) from the same vial are shown using the ELSD. The content of impurity C, as the only impurity found in this batch, was below the LOQ. All other peaks are nonreproducible ghost peaks. The chromatographic conditions are described in Section 11.3.2.

11.4.3 TLC and HPTLC Limit Test for Impurity A

As the previously described HPLC–CAD method is not sensitive enough for the detection of impurity A, we applied an HPTLC limit test for its control. The test is based on the USP monograph for topiramate [19], where TLC is employed to limit *any impurity other than* impurity A. For our purpose, the method was changed using an ethylsilylated silica stationary phase instead of bare silica (Cilag AG, personal communication, 2011).

When HPTLC plates were applied, impurity A was separated from topiramate and from all other detectable impurities. Using HPTLC plates, appropriate specificity could be demonstrated, whereas the use of TLC plates did not allow its complete separation from another impurity (data not shown).

11.5 Conclusion

An HPLC method was developed with either ELSD or CAD capable of separating five potential impurities from topiramate in a single run. The highly polar impurity E was not retained under the method conditions and eluted before the system peak. Nevertheless, accurate and precise quantification is achievable using the peak height for the calculation of the impurity content. In addition, the HPLC method was successfully applied to the content determination of topiramate.

Because of its relatively high vapor pressure, impurity A was not detectable with ELSD but gave a signal with CAD, which was significantly lower compared with all other impurities. By means of post-column addition of acetonitrile and a decreased nebulizer temperature of the Corona ultra RS, a ninefold increase in response was achieved. Still, the LOQ of impurity A is about a factor of 10^3 higher compared with all other impurities. Therefore, an HPTLC limit test has to be applied for the control of impurity A.

It was found that CAD was superior to ELSD in terms of repeatability, sensitivity, and linearity. Still, both detectors gave satisfactory recoveries for all impurities. The lack of sensitivity in ELSD could easily be compensated by increasing the concentration of the test solution. However, the occurrence of nonreproducible peaks eluting on the tail of the principle peak when using ELSD impeded its use in our study.

Acknowledgment

The authors gratefully thank the European Directorate for the Quality of Medicines & HealthCare (EDQM) for the provision of analytical equipment, consumables, and samples. Furthermore, thanks are due to the Federal Institute of Drugs and Medical Devices (BfArM, Bonn, Germany) for financial support.

References

1 Ferrari A, Tiraferri I, Neri L, Sternieri E. Clinical pharmacology of topiramate in migraine prevention. Expert Opinion on Drug Metabolism & Toxicology 2011;7(9):1169–1181.
2 Allison DB, Gadde KM, Garvey WT, Peterson CA, Schwiers ML, Najarian T, Tam PY, Troupin B, Day WW. Controlled-release phentermine/topiramate in severely obese adults: a randomized controlled trial (EQUIP). Obesity 2012;20(2):330–342.
3 Wang QP, Bai M. Topiramate versus carbamazepine for the treatment of classical trigeminal neuralgia: a meta-analysis. CNS Drugs 2011;25(10):847–857.
4 Johnson BA. Recent advances in the development of treatments for alcohol and cocaine dependence: focus on topiramate and other modulators of GABA or glutamate function. CNS Drugs 2005;19(10):873–896.
5 Shinn AK, Greenfield SF. Topiramate in the treatment of substance-related disorders: a critical review of the literature. Journal of Clinical Psychiatry 2010;71(5):634–648.
6 Popov TV, Maricic LC, Prosen H, Voncina DB. Determination of topiramate in human plasma using liquid chromatography tandem mass spectrometry. Acta Chimica Slovenica 2013;60(1):144–150.
7 Shibata M, Hashi S, Nakanishi H, Masuda S, Katsura T, Yano I. Detection of 22 antiepileptic drugs by ultra-performance liquid chromatography coupled with tandem mass spectrometry applicable to routine therapeutic drug monitoring. Biomedical Chromatography 2012;26(12):1519–1528.
8 La MG, Malvagia S, Filippi L, Fiorini P, Innocenti M, Luceri F, Pieraccini G, Moneti G, Francese S, Dani FR, Guerrini R. Rapid assay of topiramate in dried blood spots by a new liquid chromatography-tandem mass spectrometric method. Journal of Pharmaceutical and Biomedical Analysis 2008;48(5):1392–1396.
9 Contin M, Riva R, Albani F, Baruzzi A. Simple and rapid liquid chromatographic-turbo ion spray mass spectrometric determination of topiramate in human plasma. Journal of Chromatography B: Biomedical Sciences and Applications 2001;761(1):133–137.
10 Vnučec Popov T, Maričič LC, Prosen H, Vončina DB. Development and validation of dried blood spots technique for quantitative determination of topiramate using liquid chromatography–tandem mass spectrometry. Biomedical Chromatography 2013;27(8):1054–1061.
11 Bahrami G, Mohammadi B. A novel high sensitivity HPLC assay for topiramate, using 4-chloro-7-nitrobenzofurazan as pre-column fluorescence derivatizing agent. Journal of Chromatography B 2007;850(1–2):400–404.
12 Bahrami G, Mirzaeei S, Mohammadi B, Kiani A. High performance liquid chromatographic determination of topiramate in human serum using UV detection. Journal of Chromatography B 2005;822(1–2):322–325.

13 Britzi M, Soback S, Isoherranen N, Levy RH, Perucca E, Doose DR, Maryanoff BE, Bialer M. Analysis of topiramate and its metabolites in plasma and urine of healthy subjects and patients with epilepsy by use of a novel liquid chromatography–mass spectrometry assay. Therapeutic Drug Monitoring 2003;25(3):314–322.

14 Styslo-Zalasik M, Li W. Determination of topiramate and its degradation product in liquid oral solutions by high performance liquid chromatography with a chemiluminescent nitrogen detector. Journal of Pharmaceutical and Biomedical Analysis 2005;37(3):529–534.

15 Biro A, Pergel E, Arvai G, Ilisz I, Szepesi G, Peter A, Lukacs F. High-performance liquid chromatographic study of topiramate and its impurities. Chromatographia 2006;63(Suppl.):S137–S141.

16 Arvai G, Garaczi S, Mate AG, Lukacs F, Viski Z, Schneider G, inventors; Helm AG, CF Pharma Gyogyszergyarto Kft., assignee. Process for the preparation of topiramate derivatives from 2,3:4,5-bis-O-(1-methylylidene)-β-D-fructofuranose. US patent 7,414,126. 19 August, 2008.

17 Cobb Z, Shaw PN, Lloyd LL, Wrench N, Barrett DA. Evaporative light-scattering detection coupled to microcolumn liquid chromatography for the analysis of underivatized amino acids: sensitivity, linearity of response and comparisons with UV absorbance detection. Journal of Microcolumn Separations 2001;13(4):169–175.

18 Gorecki T, Lynen F, Szucs R, Sandra P. Universal response in liquid chromatography using charged aerosol detection. Analytical Chemistry 2006;78(9):3186–3192.

19 *Topiramate, United States Pharmacopoeia*, USP 38 NF 33, The United States Pharmacopeial Convention, Rockville, MD, 2015.

20 Almeling S, Ilko D, Holzgrabe U. Charged aerosol detection in pharmaceutical analysis. Journal of Pharmaceutical and Biomedical Analysis 2012;69:50–63.

21 *Technical Guide for the Elaboration of Monographs*, 6th Edition, European Directorate for the Quality of Medicines & HealthCare (EDQM), Strasbourg, France, 2011.

22 Almeling S, Holzgrabe U. Use of evaporative light scattering detection for the quality control of drug substances: influence of different liquid chromatographic and evaporative light scattering detector parameters on the appearance of spike peaks. Journal of Chromatography A 2010;1217(14):2163–2170.

23 Crafts C, Bailey B, Plante M, Acworth I. Validating analytical methods with charged aerosol detection. Available at http://www.dionex.com/en-us/webdocs/110512-PO-HPLC-ValidateAnalyticalMethods-CAD-31Oct2011-LPN2949-01.pdf (Accessed July 10, 2014).

12

Applying Charged Aerosol Detection to Aminoglycosides

Development and Validation of an RP-HPLC Method for Gentamicin and Netilmicin

Arul Joseph[1] and Abu Rustum[2]

[1] *Gilead Sciences, Inc., Foster City, CA, USA*
[2] *Merck & Co Inc., Summit, NJ, USA*

CHAPTER MENU

12.1 Introduction

Aminoglycosides are a class of antibiotics that have been used effectively against Gram-negative bacteria and to lesser extent Gram-positive bacteria for many decades. The aminoglycoside streptomycin, the first antibiotic effective against tuberculosis, was isolated from *Streptomyces griseus* in 1943, and for this discovery S.A. Waksman received the Nobel Prize in Physiology or Medicine in 1952 [1]. The aminoglycoside gentamicin was isolated from *Micromonospora purpurea* in the research laboratories of Schering-Plough Corporation [2, 3] in 1963 and is used to treat Gram-negative bacterial infections. Formulations (creams, ointments, ophthalmic solutions, etc.) containing gentamicin are still used globally to treat bacterial infections. Subsequently, other aminoglycoside antibiotics were isolated from other species of *Micromonospora*, for example, sisomicin, or produced by structural modification of aminoglycosides to yield antibiotics that are less toxic and more resistant to bacterial inactivation, for example, netilmicin. Gentamicin is derived via

Charged Aerosol Detection for Liquid Chromatography and Related Separation Techniques, First Edition. Edited by Paul H. Gamache.

fermentation and is a mixture that contains four major compounds gentamicin C1, gentamicin C1a, gentamicin C2, gentamicin C2a and a minor compound gentamicin C2b. Netilmicin is a semisynthetic aminoglycoside antibiotic that is effective against bacteria that are resistant to gentamicin, sisomicin, and tobramycin and has been found to be less toxic compared with these aminoglycoside antibiotics [4, 5]. Netilmicin is an ethyl derivative (at the 1-N position of the deoxystreptamine ring) of sisomicin and was synthesized by the alkylation of sisomicin by Wright in 1976 [6].

Aminoglycosides derived from *Micromonospora* (gentamicin, sisomicin, netilmicin, and related substances) are very water soluble, have low solubility in most organic solvents, and have only a weak UV chromophore. Thus, developing an RP-HPLC method for aminoglycosides with the sensitivity that is necessary to detect and estimate low levels of impurities and degradation products is a challenge. Due to the weak UV absorbance of aminoglycosides, direct UV detection cannot be used to develop such a sensitive method. A sensitive RP-HPLC method using charged aerosol detection (CAD) was developed and validated for the aminoglycoside gentamicin in a pharmaceutical cream [7].[1] This method was then used as a model to develop and validate a method for the aminoglycoside netilmicin [8]. The method can be applied to gentamicin and netilmicin in other formulations and can also be used as a model to develop and validate methods for other aminoglycosides.

12.1.1 Background

HPLC methods for the analysis of gentamicin and netilmicin using both direct and indirect methods have been reported in the literature [9–43]. The indirect methods involve either pre- or post-column derivatization with *o*-phthalaldehyde or dansyl chloride (5-dimethylamino-1-naphthalene sulfonyl chloride) with UV or fluorescence detection [9–20]. Direct detection methods using refractive index (RI) [21], mass spectrometry (MS) [22–26], evaporative light scattering detection (ELSD) [27, 28], and electrochemical detection (ECD) [29–31] have been reported. For netilmicin, only two direct detection methods using ECD [42] and MS [43] have been reported. A recent report compares ECD and ELSD methods using polymer and C-18 columns [32]. The methods reported in the literature have limitations with respect to selectivity and sensitivity to separate and quantitate gentamicin and its related substances, as well as netilmicin and its related deoxystreptamine, *N*-ethyl garamine, sisomicin, and ethyl sisomicin derivatives. CAD due to its greater sensitivity was chosen to develop and validate a method for the determination of the

1 The gentamicin and netilmicin methods discussed in this chapter, including all tables and figures, were published in articles 7 and 8 and are being reproduced with permission from the publishers who own the copyrights.

composition of gentamicin C1, C1a, C2, C2a, and C2b and the estimation of the related substances of gentamicin in a pharmaceutical cream.

12.2 Development and Validation of an RP-HPLC Method for Gentamicin Using Charged Aerosol Detection

12.2.1 Method Development

The structures and selected properties of gentamicin C1, C1a, C2, C2a, and C2b, deoxystreptamine, garamine, and sisomicin are shown in Figure 12.1. The chemical properties of gentamicin create several challenges for developing a suitable HPLC method. For instance, their hydrophilicity makes retaining and separating them on reversed-phase stationary phases challenging; their insolubility in most organic solvents necessitates the use of highly aqueous mobile phases. Reversed phase stationary phases under such aqueous (95% water) mobile phase conditions undergo desolvation—loss of mobile phase from stationary phase pores, causing a decrease in the retention of analytes due to the inability of the stationary phase to maintain hydrophobic interaction with the analytes [44].

12.2.1.1 Selection of Detector

Gentamicin lacks UV chromophores, making conventional UV detection unfeasible. Other detection techniques such as RI detection, ELSD, pre-column derivatization followed by UV detection, and ECD either are unsuitable or have limitations for use in the routine analysis of gentamicin. RI detection is incompatible with gradient methods necessary to separate gentamicin and related substances. ELSD has lower sensitivity (typically 10-fold compared with CAD), precision, and dynamic range. Pre-column derivatization can be variable, may not proceed to completion, and can degrade the sample. ECD is less rugged as the electrodes can foul and require frequent cleaning, and the ECD response is affected by temperature, pump pulsations, and any extraneous electrical current. As routine analytical method for gentamicin cannot be developed using UV detectors and as other conventional detection techniques are either unsuitable or cumbersome for routine analysis of gentamicins, CAD was used.

12.2.1.2 Related Substances

Despite a sustained effort to obtain garamine, deoxystreptamine, Compound D, and gentamicin B1 commercially, only garamine and deoxystreptamine were available through a contract manufacturing organization. In addition, the limited quantity and significant impurity of these compounds precluded their use in

Compound name	Structure	Identity	Properties
Deoxystreptamine sulfate	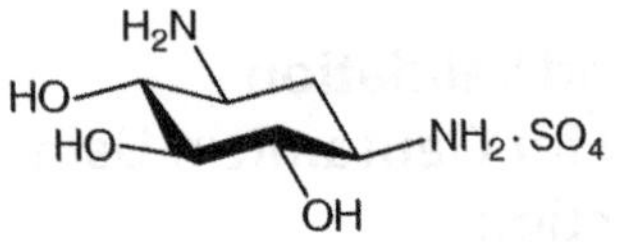 H_2N HO HO $NH_2 \cdot SO_4$ OH	Impurity/ degradation product	UV inactive, water soluble
Garamine sulfate	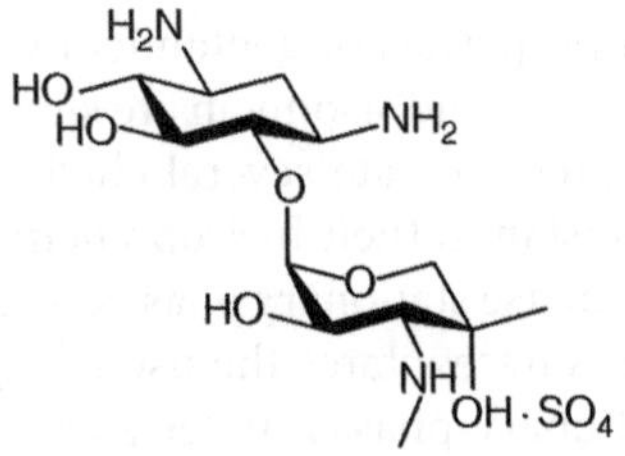 H_2N HO HO NH_2 O O HO NH $OH \cdot SO_4$	Impurity/ degradation product	UV inactive, water soluble
Sisomicin sulfate	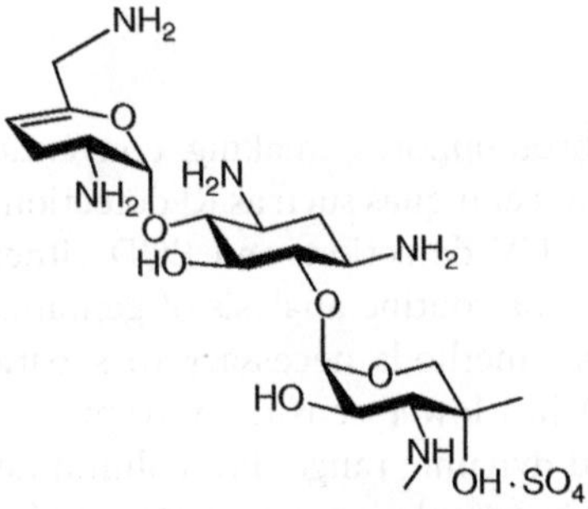 NH_2 O NH_2 H_2N O HO NH_2 O O HO NH $OH \cdot SO_4$	Impurity/ degradation product	UV inactive, water soluble

Figure 12.1 Chemical structures and selected properties of gentamicin C1, C1a, C2, C2a, and C2b, deoxystreptamine, garamine, and sisomicin.

validation studies except to prepare a specificity mixture. In addition, the precut and post-cut column fractions of the gentamicin mother liquor from the purification of gentamicin were obtained and analyzed using liquid chromatography–mass spectrometry (LC-MS) to determine if compound D and gentamicin B1 could be isolated from these fractions. The LC-MS analysis indicated that these column fractions did not contain the desired related substances.

Compound name	Structure	Identity	Properties
Gentamicin sulfate C1a	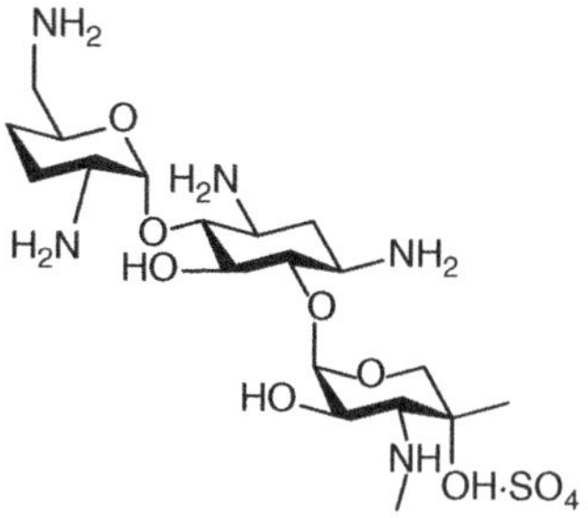	API	UV inactive, water soluble
Gentamicin sulfate C2	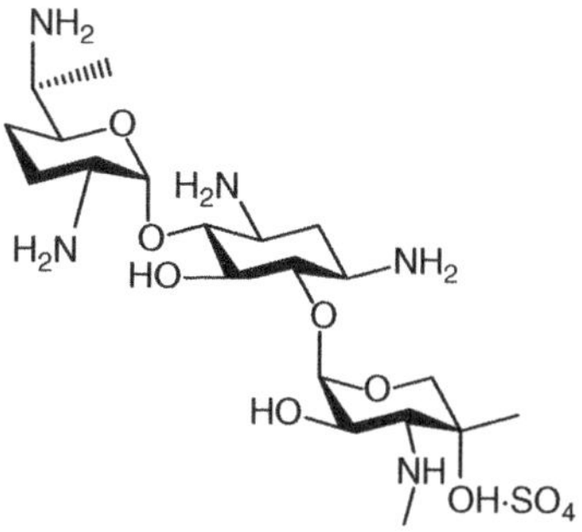	API	UV inactive, water soluble
Gentamicin sulfate C2a	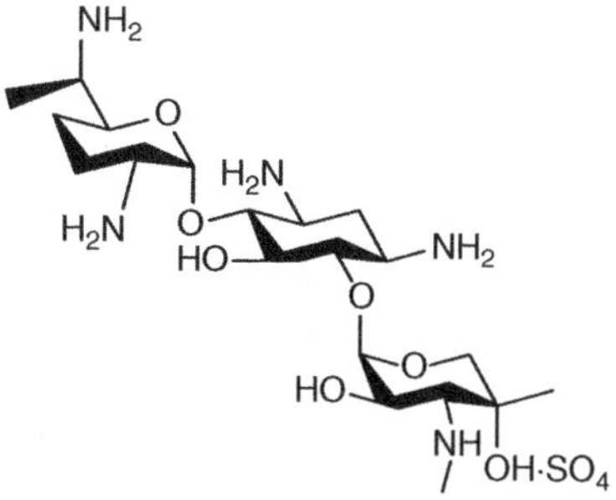	API	UV inactive, water soluble

Figure 12.1 (Continued)

Compound name	Structure	Identity	Properties
Gentamicin sulfate C2b	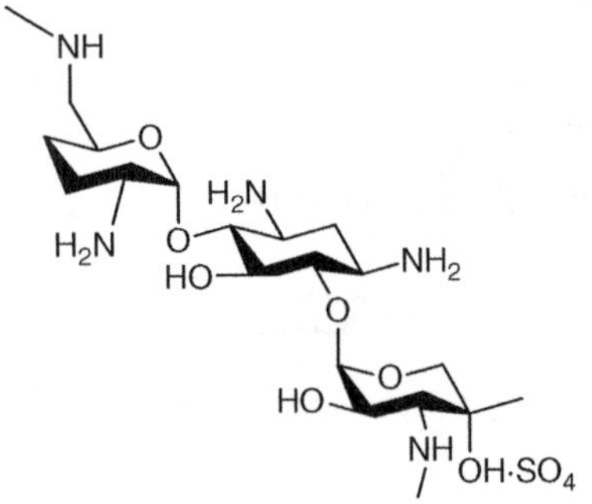	API	UV inactive, water soluble
Gentamicin sulfate C1	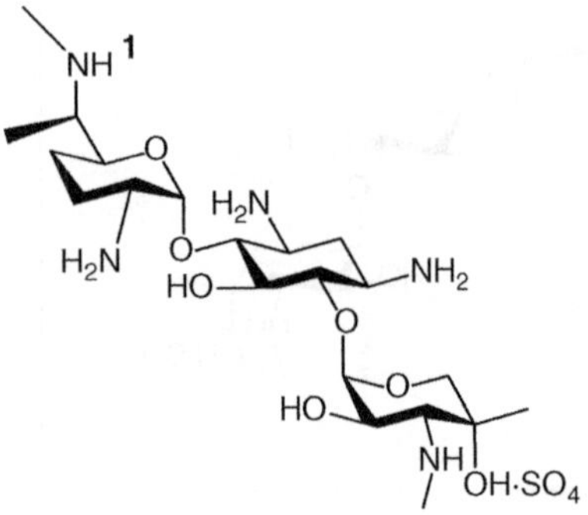	API	UV inactive, water soluble

Figure 12.1 (Continued)

12.2.1.3 Mobile Phase Composition and Column Selection

The objective was to separate gentamicin C1, C1a, C2, C2a and C2b, deoxystreptamine, garamine, and sisomicin and to retain deoxystreptamine and garamine to have sufficiently high capacity factors to enable reliable quantitation. Method development was initiated using a C18 stationary phase and a mobile phase of 50 mM trifluoroacetic acid in water : methanol (98 : 2). The composition of the mobile phase was modified to test varying proportions of ion-pairing agents (e.g., trichloroacetic acid, trifluoroacetic acid (TFA), pentafluoropropionic acid (PFPA), and heptafluorobutyric acid (HFBA)), acids (e.g., methane sulfonic acid, formic acid, acetic acid, etc.), organic modifiers (e.g., methanol, isopropanol, THF, etc.). The retention of gentamicin C1, C1a, C2, C2a and C2b, deoxystreptamine, garamine, and sisomicin could be increased by increasing the concentration of TFA or by switching from TFA to PFPA or HFBA. However, PFPA and HFBA even at low concentrations resulted in much longer retention of gentamicin C1 and C2b, leading to poor peak shapes and a long run time.

Although, deoxystreptamine and garamine could be separated from gentamicin using a higher concentration of TFA, sisomicin could not be well separated (resolution >2.0) from gentamicin C1a. Sisomicin and gentamicin C1a are structurally similar; they differ in only a ring double bond. Modifying the mobile phase composition (concentration of TFA; proportion of acetonitrile; organic modifier—methanol, isopropanol, and THF or acid—methane sulfonic acid, formic acid, and acetic acid) did not provide the desired resolution of peaks. In addition, the effect of temperature and gradient were also studied to improve separation of peaks. Theoretically, moving from a C18 stationary phase to a phenyl phase should resolve sisomicin and gentamicin C1a due to π–π interaction. However, it was found that the phenyl phase does not retain gentamicins well and the overall retention and separation of all peaks were poor. C18, phenyl, cyano, HILIC, and cyclodextrin stationary phases were screened.

The solution was to move to a novel stationary phase and mechanism to separate these compounds. This novel phase is a pentafluorophenyl stationary phase on which the dipole–dipole interaction between the C–F bonds of the ion-paired gentamicins and the C–F stationary phase provides better retention and differentiation between these closely related structures. In addition, the pentafluorophenyl stationary phase provides π–π interactions to distinguish sisomicin from gentamicin C1a. A pentafluorophenyl stationary phase was used successfully to separate gentamicin C1a from sisomicin while retaining and separating all other gentamicins (C1, C2, C2a, and C2b) and the related substances deoxystreptamine and garamine (Figure 12.3). The Restek Allure PFP column (50 mm × 4.6 mm, 3 μm) was chosen as the primary column, and the Thermo Scientific Fluophase column (50 mm × 4.6 mm, 5 μm) was chosen as the secondary column based on a column screening step.

An excellent separation of the peaks was achieved on the pentafluorophenyl columns using a mobile phase of TFA : H_2O : CH_3CN (0.025 : 95 : 5, v/v/v) under isocratic conditions. However, deoxystreptamine eluted close to the solvent front, and both deoxystreptamine and garamine had very low capacity factors. To retain and thereby provide reliable quantitation of deoxystreptamine and garamine while eluting the gentamicin peaks within a reasonable run time (33 min), an ion-pair step gradient was used. The ion-pair step gradient involves using HFBA (HFBA : H_2O : CH_3CN (0.025 : 95 : 5, v/v/v)) initially to retain deoxystreptamine and garamine and then using TFA (TFA : H_2O : CH_3CN (0.025 : 95 : 5, v/v/v)) to elute the gentamicins.

Table 12.1 Gradient program of the gentamicin method.

Time (min)	Mobile phase A (%)	Mobile phase B (%)	Mobile phase C (%)	Gradient curve	Comment
0.00	100.0	0.0	0.0	Linear	First isocratic ion-pair reagent (HFBA) elution
8.00	100.0	0.0	0.0	Linear	
12.00	80.0	20.0	0.0	Linear	
16.00	0.0	100.0	0.0	Linear	Second isocratic ion-pair reagent (TFA) elution
33.00	0.0	100.0	0.0	Linear	
Column wash and equilibrate to initial conditions					
33.50	0.0	0.0	100.0	Linear	Column wash
43.50	0.0	0.0	100.0	Linear	
44.00	100.0	0.0	0.0	Linear	Column equilibration
48.00	100.0	0.0	0.0	Linear	

The aqueous mobile phase (95% water) caused a decrease in retention time with each successive injection due to desolvation. To reduce desolvation, a column wash step using $H_2O:CH_3CN$ (20:80, v/v) was introduced. The gradient program of the final method is shown in Table 12.1. Representative chromatograms of gentamicin sulfate reference standard and a standard mixture containing gentamicin, deoxystreptamine, garamine, and sisomicin are shown in Figure 12.2. The chromatograms were obtained using $HFBA:H_2O:CH_3CN$ (0.025:95:5, v/v/v) as mobile phase A and $TFA:H_2O:CH_3CN$ (0.025:95:5, v/v/v) as mobile phase B. The chromatographic run time was 33 min and is followed by a column wash step with $H_2O:CH_3CN$ (80:20, v/v).

12.2.1.4 Sample Preparation

The water solubility of gentamicin can be advantageously used to simplify the sample preparation procedure, as the other organic compounds present in a pharmaceutical cream (cetomacrogol 1000, cetostearyl alcohol, white petrolatum, and paraffinum liquidum) are insoluble in water and the water-soluble compounds such as sodium hydroxide, phosphoric acid, and sodium dihydrogen phosphate will not be retained by the stationary phase and thus not interfere with the gentamicin peaks. The sample preparation involves dissolving 6.0 ± 0.05 g of the pharmaceutical cream in 6.0 mL of water followed by extraction of the hydrophobic components of the cream with 5.0 mL of dichloromethane. The aqueous solution, containing gentamicin, is diluted 1:2 with water and then injected into the HPLC for the determination of the

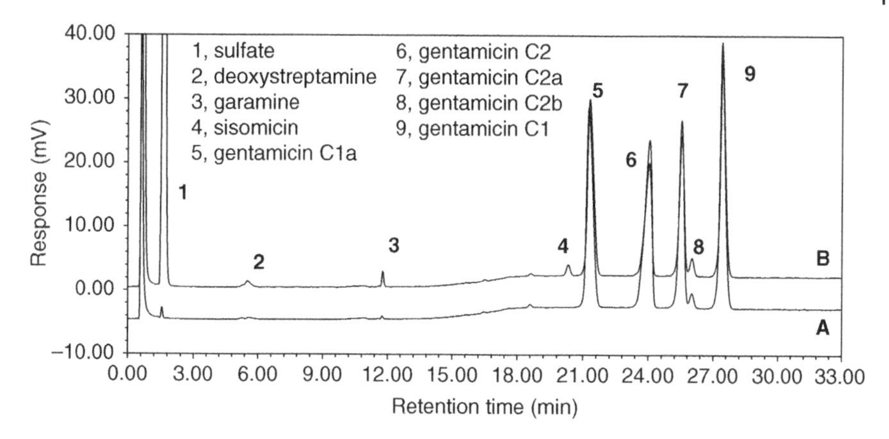

Figure 12.2 Representative chromatograms of (A) gentamicin reference standard and (B) standard mixture containing gentamicin, deoxystreptamine, garamine, and sisomicin.

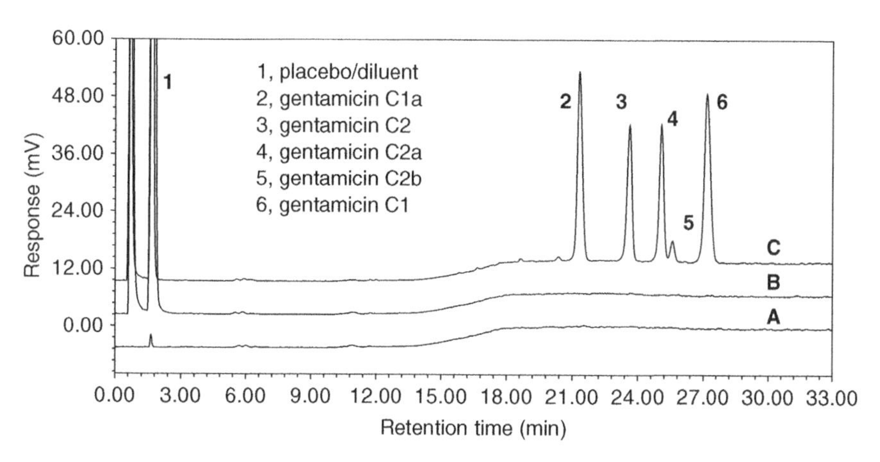

Figure 12.3 Representative chromatograms of (A) diluent, (B) pharmaceutical cream placebo, and (C) pharmaceutical cream sample.

composition of gentamicin C1, C1a, C2, C2a, and C2b and for the estimation of related substances. The analytical concentration of gentamicin in the sample solution was approximately 0.5 mg/mL, which corresponds to an analytical concentration of gentamicin C1a of approximately 0.135 mg/mL. Representative chromatograms of diluent, placebo, and pharmaceutical cream sample are shown in Figure 12.3.

Gentamicin standard solutions were prepared by dissolving the reference standard in water to have an analytical concentration of gentamicin of approximately 0.5 mg/mL (gentamicin C1a of approximately 0.135 mg/mL). Specificity mixture solutions were prepared by spiking a pharmaceutical cream sample with approximately 1% each of related substances deoxystreptamine, garamine, and sisomicin.

12.2.2 Method Validation

Two analysts from different laboratories performed the method validation of pharmaceutical cream using testing parameters such as response linearity, precision, specificity, robustness, sample stability, limit of quantitation (LOQ), and limit of detection (LOD). The analytical concentration of gentamicin C1a was approximately 0.135 mg/mL (0.5 mg/mL of gentamicin). Strategically, as the individual gentamicin C1, C1a, C2, C2a, and C2b standards were unavailable, validation was conducted using the gentamicin sulfate reference standard, a mixture of C1, C1a, C2, C2a, and C2b. The certificate of analysis of the gentamicin reference standard listed gentamicin C1 (26%), C1a (27%), and gentamicin C2 and C2a as a mixture (47%). As the percent composition of gentamicin C1a (27%) listed in the Certificate of Analysis was very close to what was observed with the current analytical method, the gentamicin reference standard could be used as a 27% pure standard of gentamicin C1a. As the percent composition gentamicin C1 (26%) was not close to the value observed with the current method and as gentamicin C2 and C2b were provided as a mixture in the certificate of analysis, these compounds could not be used in the same manner as gentamicin C1a. Therefore, the linearity and accuracy of gentamicin C1a was verified over the concentration range from 0.001 to 0.2025 mg/mL, which corresponds to QL (0.2%) to 150% of gentamicin label claim, in the presence of placebo.

12.2.2.1 Experimental

Materials Only HPLC grade solvents were used. Acetonitrile and dichloromethane were purchased from Sigma (St. Louis, Missouri, USA). HFBA ≥ 99% and TFA ≥ 99.5% were purchased from Alfa Aesar (Ward Hill, Massachusetts, USA), and PFP ≥ 98% was purchased from Sigma (St. Louis, Missouri, USA). Water (18.2 MΩcm) from a Milli-Q system (Millipore, Billerica, Massachusetts, USA) was used. Gentamicin sulfate, netilmicin sulfate, and *N*-ethyl garamine sulfate were obtained from a standards group within the company. Sisomicin sulfate USP reference standard was purchased from Fisher Scientific (USA). Cream with gentamicin samples and placebo were obtained from groups within the company. Deoxystreptamine sulfate and garamine sulfate were obtained through a contract manufacturing organization.

Instrumentation and Chromatographic Conditions A Waters 2695 Alliance HPLC system (Milford, Massachusetts, USA) equipped with a column compartment with temperature control, an online degasser, and a Corona® CAD® (ESA Biosciences, Chelmsford, Massachusetts, USA) was used. Data acquisition, analysis, and reporting were performed using Millennium32 and/or Empower chromatography software (Milford, Massachusetts, USA). Waters HPLC systems in different laboratories were used for method validation. The HPLC column used in this method (Restek Allure PFP Propyl, 50 mm × 4.6 mm I.D., 3 μm particle size) is manufactured by Restek Corporation (State College, Pennsylvania, USA) and distributed by Thermo Fisher Scientific. An alternate column, Thermo Scientific Fluophase (50 mm × 4.6 mm I.D., 5 μm particle size) manufactured by Thermo Fisher Scientific, was also validated.

12.2.2.2 Specificity

The method specificity was demonstrated by a peak identification test and by testing three representative cream samples from different manufacturing dates. The identification test was performed to show that the method is capable of resolving gentamicin and key-related substances in a specificity test mixture using the method to determine the resolution and relative retention (RR) of each related substance. To demonstrate method specificity, ICH guidelines require stress studies for new drug candidates in order to predict real-life degradation chemistry of the drug substances and formulated products under specific stability storage conditions. The pharmaceutical cream has been commercially available for many years, and its degradation chemistry is well characterized. Drug development processes related to pharmaceutical cream, such as drug formulation design, selection of storage conditions, and packaging, are well-established. Based on all the aforementioned facts, stress studies under conditions such as heat, photostability, acid, base, and oxidation were not performed. Instead, a specificity test mixture was analyzed using the method to determine the resolution factor and the RR of each gentamicin-related substance. The specificity test mixture consisted of an expired pharmaceutical cream sample spiked with approximately 1% each of deoxystreptamine, garamine, and sisomicin. The RR of gentamicin C1, C1a, C2, C2a, C2b and related substances were compared with the retention time and RR listed in the analytical method. Also, three representative cream samples from different manufacturing dates (expired and unexpired) were tested for method specificity in the method validation. These stability samples provided a true reflection of the degradation chemistry under real-life conditions. Gentamicin was adequately separated from its related substances demonstrating that the method is a stability-indicating method.

12.2.2.3 Linearity

CAD is a quadratic response detector and has a linear response over only a short concentration range and at low analyte concentrations. For this method, a linear fit using a single 100% level gentamicin standard (0.5 mg/mL) was preferred to using a quadratic fit with multiple concentration level standards and calibration curves. Also, analysts in QC analytical labs routinely use a single

100% level standard-based analytical methods and are familiar with such methods. To use CAD for quantitation of components using a linear fit over the wide concentration range of 0.2% (LOQ) to 150% of 0.135 mg/mL requires much broader method acceptance criteria than that required for an HPLC method using a linear response UV detector. Thus, this method has a broader acceptance criteria compared with typical UV detector-based HPLC methods.

For gentamicin C1a the linearity range investigated covered the concentration range from 0.001 to 0.2025 mg/mL, which corresponds to QL (0.2%) to 150% of gentamicin label claim, in the presence of placebo. For sisomicin sulfate the linearity range investigated covered the concentration range from 0.001 to 0.025 mg/mL, which corresponds to LOQ (0.2%) to 5% of gentamicin label claim, in the presence of placebo and 0.5 mg/mL of gentamicin. Although the typical concentration of the related compounds was not greater than 0.5%, we tested a wider range to cover any future potential increase in concentrations of the related compounds in pharmaceutical cream samples. The slope, *y*-intercept, and coefficient of determination (r^2) were obtained from linear regression analysis performed using the software SAS System JMP® version 4. The peak areas of each individual compound were plotted against corresponding concentrations. Linear regression analysis yielded a coefficient of determination r^2 of greater than 0.99 ($n = 18$) for gentamicin and greater than 0.98 ($n = 15$) for sisomicin.

12.2.2.4 Accuracy

The recovery/accuracy of the method was determined using the solutions used for the linearity study. This method determines the percentage composition by area normalization and is not a quantitative assay method. The percentage recovery values obtained for gentamicin C1a are listed in Table 12.2. The average % recovery from 15 preparations (three for each concentration level (50, 75, 100, 125, and 150%)) was 101.3% for Analyst 1 and 99.2% for Analyst 2. The accuracy values also reflect the quadratic response of CAD, that is, a higher response at lower concentrations and lower response at higher concentrations. Thus, a higher average accuracy value of 107.0% for Analyst 1 and 105.1% for Analyst 2 is observed for the 50% concentration level, and a lower average accuracy value of 95.4% for Analyst 1 and 94.6% for Analyst 2 is observed for the 150% concentration level. The percentage recovery of the related substance sisomicin for the 0.2–5% concentration level ranged from 109 to 72%. For recovery, the sample solutions were bracketed between two gentamicin reference standard solutions, and the experimental concentration was determined using the following equation:

$$\text{Experimental concentration} = \frac{\text{P2} \times \text{C1}}{\text{P1} \times \text{RRF}}$$

where P1 is the average peak area of gentamicin C1a in adjacent bracketing standards, P2 is the peak area of gentamicin or related substance in recovery

Table 12.2 Recovery of gentamicin C1a from triplicate preparations.

Percent label strength	Preparation	Percent recovery of gentamicin C1a[a]	
		Analyst 1	Analyst 2
50%	1	106.4	105.7
	2	106.7	105.4
	3	107.9	104.1
75%	1	102.0	101.8
	2	102.3	101.8
	3	101.3	102.9
100%	1	99.4	98.4
	2	99.4	96.8
	3	99.4	99.7
125%	1	98.8	95.9
	2	98.0	96.9
	3	97.2	95.4
150%	1	95.7	95.7
	2	96.0	95.9
	3	94.6	92.2
Average percent recovery		101.3	99.2

[a] % Recovery = Experimental concentration ÷ Theoretical concentration × 100%.

sample solution, and relative response factor (RRF) is the ratio between the response factor of each individual gentamicin compound impurity or related substance and the response factor of gentamicin C1a. The RRF was obtained by dividing the slope of the linear regression curve of gentamicin C1a by the slope of the linear regression curve of the gentamicin compound or individual related substance. The recovery of each concentration level was then determined by the following equation:

$$\text{Percent recovery} = \frac{\text{Experimental concentration}}{\text{Theoretical concentration}} \times 100$$

12.2.2.5 Limit of Detection and Limit of Quantitation

The LOD and LOQ of gentamicin C1a was 0.37% (0.00185 mg/mL gentamicin/0.0005 mg/mL gentamicin C1a) and 0.74% (0.0037 mg/mL gentamicin/0.001 mg/mL gentamicin C1a), respectively, of the gentamicin analytical concentration of 0.5 mg/mL. The LOD and LOQ of sisomicin was 0.1% (0.0005 mg/mL sisomicin) and 0.2% (0.001 mg/mL sisomicin), respectively, of the gentamicin analytical concentration of 0.5 mg/mL. Signal-to-noise (S/N) values ranging from 11 to 142 obtained for the LOQ and values ranging from 5 to 65 obtained for LOD were observed by both analysts during validation depending on the HPLC-CAD system used for the analysis. The lowest values of 5 and 11 were observed in a CAD system with low sensitivity, which was near its due date for annual performance maintenance cleaning and calibration.

12.2.2.6 Reproducibility and Precision

The method reproducibility and precision were determined using the data obtained from the linearity study. The precision was determined from the percentage relative standard deviation (%RSD) of the recoveries obtained from samples prepared in triplicate at the low (50%), middle (100%), and high (150%) gentamicin C1a concentration levels. For sisomicin, the precision was determined from the %RSD of the recoveries obtained from samples prepared in triplicate at the low (0.2%), middle (1.25%), and high (5%) concentration levels. The reproducibility was determined from the difference in the average recoveries and the difference in the %RSD of recoveries between the two analysts in different laboratories. The results for gentamicin C1a and sisomicin are listed in Table 12.3, which reveals that this method has good reproducibility and precision.

12.2.2.7 Robustness

Deliberate variations in HPLC parameters were made to demonstrate the robustness of the method. We evaluated the method robustness based on the changes in RR, the resolution between gentamicin C1a and sisomicin, and the tailing factor of gentamicin C1a under the tested conditions. Gentamicin C1a peak was used for RR calculation. The RRs of all the peaks were calculated against the gentamicin C1a peak, and the resolution between sisomicin and gentamicin C1a was found to be ≥1.5, and the tailing factor was found to be ≤2.0 under the various chromatographic conditions tested. The RRs of the tested compounds obtained under a few representative HPLC conditions are summarized in Table 12.4. It can be seen that the RRs obtained under various chromatographic conditions remain fairly close.

12.2.2.8 Alternate Column Validation

Another column, Fluophase (50 mm × 4.6 mm I.D., 5 µm particle size) manufactured by Thermo Fisher Scientific, was also validated as an alternate column for the method. This validation was conducted to have an alternate column in the event the primary column (Restek Allure PFP) becomes

Table 12.3 Reproducibility and precision of gentamicin C1a and sisomicin.

Analyst	% RSD gentamicin sulfate C1a	Difference in % RSD gentamicin sulfate C1a (Analyst 1 and Analyst 2)		
		50%	100%	150%
Analyst 1		0.1	1.5	1.4
50%	0.7			
100%	0.0			
150%	0.8			
Analyst 2				
50%	0.8			
100%	1.5			
150%	2.2			
Analyst	**% RSD sisomicin sulfate**	**Difference in % RSD sisomicin sulfate (Analyst 1 and Analyst 2)**		
		0.2%	**1.25%**	**5%**
Analyst 1		4.2	2.3	2.4
0.2%	5.7			
1.25%	0.5			
5%	0.9			
Analyst 2				
0.2%	1.5			
1.25%	2.8			
5%	3.3			

commercially unavailable or its properties undergo change to limit its suitability for the current method.

For this method the Restek Allure PFP column underwent a change in the silica particle, which affected the stability of the column for use with the method and the commercial availability of the column. The Thermo Scientific Fluophase column was successfully used to replace the Restek Allure PFP column for the routine use of the method.

12.2.2.9 Calculation

The composition of gentamicin C1, gentamicin C1a, gentamicin C2, gentamicin C2a, and gentamicin C2b is calculated as a percent area by area normalization.

Table 12.4 The relative retentions (RRs) of gentamicin and related substances under selected HPLC parameter robustness conditions.

Compound name	Method condition	Column temperature = 30°C	Column temperature = 40°C	Flow rate = 0.9 mL/min	Flow rate = 1.1 mL/min	Injection volume = 18 μL	Injection volume = 22 μL
Deoxystreptamine	0.33	0.29	0.30	0.33	0.30	0.33	0.32
Garamine	0.57	0.56	0.58	0.56	0.58	0.57	0.57
Sisomicin	0.96	0.96	0.96	0.95	0.96	0.96	0.96
Gentamicin C1a	1.00	1.00	1.00	1.00	1.00	1.00	1.00
Gentamicin C2	1.11	1.10	1.11	1.11	1.10	1.11	1.11
Gentamicin C2a	1.17	1.17	1.17	1.18	1.17	1.17	1.17
Gentamicin C2b	1.20	1.20	1.20	1.20	1.19	1.20	1.20
Gentamicin C1	1.29	1.29	1.27	1.30	1.27	1.29	1.28

The percent area of each individual gentamicin and total percent area of gentamicin can be calculated using the equations shown in the following.

$$\text{Percent area of gentamicin sulfate compound } i = \frac{\text{Peak area of gentamicin sulfate compound } i}{\text{Total peak area of gentamicin sulfate}} \times 100$$

12.2.2.10 Chromatographic Conditions of the Final Method

The composition of mobile phase A is HFBA: H_2O:CH_3CN (0.025:95:5, v/v/v) and that of mobile phase B is TFA:H_2O:CH_3CN (1:95:5, v/v/v). The gradient program is listed in Table 12.1. This method employs an isocratic mobile phase of TFA:H_2O:CH_3CN (1:95:5, v/v/v) from retention time 16–33 min for eluting the gentamicins (C1, C1a, C2, C2a, and C2b) and most of the related substances of gentamicin. In the initial 16 min, it uses an ion-pair gradient varying the proportion of the ion-pairing agents HFBA and TFA over a very small range from 0.025 to 1 (v/v) compared to the constant H_2O:CH_3CN proportion of 95:5 (v/v). Thus, the resulting ion-pair gradient involves relatively minor changes in the total organic content of the mobile phase and thereby minimizes the impact of change in the nebulization efficiency and response of CAD due to change in organic content. To prevent desolvation and maintain the retention and performance of the reversed-phase column despite use with highly aqueous mobile phase, the column was flushed with mobile phase C, H_2O:CH_3CN (20:80, v/v), for 10 min. To ensure re-equilibration after each gradient elution, the column was re-equilibrated at the gradient initial condition for 4 min. The column temperature was 35°C, and the injection volume was 20 µL. The Corona CAD was set at 100 pA gain and a medium noise filter.

12.2.3 Discussion

This HPLC method is the first known method that can separate and accurately estimate all the individual analogues of gentamicin including all the related compounds by direct detection (i.e., no derivatization) using CAD. This method was successfully validated by two analysts from two different laboratories and has been demonstrated to have good accuracy, linearity, precision, reproducibility, specificity, and robustness. The analytical method described in this paper has been successfully used to determine the composition of gentamicin and also to estimate its related substances in a pharmaceutical cream. This HPLC method is a percent composition method and not an assay method because gentamicin is an antibiotic, and a microbial assay that measures the activity of the gentamicin in a sample is required by health authorities and cannot be substituted by an HPLC assay method. This method is also a stability-indicating method for a pharmaceutical cream because it can separate all the known and unknown degradation products of gentamicin from the active pharmaceutical ingredient (API) peaks and from each other and can accurately quantitate the content of gentamicin in the samples of pharmaceutical cream. Thus, this method can be used in quality control labs for stability studies and also for routine analysis of gentamicin in commercial lots of pharmaceutical cream including release testing.

Subsequent to the publication of this method [7], the silica used in the Restek Allure PFP column was changed, and the new Restek columns have much lower column lifetime. The retention on the alternate column, Thermo Scientific Fluophase, has been found to be much superior and can be substituted for the Restek PFP column.

After the publication of this method [7], another method for gentamicin using CAD and a C18 column has been published [45]. This C18 method has the limitation that deoxystreptamine elutes at the solvent front and the capacity factor of garamine is low; thus these related compounds of gentamicin cannot be reliably quantitated. C18 columns can be used to effectively separate gentamicin (as described in publication [7]); however, an ion-pair gradient would have to be used to retain deoxystreptamine and garamine for reliable quantitation, and the resolution of sisomicin and gentamicin C1a on PFP is superior to C18 columns. A new combination C18 and PFP phase, the ACE C18-PFP column, has been commercially available from MAC-MOD Analytical Inc., (Pennsylvania, USA). It can be reasonably expected that this combination C18-PFP phase should be more stable and provide greater retention, allowing for the reduction in the TFA concentration in mobile phase B and for superior column lifetime.

12.3 Application of Strategy to Netilmicin Sulfate

12.3.1 Method Development

The objective was to leverage the existing PFP-based conditions to develop an HPLC-CAD method to separate netilmicin from its known (*N*-ethyl garamine and sisomicin) and unknown related substances (ethyl sisomicin derivatives prepared by the alkylation of sisomicin and impurities formed from stress studies). Chemical structures of netilmicin and its related substances are shown in Figure 12.4. Netilmicin and its ethyl sisomicin derivatives such as 3-*N*-ethyl sisomicin, 2′-*N*-ethyl sisomicin, 6′-*N*-ethyl sisomicin, and 3″-*N*-ethyl sisomicin differ from one another only in the position of substitution of the ethyl group on sisomicin.

Netilmicin is a semisynthetic derivative of sisomicin, so the related substances of netilmicin can be generated as by-products of the synthesis of netilmicin

Compound name	Structure	Identity	Properties
N-ethyl deoxystreptamine sulfate	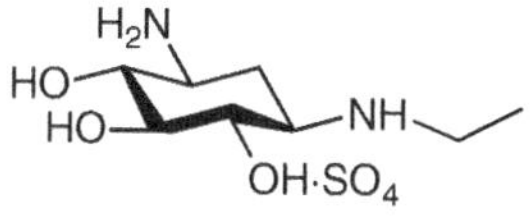	Related substance	UV inactive, water soluble
N-ethyl garamine sulfate	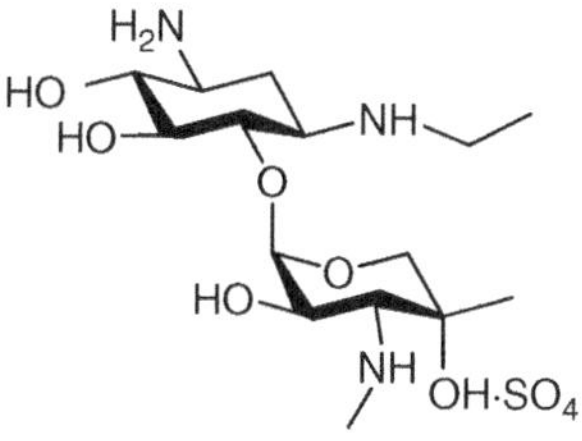	Related substance	UV inactive, water soluble
Sisomicin sulfate	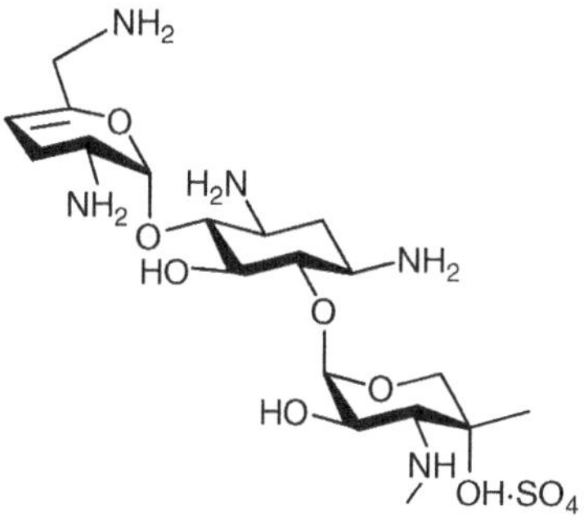	Related substance	Weak UV absorbance, water soluble

Figure 12.4 Compound name, structure, identity, and properties of netilmicin and its related substances.

from sisomicin. As we were unable to find commercial sources for the ethyl derivatives of sisomicin, we synthesized limited amounts of these compounds for the development and validation of the method using reductive alkylation of sisomicin [43]. The procedure consisted of dissolving sisomicin sulfate (500 mg) in 50 mL of water in a round bottomed flask; 100 μL of acetaldehyde was added with stirring and the solution was allowed to stir for 10 min. 50 mg of sodium

Compound name	Structure	Identity	Properties
Netilmicin sulfate	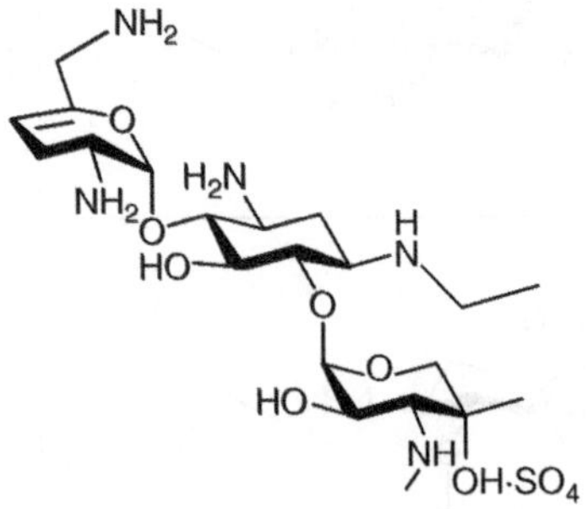 NH_2, O, NH_2, O, H_2N, H, N, HO, O, O, HO, NH, OH·SO_4	API	Weak UV absorbance, water soluble
3-*N*-ethyl sisomicin sulfate	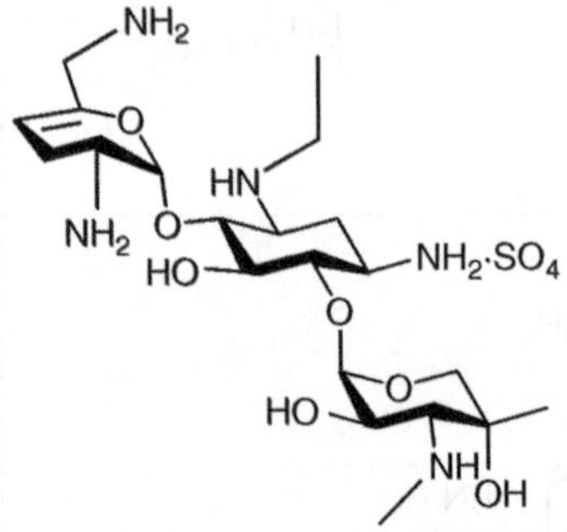 NH_2, O, HN, NH_2, O, HO, NH_2·SO_4, O, O, HO, NH, OH	Related substance	Weak UV absorbance, water soluble
6′-*N*-ethyl sisomicin sulfate	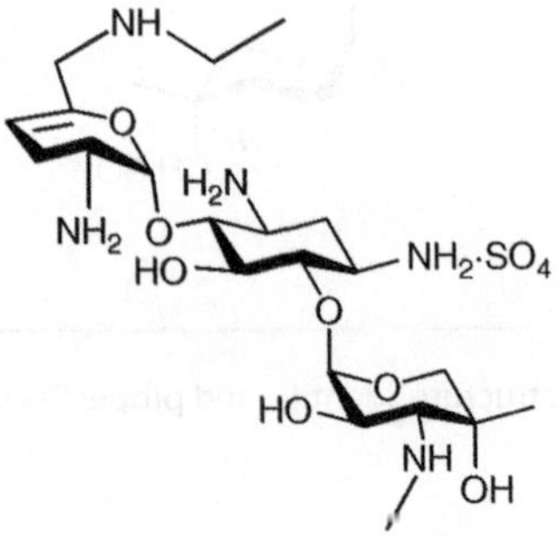 NH, O, H_2N, NH_2, O, HO, NH_2·SO_4, O, O, HO, NH, OH	Related substance	Weak UV absorbance, water soluble

Figure 12.4 (Continued)

Compound name	Structure	Identity	Properties
3″-*N*-ethyl sisomicin sulfate	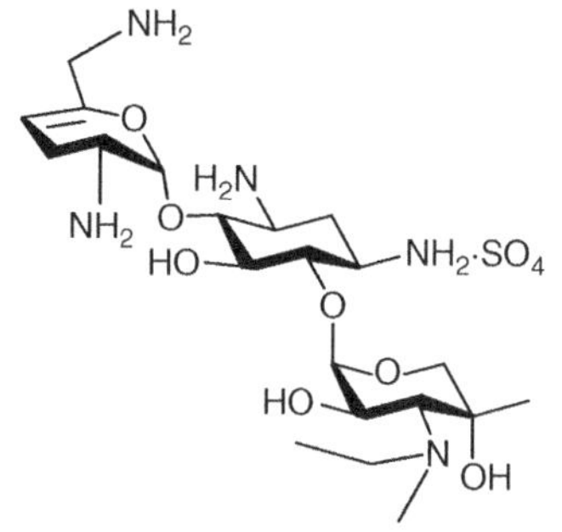	Related substance	Weak UV absorbance, water soluble
2′-*N*-ethyl sisomicin sulfate	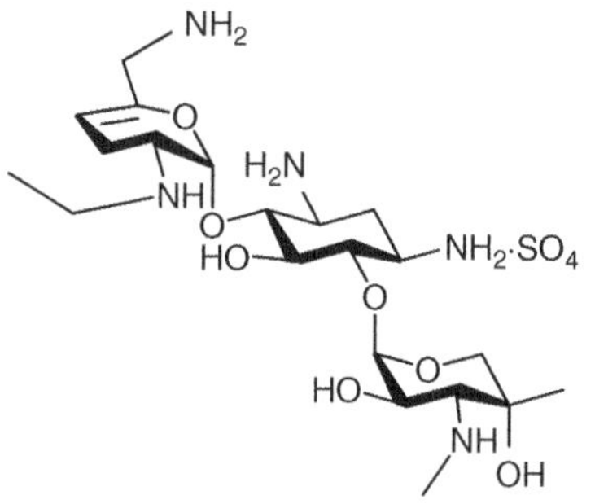	Related substance	Weak UV absorbance, water soluble

Figure 12.4 (Continued)

cyanoborohydride was added to the flask and the reaction mixture was allowed to stir for 15 min. The reductive alkylation of sisomicin yielded a mixture of ethyl derivatives of sisomicin including netilmicin sulfate.

Method development activities were initiated using a PFP column with mobile phases containing ion-pairing agents. The Restek Allure PFP column (100 mm × 4.6 mm, 5 μm particles) was tested, and this column provided good separation of all the ethyl derivatives of sisomicin from netilmicin.

The composition of the mobile phase was modified to investigate different ion-pairing agents, for example, TFA, PFPA, and HFBA. In addition, various

Table 12.5 Gradient program of the netilmicin method.

Time (min)	Mobile phase A (%)	Mobile phase B (%)	Mobile phase C (%)	Gradient curve	Comment
0.00	100.0	0.0	0.0	Linear	First isocratic ion-pair reagent (PFPA) elution
3.00	100.0	0.0	0.0	Linear	
3.50	0.0	100.0	0.0	Linear	Second isocratic ion-pair reagent (TFA) elution
30.00	0.0	100.0	0.0	Linear	
Column wash and equilibrate to initial conditions					
30.50	0.0	0.0	100.0	Linear	Column wash
40.50	0.0	0.0	100.0	Linear	
41.00	100.0	0.0	0.0	Linear	Column equilibration
45.00	100.0	0.0	0.0	Linear	

organic modifiers (e.g., methanol, isopropanol, etc.) in combination with water at different proportions were also investigated. Netilmicin was successfully separated from *N*-ethyl garamine and sisomicin using a mobile phase of TFA:H_2O:CH_3CN (1:96:4, v/v/v). To retain deoxystreptamine and elute netilmicin and related substances within a reasonable run time, a shallow ion-pair gradient was employed. The ion-pair gradient uses 0.1% PFPA from time 0 to 3 min to retain deoxystreptamine. Then 1.0% TFA is introduced to elute ethylgaramine, sisomicin, and netilmicin (Table 12.5). The PFPA concentration (0.1%) used in the ion-pair gradient was selected to elute the deoxystreptamine close to 3 min and prevented its overlap with peaks at the solvent front. The TFA concentration of 1% was ideal to resolve the sisomicin alkyl groups within a reasonable run time. To reduce the impact of the highly aqueous mobile phase, a column wash step using H_2O:CH_3CN (20:80, v/v) was introduced at the end of each gradient cycle. Representative chromatograms of diluent, deoxystreptamine, *N*-ethyl garamine, sisomicin, and netilmicin are shown in Figure 12.5.

12.3.1.1 Sample Preparation

Standard solutions of netilmicin sulfate were prepared by dissolving the compound in water to have an analytical concentration of approximately 0.25 mg/mL. A specificity mixture solution was prepared by mixing *N*-ethyl garamine sulfate with the sisomicin sulfate reductive alkylation reaction mixture containing the ethyl derivatives of sisomicin (including netilmicin).

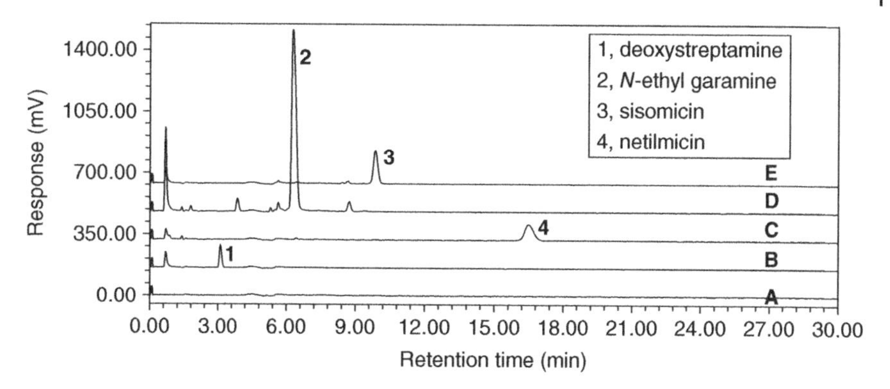

Figure 12.5 Representative chromatograms of (A) diluent, (B) deoxystreptamine, (C) netilmicin, (D) *N*-ethyl garamine, and (E) sisomicin.

12.3.2 Method Validation

Validation of this method was conducted to demonstrate the response linearity, specificity, robustness, LOQ, and LOD of this method. The results of the specificity, linearity, robustness, LOQ, and LOD experiments are described in the following. The materials, instrumentation and chromatographic conditions are described in the experimental section of the gentamicin method validation.

12.3.2.1 Specificity

The method specificity was demonstrated by successfully resolving netilmicin from its related substances in samples that were subjected to stress studies (Figure 12.6) and by successfully resolving netilmicin from sisomicin and ethyl derivatives of sisomicin in a specificity test mixture consisting of the sisomicin reductive alkylation reaction mixture spiked with ethyl garamine. The method was able to successfully resolve netilmicin, sisomicin, and *N*-ethyl garamine from all six unknown compounds as shown in Figure 12.7, demonstrating that it is a stability-indicating method.

Stress studies were conducted using heat, acid, and base. When netilmicin is subjected to heat, it undergoes degradation to form *N*-ethyl garamine and two unknown degradants. The method successfully resolved netilmicin from its three degradants as shown in Figure 12.6b. Acid treatment primarily led to the formation of *N*-ethyl garamine due to hydrolysis; the resulting chromatogram is shown in Figure 12.6c. Base-treated netilmicin solutions did not reveal any detectable peaks due to the high background noise from the base.

12.3.2.2 Linearity

The linearity range investigated for netilmicin covered the concentration range from 0.0025 to 0.5 mg/mL, which corresponds to 1 to 200% of the analytical concentration of netilmicin of 0.25 mg/mL. Triplicate preparations of netilmicin

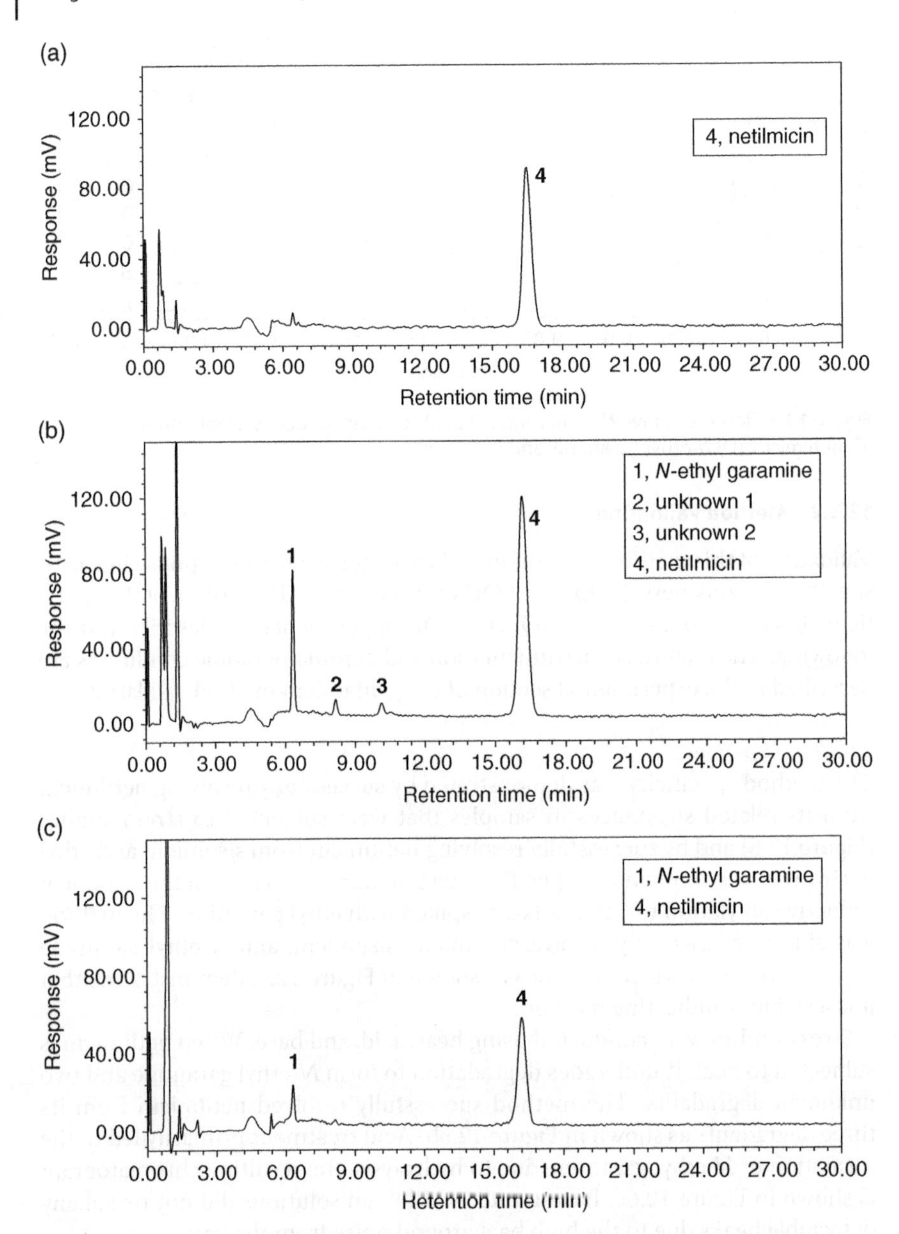

Figure 12.6 Representative chromatograms of stress study of netilmicin (a) netilmicin, (b) netilmicin heat stress, and (c) netilmicin acid stress (HCl).

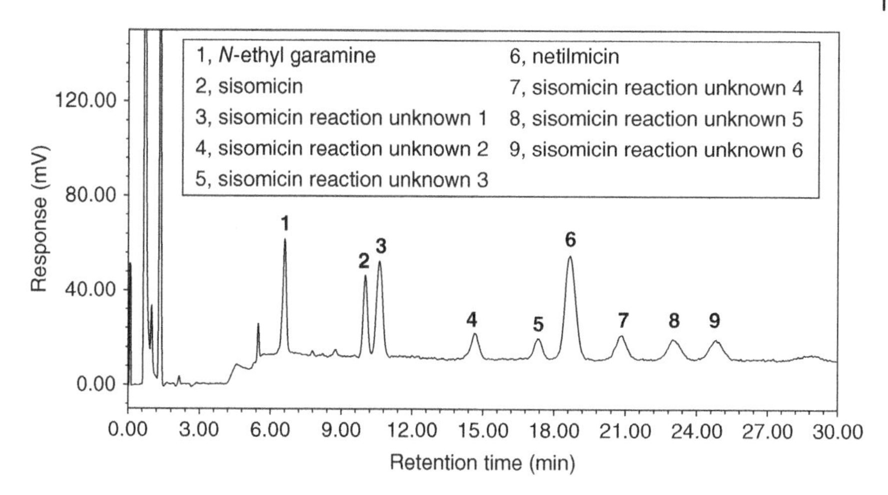

Figure 12.7 Representative chromatogram of a specificity mixture containing the ethyl derivatives of sisomicin (reaction mixture of sisomicin reductive alkylation) and *N*-ethyl garamine sulfate.

solutions at each of seven levels of sample concentration were prepared. The seven levels of sample concentration were 1, 2, 20, 40, 50, 100, and 200% of the netilmicin analytical concentration (0.25 mg/mL). For sisomicin the linearity range investigated covered the concentration range from 0.0025 to 0.025 mg/mL, which corresponds to 1–10% of netilmicin analytical concentration. The three levels of sample concentration for sisomicin were 1, 2, and 10.0% of the analytical concentration of netilmicin. The slope, y-intercept, and coefficient of determination (r^2) were obtained from linear regression analysis performed using the software SAS System JMP® version 4. The peak areas of each individual compound were plotted against corresponding concentrations. Linear regression analysis yielded a coefficient of determination r^2 of greater than 0.99 ($n = 21$) for netilmicin and greater than 0.99 ($n = 9$) for sisomicin.

12.3.2.3 Limit of Detection and Limit of Quantitation

The LOD and LOQ of netilmicin and sisomicin are 0.0025 and 0.005 mg/mL, 1 and 2%, respectively, of the netilmicin analytical concentration of 0.25 mg/mL. S/N values of greater than 10 for the LOQ and greater than 3 for LOD were routinely observed.

12.3.2.4 Robustness

Deliberate variations in HPLC parameters were made to demonstrate the robustness of the method. The parameters tested include injection volume, flow rate, temperature, and proportion of acetonitrile and ion-pairing agents in mobile phase. The method robustness was evaluated based on the changes in the resolution between the peaks in the unknown netilmicin sulfate and sisomicin sulfate and the tailing factor of netilmicin sulfate under the tested conditions. The resolution between sisomicin and sisomicin reaction unknown 1 was found to be ≥1.4, and the tailing factor was found to be ≤2.0 under the various chromatographic conditions tested. The resolution of netilmicin and its related substances in the specificity mixture obtained under a few representative HPLC robustness conditions are summarized in Table 12.6. It can be seen that netilmicin and all its related substances under various chromatographic conditions are well resolved.

12.3.2.5 Calculation

The composition of netilmicin sulfate is calculated as a percent area by area normalization. The percent area of netilmicin sulfate can be calculated using the equation that follows.

$$\text{Percent area of netilmicin sulfate} = \frac{\text{Peak area of netilmicin sulfate}}{\text{Total peak area of all peaks in sample}} \times 100$$

12.3.2.6 Chromatographic Conditions of the Final Method

Mobile phase A is PFPA:H_2O:CH_3CN (0.1:96:4, v/v/v) and mobile phase B is TFA:H_2O:CH_3CN (1:96:4, v/v/v). The gradient program is listed in Table 12.5. The column is washed with H_2O:CH_3CN (20:80, v/v), for 10 min. The column temperature is maintained at 40°C with a flow rate of 1.5 mL/min and the sample injection volume is 20 μL. The Corona CAD was set at 100 pA gain and a medium noise filter.

12.3.3 Discussion

The HPLC method described in this report is the first known method that can separate and accurately estimate all the individual analogues of netilmicin including all the related compounds by direct detection (i.e., no derivatization) using CAD. Validation of the method demonstrated linearity, specificity, and robustness of CAD that is comparable with UV detection. The analytical method described in this paper has been successfully used to determine the composition of netilmicin and also to estimate its related substances. This method is also a stability-indicating method for netilmicin because it can separate all the known and unknown degradation products of netilmicin and can accurately quantitate the content of netilmicin in any sample. Thus, this method can be used for routine analysis of netilmicin including the analysis of stability samples.

Table 12.6 The resolution of netilmicin and related substances in the specificity mixture under selected HPLC parameter robustness conditions.

Compound name	Method condition	Column temperature = 35°C	Column temperature = 45°C	Flow rate = 1.3 mL/min	Flow rate = 1.7 mL/min	Injection volume = 18 µL	Injection volume = 22 µL
N-ethyl garamine	n/a	n/a	n/a	n/a	n/a	n/a	n/a
Sisomicin	10.4	10.1	12.0	11.1	11.0	10.7	10.6
Sisomicin unknown 1	1.5	1.7	1.4	1.7	1.6	1.5	1.5
Sisomicin unknown 2	7.0	7.2	6.2	7.5	7.4	6.9	7.2
Sisomicin unknown 3	4.0	3.6	3.4	4.0	4.2	4.1	4.2
Netilmicin	1.7	1.6	1.6	1.7	1.7	1.9	1.7
Sisomicin unknown 4	2.2	2.0	2.6	2.5	2.6	2.3	2.3
Sisomicin unknown 5	2.0	2.4	2.1	4.4	2.4	2.1	2.1
Sisomicin unknown 6	1.7	1.3	1.7	2.4	1.6	1.6	1.4

12.4 Conclusion

A sensitive RP-HPLC method using CAD was developed and validated for gentamicin in a pharmaceutical cream [7]; this method was used as a model to develop and validate a method for netilmicin [8]. These methods can be applied to gentamicin and netilmicin in other formulations and can also be used as a model to develop and validate methods for other aminoglycosides.

Thermo Scientific Fluophase column can be substituted for the Restek Allure PFP column and has been found to have much superior stability. Furthermore, the new Ace C18-PFP column that is commercially available should be more stable and provide greater retention of the aminoglycosides due to multiple retention mechanisms (hydrophobic, dipole-dipole, and π–π interactions).

Acknowledgments

The author would like to thank Dr. Abu Rustum for his guidance on this project, Ms. Shrina Patel and all the analytical scientists in the Supply Analytical Sciences group for their support of this study, and the analysts from the Quality Laboratories in Heist, Belgium, for their participation in the method validation.

References

1 S. A. Waksman. Streptomycin: background, isolation, properties, and utilization. Science 1953; 118; 259–266.
2 M. J. Weinstein, G. M. Luedemann, E. M. Oden, G. H. Wagman. Gentamicin, a new broad-spectrum antibiotic complex. Antimicrob. Agents Chemother. 1963; 1; 1–7.
3 M. J. Weinstein, G. M. Luedemann, E. M. Oden, G. H. Wagman, J. P. Rosselet, J. A. Marquez, C. T. Coniglio, W. Charney, H. L. Herzog, J. Black. Gentamicin, a new antibiotic complex from micromonospora. J. Med. Chem. 1963; 6; 463–464.
4 F. C. Luft. Netilmicin: a review of toxicity in laboratory animals. J. Int. Med. Res. 1978; 6; 286–299.
5 P. Noone. Sisomicin, netilmicin and dibekacin: a review of their antibacterial activity and therapeutic use. Drugs 1984; 27; 548–578.
6 J. J. Wright. Synthesis of 1-N-ethylsisomicin: a broad-spectrum semisynthetic aminoglycoside antibiotic. J. Chem. Soc. Chem. Commun. 1976; 1976; 206–208.
7 A. Joseph, A. Rustum. Development and validation of a RP-HPLC method for the determination of gentamicin sulfate and its related substances in a

pharmaceutical cream using a short pentafluorophenyl column and a Charged Aerosol Detector. J. Pharm. Biomed. Anal. 2010; 51; 521–531.

8 A. Joseph, S. Patel, A. Rustum. Development and validation of a RP-HPLC method for the estimation of netilmicin sulfate and its related substances using Charged Aerosol Detection. J. Chromatogr. Sci. 2010; 48; 607–612.

9 J. P. Anhalt. Assay of gentamicin in serum by high-pressure liquid chromatography. Antimicrob. Agents Chemother. 1977; 11; 651–655.

10 S. K. Maitra, T. Y. Yoshikawa, J. L. Hansen, I. Nilsson-Ehle, W. J. Palin, M. C. Schotz, L. B. Guze. Serum gentamicin assay by high-performance liquid chromatography. Clin. Chem. 1977; 23; 2275–2278.

11 G. W. Peng, M. A. F. Gadalla, A. Peng, V. Smith, W. L. Chiou. High-performance liquid chromatography method for determination of gentamicin in plasma. Clin. Chem. 1977; 23; 1838–1844.

12 G. Seidl, H. P. Nerad. Gentamicin C: separation of C1, C1a, C2, C2a and C2b components by HPLC using lsocratic ion-exchange chromatography and post-column derivatization. Chromatographia 1988; 25; 169–171.

13 P. Gambardella, R. Punziano, M. Gionti, C. Guadalupi, G. Mancini, A. Mangia. Quantitative determination and separation of analogues of aminoglycoside antibiotics by high-performance liquid chromatography. J. Chromatogr. 1985; 348; 229–240.

14 L. O. White, A. Lovering, D. S. Reeves. Variations in gentamicin C1, C1a, C2, and C2a content of some preparations of gentamicin sulphate used clinically as determined by high-performance liquid chromatography. Ther. Drug Monit. 1983; 5; 123–126.

15 P. J. Claes, R. Busson, H. Vanderhaeghe. Determination of the component ratio of commercial gentamicins by high-performance liquid chromatography using pre-column derivatization. J. Chromatogr. 1984; 298; 445–447.

16 J. H. Albracht, M. S. De Wit. Analysis of gentamicin in raw material and in pharmaceutical preparations by high-performance liquid chromatography. J. Chromatogr. 1987; 389; 306–311.

17 M. Freeman, P. A. Hawkins, J. S. Loran, J. A. Stead. The analysis of gentamicin sulphate in pharmaceutical specialities by high performance liquid chromatography. J. Chromatogr. 1979; 2; 1305–1317.

18 K. Kraisintu, R. T. Parfitt, M. G. Rowan. A high-performance liquid chromatographic method for the determination and control of the composition of gentamicin sulphate. Int. J. Pharm. 1982; 10; 67–75.

19 P. J. Claes, Y. Chaerani, H. Vanderhaeghe. Differentiation of the C2 and C2a by paired-ion high performance liquid chromatography of underivatized gentamicin. J. Pharm. Belg. 1985; 40; 95–99.

20 E. Kaale, S. Leonard, A. Van Schepdael, E. Roets, J. Hoogmartens. Capillary electrophoresis analysis of gentamicin sulphate with UV detection after pre-capillary derivatization with 1,2-phthalic dicarboxaldehyde and mercaptoacetic acid. J. Chromatogr. A 2000; 895; 67–79.

21 D. Samain, P. Dupin, P. Delrieu, G. Inchauspe. Multidimensional ion-pair HPLC for the purification of aminoglycoside antibiotics with refractive index detection. Chromatographia 1987; 24; 748–752.

22 R. T. Parfitt, D. E. Games, M. Rossiter, M. S. Rogers, A. Weston. Chemical ionization and field desorption mass spectrometry of the gentamicins. Biomed. Mass Spectrom. 1976; 3; 232–234.

23 T. A. Getek, M. L. Vestal, T. G. Alexander. Analysis of gentamicin sulfate by high-performance liquid chromatography combined with thermospray mass spectrometry. J. Chromatogr. 1991; 554; 191–203.

24 M. Cherlet, S. De Baere, P. De Backer. Determination of gentamicin in swine and calf tissues by high-performance liquid chromatography combined with electrospray ionization mass spectrometry. J. Mass Spectrom. 2000; 35; 1342–1350.

25 D. Loffler, T. A. Ternes. Analytical method for the determination of the aminoglycoside gentamicin in hospital wastewater via liquid chromatography–electrospray-tandem mass spectrometry. J. Chromatogr. A 2003; 1000; 583–588.

26 C. Lecaroz, M. A. Campanero, C. Gamazo, M. J. Blanco-Prieto. Determination of gentamicin in different matrices by a new sensitive high-performance liquid chromatography-mass spectrometric method. J. Antimicrob. Chemother. 2006; 58; 557–563.

27 N. C. Megoulas, M. A. Koupparis. Development and validation of a novel LC/ELSD method for the quantitation of gentamicin sulfate components in pharmaceuticals. J. Pharm. Biomed. Anal. 2004; 36; 73–79.

28 I. Clarot, P. Chaimbault, F. Hasdenteufel, P. Netter, A. Nicolas. Determination of gentamicin sulfate and related compounds by high-performance liquid chromatography with evaporative light scattering detection. J. Chromatogr. A. 2004; 1031; 281–287.

29 E. Adams, W. Roelants, R. De Paepe, E. Roets, J. Hoogmartens. Analysis of gentamicin by liquid chromatography with pulsed electrochemical detection. J. Pharm. Biomed. Anal. 1998; 18; 689–698.

30 T. A. Getek, A. C. Haneke, G. B. Selzer. Determination of gentamicin sulfate liquid chromatography with electrochemical detection. J. Assoc. Off. Anal. Chem. 1983; 66; 172–179.

31 V. Manyanga, K. Kreft, B. Divjak, J. Hoogmartens, E. Adams. Improved liquid chromatographic method with pulsed electrochemical detection for the analysis of gentamicin. J. Chromatogr. A 2008; 1189; 347–354.

32 V. Manyanga, O. Grishina, Z. Yun, J. Hoogmartens, E. Adams. Comparison of liquid chromatographic methods with direct detection for the analysis of gentamicin. J. Pharm. Biomed. Anal. 2007; 45; 257–262.

33 E. Adams, D. Puelings, M. Rafiee, E. Roets, J. Hoogmartens. Determination of netilmicin sulfate by liquid chromatography with pulsed electrochemical detection. J. Chromatogr. A 1998; 812; 151–157.

34 G. W. Peng, G. G. Jackson, W. L. Chiou. High-pressure liquid chromatographic assay of netilmicin in plasma. Antimicrob. Agents Chemother. 1977; 12; 707–709.

35 S. E. Back, I. Nilsson-Ehle, P. Nilsson-Ehle. Chemical assay, involving liquid chromatography, for aminoglycoside antibiotics in serum. Clin. Chem. 1979; 25; 1222–1225.

36 J. Marples, M. D. G. Oates. Serum gentamicin, netilmicin and tobramycin assays by high performance liquid chromatography. J. Antimicrob. Chemother. 1982; 10: 311–318.

37 L. Essers. An automated high-performance liquid chromatographic method for the determination of aminoglycosides in serum using pre-column sample clean-up and derivatization. J. Chromatogr. 1984; 305; 345–352.

38 M. Santos, E. Garcia, F. G. Lopez, J. M. Lanao, A. Dominguez-Gil. Determination of netilmicin in plasma by HPLC. J. Pharm. Biomed. Anal. 1995; 13; 1059–1062.

39 S. Dionisotti, F. Bamonte, M. Gamba, E. Ongini. High-performance liquid chromatographic determination of netilmicin in guinea-pig and human serum by fluorodinitrobenzene derivatization with spectrophotometric detection. J. Chromatogr. 1988; 434; 169–176.

40 H. Kubo, T. Kinoshita, Y. Kobayashi, K. Tokunaga. Micro-scale method for determination of tobramycin in serum using high-performance liquid chromatography. J. Liq. Chromatogr. 1984; 7; 2219–2228.

41 H. Fabre, M. Sekkat, M. D. Blanchin, B. Mandrou. Determination of aminoglycosides in pharmaceutical formulations—II. High-performance liquid chromatography. J. Pharm. Biomed. Anal. 1989; 7; 1711–1718.

42 D. C. Rigge, M. F. Jones. Shelf lives of aseptically prepared medicines—stability of netilmicin injection in polypropylene syringes. J. Pharm. Biomed. Anal. 2004; 35; 1251–1256.

43 B. Li, A. V. Schepdael, J. Hoogmartens, E. Adams. Characterization of impurities in sisomicin and netilmicin by liquid chromatography/mass spectrometry. Rapid Commun. Mass Spectrom. 2008; 22; 3455–3471.

44 N. Nagae, T. Enami, S. Doshi. The retention behavior of reversed-phase HPLC columns with 100% aqueous mobile phase. LCGC 2002; 20; 964–972.

45 K. Stypulkowska, A. Blazewicz, Z. Fijalek, K. Sarna. Determination of gentamicin sulphate composition and related substances in pharmaceutical preparations by LC with charged aerosol detection, Chromatographia 2010; 72; (11/12); 1225–1229.

13

Determination of Quaternary Ammonium Muscle Relaxants with Their Impurities in Pharmaceutical Preparations by LC-CAD

Agata Blazewicz[1], Magdalena Poplawska[2], Malgorzata Warowna-Grzeskiewicz[1,2], Katarzyna Sarna[1], and Zbigniew Fijalek[1,2]

[1] *Pharmaceutical Chemistry Department, National Medicines Institute, Warsaw, Poland*
[2] *Warsaw Medical University, Warsaw, Poland*

CHAPTER MENU

13.1 Summary

Quaternary ammonium muscle relaxants are non-depolarizing neuromuscular blocking drugs used for muscle relaxation, most commonly in anesthesia. They inhibit neuromuscular transmission by competing with acetylcholine for the cholinergic receptors of the motor end plate, thereby reducing the response of the end plate to acetylcholine. Drugs in this category include well-known compounds such as pancuronium, atracurium, mivacurium, vecuronium, rocuronium and pipecuronium. In terms of toxicity and forensics, a clinically effective dose of one of these drugs can be also a lethal dose if respiratory assistance is not provided. Thus, neuromuscular blocking agents are potential murder weapons, as described in many papers concerning forensic medicine (in cases of suicide and murder).

Charged Aerosol Detection for Liquid Chromatography and Related Separation Techniques, First Edition. Edited by Paul H. Gamache.

Because of the lack of chromophore and thermal stability and also because of the presence of a permanent positive charge, quaternary ammonium muscle relaxants are difficult to analyze by conventional analytical methods, and therefore they are analytically challenging compounds. For the determination of muscle relaxants with or without their impurities or metabolites, various methods have been reported: fluorimetric, electrochemical, and liquid chromatographic (LC) with the following types of detection—electrochemical, fluorimetric, NMR or ESI-MS(MS). For pancuronium, European Pharmacopoeia 7.1 recommends TLC due to the presence of quaternary amine moieties and lack of UV chromophores. Moreover, impurities of pancuronium co-migrate and give one spot in TLC.

The purpose of this work was to develop quick and sensitive LC-CAD methods for the simultaneous determination of quaternary ammonium muscle relaxants with their impurities, in substance and in pharmaceutical preparations. It is very important to identify and determine impurities present in medicinal products because of the risk of their toxicity. The concentration of impurities in medicines is very low, so it is still necessary to look for more sensitive and selective analytical methods suitable for their determination.

Novel, sensitive, and simple LC-CAD methods for the determination of three isomers of atracurium, three isomers of mivacurium, cisatracurium, and pancuronium with their impurities were developed. The LC-CAD method was also optimal for the identification of seven muscle relaxants (rocuronium, pipecuronium, pancuronium, vecuronium, atracurium, cisatracurium, and new potential non-depolarizing neuromuscular blocking agent SZ1677) in the mixture. The conditions were optimized to obtain the best signal and stability of the measurement with the highest sensitivity. The preparation of samples and their analysis were performed in a relatively short time. All impurities were identified using time-of-flight mass spectrometry with electrospray ionization. The quantitative aspects of the proposed methods were examined according to ICH guidelines; the elaborated method was validated for LOD, LOQ, linearity, precision, accuracy, stability, and robustness.

The elaborated methods for the analysis of muscle relaxants and their impurities proved to be fast, precise, accurate, and sensitive and could be applied for routine analysis in substances and in pharmaceutical preparations.

13.2 Introduction

Atracurium besilate and its 1R-*cis*, 1R'-*cis* isomer, cisatracurium besilate, which are non-depolarizing neuromuscular blocking drugs of intermediate duration of action, are widely used clinically [1]. Due to four chiral atoms and symmetric structure, atracurium possesses 10 stereoisomers. Cisatracurium is a more potent (3–5-fold) isomer of atracurium and has lower histamine-releasing potential in clinical doses. Both drugs undergo Hofmann elimination [2, 3], a nonenzymatic process dependent on pH and temperature, yielding laudanosine and quaternary monoacrylate. In acidic solutions (pH < 3) atracurium degrades also by ester hydrolysis, where the monoquaternary alcohol and monoquaternary acid are the primary products (Figure 13.1). Unlike atracurium, cisatracurium does not appear to be degraded directly by ester hydrolysis.

Mivacurium is a mixture of three isomers (trans–trans, cis–trans, cis–cis) [4]. It has a short to intermediate duration of action and is hydrolyzed to monoquaternary alcohol and acid (Figure 13.2). Opposite to atracurium, its trans–trans isomer is more potent than the other ones.

As pancuronium is mainly used in relaxation of respiratory muscles, it is very important to manage patients with a mechanical ventilator. In other cases, it can be lethal, as described in many papers concerning forensic medicine (in cases of suicide [5–7] and murder [8, 9]).

Due to deacetylation, pancuronium may be metabolized to various compounds. Deacetylation proceeds at the 3-position and leads to a 3-hydroxy derivative of pancuronium (impurity B acc. to Ph. Eur.) and at the 17-position to give 17-hydroxy derivative of pancuronium (impurity A acc. to Ph. Eur.). Vecuronium (impurity D acc. to Ph. Eur.) is a monoquaternary ammonium salt of pancuronium.

Various methods have been reported for the determination of muscle relaxants alone without their impurities or metabolites: fluorimetric [10], electrochemical [11], and liquid chromatographic (LC) with the following types of detection: UV [12], fluorimetric [13], NMR [14], and recently electrospray ionization (ESI)-MS/MS [15–17]. Cirimele *et al.* [15] developed a procedure for eight quaternary nitrogen muscle relaxants in blood using LC-MS. A general screening method of 20 quaternary ammonium drugs in equine urine was worked out after solid-phase extraction (SPE) by LC-MS/MS [16]. A similar LC-MS/MS procedure for the determination of eight quaternary ammonium drugs and herbicides in human whole blood after weak cation-exchange SPE was developed by Ariffin and Anderson [17]. Although the previous methods can be used for simultaneous determination of quaternary ammonium drugs, it is necessary to detect and determine their degradants and metabolites in the same procedure. Of the published methods, LC with fluorimetric detection was mainly used for the determination of mivacurium in combination with their metabolites in human plasma after SPE [18] and without extraction [19], of atracurium with laudanosine as the main degradant [20], and of cisatracurium with laudanosine [21] and moquaternary alcohol [22]. Atracurium and cisatracurium with their metabolites were also determined using LC-MS [23–26]. However, some of the published methods either did not report assay validation [16, 21, 23, 25] or else reported incomplete assay validations—only for the main compound [26]. Most of them were unable to perform simultaneous determination of isomers of atracurium and its metabolites. The published stereoselective assays for the determination of isomers of atracurium without its metabolites used LC–NMR [14], but the procedure is time consuming and expensive. When diastereoisomers differ in pharmacological properties, it is very important

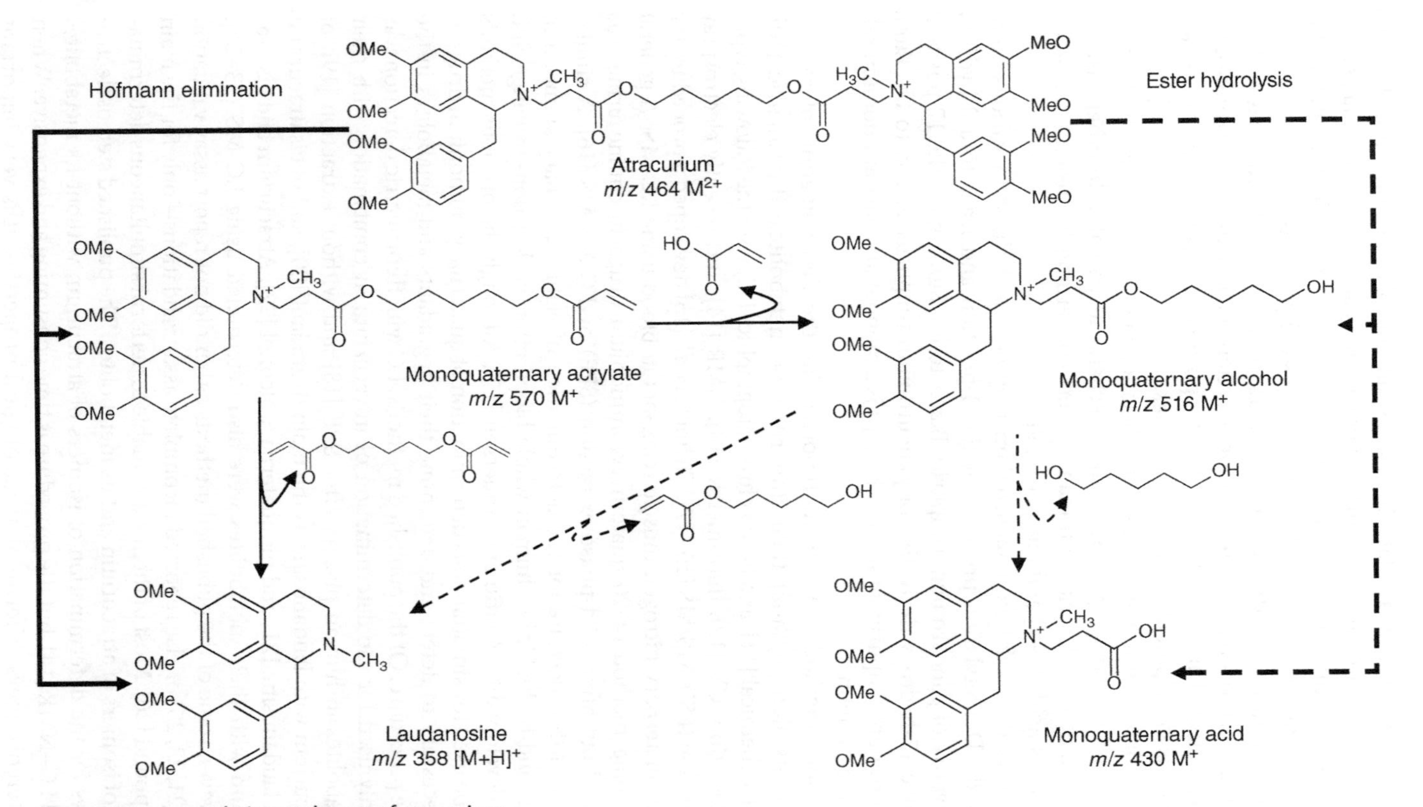

Figure 13.1 Degradation pathways of atracurium.

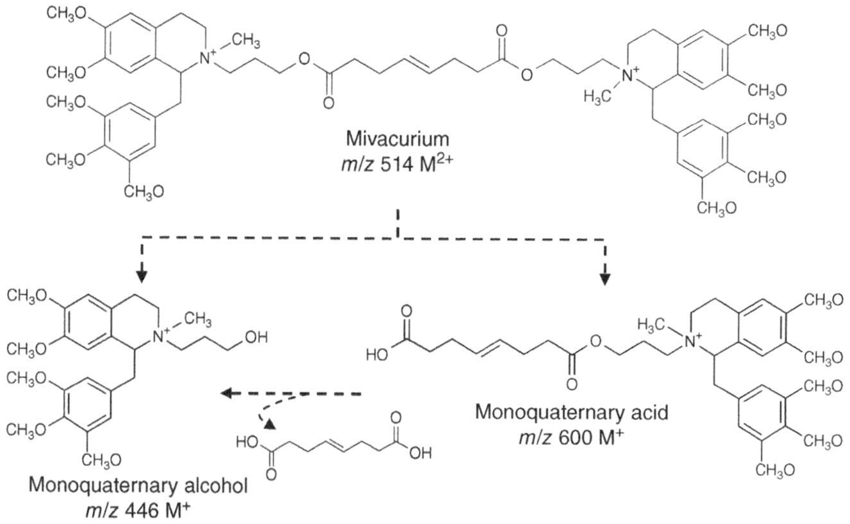

Figure 13.2 Degradation pathways of mivacurium.

to have a stereoselective method for analysis. European Pharmacopoeia 7.1 (Ph. Eur.) recommends LC-UV for the determination of the isomers of atracurium and its impurities; however, the time of analysis is very long—50 min [27].

For the determination of pancuronium with impurities or metabolites, only few methods have been mentioned. Ph. Eur. recommends TLC due to the presence of quaternary amine moieties and lack of UV chromophores [28]. Moreover, impurities of dacuronium (impurity A) and vecuronium (impurity D) co-migrate and give one spot in TLC. Although, according to Ph. Eur., pancuronium bromide and its impurities cannot be analyzed by LC-UV because of insufficient sensitivity (little or no chromophores), methods using this approach have been published [29, 30]. Other methods for the determination of pancuronium and impurities involve mainly MS [15, 31, 32] or MS/MS [7, 25] coupled to LC system.

The purpose of this work was to develop quick and sensitive LC methods with charged aerosol detection (CAD) for the determination of the three isomers of atracurium, cisatracurium, the three isomers of mivacurium with their degradants, and also pancuronium and its impurities, which could be applied to their analysis in substances and pharmaceutical preparations.

13.3 Experimental

13.3.1 Equipment and Conditions

A Corona® CAD® instrument (ESA, Chelmsford, MA, USA) was equipped with an LC UltiMate 3000 system (Dionex, Germering, Germany) consisting of a pump, a degasser, an autosampler, a column heater, and a pulse damper. Data processing was carried out with Chromeleon 6.8 software (Dionex). Nitrogen gas from nitrogen generator N2-MISTRAL-4 (Schmidlin-DBS, Switzerland), regulated at 35 psi, was introduced to the detector, and the resultant gas flow rate was regulated automatically and monitored by the CAD device. Response range was set to 50 pA full scale. Chromatographic analysis was carried out at 25°C. The analysis was performed on a C_{18} analytical column (Hypersil GOLD, 150 mm × 4.0 mm; 3 μm particle size; Thermo Fisher Scientific, Waltham, MA, USA) with a guard column (Hypersil GOLD, 10 mm × 4.0 mm; 3 μm particle size; Thermo Fisher Scientific).

The mass spectrometer micrOTOF-Q II from Bruker Daltonics (Billerica, MA, USA) was used to obtain the electrospray ionization time-of-flight mass spectra (ESI-TOF-MS), when peak identifications were required.

13.3.2 Material Studied

Reference standards:

Atracurium besilate (Abbott Laboratories), which is a mixture of the cis–cis, cis–trans, and trans–trans isomers of 2,2′-[pentane-1,5-diylbis[oxy(3-oxopropane-1,3-diyl)]]bis[1-(3,4-dimethoxybenzyl)-6,7-dimethoxy-2-methyl-1,2,3,4-tetrahydroisoquinolinium] dibenzenesulfonate

Cisatracurium besilate (GlaxoSmithKline)—1R-*cis*, 1R′-*cis* isomer of atracurium

Laudanosine (LGC Standards)—1-(3,4-dimethoxybenzyl)-6,7-dimethoxy-2-methyl-1,2,3,4-tetrahydroisoquinoline

Pancuronium bromide—1,1′-[3α,17β-bis(acetyloxy)-5α-androstane-2β,16β-diyl]bis(1-methylpiperidinium) dibromide (Sigma)

Dacuronium bromide (impurity A)—1,1′-[3α-(acetyloxy)-17β-hydroxy-5α-androstane-2β,16β-diyl]bis(1-methyl-piperidinium) dibromide (Organon)

Vecuronium bromide (impurity D)—1-[3α,17β-bis(acetyloxy)-2β-(piperidin-1-yl)-5α-androstan-16β-yl]-1-methylpiperidinium bromide (Organon)

Pipecuronium bromide—4,4′-[3α,17β-bis(acetyloxy)-5α-androstane 2β,16β-diyl]bis[1,1-dimethylpiperazinium]dibromide (Gedeon Richter)

SZ1677—1-[17β-(acetyloxy)-2β-(1,4-dioxa-8-azaspiro[4,5]dec-8-yl)-3α-hydroksy-5α-androstan-16β-yl]-1-(prop-2-enyl) pyrrolidinium bromide (Gedeon Richter)

Rocuronium bromide—1-[17β-(acetyloxy)-3α-hydroxy-2β-(morpholin-4-yl)-5α-androstan-16β-yl]-1-(prop-2-enyl) pyrrolidinium bromide (Organon)

Pharmaceutical preparations:

Nimbex, a solution for injections and infusions containing 2 mg mL^{-1} of cisatracurium (GlaxoSmithKline)

Tracrium, a solution for injections and infusions containing 10 mg mL^{-1} of atracurium besilate (GlaxoSmithKline)

Mivacron, a solution for injections containing 2 mg mL^{-1} of mivacurium chloride (GlaxoSmithKline)
Pancuronium, solution for injections containing 2 mg mL^{-1} of pancuronium from Jelfa (Jelenia Góra, Poland)

Methanol from Labscan (Dublin, Ireland), formic acid (FA) from Park Scientific Limited (Northampton, UK), and trifluoroacetic acid (TFA) from Biosolve (Valkenswaard, The Netherlands) are of purity suitable for LC; doubly distilled water additionally purified in the NANOpure DIamond UV Deionization System from Barnstead (Dubuque, IA, USA) was used throughout.

13.3.3 Standard Solutions

Stock standard solutions were prepared weekly. Approximately 10 mg of each active substance and 5 mg of the impurity of laudanosine were weighed accurately into a 10-mL volumetric flask and dissolved with the solvent containing 0.1% (v/v) FA in water (for atracurium, cisatracurium, and laudanosine) or with the solvent containing 0.1% (v/v) TFA in water (for pancuronium, vecuronium, and dacuronium). These solutions were further successively diluted with the solvent (0.1% TFA or 0.1% FA, respectively) to obtain the required concentrations. All solutions were stored in a cool, dark place when not in use.

13.4 Results and Discussion

13.4.1 Selection of Chromatographic Conditions

13.4.1.1 LC-CAD Method for Atracurium, Cisatracurium, and Mivacurium and Their Impurities

To obtain optimal chromatographic separation, different mobile phases in isocratic and gradient elution were evaluated. For analytical purposes, the gradient elution was much better for separation of atracurium or mivacurium isomers and impurities from each other. Mobile phases with 0.1% TFA (pH = 2) and 0.1% FA (pH = 3) were evaluated. However, when TFA was used, the retention times were longer, the noise was higher, and the signal-to-noise (S/N) ratio was lower (S/N was equal to 4.8 and 33.2 for TFA and FA, respectively).

The best response was obtained with a C_{18} analytical column (150 mm × 4.0 mm, 3 μm particle size; Thermo) and the mobile phase containing 0.1% FA in water (solvent A) and 0.1% FA in methanol (solvent B) in gradient elution. The following gradient was used before returning to the initial conditions at 17 min: 0–2 min 30% B, 2–10 min 30–50% B, and 10–15 min 50% B. The column was equilibrated (in 30% B) for 3 min at the end of each run.

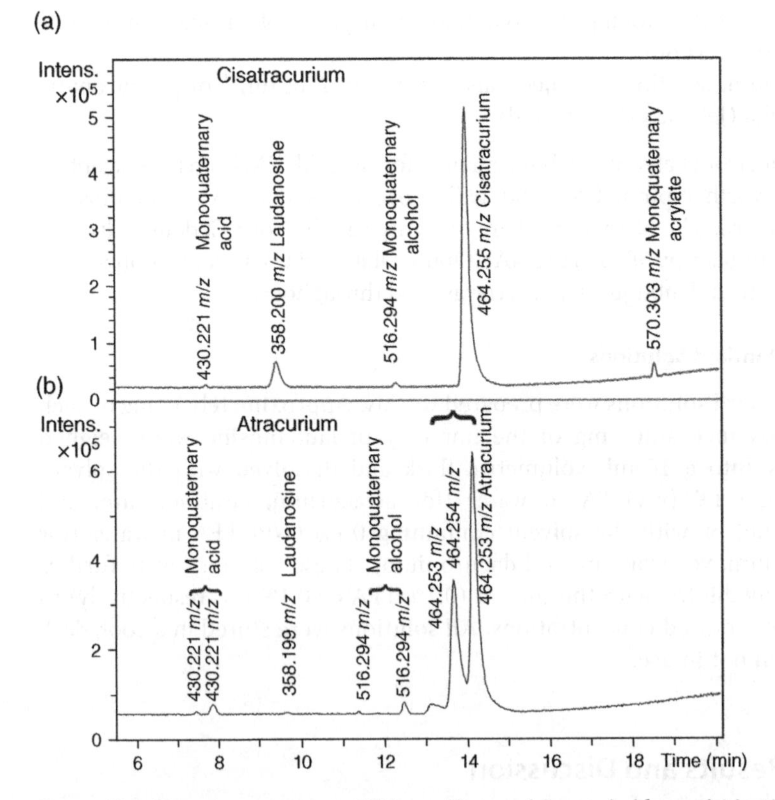

Figure 13.3 Mass chromatograms (ESI in positive mode) recorded for peak identification from the solutions containing 10 μg mL^{-1} of cisatracurium (a) and 10 μg mL^{-1} of atracurium (b); injection volume 1 μL.

The effect of different flow rates (from 0.4 to 0.6 mL min^{-1}) and column temperature (from 20 to 30°C) on the peak area and resolution was studied. A flow rate was set at 0.5 mL min^{-1}, 25°C was used throughout, and the injection volume was 10 μL. These conditions were optimal for the separation of atracurium (Figure 13.3) and mivacurium (Figure 13.4) isomers and their degradants from each other with the resolution values ranging from 2 to 18. All retention times, resolution values, tailing factors, and numbers of theoretical plates are presented in Table 13.1.

The aforementioned chromatographic conditions were finally chosen for further investigations.

13.4.1.2 LC-CAD Method for Pancuronium and Its Impurities

To obtain optimal chromatographic separation, different mobile phases in isocratic and gradient elution modes were evaluated. For analytical purposes,

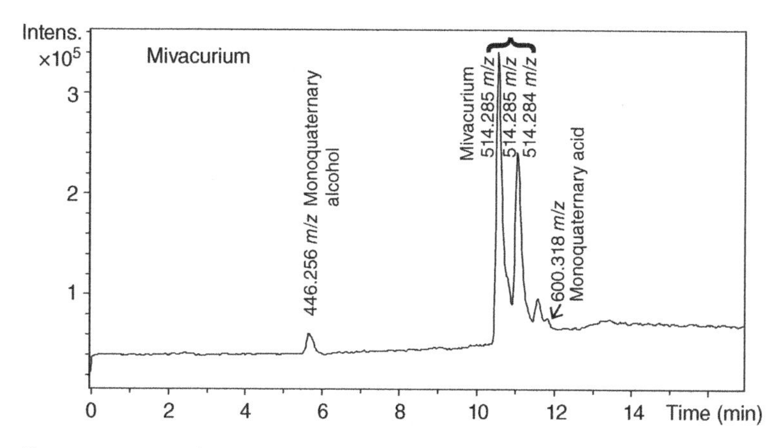

Figure 13.4 Mass chromatogram (ESI in positive mode) recorded for peak identification from the solutions containing 10 μg mL^{-1} of Mivacron; injection volume 1 μL.

Table 13.1 Observed *m/z* values and chromatographic (LC-CAD) parameters for atracurium and its impurities.

	m/z	Retention time (min)	Resolution	Tailing factor	Plates
Monoquaternary acid trans	430 M$^+$	4.56	2.28	1.38	11,835
Monoquaternary acid cis	430 M$^+$	4.79	8.70	0.97	10,427
Laudanosine	358 [M + H]$^+$	5.87	7.57	1.20	31,762
Monoquaternary alcohol trans	516 M$^+$	7.19	5.26	1.14	49,945
Monoquaternary alcohol cis	516 M$^+$	8.31	2.19	1.07	67,572
Atracurium trans–trans	464 M^{2+}	9.18	2.16	1.69	34,634
Atracurium cis–trans	464 M^{2+}	9.83	2.02	1.80	56,687
Atracurium cis–cis	464 M^{2+}	10.36	18.11	1.55	68,028
Monoquaternary acrylate trans	570 M$^+$	13.94	3.21	1.05	119,170
Monoquaternary acrylate cis	570 M$^+$	14.49	—	1.00	101,402

the gradient elution was much better for the separation of pancuronium and its impurities from each other. Mobile phases with 0.1% (v/v) TFA or 0.1% (v/v) FA in water (solution A) and in methanol (solution B) were evaluated. However, when FA was used, the resolution was not sufficient.

The best response was obtained with a C_{18} analytical column (150 mm × 4.0 mm, 3 μm particle size; Thermo Fisher Scientific, Waltham, MA, USA) with a guard column (Hypersil GOLD, 10 mm × 4.0 mm; 3 μm; Thermo Fisher Scientific) and the mobile phase containing 0.1% (v/v) TFA in water (solvent A) and in methanol (solvent B) in gradient elution. The following gradient was used before returning to the initial conditions at 15 min: 0–5 min 35% B, 5–12 min 35–57% B, 12–13 min 57% B. The column was equilibrated (in 35% B) for 2 min at the end of each run. An effect of different flow rates (from 0.4 to 0.6 mL min^{-1}) and column temperature (from 20 to 30°C) on the peak area and resolution was studied. A flow rate was set at 0.5 mL min^{-1}, 25°C was used throughout, and the injection volume was 10 μL. These conditions were optimal for the separation of pancuronium and its impurities from each other with good resolution (Figure 13.5) and were finally chosen for further investigations.

Similar conditions (0–5 min 35% B, 5–12 min 35–57% B, 12–15 min 57% B, 15–17 min 57–35% B, 17–20 min 57% B) were also optimal for the separation of seven muscle relaxants in the mixture (Figure 13.6). As three of these muscle relaxants (rocuronium, pipecuronium, and new potential non-depolarizing neuromuscular blocking agent SZ1677) with their impurities were determined using LC with electrochemical detection and described in our previous papers [33–35], they were not quantitatively determined by this LC-CAD method.

13.4.2 Identification of Analytes

Standards of mivacurium and degradants of atracurium, cisatracurium, and mivacurium like monoquaternary acids, monoquaternary alcohols (of ester hydrolysis), or monoquaternary acrylates (of Hofmann elimination) were not available for us to prepare standard stock solutions for identification and quantitation purposes. However, because in LC-CAD and LC-MS the same volatile mobile phases can be used, the elaborated LC-CAD method for chromatographic separation was transferred into LC-ESI-TOF-MS. By this analysis unknown peaks were identified as products corresponding to degradation pathways of cisatracurium (Figure 13.3a), atracurium (Figure 13.3b), and mivacurium (Figure 13.4), and similar chromatograms were obtained by LC-CAD (Figure 13.7).

Except dacuronium and vecuronium, standards of other impurities described in European Pharmacopoeia for pancuronium bromide were not available for us to prepare standard stock solutions for identification and quantitation purpose. The molecular ions of the target analytes have been verified through LC-ESI-TOF-MS, and an unknown peak was identified as impurity B of pancuronium (Figure 13.5).

13.4.3 Validation of the Methods

The quantitative aspects of the proposed methods were examined according to ICH guidelines [36]. The statistical evaluation for all analyzed substances was calculated using Chromeleon Validation ICH software. The data concerning method validation are summarized in tables. Peak areas were evaluated in the whole validation.

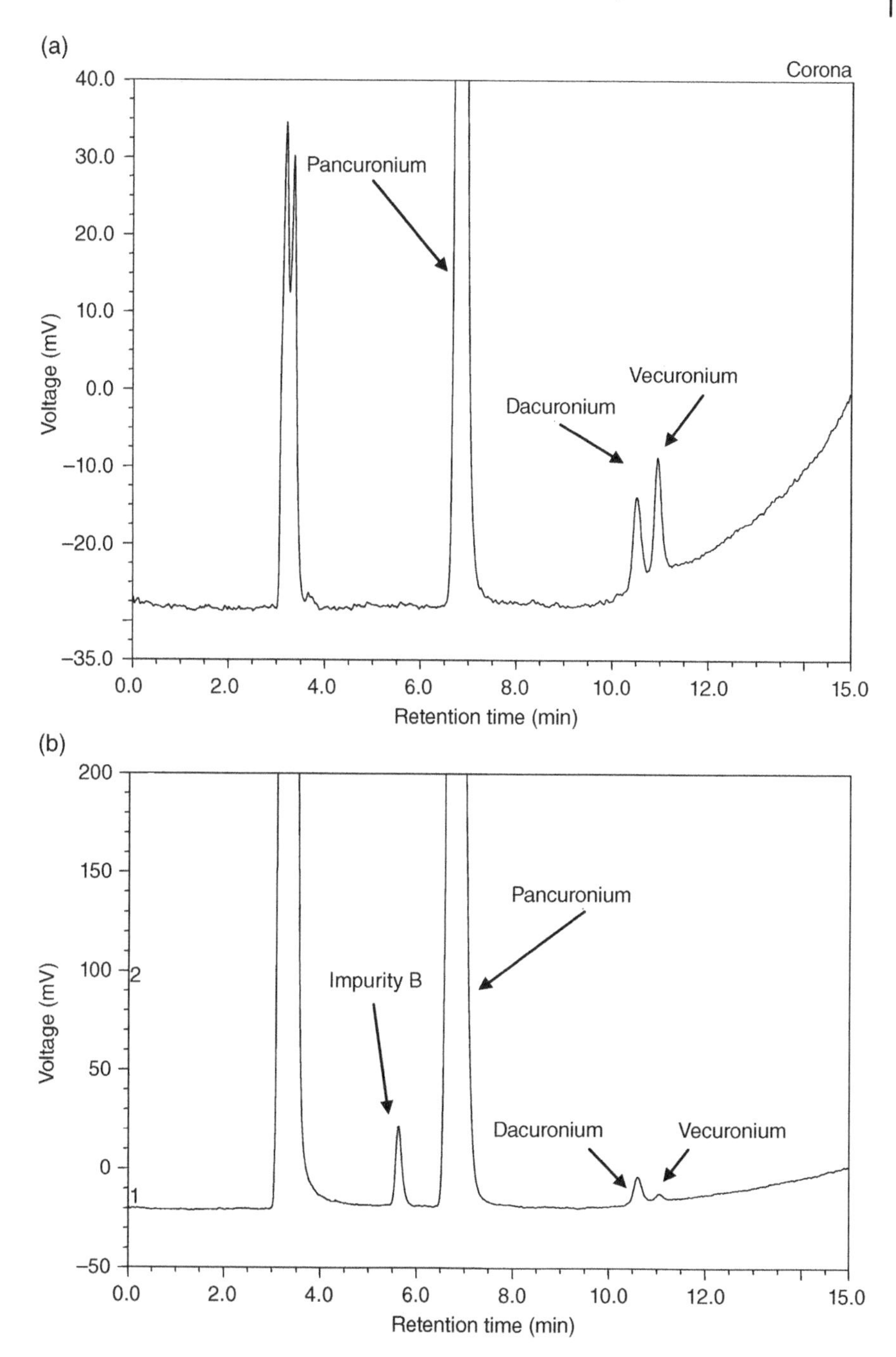

Figure 13.5 LC-CAD chromatogram recorded from the solutions containing 100 μg mL^{-1} pancuronium and 3 μg mL^{-1} of dacuronium and vecuronium in 0.1% TFA (a) and from the solution containing 800 μg mL^{-1} pancuronium from pharmaceutical preparation in 0.1% TFA (b).

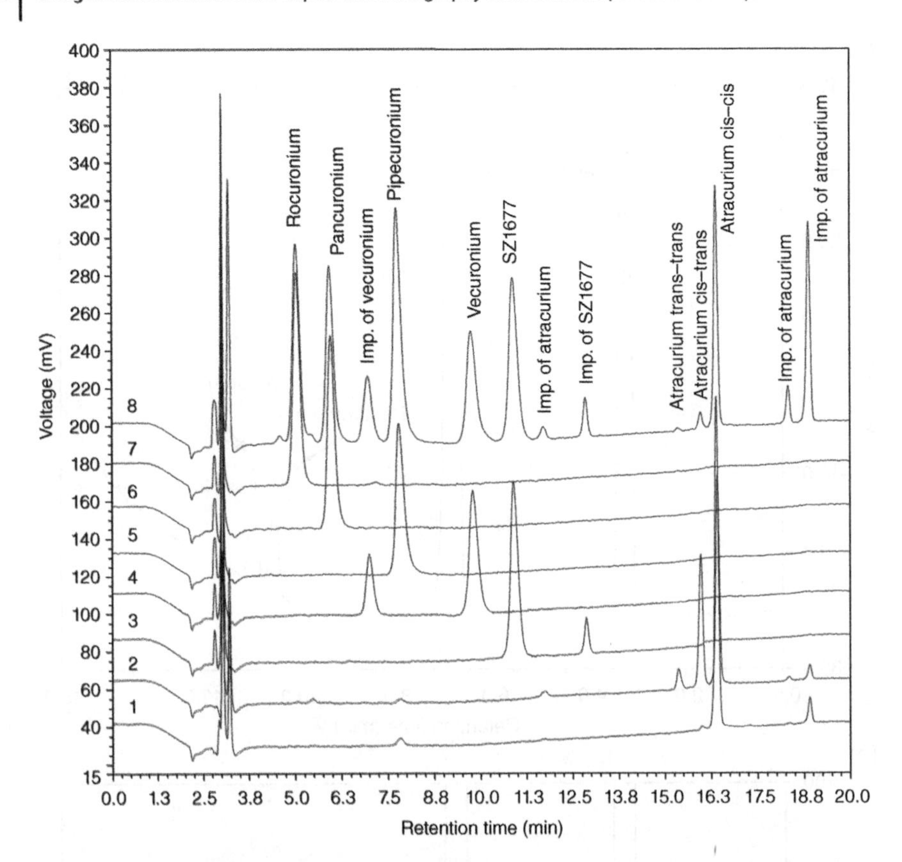

Figure 13.6 LC-CAD chromatograms recorded from the solution containing 100 μg mL^{-1} of: cisatracurium (1), atracurium (2), SZ1677 (3), vecuronium (4), pipecuronium (5), pancuronium (6), and rocuronium (7) and the mixture of seven muscle relaxants (8) in 01% TFA.

Except laudanosine, the other degradants of atracurium, cisatracurium, and mivacurium were not available for us to prepare standard stock solutions for calibration purposes. They could not, therefore, be quantitatively determined. Because of universal response of Corona CAD to almost all nonvolatile species, independently of their nature and spectral or physicochemical properties, even unidentified impurities or the analytes, for which no pure standards are available, can be quantified in isocratic elution.

13.4.3.1 Linearity

Linearity was estimated by analyzing atracurium, cisatracurium, laudanosine, pancuronium, vecuronium, and dacuronium standards. Several (at least five) concentrations of the analyzed substances ranging from 5 to 150 μg mL^{-1}

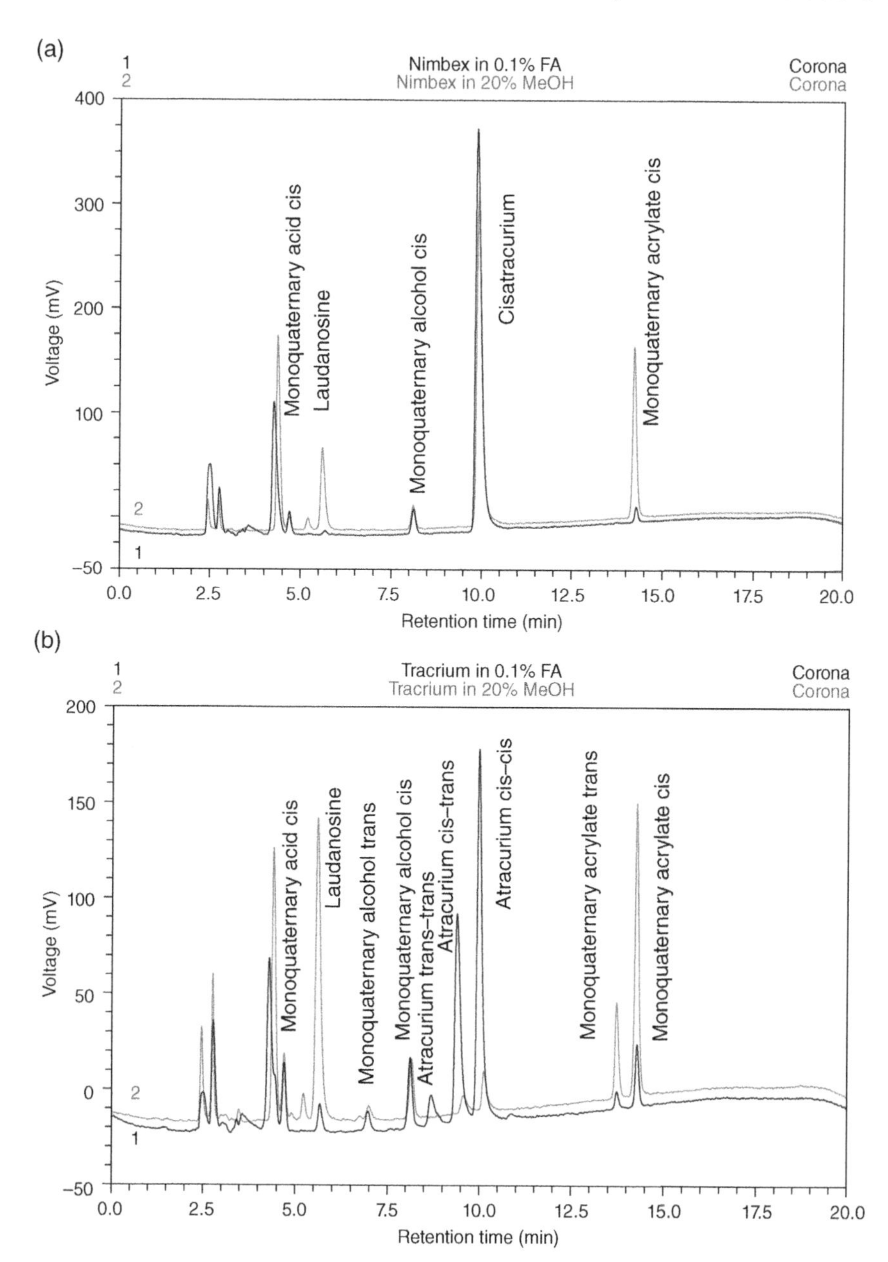

Figure 13.7 LC-CAD chromatograms recorded from pharmaceutical preparations containing ~100 μg mL^{-1} of atracurium from Tracrium (a), cisatracurium from Nimbex (b), and mivacurium from Mivacron (c) in 0.1% FA (black) and in 20% methanol (grey); injection volume 10 μL.

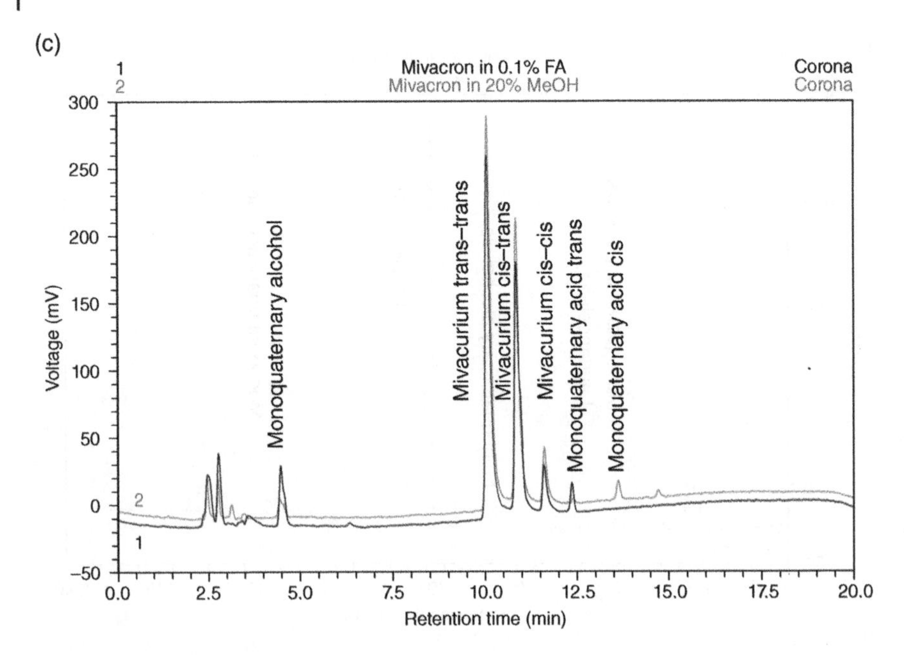

Figure 13.7 (Continued)

(for active substances), from 1 to 15 μg mL^{-1} for laudanosine, and from 1 to 10 μg mL^{-1} for vecuronium and dacuronium were used to obtain calibration curves. The 1 μg mL^{-1} concentration standard of atracurium and cisatracurium was excluded, because peak areas were lower than LOQ. A linear response was not expected since aerosol charging does not depend directly on the aerosol mass [37]. For cisatracurium the response of the Corona CAD was not linear (Table 13.2), but good linearity was obtained when the curve was plotted as a power function $y = ax^b$, where y is the response of the Corona CAD (peak area), x is the concentration of the sample, and a and b are coefficients that depend on droplet size, nature of solute, gas and liquid flow rates, molar volatility, etc. [38]. A double logarithmic coordinate system was used to obtain a linear calibration curve $\log y = \log a + b \log x$, with a good linear fit (Table 13.2).

Although it is known that Corona CAD response is nonlinear within the range of four orders of magnitude, we found that the signal is nearly linear in the examined ranges of other analyzed compounds [38].

13.4.3.2 Detection and Quantitation Limits

S/N ratio was determined by comparing measured signals from samples of known low concentrations of the analyte with those of blank samples and establishing the minimum concentration at which the analyte can be reliably detected. The detection limit (LOD) and quantitation limit (LOQ) were defined

Table 13.2 Linear and exponential fit equations for calibration curves for atracurium, cisatracurium, and laudanosine determined by LC-CAD.

		Linear		Power		Log–log	
	Range ($\mu g\,mL^{-1}$)	$y = ax + b$	R^2	$y = ax^b$	R^2	$Log\ y = b\ log\ x + log\ a$	R^2
Atracurium (three isomers)	1–150						
cis–cis		$0.3232x - 0.6894$	0.9986	$0.3740x^{0.98}$	0.9996	$0.9803x - 0.4366$	0.9996
cis–trans		$0.2158x - 0.0368$	0.9997	$0.1761x^{1.05}$	0.9991	$1.050x - 0.7542$	0.9991
trans–trans		$0.0381x - 0.7320$	0.9986	$0.0068x^{1.32}$	0.9961	$1.324x - 2.170$	0.9961
Cisatracurium	1–150	$0.5172x + 2.6031$	0.9926	$1.032x^{0.87}$	0.9995	$0.8725x + 0.0139$	0.9995
Laudanosine	1–10	$0.7362x - 0.0351$	0.9992	$0.6749x^{1.04}$	0.9986	$1.043x - 0.1708$	0.9986

Table 13.3 Limits of detection (LOD) and limits of quantitation (LOQ) for atracurium, cisatracurium, and laudanosine determined by LC-CAD.

				SD of the response and the slope					
		S/N		SD of the blank		Calibration curve		Experimentally verified	
	Range ($\mu g\,mL^{-1}$)	LOD	LOQ	LOD	LOQ	LOD	LOQ	LOD	LOQ
Atracurium (three isomers)									
cis–cis	1–150	1.496	4.987	0.647	1.961	0.780	2.363	1.0	3.0
cis–trans	5–150	3.912	13.04	1.460	4.425	1.209	3.662	2.0	6.0
trans–trans	50–150	19.50	65.01	10.81	32.75	5.139	15.57	15.0	50.0
Cisatracurium	1–15	0.660	2.198	0.387	1.172	0.779	2.360	0.8	2.5
Laudanosine	1–10	0.499	1.664	0.267	0.810	0.305	0.923	0.3	1.0

as S/N ratio of 3:1 and 10:1, respectively. The lowest LOD and LOQ were obtained for the impurity—laudanosine—and were equal to 0.50 $\mu g\,mL^{-1}$ (LOD) and 1.66 $\mu g\,mL^{-1}$ (LOQ). For pancuronium and its impurities, the lowest LOD and LOQ were obtained for vecuronium and were equal to 0.50 $\mu g\,mL^{-1}$ (LOD) and 1.64 $\mu g\,mL^{-1}$ (LOQ). Estimates of LOD and LOQ are presented in Table 13.3 for atracurium and cisatracurium and their impurities and in Table 13.4 for pancuronium and its impurities.

Detection and quantitation limits were also determined as 3.3 σ/S (LOD) and 10 σ/S (LOQ), where σ is the standard deviation (SD) of the response and S is the slope estimated from the calibration curve of the analyte. The estimation of σ was carried out in two ways: based on SD of the blank and based on the calibration curve. For the first of them, the measurement of the magnitude of analytical background response was performed by analyzing numerous blank samples and calculating the SD of these responses. For the other one, a specific calibration curve was studied using samples containing an analyte in the range of LOQ.

Several approaches for the determination of the detection and quantitation limits are possible [36], but the results obtained in our study vary (Table 13.3). The approach based on SD of the blank proved to be the most sensitive, while S/N approach was the least sensitive. The smallest LOD/LOQ values calculated for the approach based on SD of the blank could not be in most cases verified experimentally, so visual evaluation was done for justification and the data were close to those obtained from the S/N approach. 1 mV noise was detected in the study.

Table 13.4 Summarized validation parameters and determination of content for pancuronium and its impurities in PANCURONIUM *injection* 2 mg mL^{-1} by LC-CAD.

Parameters	Pancuronium	Dacuronium (impurity A)	Vecuronium (impurity D)
Concentration range (μg mL^{-1})	5–150	1–10	1–10
Regression equation	$y = 0.6304x + 3.2108$	$y = 0.6188x - 0.2041$	$y = 0.8997x - 0.3051$
R^2	0.9940	0.9948	0.9958
LOD	0.55	1.02	0.50
LOQ	1.82	3.42	1.64
Concentration found ± confidence level (μg mL^{-1})	(d) 81.34 ± 0.86	(a) 4.92 ± 0.24	(a) 4.85 ± 0.06
	(e) 99.07 ± 0.91	(b) 8.08 ± 0.18	(b) 8.00 ± 0.04
	(f) 114.41 ± 0.43	(c) 9.67 ± 0.24	(c) 10.51 ± 0.01
RSD intra-day (%)	(d) 0.430	(a) 1.940	(a) 0.546
	(e) 0.370	(b) 0.895	(b) 0.191
	(f) 0.151	(c) 1.002	(c) 0.055
RSD inter-day (%)	(e) 0.383	(a) 2.814	(a) 0.362
Recovery (%)	(d) 103.25	(a) 99.07	(a) 99.59
	(e) 99.16	(b) 103.43	(b) 101.12
	(f) 96.08	(c) 97.37	(c) 105.82
Content in pharmaceutical preparation ± confidence level (RSD)	2.10 ± 0.02 mg mL^{-1}* (0.74%)	0.36 ± 0.01%** (2.29%)	0.13 ± 0.01%** (1.38%)

Concentrations: (a) 5.0 μg mL^{-1}, (b) 8.0 μg mL^{-1}, (c) 10.0 μg mL^{-1}, (d) 80.0 μg mL^{-1}, (e) 100.0 μg mL^{-1}, and (f) 120.0 μg mL^{-1}.

* According to manufacturer's specification the limit of acceptance for pancuronium is 1.80–2.20 mg mL^{-1}, for dacuronium max 3.0%, and for vecuronium max 1.0%.

** Percentage of pancuronium content.

These results are in accordance with the literature [36], where the approach based on S/N was suggested to analytical procedures that exhibit constant baseline noise, that is, in chromatographic methods. In the case of spectrophotometric methods, LOD and LOQ are determined using the approaches based on SD.

13.4.3.3 Precision and Accuracy

Repeatability was assessed using three concentrations covering the specified range for the procedure. Precision was calculated from three consecutive injections for each concentration, and the observed RSD ranged from 0.10 to 1.60%

Table 13.5 Precision and accuracy for atracurium, cisatracurium, and laudanosine determined by LC-CAD.

Substance	Concentration added (μg mL^{-1})	Concentration found ± confidence level (μg mL^{-1})	SD	RSD intra-day (%)	RSD inter-day (%)	Recovery (%)
Atracurium (three isomers)						
cis–cis isomer	78.24	79.56 ± 0.20	0.444	0.558	0.808	101.63
	97.80	97.67 ± 0.58	0.233	0.239	0.539	99.86
	117.36	116.36 ± 0.44	0.179	0.154	0.315	99.06
cis–trans isomer	78.24	78.23 ± 1.87	0.751	0.960	1.034	100.80
	97.80	96.80 ± 1.62	0.652	0.673	0.526	100.00
	117.36	116.13 ± 1.21	0.487	0.419	0.355	99.84
trans–trans isomer	78.24	76.32 ± 3.04	1.224	1.604	0.340	96.50
	97.80	97.24 ± 3.41	1.373	1.412	1.756	99.23
	117.36	118.59 ± 4.21	1.694	1.429	2.092	100.25
Cisatracurium	77.60	78.12 ± 1.09	0.848	0.560	0.624	99.71
	96.99	95.84 ± 0.24	0.452	0.100	0.347	100.45
	116.39	116.91 ± 1.70	0.446	0.584	0.373	99.90
Laudanosine	2.99	3.07 ± 0.08	0.034	1.106	2.737	100.25
	4.98	5.18 ± 0.09	0.037	0.715	1.267	100.26
	7.97	8.06 ± 0.14	0.055	0.681	1.279	98.87

(Table 13.5) for atracurium, cisatracurium, and laudanosine and from 0.06 to 2.81% for pancuronium and impurities (Table 13.4). Intermediate precision was calculated from 2 days. The accuracy of the recovery was evaluated at three concentrations. The mean recoveries for all samples from each run were in the range of 96.5–101.6% for atracurium, cisatracurium, and laudanosine (Table 13.5) and in the range of 96.1–105.8% for pancuronium and impurities (Table 13.4).

13.4.3.4 Range

The calibration curves for the response of atracurium, cisatracurium, and pancuronium in the range from 80 to 120 μg mL^{-1} corresponding to 80–120% of the assay concentration level of 100 μg mL^{-1} were performed. Responses obtained in the examined range can be expressed by a linear equation $y = ax + b$ with good R^2 correlation coefficient values. It was confirmed that when the level is very low or when the range is small, the calibration curve is close to a linear curve [37].

13.4.4 Determination of Active Substances and Impurities in Pharmaceutical Preparations

0.8 mL of Nimbex solutions containing cisatracurium and 2.0 mL of Tracrium solutions containing atracurium were transferred into 20-mL volumetric flasks and diluted with 0.1% FA and were further successively diluted with 0.1% FA to the concentration of approximately 100 μg mL^{-1} of the active substance. For laudanosine determination, 5 mL of Nimbex and 1.0 mL of Tracrium were transferred into 10-mL volumetric flasks and diluted with 0.1% FA to the concentration of approximately 1 mg mL^{-1} of the active substance.

Thus prepared sample solutions were used for qualitative studies. In both determinations six prepared samples were used. Then 10 μL of solutions was introduced into the column, and the chromatograms were recorded for 20 min. The contents were determined by the calibration curve method. All data are summarized in Table 13.6. The cis–cis isomeric group constituted 58.28%, cis–trans 36.66%, and trans–trans 5.06% of the atracurium mixture.

The other impurities could not be determined due to lack of standards; however, the determination should be possible when using a second pump for mobile phase compensation.

Half milliliter of Pancuronium preparation was transferred into 10-mL volumetric flasks and diluted with 0.1% TFA and was further successively diluted with 0.1% TFA to the concentration of approximately 100 μg mL^{-1} of the active substance. For the determination of impurities, 2 mL of Pancuronium preparation was transferred into 5-mL volumetric flasks and diluted with 0.1% TFA to the concentration of approximately 0.8 mg mL^{-1} of the active substance.

Thus prepared sample solutions were used for quantitative studies. In both determinations six prepared samples were used. Then 10 μL of solutions was introduced into the column, and the chromatograms were recorded for 17 min. Three impurities were found in the pharmaceutical preparation of Pancuronium (Figure 13.5b). The assay and content of dacuronium and vecuronium were determined by the calibration curve method. The content of dacuronium and vecuronium was 0.36 and 0.13%, respectively, when pancuronium in pharmaceutical preparation was quantified at concentration of 2.10 mg mL^{-1}.

13.4.5 Stability

In vitro investigations in various buffers and plasma [2, 3] suggest that cisatracurium and atracurium undergo temperature- and pH-dependent Hofmann elimination, yielding laudanosine and a quaternary monoacrylate (Figure 13.1). Due to ester hydrolysis of atracurium and mivacurium, monoquaternary acids and alcohols are formed (Figures 13.1 and 13.2) as confirmed in our study. The analyzed active substances and pharmaceutical preparations were also dissolved in 20% methanol. The degradation was very fast. As we expected, for atracurium the contents of all possible degradants from Hofmann elimination

Table 13.6 Determination of active substances and impurities in pharmaceutical preparations by LC-CAD.

		Declared amount	Found amount	Percent of declared amount	SD	RSD (%)
TRACRIUM	Atracurium (three isomers)	$10.0\ mg\ mL^{-1}$	$10.83 \pm 0.05\ mg\ mL^{-1}$	108.35	0.044	0.407
	cis–cis isomer	55.0–60.0%	58.28 ± 0.17%		0.166	0.285
	cis–trans isomer	34.5–38.5%	36.66 ± 0.14%		0.134	0.366
	trans–trans isomer	5.0–6.5%	5.06 ± 0.10%		0.097	1.918
	Laudanosine	Max 1.0%*	0.50 ± 0.01%		0.334	0.677
NIMBEX	Cisatracurium	$2.68\ mg\ mL^{-1}$	$2.87 \pm 0.03\ mg\ mL^{-1}$	107.14	0.017	0.595
	Laudanosine	Max 1.0%*	0.56 ± 0.05%		0.108	0.969

* For substance atracurium besilate, not for pharmaceutical preparation.

and ester hydrolysis increased in Tracrium sample, whereas the content of atracurium sharply decreased (Figure 13.7a). In the case of cisatracurium (Nimbex), only the contents of laudanosine and monoquaternary acrylate slightly increased, while the content of cisatracurium decreased (Figure 13.7b). For the Mivacron sample, higher contents of monoquaternary acid and alcohol were observed (Figure 13.7c). The formation of monoquaternary acid, alcohol, and monoacrylate was estimated by comparing chromatographic peak areas.

It is crucial, therefore, to prepare the stock solutions of atracurium, cisatracurium, and mivacurium in a slightly acidic solution (pH = 3), in our study in 0.1% FA, to avoid acidic hydrolysis and Hofmann elimination. Thus prepared solutions were stable for at least 2 weeks when stored at +6°C (±2°C). After 2 months of storage the content of atracurium decreased by almost 10%, while the content of monoquaternary acid and alcohol slightly increased.

13.5 Conclusion

The LC-CAD method for the determination of the three isomers of atracurium, cisatracurium, and the three isomers of mivacurium with their impurities and the LC-CAD method for the determination of pancuronium and its impurities were developed to provide a sensitive, quantitative assay of active substances and their impurities in substance and in pharmaceutical preparations. The conditions were optimized so as to obtain the best signal and stability of the measurement with the highest sensitivity. The preparation of samples and their analysis were performed within a relatively short time. These elaborated methods for the analysis of atracurium and cisatracurium with their impurities and for the analysis of pancuronium and its impurities proved to be fast, precise, accurate, and sensitive and could be applied to routine analysis in substances and in pharmaceutical preparations. Mivacurium and all impurities were identified using time-of-flight mass spectrometry with electrospray ionization.

Acknowledgments

This work is based on two combining papers (with copyright permission):

- Blazewicz A, Fijalek Z, Warowna-Grzeskiewicz M, Jadach M. Determination of atracurium, cisatracurium and mivacurium with their impurities in pharmaceutical preparations by liquid chromatography with charged aerosol detection. J. Chromatogr. A, 2010; 1217; 1266–1272.
- Blazewicz A, Fijalek Z, Sarna K, Warowna-Grzeskiewicz M. Determination of pancuronium and its impurities in pharmaceutical preparation by liquid chromatography with charged aerosol detection. Chromatographia, 2010; 72; 183–186.

References

1 Kisor DF, Schmith VD. Clinical pharmacokinetics of cisatracurium besilate. Clin. Pharmacokinet., 1999; 36; 27–40.
2 Weindlmayr-Goettel M, Kress HG, Hammerschmidt F, Nigrovic V, In vitro degradation of atracurium and cisatracurium at pH 7.4 and 37°C depends on the composition of the incubating solutions. Br. J. Anaesth., 1998; 81; 409–414.
3 Weindlmayr-Goettel M, Gilly H, Kress HG. Does ester hydrolysis change the in vitro degradation rate of cisatracurium and atracurium? Br. J. Anaesth., 2002; 88; 555–562.
4 Atherton DP, Hunter JM. Clinical pharmacokinetics of the newer neuromuscular blocking drugs. Clin. Pharmacokinet., 1999; 36; 169–189.
5 Martinez MA, Ballesteros S, Almarza E. Anesthesiologist suicide with atracurium. J. Anal. Toxicol., 2006; 30; 120–124.
6 Kintz P, Tracqui A, Ludes B. The distribution of laudanosine in tissues after death from atracurium injection. Int. J. Legal Med., 2000; 114; 93–95.
7 Kłys M, Białka J, Bujak-Giżycka B. A case of suicide by intravenous injection of pancuronium. Leg. Med., 2000; 2; 93–100.
8 Andresen BD, Alcaraz A, Grant PM. The application of pancuronium bromide (Pavulon) forensic analyses to tissue samples from an "Angel of Death" investigation. J. Forensic Sci., 2005; 50; 215–219.
9 Maeda H, Fujita MQ, Zhu BL, Ishidam K, Oritani S, Tsuchihashi H, Nishikawa M, Izumi M, Matsumoto F. A case of serial homicide by injection of succinylcholine. Med. Sci. Law, 2000; 40; 169–174.
10 Fernández R, Bello MA, Callejón M, Jiménez JC, Guiraúm A. Spectrofluorimetric determination of cisatracurium and mivacurium in spiked human serum and pharmaceuticals. Talanta, 1999; 49; 881–887.
11 Torres RF, Mochón MC, Sánchez JCJ, López MAB, Pérez AG. Electrochemical oxidation of cisatracurium on carbon paste electrode and its analytical applications. Talanta, 2001; 53; 1179–1185.
12 Bjorksten AR, Beemer GH, Crankshaw DP. Simple high-performance liquid chromatographic method for the analysis of the non-depolarizing neuromuscular blocking drugs in clinical anaesthesia. J. Chromatogr. B, 1990; 533; 241–247.
13 Lugo SI, Eddington ND. Rapid method for the quantitation of mivacurium isomers in human and dog plasma by using liquid chromatography with fluorescence detection: Application to pharmacokinetic studies. J. Pharm. Biomed. Anal., 1996; 14; 675–683.
14 Lindon C, Nicholson JK, Wilson ID. Directly coupled HPLC-NMR and HPLC-NMR-MS in pharmaceutical research and development. J. Chromatogr. B, 2000; 748; 233–258.

15 Cirimele V, Villain M, Pepin G, Ludes B, Kintz P. Screening procedure for eight quaternary nitrogen muscle relaxants in blood by high-performance liquid chromatography-electrospray ionization, mass spectrometry. J. Chromatogr. B, 2003; 789; 107–113.
16 Yiu KCH, Ho ENM, Wan TSM. Detection of quaternary ammonium drugs in equine urine by liquid chromatography—mass spectrometry. Chromatographia, 2004; 59; 45–50.
17 Ariffin MM, Anderson RA. LC/MS/MS analysis of quaternary ammonium drugs and herbicides in whole blood. J. Chromatogr. B, 2006; 842; 91–97.
18 Lacroix M, Tu TM, Donati F, Varin F. High-performance liquid chromatographic assays with fluorometric detection for mivacurium isomers and their metabolites in human plasma. J. Chromatogr. B, 1995; 663; 297–307.
19 Biederbick W, Aydinciouglou G, Diefenbach C, Theisohn M. Stereoselective high-performance liquid chromatographic assay with fluorometric detection of the three isomers of mivacurium and their *cis*- and *trans*-alcohol and ester metabolites in human plasma. J. Chromatogr. B, 1996; 685; 315–322.
20 Ferenc C, Audran M, Lefrant JY, Mazerm I, Bressolle F. High-performance liquid chromatographic method for the determination of atracurium and laudanosine in human plasma. Application to pharmacokinetics. J. Chromatogr. B, 1999; 724; 117–126.
21 Reich D, Hollinger I, Harrington D, Seiden H, Chakravorti S, Cook R. Comparison of cisatracurium and vecuronium by infusion in neonates and small infants after congenital heart surgery. Anesthesiology, 2004; 101; 1122–1127.
22 Welch RM, Brown A, Ravitch J, Dahl R. The in vitro degradation of cisatracurium, the R, cis-R′-isomer of atracurium, in human and rat plasma. Clin. Phamacol. Ther., 1995; 58; 132–142.
23 Kerskes CHM, Lusthof KJ, Zweipfenning PGM, Franke JP. The detection and identification of quaternary nitrogen muscle relaxants in biological fluids and tissues by ion-trap LC-ESI-MS. J. Anal. Toxicol., 2002; 26; 29–34.
24 Sayer H, Quintela O, Marquet P, Dupuy JL, Gaulier JM, Lachâtre G. Identification and quantitation of six non-depolarizing neuromuscular blocking agents by LC-MS in biological fluids. J. Anal. Toxicol., 2004; 28; 105–110.
25 Ballard KD, Vickery WE, Nguyen LT, Diamond FX. An Analytical strategy for quaternary ammonium neuromuscular blocking agents in a forensic setting using LC-MS/MS on a tandem quadrupole/time-of-flight instrument. J. Am. Soc. Mass Spectrom., 2006; 17; 1456–1468.
26 Zhang H, Wang P, Bartlett MG, Stewart JT. HPLC determination of cisatracurium besylate and propofol mixtures with LC-MS identification of degradation products. J. Pharm. Biomed. Anal., 1998; 16; 1241–1249.
27 European Pharmacopoeia Supplement 7.1 to the 7th Edition, Monograph 01/2008:1970, Council of Europe, Strasbourg, France, 2011.

28 European Pharmacopoeia Supplement 7.1 to the 7th Edition, Monograph 01/2008:0681, Council of Europe, Strasbourg, France, 2011.

29 Zecevic M, Zivanovic L, Stojkovic A. Validation of a high-performance liquid chromatography method for the determination of pancuronium in Pavulon injections. J. Chromatogr. A, 2002; 949; 61–64.

30 Lopez Garcia P, Gomes F, Santoro M, Hackmann E. Validation of an HPLC analytical method for determination of pancuronium bromide in pharmaceutical Injections. Anal. Lett., 2008; 41; 1895–1908.

31 Nishikawa M, Nishioka H, Katagi M, Tsuchihashi H. Analysis of quaternary ammonium neuromuscular blocking agents, pancuronium and vecuronium by ESI-LC/MS. Jpn. J. Forensic Toxicol., 1999; 17; 116–117.

32 Usui K, Hishinuma T, Yamaguchi H, Saga T, Wagatsuma T, Hoshi K, Tachiiri N, Miura K, Goto J. Simultaneous determination of pancuronium, vecuronium and their related compounds using LC-ESI-MS. Leg. Med., 2006; 8; 166–171.

33 Blazewicz A, Fijalek Z, Warowna-Grzeskiewicz M, Boruta M. Simultaneous determination of rocuronium and its eight impurities in pharmaceutical preparation using high-performance liquid chromatography with amperometric detection. J. Chromatogr. A, 2007; 1149; 66–72.

34 Blazewicz A, Fijalek Z, Samsel K. Determination of pipecuronium bromide and its impurities in pharmaceutical preparation by high-performance liquid chromatography with coulometric electrode array detection. J. Chromatogr. A, 2008; 1201; 191–195.

35 Blazewicz A, Fijalek Z, Warowna-Grzeskiewicz M, Banasiuk J. Application of high-performance liquid chromatography with amperometric and coulometric detection to the analysis of SZ1677, a new neuromuscular blocking agent, and its two derivatives. J. Chromatogr. A, 2008; 1204; 114–118.

36 International Conference on Harmonization (ICH), Topic Q2 (R1): Validation of Analytical Procedures: Text and Methodology, EMEA, Geneva, 2005, www.ich.org (accessed March 6, 2017).

37 Gamache PH, McCarthy RS, Freeto SM, Asa DJ, Woodcock MJ, Laws K, Cole RO. HPLC analysis of nonvolatile analytes using charged aerosol detection. LCGC North Am., 2005; 23; 150–155.

38 Forsatz B, Snow NH. HPLC with charged aerosol detection for pharmaceutical cleaning validation. LCGC North Am., 2007; 25; 960–968.

14

Charged Aerosol Detection of Scale Inhibiting Polymers in Oilfield Chemistry Applications

Alan K. Thompson

Nalco Champion, an Ecolab Company, Aberdeen, UK

CHAPTER MENU

14.1 Summary

The use of HPLC with charged aerosol detector (CAD) has opened a new field in the area of polymeric scale inhibitor analysis at residual concentrations. The sensitivity and robustness of the detector has allowed analysis to be performed that was not possible using the classical polymeric scale inhibitor analytical techniques. To be able to "see" the polymer of interest reliably at sub 10 ppm levels has taken the analysis into new areas over the past 5 years and has sparked a large amount of new developmental work in the area of instrumental analysis of polymeric scale inhibitor.

The new developments are going in two related but different directions. The first is to improve the chromatographic part of the analysis. The use of GPC columns will always run up against the limiting factor that similar sized materials will elute at similar retention times, thus limiting the use of the HPLC-CAD to the resolution of the closest peak. The use of more novel HPLC columns will likely widen the use of the system as better chromatographic separations are obtained.

Charged Aerosol Detection for Liquid Chromatography and Related Separation Techniques, First Edition. Edited by Paul H. Gamache.

The use of polymeric scale inhibitor is widely prevalent in oilfield chemistry, but in absolute terms, these are dwarfed by the use of corrosion inhibitors. Unlike polymeric scale inhibitors that tend to be a binary or ternary mix of materials, corrosion inhibitors can be a complex mix of chemistries. These can be analyzed by dye extraction techniques and by LCMS, but there is a need for methods that sit between the two, more specific and sensitive than the dye transfer but simpler and cheaper than the LCMS. The use of HPLC-CAD has been investigated and shown to have potential utilization in this area.

Note: It is anticipated that not all the readers will be familiar with the offshore oil and gas industry nor the use of chemicals in that industry. The following text is therefore recommended for further reading: *Production Chemicals for the Oil and Gas Industry*, Malcolm Kelland, CRC Press, Boca Raton, FL, 2009.

14.2 Background to Scale Inhibition in Oilfields

14.2.1 General Background

Two hundred and fifty million years ago, the earth was warmer and more humid than today. Great forests grew around inland seas, and the land and sea teemed with life; this was the age of the dinosaur, the Jurassic and Triassic periods in the earth's history, but it was also the beginning of the creation of the most valuable natural resources the earth has ever seen—fossil fuels: coal, oil, and natural gas.

As the different forms of life died, they collected in the sea and river beds. Over geological time frames these beds of organic matter disappeared under mud and sand creating pressure and therefore temperature. In the pressure cooker environment, the organic matter underwent a chemical change to form solid, liquid, and gaseous hydrocarbons, which became better known as coal, oil, and natural gas. Coal is ignored for the purposes of this chapter but is of vital importance to the economies of the world; instead we will concentrate on the oil and natural gas formations (hydrocarbons). As the Pangea continents moved into their current places over millennia of time, the hydrocarbons moved with them. In some areas, as the rock formations failed to trap the hydrocarbons, they evaporated and were lost to the atmosphere, while in other areas, the hydrocarbons became trapped under sheets of impervious rock strata resulting in today's oilfields. These fields can occur anywhere on earth where suitable rock formations exist and have been identified from Greenland to the Falkland Islands and from the Gulf of Mexico to the Sakhalin Islands, Vietnam, India, Angola, Sudan, Australia, Austria, Romania, and, of course, the regions that most people identify first with hydrocarbon production, the middle east and north sea regions.

Some of these fields have been exploited by man for many centuries: in Arabia "naphtha" has been used as a fuel and light source since before recorded history, while Marco Polo wrote of oil lamps being used in Baku on the shores of the Caspian Sea during his voyages along the old Silk Road to China in 1272. However, it was not until midway through the nineteenth century that hydrocarbons became universally important.

In 1859, a well was drilled at Titusville, Pennsylvania, USA by Edwin Drake; this was the first recorded well drilled deliberately for oil and sparked the worldwide exploration and production that is still ongoing today. The century from 1860 to 1960 gave the world the scientific and technical basis for extracting oil from onshore oilfields and from fields close to shore. It gave the world the old, familiar oil company names, Shell, British Petroleum (BP), Esso, Elf, Mobil, and Exxon, and bound the first world countries to a dependence on oil that continues to this day.

The North Sea is a shallow sea, connected to the Atlantic Ocean by the English Channel to the south, the Norwegian Sea to the north, and the Baltic Sea by the Kattegat. Approximately 600 miles long by 360 miles wide, it principally lies between the UK and the Scandinavian countries of Norway and Denmark but is also bounded by Germany and the Netherlands to the south and Shetland to the north and west.

Oil and gas have been produced around the edges of the North Sea since 1851 when the oil shale in Scotland was mined and the oil extracted by James "Paraffin" Young, but the evaluation of the North Sea itself became more urgent in 1964 following a seismic exploration by BP. A number of commercial gas fields were developed through the 1960s (West Sole, Hewitt, and Indefatigable), but the real explosion happened when the Ekofisk field in the Norwegian sector was discovered by Phillips in 1969. In 1970, BP discovered the Forties field, while the Brent field followed in 1971. Commercial hydrocarbons were first produced from the Forties field in 1975. Over the next 25 years, millions of tonnes of oil have been produced not only from the North Sea, primarily from the UK and Norwegian sectors, but also from the Danish, Dutch, and German sectors, from in excess of 150 fields.

Oil is a complex mix of hydrocarbons, from gaseous components such as methane and ethane through the commercially important low carbon number liquids to tars and napthanates, the ratio of which changes from field to field. The changing chemical nature of the produced fluids requires a complex mix of production chemicals to make production safe and reliable, while the environmental conditions under which the oil platforms are used require a further mix of chemicals to prevent corrosion of the metalwork and to keep the fluids liquid even in the depth of a north sea winter storm where temperatures can fall below −10°C, even without the wind chill factor. These chemicals include corrosion inhibitors, wax inhibitors, defoaming chemicals, and demulsifying chemicals. All of which are discrete chemical families with their own requirements for supply, storage, usage, and analysis.

In this chapter, however, we are only concerned with the problem of scale. Scale is important because, when oil and gas are produced from a reservoir, water is also produced. This water can come from the oilfield formation rock (connate water) or can be a result of water injection (generally seawater in the North Sea) to stimulate oil production. This connate water is always saturated with inorganic ions, sodium and chloride with smaller amounts of calcium, magnesium, barium, and so on, and can actually be supersaturated as it is held under high pressure and high temperature conditions in the oilfield rock formation. As the connate water cools, reduces in pressure, and mixes with the injected water, there is a significant risk of the formation of inorganic scale, generally barium sulfate and calcium carbonate in the North Sea environments. This scale can form in large amounts over a short time period and can shut down production as safety valves fail and pipes block. The classical way to control the formation of scale is to have a regular and ongoing squeeze program that uses a chemical to prevent the formation of scale.

14.2.2 Squeeze Programs

A classical squeeze program is designed to provide the maximum period of protection from scale to the well within the operational constraints of the platform.

A squeeze program is based on developmental lab work that allows the rate of absorption of the inhibitor to a typical rock surface (a short core of rock from the original well drilling is used), the rate of desorption from the same rock surface, and the minimum effective concentration of the inhibitor to be calculated. These calculations result in an isotherm that plots the fall in theoretical concentration of the inhibitor against water volume produced. This is then extrapolated by commercial computer programs, for example, Squeeze VI© to take into account the well depth, the well orientation, the perforation intervals of the well, the temperature and pressure of the well, and the permeability of the different types of rock from which the well produces oil. The final outcome is a squeeze design that specifies the amounts of chemical and other fluids used that are needed to give the desired protection to the well.

Performance of the squeeze protocol requires that production is halted from the well and a "pre-flush" package is pumped down the well bore and out into the rock formation. This pre-flush package is generally an organic solvent, for example, diesel, which cleans and dries the surface of the rock prior to the scale inhibitor being applied. Once the pre-flush is complete, the "main pill" is pumped down the well in the same manner. The main pill is a solution of the desired scale inhibitor, generally in an aqueous solvent, for example, seawater. The volume pumped has been calculated to give the desired lifetime to the squeeze program. Following the main pill treatment, an "overflush" is pumped down the well. The purpose of this is to drive the scale inhibitor the desired distance into the rock formation, typically 1–2 m. The well is then "shut in" for a period of 12–24 h; this period where the fluids are not moving gives the scale inhibitor time to adsorb to the rock formation and is calculated from the isotherms derived from the theoretical work performed previously. At the end of the shut in period, the well is slowly brought back into production. Samples of the produced fluids are taken after regular time intervals, typically every hour for the first 24 h of production then in a gradually expanding frequency until they are being taken once per month as the squeeze extends toward its lifetime. These samples require analysis for the residual inhibitor concentration present and also for the inorganic ions in solution. The residual inhibitor data is plotted against water production from the well to give a return curve, which indicates when the well requires to be re-squeezed.

A typical squeeze lifetime would be 6–12 months, but the lifetime is unique to the operational parameters of the well, which will include the ratio of water to oil being produced, the uniformity of the rock formation, the severity of the scaling risk, and the minimum effective concentration of the particular scale inhibitor used in the application. An example return curve is given in Figure 14.1 (plotted against time rather than water production), which illustrates some of the technical challenges in the field of residual analysis.

The concentration range of the inhibitor can be, typically, from close to 100,000 ppm (10%) down to sub 10 ppm, resulting in the need for analytical methodology with a dynamic range covering 4 orders of magnitude.

The concentrations of less than 10 ppm require a very sensitive detection method to accurately and reliably quantify polymers at these concentrations in complex matrices.

The sample concentrations may vary considerably with time even once the plateau region of the return curve has been reached (<100 ppm) due to the nonhomogeneous nature of the rock formation into which the chemical has been pumped; different rock structures will adsorb and desorb the chemical at

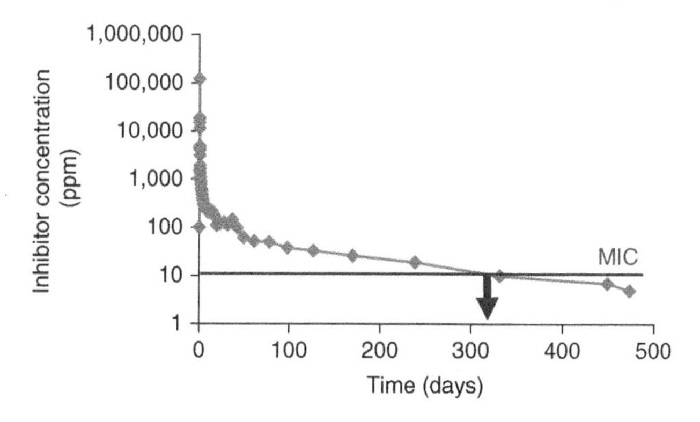

Figure 14.1 Example scale inhibitor return curve.

different kinetic rates. If the rock formation has fractures, then some chemical will not return and will be lost to the rock formation. In classical squeeze technology, it is estimated that approximately 30% of the chemical will return immediately due to not being adsorbed to the rock face, 30% will be lost to the rock formation, and 40% will return as the effective squeeze lifetime.

The chemicals used for scale prevention by squeezing must conform to a number of stringent requirements. These include that they must be stable under the temperature and pressures experienced in the well and rock formation, they must be effective scale inhibitors at parts per million levels, they must be relatively cheap to produce and deploy, they must be able to be monitored at parts per million levels in oilfield-produced fluids, and they must be relatively nontoxic to the environment. Unsurprisingly, there are a limited number of candidates that fill all these requirements, the two major classes being phosphonate esters and a small number of polymeric materials.

The phosphonate esters have been used in offshore scale squeeze programs for many years. They are both effective inhibitors of common scale and relatively cheap to manufacture and use. In recent years however, as environmental legislation has tightened (post ~2000) and concern as to the environmental impact of this class of chemical has increased, their use has rapidly declined in the North Sea, particularly in the Norwegian sector, and they are now only employed when there is an overriding operational reason to do so. Their mode of action and their analysis, typically by ICP-OES, are both ignored for the purpose of this chapter as we will be concentrating on Charged Aerosol Detection, which is particularly useful for the polymeric scale inhibitors rather than the phosphate esters, but a significant body of literature exists on the properties and analysis of phosphonate esters down to sub ppm levels [1–7].

14.2.3 Polymeric Inhibitors

All commercially produced polymeric scale inhibitors have a number of similarities in their chemical make-up and structural parameters. The polymer backbone is generally a nonpolar polymer with polar groups suspended from the backbone. These polar moieties are generally phosphino, sulfonate, or carboxyl groupings, but more recently, other chemistries such as amino groups have been employed to give different modes of operation and thus different specificities in their applications. The mode of action of the polymeric scale inhibitors is primarily crystal growth inhibition where the presence of the polymer inhibits the growth of the scale crystal on one or more faces, thus leading to deformed crystals and a limit to the amount of scale formed from the available inorganic ions.

Typical structures of some of the polymeric scale inhibitors that are used as part of downhole scale control squeezes are given in Figures 14.2, 14.3, 14.4, and 14.5.

In each case the molecular weight envelope of the polymeric scale inhibitor is in the 3–9000 MW range with the associated polydispersity of each type of polymer being a function of the synthetic pathway, the purity of the monomers used, and the extent of the purification processes involved. Literature suggests that the PVS and VS-Co type polymers are most suited to the prevention of barium sulfate scale, while the PAA type polymers are more suited to the prevention of Carbonate scale, for example, calcium carbonate, but the application aspects of polymeric scale inhibitors are not covered here. Other polymers are available that are good preventers of scale but are not suitable for squeeze applications; these are generally used in "topside" applications where they can be applied at a constant dose and are not required to adsorb to a rock formation. These polymeric scale inhibitors may still require analysis but not generally at the residual parts per million levels of the residual squeeze inhibitor polymers.

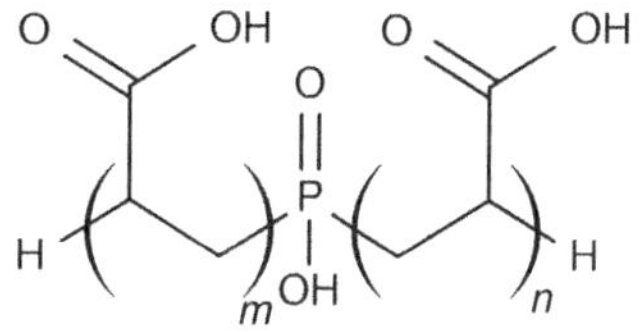

Figure 14.2 Polyphosphino polycarboxylic acid (PPCA).

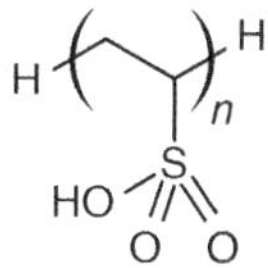

Figure 14.3 Polyvinyl sulfonate (PVS).

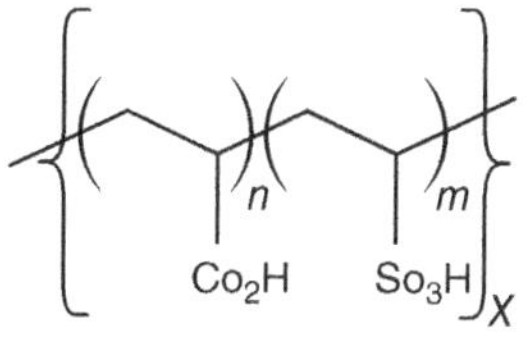

Figure 14.4 Sulphonated polyacrylic acid copolymer (VS-Co).

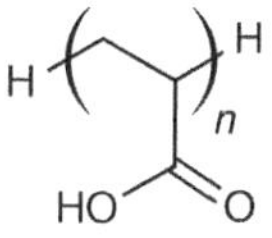

Figure 14.5 Polyacrylic acid (PAA).

14.3 Historical Methods of Analysis

Phosphorous inductively Coupled Plasma (P-ICP) is the classical method of analysis used to quantify phosphonate ester scale inhibitors; the methods are well characterized, accurate, and specific with a lengthy list of literature available to investigate the technique further [1–5]. When the phosphonate ester scale inhibitors began to be replaced with polymeric scale inhibitors, P-ICP continued to be used, as many of the polymeric scale inhibitor had concentrations of phosphorous designed in as part of the polymer to aid analytical detection using this technique.

The continued presence of phosphorous in scale inhibitors was regarded as environmentally undesirable, therefore manufacturers developed and commercialized polymeric scale inhibitors with little, or no, phosphorous present. This presented a challenge to the analytical chemists of the oilfield chemistry industry as a different type of analysis was required for these materials.

The initial chosen method of analysis was Hyamine 1622, which is routinely used within the industry to quantify polymeric scale inhibitor species. In clean water systems the Hyamine 1622 method is demonstrably accurate and precise down to low ppm levels of polymer.

However, in samples of produced water from real oilfields, both the P-ICP and the Hyamine 1622 methods suffer from an unavoidable deficiency in that neither method is specific for the different polymeric species or even for polymers as a chemical species.

P-ICP will detect and quantify all phosphorous present in the sample being analyzed, without differentiating the source of the phosphorous. In oilfield-produced waters, there are a number of potential sources of phosphorous including background phosphorous from the rock formation and phosphorous from other production chemicals such as corrosion inhibitors and topside scale inhibitors. The assumption is made during analysis that all the phosphorous quantified in the sample is attributable only to the polymeric scale inhibitor, which, in some instances, can give an erroneous data set that overestimates the polymeric scale inhibitor residuals and therefore the efficacy and lifetime of the scale inhibitor squeeze treatment.

Hyamine 1622 is a quaternary amine salt that will form a complex and give a turbid solution, that is, a response, with any anionic material that reacts with the quaternary amine species. This turbidity is then measured by ultraviolet (UV) Spectroscopy to quantify the polymeric scale inhibitor present, the assumption being made that all of the observed turbidity is a result of the polymeric scale inhibitor. The response of the amine, however, is not limited to the polymer but can include other polymeric species, other species, which can carry a negative charge, and, most critically in brine samples, the chloride ions. This nonspecificity of response demands that a complex clean-up procedure is used when field samples are analyzed by Hyamine 1622 for residual polymeric inhibitor to remove chloride ions (typically at >20,000 ppm) in the presence of less than 20 ppm of inhibitor.

The subsequent detection of the complex species by UV Spectroscopy is not straightforward: the Hyamine 1622 complex exhibits a third order polynomial response rather than a linear response, which gives rise to inaccuracies in the calibration and subsequent determination of sample results.

A typical Hyamine 1622 calibration graph is given in Figure 14.6.

As both historical method types had serious defects, considerable investigative and developmental work was performed in the industry to develop and validate chromatographic methods of analysis for the polymeric scale inhibitors.

Classical HPLC has many applications in many industries but, up until 2005, was little used in the oilfield chemicals industry for the quantification of scale inhibitors, primarily due to the lack of a suitable detector for the polymeric scale inhibitors. Classical HPLC technology utilizes UV detectors as the

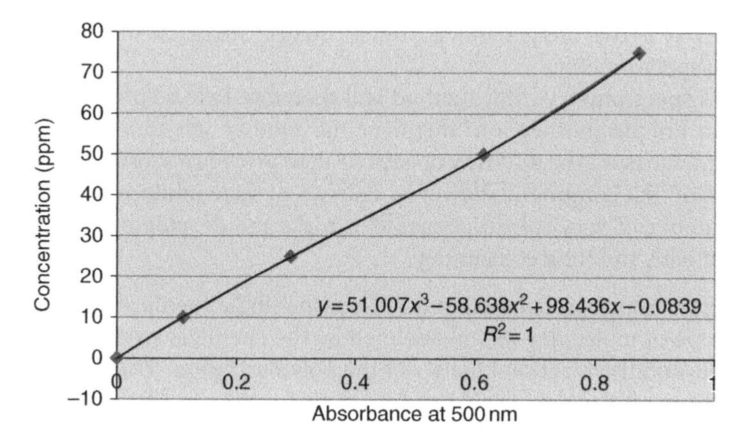

Figure 14.6 Typical Hyamine 1622 calibration curve.

primary detection systems (either as single wavelength detectors or as diode array detectors) with typical options for refractive index, electrochemical detection, and Mass Spectrometric detection.

The use of UV detection for the quantification of polymeric scale inhibitor has major problems in that the majority of the polymeric scale inhibitors used in the oilfield chemistry industry do not have strong chromophores, and the chromophores that they do have tend to be at low wavelengths (<210 nm). Most oilfield chemical samples are also contaminated, to a greater or lesser degree, with crude oil and other oilfield chemicals, thus the background absorbance from the samples frequently swamped the signal from the polymeric scale inhibitor even at relatively high concentrations. At low concentrations (<100 ppm), most polymeric scale inhibitors are undetectable by UV detection.

Refractive index detection is useful for the detection of polymers at high concentrations in very clean samples, but in samples that contain salt or low levels of the polymeric scale inhibitor, the detection levels for the technique are too high to be useful.

The use of HPLC-MS for the quantification of polymeric scale inhibitors is a complex area; in simple terms, the most readily available type of mass spectrometer works by creating an ion from the analyte then measuring the mass/charge (m/z) ratio of that ion. Specificity can be obtained either by the straightforward intensity measurement of the m/z (usually referred to as the mass as the charge is generally 1) or by applying a second voltage and creating a fragment (daughter ion) from the original ion. This mother–daughter transition is very specific for individual compounds and is accepted in most types of analysis requiring high levels of proof of identification, for example, forensics and pharmaceutical metabolite studies. The matrix that the sample is in does not usually interfere as the ions generated from the matrix are below the detection limit of the spectrometer.

An ideal mass spectrometry (MS) method will therefore take a compound, create a pseudo unique mother ion, fragment the mother ion to a unique daughter ion, and measure the intensity of each peak to provide a sensitive and specific analysis of that compound; however, analyses of scale inhibiting polymers have a number of very specific issues that make the development of a method very difficult and time consuming.

Issue 1: Polymers are, by their chemical nature, not one single chemical species but an envelope of molecular weights defined by the chemical structure of the monomers and the reaction kinetics of the polymerization. This means that any one single molecular weight species is only a small percentage of the total polymer, thus reducing the sensitivity of any method that isolates single masses. In analytical terms this means that no one single ion is generated by the mass spectrometer but instead numerous ions can be observed, each with a low intensity.

Issue 2: Multiple charges. Unlike "small" molecules, polymers can carry more than one electronic charge spread over the polymer moiety. In practice, this can result in changes in apparent mass values (as described before, what is measured is mass/charge), making the identification of which peaks are from the polymer very confusing and difficult.

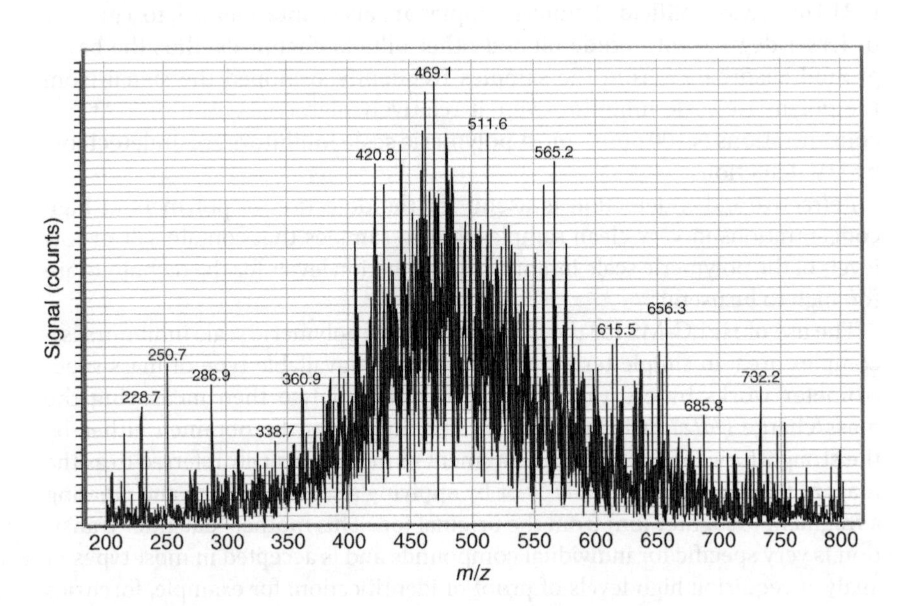

Figure 14.7 Typical mass spectrum of a polymeric scale inhibitor.

Issue 3: Difficulty in ionization. Ionization is a dynamic process rather than a fixed one. Areas that can influence ionization include the spectrometer voltages, flow rates, matrix effects and mobile-phase effects plus the chemical structure of the species being ionized (polymers are particularly difficult). The mobile phase can suppress, enhance or have no effect on the ionization potential.

Figure 14.7 demonstrates the complexity of the mass spectra of a polymeric scale inhibitor that are obtained during this type of analysis.

As a result of the issues and complexities of the analysis, LCMS is of little use in the oilfield chemistry industry to quantify scale inhibiting polymers. Some very recent work indicates that some of the issues are in the process of being solved, but it may be some time yet before the technique is used for routine analysis [8–10].

14.4 Charged Aerosol Detection for Polymeric Scale Inhibitors

14.4.1 Theoretical Application of CAD

As described in Section 14.3, analysis of polymeric scale inhibitors is made difficult by the structure of the polymers; the majority of the polymeric materials used in the oilfield industry for scale inhibition are not heavily functionalized with functional groups; thus detection by classical HPLC UV detectors is often not practical. The low concentrations required (<10 ppm) and the presence of brine salts and other oilfield chemicals make use of classical polymer detection systems such as Refractive Index detection unreliable. Use of MS in the field of oilfield scale inhibitor analysis is a developing field but is restricted by the chemical structure of the polymers, the polydispersive nature of the formulations used, and the cost of the instrumentation.

These difficulties did not remove the need for the analysis but opened the door to the application of a novel detection technique, charged aerosol detector (CAD). In principle, CAD will detect any chemical entity that will accept and carry a charge, thus removing the requirement that the analyte has a chromophore. The CAD has, in principle, a very broad dynamic range and a very sensitive limit of detection, both of which are important for the proposed application. This detection is independent of the chemical class that is being analyzed, which, in this specific application, allows the possibility of CAD giving a response to many of the specific polymeric scale inhibitors without having to tune the detector for each individual polymer as would have been required for UV detectors or MS detection. This universality of detection was an important consideration in the application as it allowed the possibility of generic methods, with all the time, resource, and efficiency advantages inherent in that, rather than a specific method for each polymeric scale inhibitor with the requirement to optimize and stabilize the HPLC system between each method.

The hardware of the detector mixes the flowing HPLC mobile phase with a dried gas stream, which causes the liquid flow to be nebulized and dried. The resulting particles are then charged and pass into a collector where a sensitive electrometer detects the charge and produces a signal that can be read by the data handling system.

The intended application of the analysis of scale inhibiting polymers appeared to fit the technical aspects of the detector; the samples being analyzed in routine applications are aqueous based; therefore a reverse-phase HPLC method was applicable to both the samples and the detector. The polymeric scale inhibitors can carry a charge due to the functionality of the polymer and therefore should show a response to the detector at low concentrations, and the polymers are not volatile and therefore should not be lost to the waste during the nebulizing process, thus maximizing the potential sensitivity of the application.

14.4.2 Practical Application of CAD

The use of CAD as a detection system was pioneered by Champion Technologies from 2006 onwards as the detector met all the analysis requirements following a trial period of work within the analytical laboratories. This trial period involved testing the detector in a number of critical application areas to investigate its operational success against what were the industry standard methods at the time. The application areas looked at were the following:

Specificity: CAD is a nonspecific detection system and gives a positive response to all and any materials that pass through the corona. This allows a generic instrumental analytical method to be set up and used for multiple polymeric materials in multiple matrices. To date in excess of 10 different polymeric scale inhibitors have been quantified in brine samples using a CAD, all of which give similar levels of sensitivity.

Sensitivity: While CAD is not as sensitive for ranges of molecules as, for example, MS nor as sensitive for specific molecules as UV, the sensitivity is, in general terms, very similar for each of the polymeric scale inhibitors tested, thus removing the need to tune the detector for each molecule being analyzed. In general terms, CAD will reliably detect polymeric scale inhibiting molecules down to approximately 1 ppm of active polymer, which will equate to 2–5 ppm of formulated polymer. In most cases, this sensitivity is appropriate to the analysis being performed.

Robustness: In an ideal world, samples from off shore would be injected directly into the HPLC system and the polymer quantified. In practice, however, the presence of high levels of salt and the potential presence of other oilfield chemicals mean that some form of clean-up is required. The resulting samples have a much reduced salt content but this is still appreciable; therefore the detector has to be able to cope with the levels of sodium and chloride in the sample without salting up. Initial work performed demonstrated that the detector could cope with one injection of uncleaned sample then the response would dramatically decrease as the salt affected the detector; however, a short flush or approximately 2 h with mobile phase was sufficient to remove all the salt and return the detector baseline to an acceptable level in terms of sensitivity and reproducibility. The use of cleaned samples reduced the salt content to a manageable level, but the chromatograms still showed response to the sodium and chloride in the samples. This resulted in the use of a flow gradient during routine operation to act as a flush for the salt. This flow gradient increased the flow from the standard 0.7 $mL.min^{-1}$ upwards to 2.0 $mL.min^{-1}$ for a 10 min period. This was sufficient to maintain a stable detector baseline and, subsequently, a stable response to the polymeric materials.

Once CAD was fully introduced to the analytical laboratories, the systems used in routine analysis have typically been at greater than 80% utilization over extended periods of time for a number of years and have proven to be reliable with only annual service being generally required to maintain operation.

14.4.3 Typical Validation of Methodology

In common with most other types of analytical methods, the use of the HPLC-CAD methods in oilfield chemistry analysis requires validation to demonstrate that the data generated are accurate, precise, and reproducible.

The major difference between this field and others, for example, the pharmaceutical or environmental fields are the requirements for validation. The oilfield chemistry industry does not have an international organization, the equivalent of the pharmaceutical International Conference on Harmonisation (ICH) to set industry standards for validation; therefore the approach taken by individual companies may vary, and be equally valid, to that given in this section.

A typical validation of a residual method using HPLC-CAD will follow the guidelines given in ICH Q2B, validation of methodology [11], to the extent that it is practical and cost effective, although the pesticide residue guidelines may have more read over in terms of quantifying a residual concentration in a complex matrix. It is likely that any typical validation protocol will use a mixture of both guidelines. In general, validation of HPLC-CAD methods consists of the standard sections of linearity of detection, assay accuracy, assay precision, injection precision, limit of detection (LOD) and quantification, specificity, ruggedness, and robustness. In each case the acceptance limits are defined before the work is commenced with the understanding that the requirements of either ICH Q2B or pesticides are not strictly required and a "fit for purpose" method that may fail in some areas is more useful than a strict requirement to meet every aspect of either guideline.

The data given in this section are the result of a typical validation performed by the author [12]. The chemical used was a sulfonated copolymer, which was spiked at various levels into produced water that had been supplied from a producing oilfield in the North Sea. This water was spiked before sample clean-up (thus the spiked inhibitor would pass through the clean-up process in mimicry of real samples) and was expected to contain a mix of standard oilfield chemicals but to be blank for scale inhibiting polymeric species.

The samples were all individually cleaned up using a proprietary solid-phase extraction (SPE) type of clean-up and all were injected onto the HPLC using an autosampler. The HPLC method utilized a standard GPC column, an organic/aqueous buffer mobile phase, and a CAD detector linked in series with a UV detector. No column heating was employed, and a flow gradient was utilized to flush the inorganic salts from the column as part of each injection.

14.4.3.1 Linearity of Detection

A series of standards containing scale inhibitor were prepared in blank field brine by spiking in known amounts of scale inhibitor (by weight) covering the range 1–100 ppm. Each standard was injected, and the peak areas used to generate linearity curves. Three separate calibration curves were generated from the data, these being a full range curve (1–100 ppm), which showed unacceptable linearity (not shown), a low range curve (1–10 ppm, Figure 14.8), and a high range curve (10–100 ppm, Figure 14.9), both of which showed good linearity. In practical terms, a set of linearity curves like these will allow the scale inhibiting polymer to be calculated from an extrapolated single

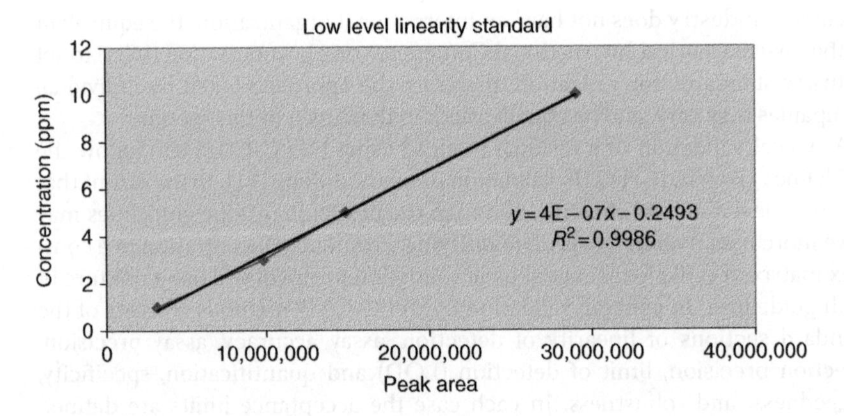

Figure 14.8 Typical low level linearity curve.

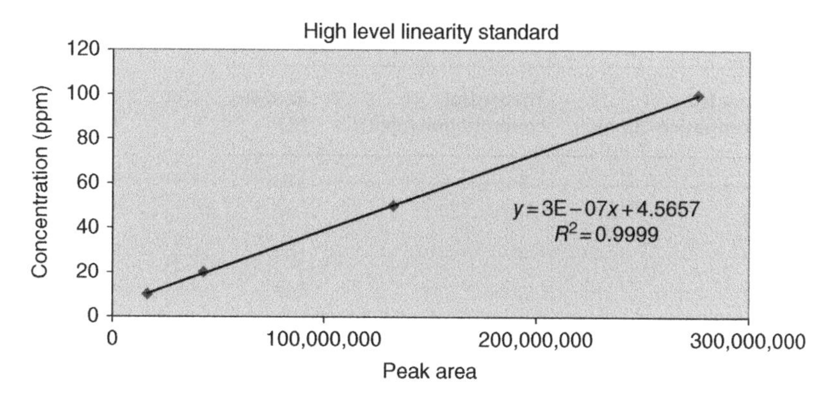

Figure 14.9 Typical high level linearity curve.

point standard at 5 ppm for the concentration range 1–10 ppm and at 30 or 40 ppm for the concentration range 10–100 ppm.

14.4.3.2 Precision of Injection

A set of 10 injections were made of a standard preparation at four concentrations (Table 14.1). The concentrations were chosen to cover the concentration range expected in routine analysis. The peak areas were recorded and a coefficient of variation was calculated for each set of injections.

Table 14.1 Precision of injection.

	Peak area (ppm)			
Injection	**3**	**5**	**10**	**100**
1	7,920,314	11,903,475	31,414,721	465,641,427
2	7,637,915	13,301,619	30,157,241	488,252,117
3	7,159,314	13,521,873	30,325,181	487,900,368
4	7,555,224	13,557,634	30,562,137	489,629,455
5	7,227,357	13,528,999	30,571,113	488,020,943
6	7,769,816	13,417,418	29,067,775	490,091,233
7	8,357,226	12,719,449	29,015,640	490,504,877
8	7,864,423	13,397,391	29,390,983	468,663,931
9	Poor injection	13,212,930	29,218,897	475,723,907
10	7,910,093	12,896,096	28,794,189	470,206,840
Mean	7,711,298	13,145,688	29,851,788	480,709,220
CV (%)	4.8	3.9	2.9	2.2

Table 14.2 Assay accuracy and precision.

QC	Calculated concentration (ppm)	Theoretical concentration (ppm)	Recovery (%)
1	3.3	3	110
2	2.2	3	73
3	3.2	3	107
4	3.2	3	107
5	9.8	10	98
6	9.6	10	96
7	6.8	10	68
8	10.8	10	108

The data show good injection precision when using CAD down to less than 3 ppm. These data demonstrate the stability of the CAD detection system when performing multiple injections of samples during routine analysis.

14.4.3.3 Assay Accuracy and Precision

A set of four samples were prepared individually at 3 and 10 ppm by spiking the polymeric scale inhibitor onto oilfield-produced water from a typical North Sea oilfield. The samples were individually cleaned using SPE and quantified against separately prepared standards that were matrix matched.

The data set (Table 14.2) shows an expected spread of results, which in most cases show less than 10% difference from the nominal concentrations. Sample 2 showed a larger difference, but the concentration spiked and being analyzed is so low that the loss of a fraction of a ppm will have a significant effect on the percentage recovery reported. The low value obtained for sample 7 is anomalous in the result set given the high recovery of the remainder of the samples at this concentration. This sample was injected again as part of the assay ruggedness, Section 14.4.3.4, and gave a value of 9.0 ppm, which would be more in trend for the data. Therefore the original analysis was taken as a poor injection on the instrument and led to a recommendation that routine samples be analyzed in duplicate to give confidence in the obtained result.

14.4.3.4 Assay Ruggedness

The four spiked samples at 3 and 10 ppm prepared for the previous section were analyzed on one HPLC system on two separate days and quantified against the appropriate, freshly prepared standards. This gave an indication of the ruggedness of the method over the typical analysis run time.

Table 14.3 Assay ruggedness.

QC	Calculated concentration (ppm)		Δ With time (ppm)
	Day 1	Day 2	
1	3.3	3.1	0.2
2	2.2	2.8	0.6
3	3.2	3.1	0.1
4	3.2	2.2	1.0
5	9.8	9.9	0.1
6	9.6	10.9	1.3
7	(6.8)	9.0	—
8	10.8	13.3	2.5

The data set (Table 14.3) indicate good stability and reproducibility of the detector system over time, indicating a high level of confidence that can be given in data generated using the detection system, even if samples are analyzed multiple times over a longer timeframe.

14.4.3.5 Assay Ruggedness 2: Inter-instrument Variability

The four spiked samples at 3 ppm prepared for the previous section were analyzed on the two HPLC systems available with different CAD detectors and quantified against the appropriate standards. This gave an indication of the ruggedness of the method across the two available instruments and also served as a cross validation to allow either instrument to be used for routine analysis.

The obtained data (Table 14.4) demonstrate the reproducibility of the CAD detector system across hardware systems. The two CAD detection systems

Table 14.4 Assay ruggedness 2.

QC	Calculated concentration (ppm)		Δ Instrument(ppm)
	Instrument 1	Instrument 2	
1	3.3	3.1	0.2
2	2.2	[a]	
3	3.2	3.3	0.1
4	3.2	3.3	0.1

[a] Vial empty.

used were different in age and attached to different models of HPLC systems but gave data that will statistically overlap at any required confidence level.

14.4.3.6 Limit of Detection and Limit of Quantification

The LOD/limit of quantification (LOQ) was calculated for the polymeric scale inhibitor using a definition of a signal-to-noise ratio of 3 to be equivalent to LOD and a signal-to-noise ratio of 10 to be equivalent to LOQ.

The obtained values (Table 14.5) demonstrate the working of the CAD detector at the absolute limits of its sensitivity range; the calculated LOQ of the particular scale inhibitor used was 3 ppm, while the samples spiked and analyzed as part of the various validation sections were at 3 and 10 ppm. Despite this, the data obtained from the CAD detector suggest that the practical LOD of the method are slightly lower than the theoretical limit of detection of the method.

14.4.3.7 Analysis of Routine Oilfield Brine Samples for Polymeric Scale Inhibitor Using HPLC-CAD

The end result of the development and validation work on the methods and the equipment was the ability to routinely analyze samples provided from offshore oilfields for the residual polymeric scale inhibitor using CAD detection. A typical set of data are given in Table 14.6; these were generated on a single sample that

Table 14.5 Limit of detection/quantification.

Compound	LOQ	LOD
Polymeric scale inhibitor	3 ppm	1 ppm

Table 14.6 Analysis of routine sample.

	Calculated concentration (ppm)		
Sample	Analyst 1	Analyst 2	Analyst 2, day 2
1	11	9	10
2	10	11	9
3	11	11	9
4	8	9	10
5	10	11	9
6	10	10	9
Mean	10	10	9
CV (%)	11	8	5

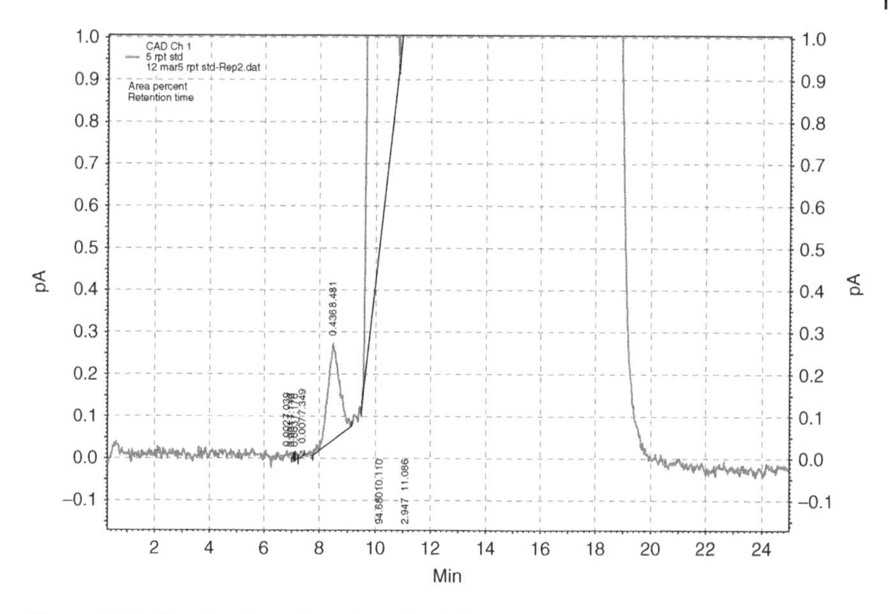

Figure 14.10 Showing the polymeric scale inhibitor peak at approximately 9 min.

was analyzed multiple times by two different analysts over 2 days using two different instruments and two different CADs. The data therefore demonstrate the reproducibility of the method and the instrumentation when used by different operators.

The data obtained (Table 14.6) for all 18 samples show good reproducibility with analyst 1 and analyst 2 data showing statistical overlap at 95% confidence levels. The range of the data over all 18 analyses is 8–11 ppm. When calculated as statistical errors at 99% confidence levels, the calculated result for this sample is 9.8 ± 0.6 ppm.

Typical chromatography for the sample is given in Figure 14.10.

14.4.4 Limits of Methodology

The use of HPLC with CAD has opened a new field in the area of polymeric scale inhibitor analysis at residual concentrations; however, the methodology is not a complete solution to the problems of the analysis. There are a number of important limitations to the use of the technology in the oilfield chemistry industry that may be resolved in the future by development of the system or application or may end up limiting the application of the technology in this field at an absolute level.

The Major Limitations Include Robustness: The detector is an exceptionally robust piece of engineering but cannot reliably analyze samples that contain significant concentrations of inorganic ions. In practical applications this requires that the oilfield sample be either diluted in distilled water (~×100), which limits the sensitivity of the method, or that the ions are removed by some chemical or physical means (e.g., SPE or desalting), which can add a degree of error to the obtained data and a level of complexity to the analytical procedure.

Linearity of Detection: The typical linearity of detection curve covers one or two orders of magnitude. This is an inconvenience rather than a major limitation but does require that a number of standards are prepared for each analysis performed to cover the expected concentration range of the samples and may result in a requirement to reanalyze samples if the actual sample concentrations are outside the linear range of the standards employed.

Sensitivity: In common with all other fields of analytical chemistry, the users of the data in the field of oilfield chemistry always want to see lower LOD and more robust data at, or close to, the LOD. For the CAD, the LOD for scale inhibition polymers in field brine looks to be approximately 1 ppm of pure polymer. The resulting LOD of the formulated products will, of course, vary from application to application depending on the concentration of the active polymer in the formulation but are typically 3–5 ppm of a formulated product. Under normal, current field operating conditions, these limits are appropriate, but, as longer squeeze lifetimes are looked for to reduce operating costs and chemical usage, the requirement for a lower limit of detection will become much more important as scientists look for the actual chemical limit of the inhibitor not just the analytical limit of detection of the method.

14.5 Conclusions and Further Work

The use of HPLC with CAD has opened a new field in the area of polymeric scale inhibitor analysis at residual concentrations. The sensitivity and robustness of the detector has allowed analysis to be performed that was not possible using the classical polymeric scale inhibitor analytical techniques. To be able to "see" the polymer of interest reliably at sub 10 ppm levels has taken the analysis into new areas over the past 5 years and has sparked a large amount of new developmental work in the area of instrumental analysis of polymeric scale inhibitor.

The new developments are going in two related but different directions. The first is to improve the chromatographic part of the analysis. The current methods use classical gel permeation chromatography columns, which perform the majority of their separation by physical size. This is useful when there is only one polymer present in the sample being analyzed as only one peak is observed and quantified. However, as many of the scale inhibiting polymers used in the oilfield chemistry industry cover a very similar size envelope, the use of GPC columns gives merged peaks when more than one polymer is present, for example, from comingled wells or from a topside inhibitor. Use of more novel HPLC columns will likely widen the use of the system as better chromatographic separations are obtained, but the practice of obtaining retention of scale inhibiting polymers on non-GPC columns is nontrivial and has not yet been conclusively demonstrated.

The use of polymeric scale inhibitor is widely prevalent in oilfield chemistry, but in absolute terms, these are dwarfed by the use of corrosion inhibitors. Unlike polymeric scale inhibitor that tends to be a simple mix of materials, corrosion inhibitors can be a complex mix of chemistries including amphoteric, quaternary amine, and surfactant chemicals. These can be analyzed by dye extraction techniques and by LCMS, but there is a need for methods that sit between the two, more specific and sensitive than the dye transfer but simpler and cheaper than the LCMS. The use of HPLC-CAD has been investigated and shown to have potential utilization in this area.

References

1 GM Graham, KS Sorbie, LS Boak, Development and Application of Accurate Detection and Assay Techniques for Oilfield Scale Inhibitors in Produced Water Samples; SPE Paper No. 28997, 1995.
2 GM Graham, KS Sorbie, LS Boak, Development and Accurate Assay Techniques for Poly Vinyl Sulphonate (PVS) and Sulphonated Co-polymer (VS-Co) Oilfield Scale Inhibitor, presented at the 6th International Symposium on Oil Field Chemicals, Geilo, March 1995.
3 GM Graham *et al.*, Complete Chemical Analysis of Produced Water by Modern Inductively Coupled Plasma Spectroscopy (ICP), presented at the 7th International Symposium on Oil Field Chemicals, Geilo, March 1996.
4 CE Anderson, G Ross, GM Graham, Automated Instrumental Analysis of Residual Polymeric Scale Inhibitor Species, presented at the 12th International Symposium on Oil Field Chemicals, Geilo, April 2001.
5 NP Chilcott *et al.*, The Development and Application of an Accurate Assay Technique for Sulphonated Polyacrylate Co-Polymer Oilfield Scale Inhibitors, SPE Paper No. 60194, 2000.
6 P Chen *et al.*, Extending Squeeze Life with a Novel Scale Inhibitor Squeeze Additive, presented at the 15th International Symposium on Oil Field Chemicals, Geilo, March 2004.
7 P Chen *et al.*, Field Experiences in the Application of an Inhibitor/Additive Interaction Package to Extend an Inhibitor Squeeze Life, SPE Paper No. 100466, 2006.
8 Internal Champion Technologies Report from Technium OpTIC, January 2007.
9 Internal Champion Technologies Report No. AS-CR-790.1.6 from CSS, May 2006.

10 Internal Champion Technologies Report from University of Strathclyde, February 2006.
11 International Conference on Harmonisation Guidelines Q2(R1): Validation of Analytical Procedures: Text and Methodology, November 2005.
12 A Thompson, K Burnett, Development and Validation of a Novel Method for Quantification of Polymeric Scale Inhibitors and a Comparison of Obtained Data with Current Commercial Techniques, presented at Chemistry in the Oil Industry X, Manchester, November 2007.
13 A Thompson *et al.*, Oil Field Data/Return Analysis: A Comparison of Scale Inhibitor Return Concentrations Obtained with a Novel Analytical Method and Current Commercial Techniques, SPE Paper No. 114049, 2008.

15

Applications of Charged Aerosol Detection for Characterization of Industrial Polymers

Paul Cools and Ton Brooijmans*

DSM Coating Resins, Waalwijk, The Netherlands
** Current address: The Dow Chemical Company, Freeport, TX, USA*

CHAPTER MENU

15.1 Introduction

Molecular analysis of macromolecules and polymers is an intriguing field within the analytical world. In general, a synthetic polymer does not only consist of the intended macromolecules but is also often a complex mixture of the macromolecules combined with additives, polymerization aids, and residual amounts of raw materials and their corresponding impurities. Especially in industrial polymer grades, the analysis of the additives, residuals, and impurities is a complicated task. In addition, the analysis of the polymer molecules is not straightforward. Various types of sample preparation and separation techniques are necessary to obtain information on the general composition of polymers.

Charged Aerosol Detection for Liquid Chromatography and Related Separation Techniques,
First Edition. Edited by Paul H. Gamache.

In this chapter the focus will be put on the application of charged aerosol detection (CAD) for the characterization of industrial synthetic oligomers (molecules consisting of only a small number of repeating units) and larger polymer molecules and their corresponding distributions. The analysis of the additives and others, including sample preparation, will not be addressed in this discussion.

For the characterization of polymer molecules, liquid chromatography (LC) is used to separate the molecules based on molar mass and chemical composition. Size exclusion chromatography (SEC) [1] is used to obtain information on molar size distributions or molar mass distributions (MMDs), and gradient polymer elution chromatography (GPEC) [2–8], or gradient LC [9], is used to obtain information on chemical composition. Another LC technique applied especially for oligomer/polymer distributions is liquid chromatography at critical conditions (LCCC) [10–12]. With LCCC polymer molecules can be separated according to end-group functionality fairly independent of the molar mass. For block copolymers, LCCC can be used to obtain information on block length and block length distribution [13, 14].

Detection in general and in particular for polymer molecules is a key concern. Especially in combination with gradient techniques, detection of polymers is not straightforward. The use of solvent displacement detection techniques, such as evaporative light scattering detection (ELSD) [3, 15] and CAD [16–18], is essential for the application of GPEC. For isocratic techniques, such as SEC, refractive index (RI) is a generally applied solution; nevertheless, evaporative detectors can also be used.

This chapter will describe the use of CAD with SEC and GPEC. Qualitative as well as quantitative aspects will be discussed. Polymers with various composition (various styrene–methyl methacrylate ratios) synthesized by emulsion polymerization will be used to discuss qualification and quantification with CAD in comparison to the ultraviolet (UV) detector and differential refractive index (dRI) detector. Besides, emulsion polymers, polymer standards with low polydispersity were applied. Finally the importance and applications of CAD in combination with mass spectrometry (MS) will be discussed in particular for the characterization and interpretation of polymer and oligomer distributions.

15.2 Liquid Chromatography of Polymers

In general, LC of polymers is divided in two types, gradient LC and isocratic LC. Under isocratic conditions, the eluent conditions do not change in time. SEC is the traditional isocratic LC method used to characterize MMDs [1, 14]. Another isocratic LC technique for polymers is LCCC [10–14]. Solvent gradient LC or GPEC [2–9] is mainly used for the separation of polymer molecules according to polarity. The polarity of a polymer molecule depends on the type (or types) of repeating unit, and the amount of repeating units or molar mass [8]. Besides solvent gradients, temperature gradients can be used [19].

Under isocratic conditions, the retention of polymers can be divided into three modes: the mode in which separation is governed by adsorption (adsorption mode), a mode where the separation is governed by exclusion (exclusion mode), and an intermediate mode called the critical condition mode where the exclusion contribution is exactly suppressed by the adsorption contribution [10, 13]. The adsorption contribution and the critical conditions depend on the eluent strength of the mobile phase, the chemistry of the repeating unit of the polymer molecule, the chemical type of column material, and temperature. Adsorption of a molecule will result in an additional retention. In comparison with regular organic molecules, polymer molecules are significantly larger in size (in solution), and the amount of functional groups (repeating units) available for adsorption is considerably larger. Hence, the adsorption of a polymer molecule is stronger. In addition, a polymer molecule in solution can have dimensions larger than the biggest pore diameter present in the column, especially when standard analytical column particles are used. Consequently, a polymer molecule in solution has considerably less column volume and surface available than those available for the mobile phase molecules. The mobile phase molecules can enter all the pores, and therefore with constant mobile phase flow, the polymer molecule will elute much sooner, depending on the pore size distribution of the used column. Whether adsorption occurs or not, exclusion of polymer molecules from the pores of the chromatographic column will always take place. In other words, if the contribution of adsorption in isocratic LC is secondary to exclusion, the polymer molecules will elute before the injected solvent molecules, and separation takes place according to size. The larger molecules will elute first, followed by smaller molecules. When the adsorption contribution is significant, for example, at weak eluent conditions, the adsorption of the repeating units will retain the polymer molecules to such an extent that the polymer molecules will elute later than the simultaneously injected solvent molecules. In this mode, the adsorption mode, the polymer molecules will be separated according to the amount of repeating units; with the increasing number of repeating units, stronger adsorption will occur and therefore the retention time will increase (as can be explained by Martin's rule [20]). The retention sequence in adsorption mode is opposite to the retention sequence in exclusion mode; small molecules will elute before the larger molecules. By controlling the mobile phase composition in combination with temperature, the adsorption can be controlled in such a way that the adsorption effect completely compensates the exclusion effect, resulting in a repeating unit independent of retention, or in other words retention independent of molar mass. This mode is also called LCCC [11].

The different retention modes are visualized in Figure 15.1. Three polystyrene standards normally used for conventional SEC calibration with different molar mass ($M_A > M_B > M_C$) are injected on a silica C18 analytical column at three different conditions: exclusion mode, 100% tetrahydrofuran THF; adsorption mode, 75/25% THF/water; and critical mode, 87/13% THF/water at 30°C [11].

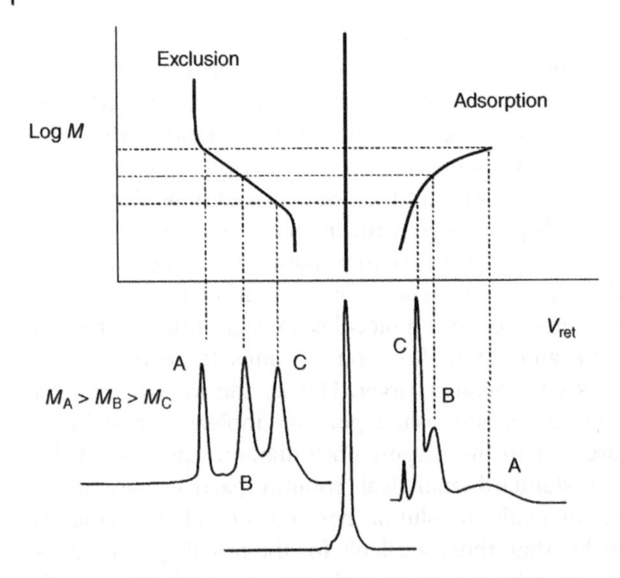

Figure 15.1 Isocratic retention modes. The retention behavior of three different polystyrene SEC standards ($M_A > M_B > M_C$) at three different conditions on analytical silica C18 column is shown: (left) eluent tetrahydrofuran, (middle) tetrahydrofuran/water 87 : 13 (at 30°C), and (right) eluent tetrahydrofuran/water 75 : 25 [11]. *Source:* Cools *et al.* [11]. Reproduced with permission of Taylor & Francis.

The extent of adsorption depends on the eluent strength of the mobile phase in combination with the type of column and repeating unit. By changing the eluent strength from weak to strong, adsorption can be controlled. The conditions in which the adsorption is considered to be absent are the conditions ideal for SEC. With SEC, polymer molecules are separated by molecular size and not necessarily molecular weight. The size depends on molar mass via the Mark–Houwink and Newton hard sphere theory [1]. When a homopolymer (consisting of only one type of repeating unit) is considered, the size can be directly calculated into molar mass. Nevertheless, for copolymers (consisting of at least two types of repeating unit), the size depends not only on the number of repeating units but also on the composition (ratio of the different units in combination with the total number of repeating units) [1].

A small adsorption contribution of repeating units can result in (isocratic) irreversible adsorption of the total polymer molecule. Irreversible adsorption is a common phenomenon in the LC of polymer molecules if weak eluent conditions are applied. Consequently, by controlled change of the eluent strength to stronger conditions (e.g., going from MeCN to THF on a silica C18 column), the polymer molecules can be desorbed, and separation can occur according to the type of repeating unit or polymer. This is actually the separation mechanism of gradient LC or GPEC [5–7].

15.3 Solvents

Since polymer molecules are difficult to dissolve, various organic solvents have to be used as eluent. The solubility of polymers is extensively described by the Flory–Huggins Theory [21, 22]. The organic solvents used for polymer analysis are not commonly applied LC eluents like water, MeCN, or methanol (MeOH). Mobile phase or solvents necessary to analyze polymers with LC are shown in Table 15.1.

In order to separate polymer molecules, the molecules need to be dissolved on a molecular basis. The molecules need to be uncoiled and this can only be achieved under extremely dilute conditions. The solvent needs to be sufficiently strong not only for solvation but also for uncoiling of the molecules. Using a good solvent will result in a solution of separate swollen molecules (e.g., THF and polystyrene). Weak solvents enable solvation of the polymer molecules but cannot bring the molecule in a complete uncoiled condition (e.g., MeCN and polystyrene). This does not result in single molecules but a cluster of molecules. For weak solvents even solvation does not occur (e.g., water and polystyrene). The solubility of a polymer in a solvent or solvent combination depends on the molar mass and the chemical type of the polymer molecules. High molar mass needs stronger solvents or solvent compositions. Hence, depending on the chemical type of the polymer molecules, different solvents have to be used.

Table 15.1 Overview of used solvents/eluents for LC polymers applications.

Solvent	LC application	Polymer application	Remarks
Water	SEC [1]	Peptides, cellulose, polyacrylic acid	Polar and/or ionic polymers
Tertrahydrofuran (THF)	SEC [1]/ GPEC [3]	Polystyrene, polyacrylates	Non-polar to medium polar polymer
Dimethyl acetamide (DMAc)	SEC [1, 23]	Polyurethanes, polysaccharides	Polar polymers
Hexafluoro isopropanol (HFIP)	SEC [7]/ GPEC [7]	Polyurethanes, polyesters, polyamides, polycarbonates	Polar, hydrogen bonding, crystalline polymers
Dimethyl sulfoxide (DMSO)	SEC [1, 23]	Dextranes starch (polysaccharides)	
Chloroform/ dichloromethane	SEC [1]/ GPEC [6]	Polystyrene	Non-polar, non- crystalline
Trichloro benzene (TCB)	SEC [1]/ GPEC [24, 25]	Polyolefins	High temperature applications (>100°C) Non-polar crystalline polymers

Polystyrene and most poly(meth)acrylate polymers dissolve in the most commonly used solvents like THF. Polar polymers such as polyurethanes and (crystalline) polyesters need more polar solvents such as DMAc and HFIP. Polysaccharides need DMSO to dissolve. Polyolefins need nonpolar organic solvents to dissolve and, in all cases, elevated temperatures higher than 150°C. The applicability of organic solvents suitable for polymer analysis with CAD needs to be studied in more detail. It is unknown whether CAD is resistant to solvents like HFIP, DMAc, and DMSO. For THF, CAD has already been optimized/modified and filtration of the solvents with a 0.1 μ filter is necessary. This will be discussed further in the succeeding sections.

15.4 Quantitative Detection of Polymer Molecules

The most preferred concentration detector is universal in selectivity and sensitivity and therefore useful for quantifying a broad range of components. With a universal detector, the qualitative and quantitative comparison of complex samples can be performed. In LC, a universal detector is missing in general. The mobile phase of LC regularly contributes to the detector signal and can block the signal entirely.

15.4.1 Ultraviolet Detection

The most commonly used LC detector is the UV detector. UV detection can detect molecules that contain chromophore groups, and in many cases this includes the LC mobile phase. Consequently, the UV extinction of the analyte present in low concentrations will be surpassed by the extinction of eluent. The baseline signal of the UV detector is always auto-zeroed in order to detect the difference between the dissolved analyte and the used mobile phase. The background extinction of the mobile phase should not be too high. UV detection can be used in isocratic methods as well as in gradient elution. In gradient elution the mobile phase gradient is noticeable over the whole range of wavelengths in the baseline. Not all types of polymers contain chromophore groups that can be detected in solvents like THF. Most polymers absorb UV light below 230 nm, and the UV extinction of THF will hamper the detection. This makes the UV detector not suitable to detect, for instance, acrylic polymers in THF. HFIP is more UV transparent and could be used to analyze polyacrylates. Unfortunately, HFIP is not a good solvent for less-polar polymers like polystyrene. Consequently for polymers, UV detection is generally not as suitable as universal detection.

15.4.2 Differential Refractive Index Detection

Another commonly used LC detector is the dRI detector. By monitoring the change of the RI of the solution in the sample flow cell compared with the RI of the pure eluent in the reference cell, small differences in RI can be used to detect polymer molecules in solution. The dRI can be used as universal detector when the solvent is chosen wisely (big RI difference between analyte in solution and eluent) and isocratic conditions are applied [16]. Consequently, the dRI is most suitable for SEC separations. Unfortunately, the dRI cannot be used in gradient applications; the change in RI caused by the eluent gradient will hamper the detection of polymer molecules in solution. In the past, efforts were made to modify the dRI optics to use the RI detector in gradients [26], unfortunately without commercial success.

15.4.3 Evaporative Detection

Since the LC eluent is causing most of the problems during the detection, eluent evaporation detectors seem to be the solution. In the past, the evaporation techniques mainly focused on the evaporation of the eluent by spraying and applying heat. The moving belt flame ionization detector (FID) for LC has been studied, but without commercial success [27]. Another technique is evaporative light scattering (ELS) detection, introduced successfully in the late 1980s. With evaporative detection, the eluent is evaporated and the analytes preferably form particles that can either be detected by light scattering or by corona charging [16]. The corona charging technique is still undergoing optimization and development, and the introduction of the Nano Quantity Analyte Detection (NQAD) [28] also shows that the development on laser detection in combination with evaporation is not mature. The application of NQAD will not be part of the discussion.

As expected, evaporation of the eluent also has drawbacks. By evaporating the eluent, low molar mass components will also evaporate, and this hinders the detection as particles. Consequently, low boiling components cannot be detected by these techniques. By applying lower temperature and better nebulizing techniques, the application temperatures of evaporative detectors have decreased significantly (from boiling point of eluents to about 30°C).

Polymer molecules are extremely suitable in combination with these detectors. Oligomers (molecules consisting of only a few number of repeating units) might still be affected by the evaporation steps, but the main part of a polymer distribution should be forming particles and therefore be detected. The ELS detector turned out to be a useful detector and definitely contributed to the success of GPEC [3, 7]. Nevertheless, the ELS detector was not perfect with respect to sensitivity and universal applicability. Especially for quantification of distributions, the early ELS detectors were limited. The new ELS detectors with optimized evaporation principles at lower temperatures are significantly more sensitive; however, the new generation of ELS detector still results in nonlinear concentration dependency [15].

15.4.4 Charged Aerosol Detection

By using CAD, combining evaporation at low temperature and corona charging of the formed particles, analytes can be detected [17, 18]. CAD is considered to be universal [16] and can be used under isocratic and gradient elution conditions.

CAD has three major drawbacks. Similar to other evaporative detectors, volatile components will not be detected due to the evaporative conditions, and the sensitivity significantly depends on the eluent solvent composition. Another drawback is the high sensitivity toward particles present in the eluents. Accordingly, the quality of the eluents should be monitored wisely. These drawbacks are similar to the modern ELS detectors. Fortunately, with the introduction of μLC equipment, the need for high-purity eluents becomes more important. As experienced with μLC, the presence of particles in the mobile phase causes problems; the presence of particles can induce clogging of the system due to the small internal diameters of the capillary, low dead volume, and small column particle sizes. For the generally applied eluents, like water and MeCN, μLC quality grades are available. In most cases, the eluent is filtered with a 0.1 μm filter instead of a 0.2 μm filter. THF at first was not applicable with μLC equipment due to incompatibility issues. With the introduction of CAD, the quality of the THF was not suitable for μLC and incurred problems with CAD. This issue has been solved by an extra filtration step over 0.1 μm filters [29]. However, impurities, erosion of the equipment, and aging of the THF can have a significant effect on CAD signal. Besides the eluent, column bleeding can also give rise to baseline effects. Therefore, the use of good quality columns with low column bleed, for example, Symmetry® (Waters) or Chromolith® (Merck Millipore), is essential. In practice, cleaning of the CAD detector needs to be performed on a regular basis to avoid malfunction of the flow regulation.

15.4.5 Molar Mass Dependent Detection

Besides concentration detection, molar mass-dependent detectors can be used. The most generally applied detectors are viscosity and light scattering. Both detection techniques have to be used under isocratic conditions and, from the signal, molar mass information can be extracted. Nevertheless, in order to calculate molar masses, a concentration detector, like the dRI, is necessary.

15.4.6 Mass Spectrometry

Another technique that has been introduced for the detection of polymer molecules is MS [30]. Of course, MS is not only a detector as such but is also considered as a separate analytical technology. Nevertheless, MS can be used to quantify components, especially in combination with LC. The issues with MS on selectivity and sensitivity are significant but can be overcome by hyphenation with LC. The detection of polymers or polymer molecules has one main difference in comparison with the classic organic compound detection: polymers emerge as a distribution. As a result, the discussion on having a universal detector becomes even more important. With LC the molecules are separated according to polarity or molecular size; hence, in all cases a distribution is obtained.

In the next paragraphs, the application of CAD in SEC and GPEC will be discussed with special attention to measuring distributions. CAD will be compared with UV and dRI detections. Some quantitative and qualitative aspects will be described with respect to chemical type and molar mass of the polymers. Furthermore, in the last paragraphs, the combination of LC with MS and CAD will be discussed in more detail.

15.5 Size Exclusion Chromatography and Charged Aerosol Detection

To compare the different signals (UV, CAD, dRI) in SEC, model copolymers were synthesized with different methyl methacrylate (MMA) and styrene (S) ratios. The polymers are made by emulsion polymerization using sodium lauryl sulfate (surfactant), ammonium persulfate (initiator), and lauryl mercaptane (chain transfer agent). Different copolymers were synthesized: pMMA, pS-co-pMMA 25:75, pS-co-pMMA 50:50, pS-co-pMMA 75:25, and pS. Well-defined pS SEC standards are used for conventional calibration. pS and pMMA standards ($M_w \sim 5$ kg/mol) are also used for illustration purposes.

All polymers were dissolved in THF (unstabilized, Biosolve, μLC-MS quality) at various concentrations: approximately 0.01 mg/mL—0.1 mg/mL—1 mg/mL—10 mg/mL. The SEC experiments were performed with two SDVB Mixed B columns (Agilent) with guard column, on an Alliance 2695 separations module (Waters) with a 2998 Photo Diode Array detector (PDA) (Waters), a 2410 dRI (Waters) and the Corona® *Plus* (Dionex). The UV signal at 254 nm was used. THF (unstabilized, Biosolve, μLC-MS quality) was used as eluent without modifiers and a flow of 1 mL/min, and an injection volume of 100 μL.

In Figure 15.2, overlays of the chromatograms of low molar mass pS (white) and pMMA (grey) standards with approximately similar concentration (10 mg/mL) obtained with CAD, UV, and dRI detectors are shown. If the signal is independent on the composition, the peak areas should be comparable. In the CAD chromatograms, typical baseline effects are visible. As discussed before, these artifacts can be caused by the quality of the THF or column bleeding. In this case the artifacts are probably caused by the virgin SDVB columns. A clear tailing also occurs in the CAD signal for both pMMA and pS. In comparison, the UV and dRI show less or no tailing.

The tailing in the CAD signal can be explained by broadening effects and dead volume probably related to the hardware of the CAD. The evaporation step and the corona charging might cause extra broadening, making the early model CAD (Corona Plus) less suitable for use in UHPLC applications. All experiments described in this chapter were performed with a Corona Plus version. To solve the broadening effect, the Corona® ultra was introduced. However, this needs to be studied further. In addition to the dead volume,

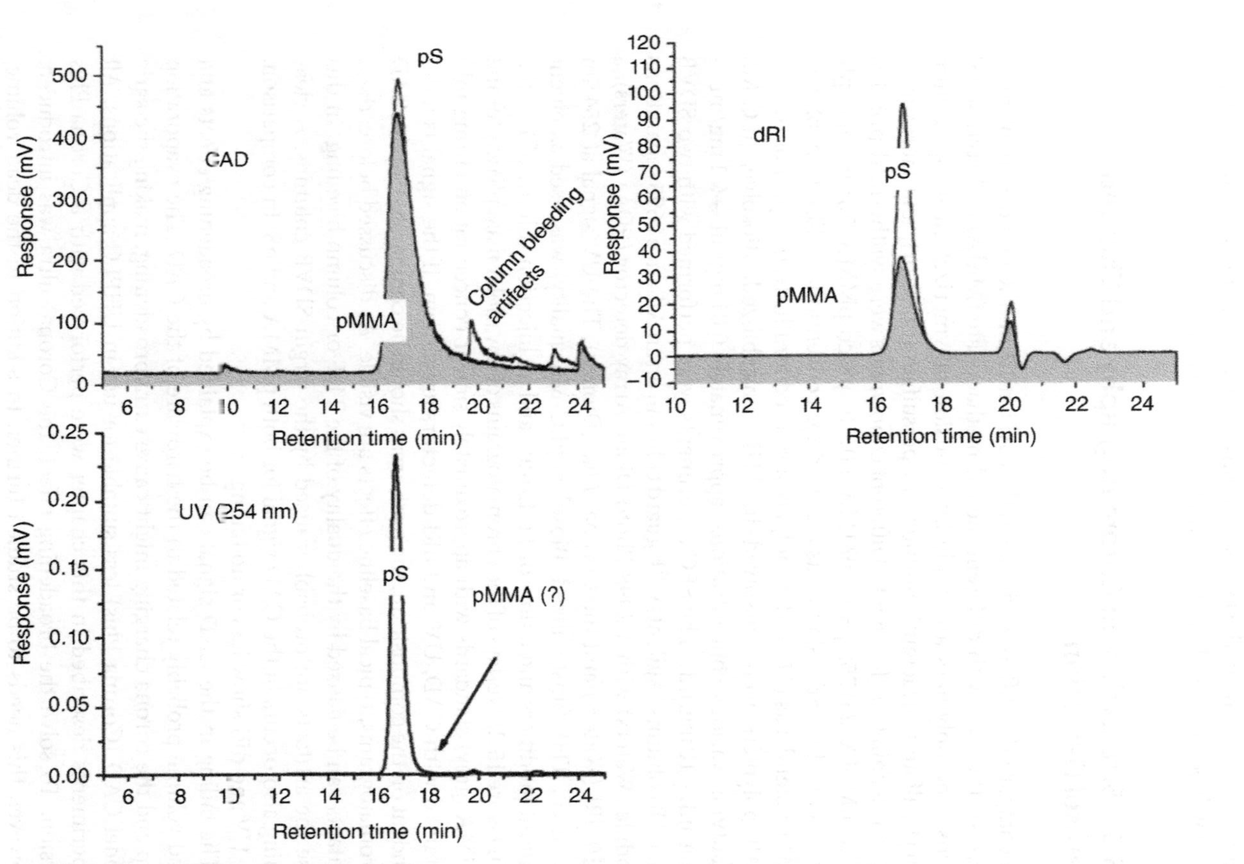

Figure 15.2 Comparison of CAD, dRI, and UV signals of homopolymers, a pS (white) standard ($M_w \sim 5$ k/mol) and a pMMA (grey) standard ($M_w \sim 5$ k/mol) at similar concentration.

the tailing can also be a result of a contaminated detector. Cleaning the detector and the particle filters on a regular basis is advised.

In Figure 15.2, it is clearly shown that the areas obtained with CAD, UV (254 nm), and dRI signals are different in pS and pMMA standards. The UV detector gives no response for the pMMA standard, while the peak area obtained with dRI is significantly larger for the pS standard than for the pMMa standard. CAD shows the most universal behavior; the areas are comparable for the copolymers and the homopolymers pMMA and pS, despite the tailing. The peak tailing is less pronounced due to the broader distribution compared to both standards. Similar findings are obtained when the copolymers are analyzed and the results are shown in Figure 15.3.

The UV response (measured at 254 nm) is highly dependent on styrene content, since MMA has no UV contribution. The dRI signal is more universal but still shows a significant dependency based on the composition. This is attributed to the difference in dn/dc of S and MMA units; the dn/dc of pS is approximately twice as large compared with that of pMMA (0.17 versus 0.08). In general the dRI is used as a universal detector. However, the dRI signal is dependent on composition, which could result in problems with respect to the quantitative analysis of heterogeneous copolymers, especially when distributions are concerned. CAD response appears to be independent of the composition. The dependency of concentration on the response of the different detectors is studied by observing total peak areas. The peak areas for the different copolymers and detectors are determined for various concentrations (0.01, 0.1, 1.0, and 10 mg/mL) and are depicted in Table 15.2.

First of all, it should be noticed that the concentrations are of the total polymer distributions and not of a single (polymer) molecule. In others words, due to the broad distribution, the concentration of a single molecule is decreased significantly and might be a factor of 100 less concentrated. Therefore, the detection levels appear to be limited. At the lowest concentration, none of the detectors showed a response due to the previously discussed distribution dilution effect. In Figure 15.4, the data depicted in Table 15.2 are plotted (peak area versus the concentration) for the different detectors and copolymers.

Both UV and dRI results show a linear dependency across the measured concentration range; CAD shows a skewed curve (probably logarithmic dependency). Due to the limited data points, an exact fit was not performed. For the discussion this is not relevant; the fact is that the relation is not linear, which poses as a major drawback of CAD. A possible correction on these types of data could be the use of log-log scales, but this is not present in commercial SEC software. For single components, using a log-log scale might be a solution, but for distribution this is less suitable. The UV and dRI detectors show a significant dependence on the composition, but for CAD this dependency is apparently absent.

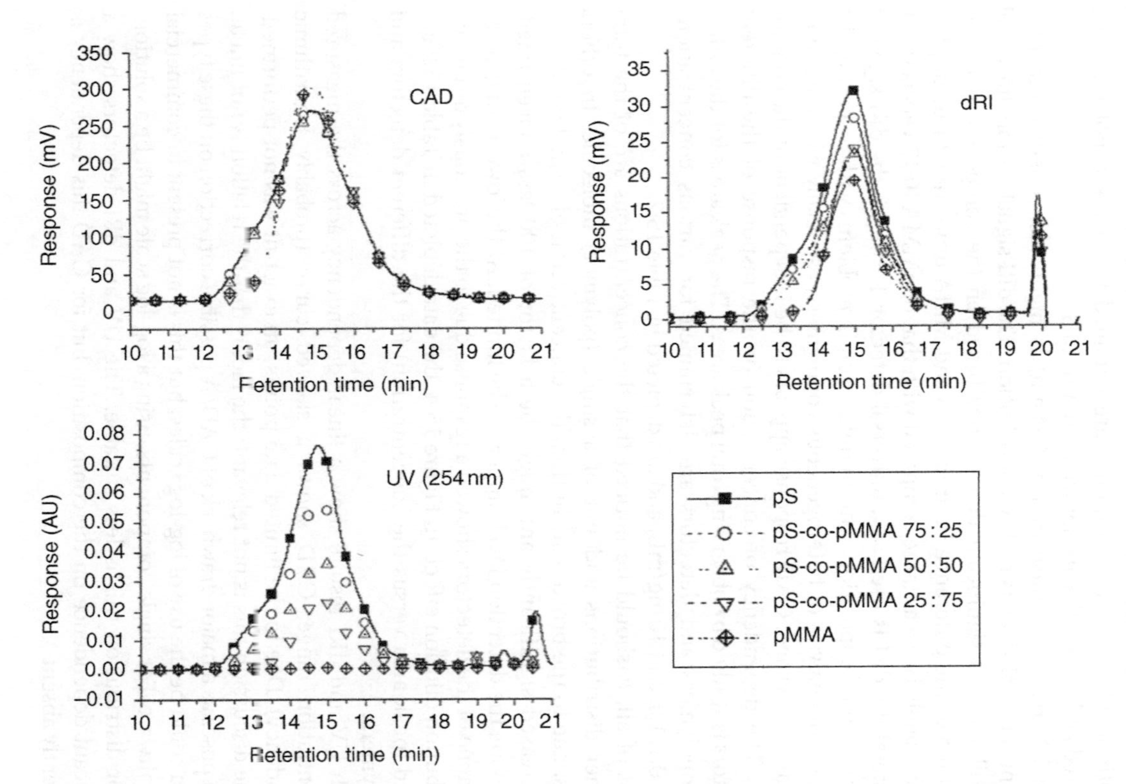

Figure 15.3 Chromatograms of copolymers using CAD, UV, and dRI. The UV and dRI clearly show dependency on sample composition.

Table 15.2 Peak areas of different polymer samples obtained with UV, CAD, and dRI detection.

	Load (mg)	Area UV254	Area CAD	Area dRI
pS	1	9.3E6	4.0E7	3.8E6
	0.1	9.2E5	1.0E7	4.0E5
	0.01	1.3E5	1.5E6	6.1E4
	0.001	*n.d.*	*n.d.*	*n.d.*
pS-co-pMMA 75:25	1	7.0E6	3.9E7	3.4E6
	0.1	6.7E5	1.0E7	3.8E5
	0.01	5.6E4	9.8E5	9.2E4
	0.001	*n.d.*	*n.d.*	*n.d.*
pS-co-pMMA 50:50	1	4.6E6	3.9E7	2.7E6
	0.1	4.6E5	1.0E7	2.7E5
	0.01	3.9E4	9.8E5	6.1E4
	0.001	*n.d.*	*n.d.*	*n.d.*
pS-co-pMMA 25:75	1	2.4E6	3.6E7	2.3E6
	0.1	2.4E5	9.9E6	2.2E5
	0.01	1.9E4	1.0E6	4.1E4
	0.001	*n.d.*	*n.d.*	*n.d.*
pMMA	1	*n.d.*	3.6E7	1.8E6
	0.1	*n.d.*	9.5E6	1.5E5
	0.01	*n.d.*	1.0E6	2.7E4
	0.001	*n.d.*	*n.d.*	*n.d.*

n.d., not detected.

A big concern for CAD results is the nonlinear response. This means that molecules present in a sample at low concentrations get overestimated compared with the major constituents. This can have a big influence on the MMD calculation since the peak tails are overestimated. In addition, the tailing effect as can be seen in Figure 15.2 also has a negative contribution. This can be observed in Figure 15.5 and Table 15.3. The MMD calculated by the dRI and CAD are different, resulting in deviating average molecular weight numbers. The calculated molar masses are shown in Table 15.3.

The MMDs are calculated for the dRI and the CAD signals via conventional calibration with pS standards. The MMD for the UV signal was not included for practical reasons. The number average molecular weight (M_n) is significantly lower for CAD results for all polymer samples, due to the tailing of the CAD signal. Tailing in the chromatogram will result in a lower M_n. The effect of the tailing is also shown in the M_w and M_z; however the effect is less pronounced.

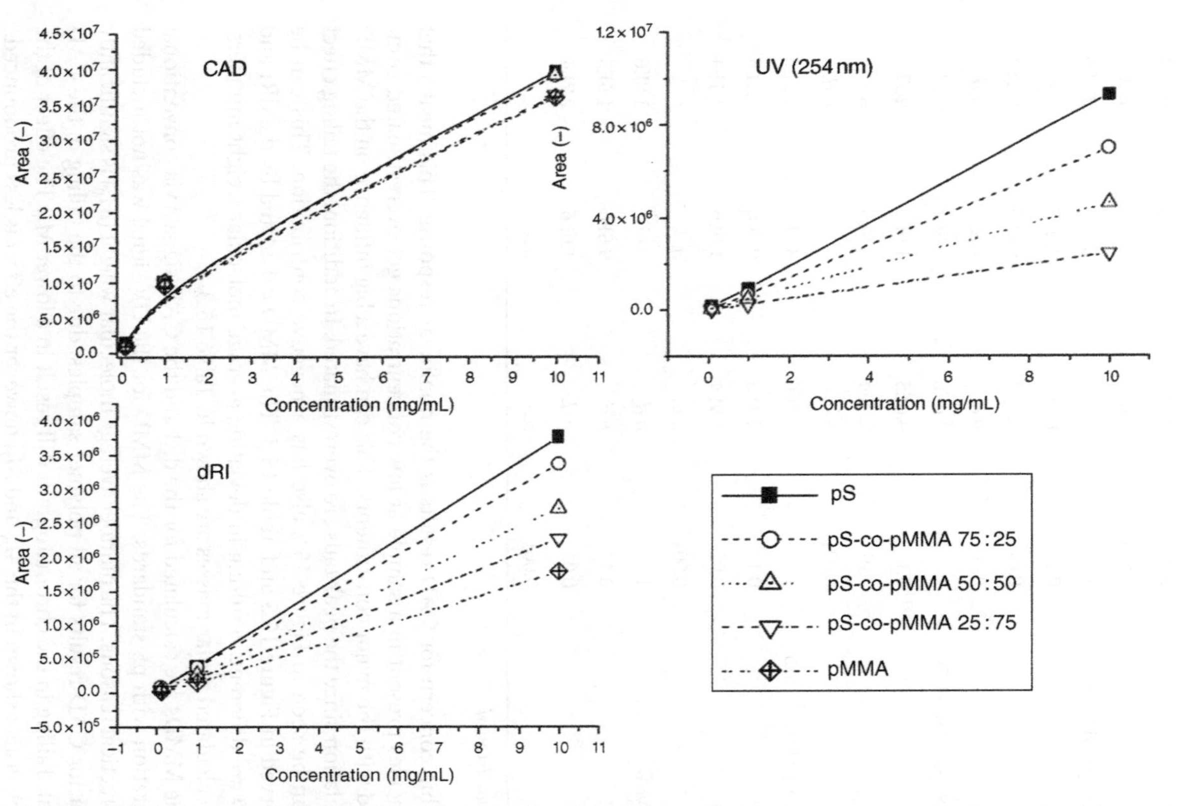

Figure 15.4 Concentration dependency of different copolymers. Different concentrations are used 0.01, 0.1, 1.0, and 10 mg/mL. Samples are analyzed using CAD, UV, and dRI detectors.

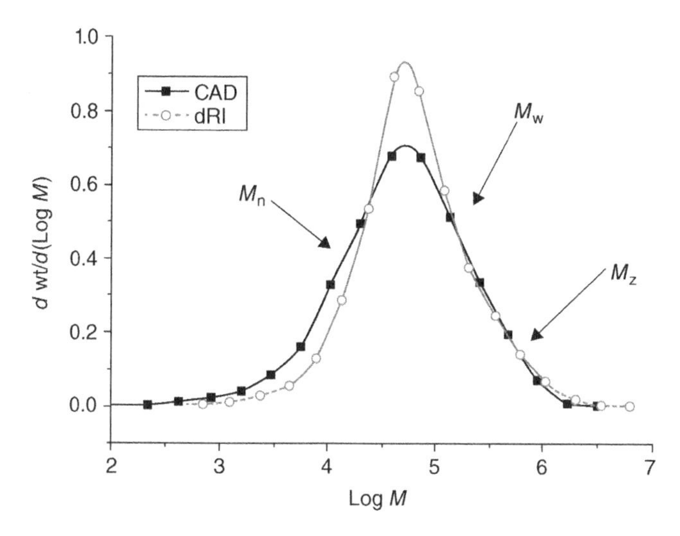

Figure 15.5 Comparison of molar mass distributions (MMD) of pS-co-pMMA 50:50 copolymer as determined by CAD and dRI.

Table 15.3 Average molecular weights of copolymers determined by CAD and dRI.

	CAD (kg/mol)			dRI (kg/mol)		
Samples	M_n	M_w	M_z	M_n	M_w	M_z
pS	15	127	446	29	139	541
pS-co-pMMA 75:25	17	129	457	30	141	556
pS-co-pMMA 50:50	14	115	394	28	128	509
pS-co-pMMA 25:75	15	68	168	27	71	175
pMMA	16	79	225	32	85	257

In conclusion, the SEC results show that all detectors have their drawbacks, UV is the most selective (the least universal) detector, and dRI shows some compositional dependency. This is especially an issue when chemically heterogeneous distributed copolymers are analyzed. In many industrial polymer grades, the molecules are built up with several monomers. CAD appears more universal with respect to composition but overestimates the tail of the distributions. Eventually, in all cases, the calculated MMDs are not the correct representation of the reality, and the MMDs should be interpreted with caution. This problem cannot be overcome by using viscosity and/or light scattering detection, since the application of these molar mass-sensitive detectors has to make use of a concentration signal in order to calculate "absolute" molar masses [1]. Ultimately, the application is and should be based on which detector is preferable.

15.6 Gradient Polymer Elution Chromatography and CAD

As described in Section 15.2, gradient LC (GPEC) is commonly used to analyze polymer compositions. GPEC can be used to separate a polymer sample into monomers, additives, oligomers, and polymer molecules, and therefore specific information on the polymer composition can be obtained. GPEC can be used to analyze specific target components, such as additives and residuals, as well as the main polymer molecules [2]. The separation mechanism is different from SEC, since polymer–column interactions, undesired in SEC, are used to separate polymer molecules based on monomer type or functional groups.

For the experiments described in the next paragraphs, GPEC is used in reversed phase mode (polar eluents in combination with a nonpolar column). A sample is injected into the column in weak eluent/solvent conditions, for example, aqueous conditions on Si C18 column. The sample will precipitate on top of the column/frit. Successively, a controlled change of eluent composition/strength is applied, resulting in the dissolution and desorption of the compounds and polymer molecules. Commonly used gradients are water/THF or water/MeCN/THF gradients. Other solvents such as methanol, isopropanol, and HFIP can also be used. Under normal phase conditions, solvents such as chloroform, heptane, and TCB are used.

As already discussed, under gradient conditions dRI detection is not an option. In general, UV detection and/or solvent evaporative detectors, for example, ELS detector and CAD, are used. The polymers described in Section 15.5 have also been analyzed by GPEC. The used system is a 2695 separation module (Waters), with a Symmetry C18 150 × 4.6 mm 3.5 μ column (Waters). Injection volume was 5 μL. A binary gradient (see Table 15.4) was applied (A = water + 0.1 v% trifluoroacetic acid (TFA), B = THF unstabilized) at a flow rate of 0.5 mL/min.

In Figures 15.6 and 15.7, the GPEC results of the various copolymers are shown. With GPEC the copolymers are separated according to composition and, to lesser extent, molar mass. The retention time increases with increasing amount of S units (higher S : MMA ratio). The UV signal is only depending on the styrene content, resulting in decreasing area with decreasing styrene content. Consequently, UV 254 nm cannot be used for qualification and quantification of these copolymers. Even when other UV wavelengths are used, the composition dependency on the response remains significant. Theoretically, at lower wavelengths, MMA units should be detectable, but the different

Table 15.4 Gradient conditions in GPEC mode.

Time	%A	%B
0	100	0
40	0	100
45	0	100
50	100	0
60	100	0

A, water + 0.1 v% TFA; B, THF.

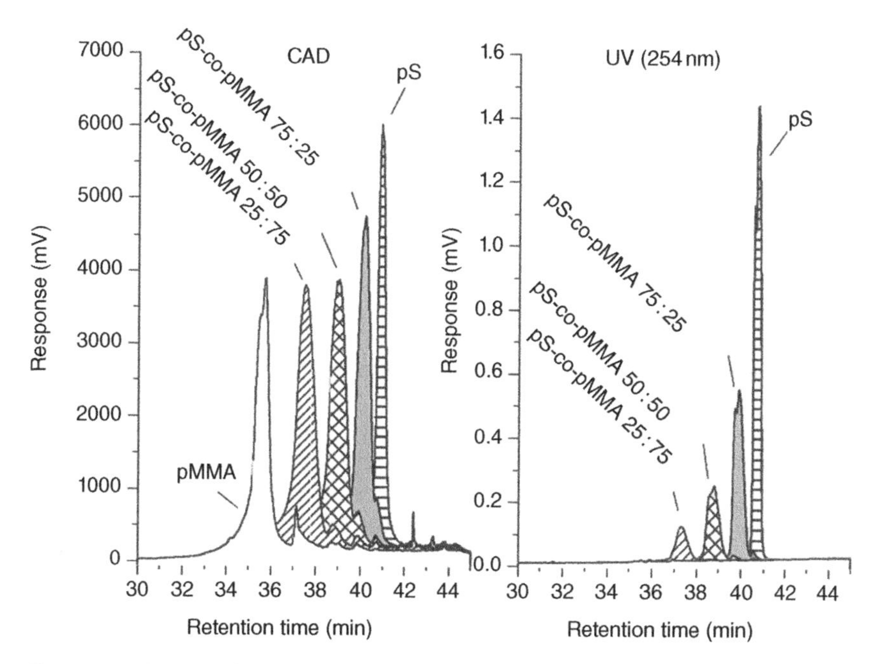

Figure 15.6 Overlay of chromatograms of various copolymers and homopolymers of styrene and MMA obtained under gradient conditions shown in Table 15.4.

extinction coefficients of styrene and MMA and the UV absorbance of THF still result in inadequate quantification of the distribution. Obviously, the response of the UV can be corrected for the amount for the S:MMA ratio, but this is less favorable. First observation of CAD chromatograms shows that the responses of the various copolymers are comparable. The response of evaporative detectors is known to be dependent on the eluent composition [15]. For the used gradient and the copolymer compositions, this dependency is not evident.

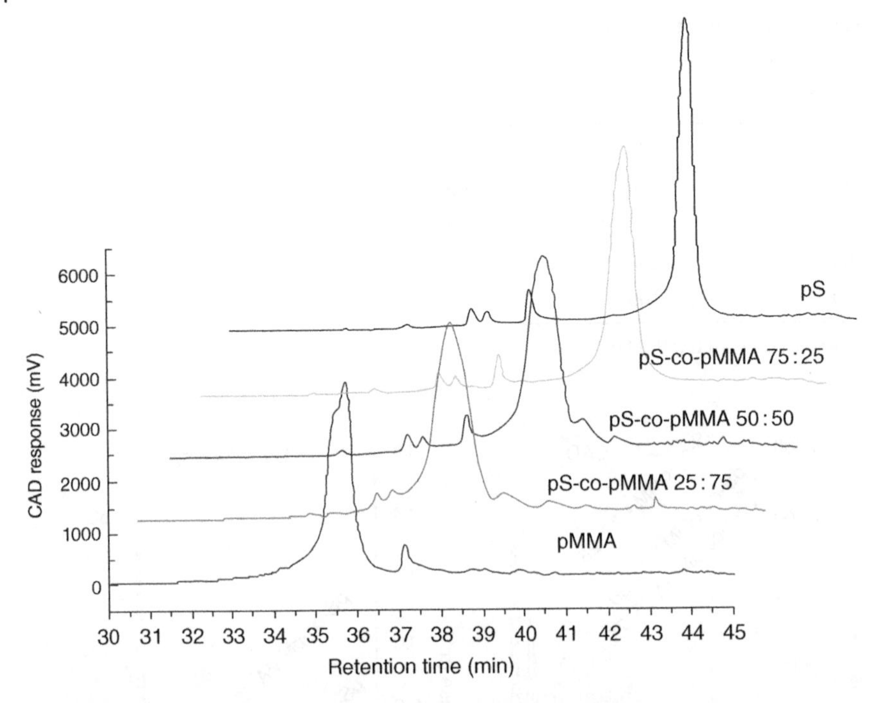

Figure 15.7 Overlay of CAD-GPEC chromatograms of the copolymers and corresponding homopolymers. The copolymers clearly show a broader distribution, resulting in a larger peak area due to the nonlinear concentration dependency of CAD.

The peaks of the homopolymers pS and pMMA appear to be narrower than the peaks of the copolymers. The homopolymers pS and pMMA do not have a chemical composition distribution, which is the case with the copolymers. For clarity reasons, the chromatograms at highest concentration of the various copolymers are shown in a different perspective in Figure 15.7.

The broadening of the peaks is mainly caused by system broadening and separation based on molar mass. The fronting of the peaks can be explained by a broad MMD. The tailing effect as detected with SEC cannot be observed in the GPEC–CAD chromatograms. The extra peaks on the tail can be explained by the monomer feeding conditions during the polymerization. Since all polymers are dissolved at similar concentrations, the peak areas should be comparable. In Figure 15.8, the peak areas of CAD and UV chromatograms for the copolymers and homopolymers are plotted versus concentration of the distribution.

With respect to the comparison of the detectors, similar results are obtained compared to SEC. The UV signal obviously showed a strong compositional dependency, and CAD detector shows a nonlinear dependency on the

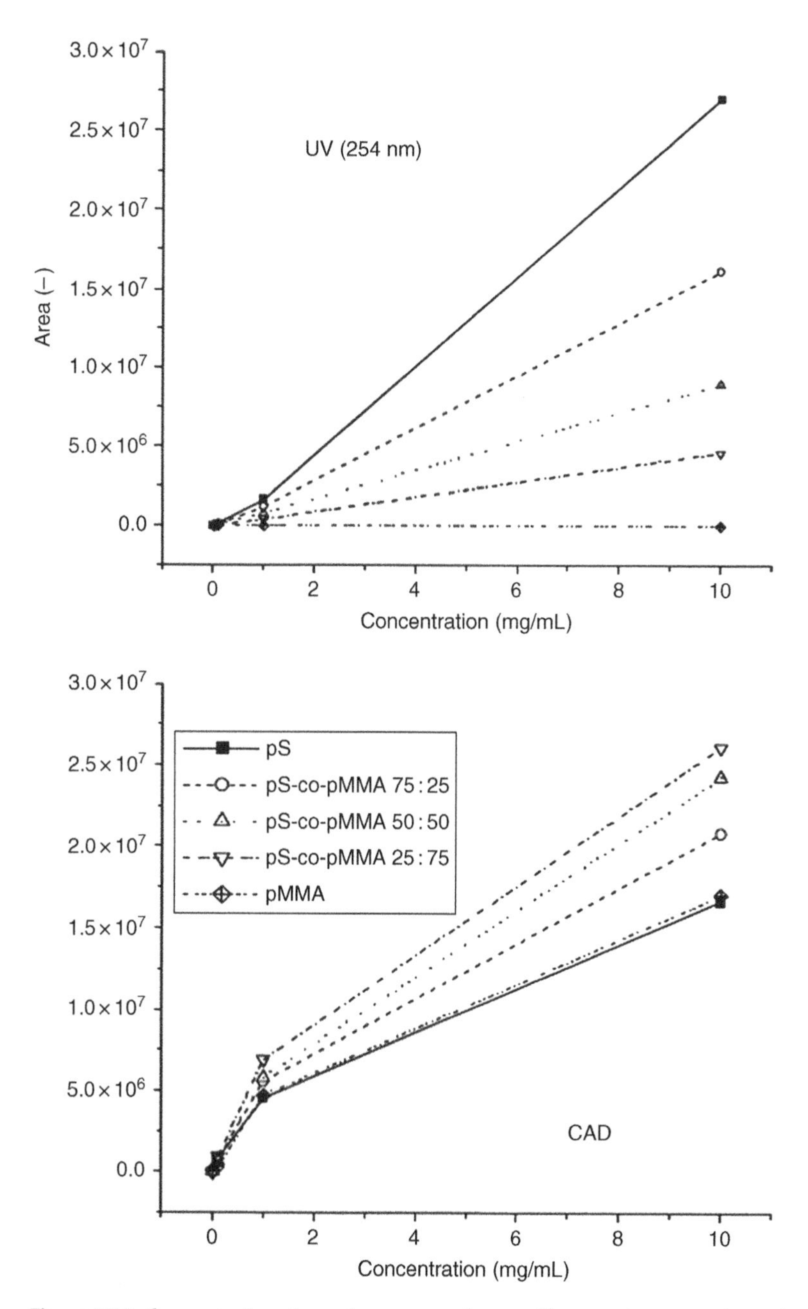

Figure 15.8 Concentration dependency on peak area of homopolymers and copolymers of the UV (254 nm) and CAD.

concentration. In the case of GPEC, the CAD signal is less universal. For the homopolymers pS and pMMA, the peak areas are comparable, indicating universal behavior. The peak areas obtained for the copolymers show a nonuniversal behavior, perhaps due to the nonlinear concentration behavior. In addition, the gradient is always visible in the baseline. The system peaks are present due to eluent and column contaminants. With UV the background extinction increases with increasing THF composition, and the response of CAD changes with the eluent composition.

The drawbacks mentioned previously hamper the universal application of UV and CAD in combination with gradient elution. However, CAD (or other evaporative detectors) is an essential detector to be used with gradient elution. Especially for the analysis of copolymers, the UV cannot be used, therefore evaporative detectors are crucial, although not ideal. As mentioned in the SEC section, a log-log scale could resolve the nonlinear behavior of CAD, but again this is more appropriate for single component analysis, and this is out of the scope of this discussion.

15.7 Liquid Chromatography Combined with UV, CAD, and MS Detection

Besides using UV, dRI, and evaporative detection, MS can also be used to detect the separated polymer molecules. LC-MS can rapidly provide information on several types of issues, such as raw material comparison, contaminations, or compositional analysis. In combination with LC, electrospray ionization (ESI) [30] is the most used system. Matrix-assisted laser desorption ionization (MALDI) [31] is applied for polymer molecules, but due to the challenging sample preparation, MALDI cannot be used in line with LC [31]. Due to the limitations of MS, polymer analysis is primarily focused on the compounds in the oligomer range ($M < 10$ kDa). In general, MS can be divided in three steps: the ionization step, the analyzing/separation step, and the detection step. The time-of-flight (TOF) analyzer is the most common for polymer analysis due to the broad application range (=1,000,000 Da) [30, 31]. Higher molar masses are applicable, but at above 10 kDa, only broad peaks are observed and no information can be obtained anymore on the repeating unit or end-groups. Additionally broad distributions are difficult to analyze directly with MS.

A large range of LC techniques is coupled with ESI-TOF-MS, the main technique being reversed phase gradient elution LC. But SEC can be coupled with MS. The separation power of SEC is limited compared with the polarity-based separation such as reversed phase LC, but it does provide direct information on the distribution since the molecules are separated according to molar size, making it significantly more straightforward.

15.7.1 LC-ESI-TOF MS System at DSM Coating Resins

The LC-ESI-TOF MS system is a Dionex U3000 system equipped with two gradient elution pumps, two isocratic pumps, one photo diode UV detector, one CAD detector, and one ESI-TOF mass spectrometer (MicroTOF, Bruker). ESI is used for the introduction from the LC to MS and the ionization. The analyzer is a TOF analyzer, and the detector is an analog microchannel plate detector. With ESI many polar and slightly polar molecules can be ionized. With TOF the molar mass range is broad, and in reflectron mode, detailed information of separate molar masses and isotope patterns can be obtained up to 10 kDa. The detector enables quantitative analysis of the formed adducts with the constraint that a reference sample of the single component/adduct must be analyzed to establish a concentration calibration curve. A schematic representation of the combined LC-MS and SEC-MS systems is shown in Figure 15.9.

In practice, the MS and LC acquisition is controlled by Bruker software, but the LC modules are controlled by Chromeleon software in order to obtain full control of the LC possibilities. In this way, SEC-MS and gradient LC-MS can be performed on one single system. Besides the efficient use of the LC system, the software-controlled performance of the TOF calibration with an HCOONa solution at any chosen moment (including at the start of the gradient LC or SEC run) is unique. The use of alternate adduct salts delivered by a gradient pump, post column added after the splitting, enhances the possibilities of the MS identification.

In general, LC-MS systems are dominated by the MS part with low flexibility in the LC configuration. In this case the LC configuration is leading. In other words, all feasible detection techniques are used in combination with MS.

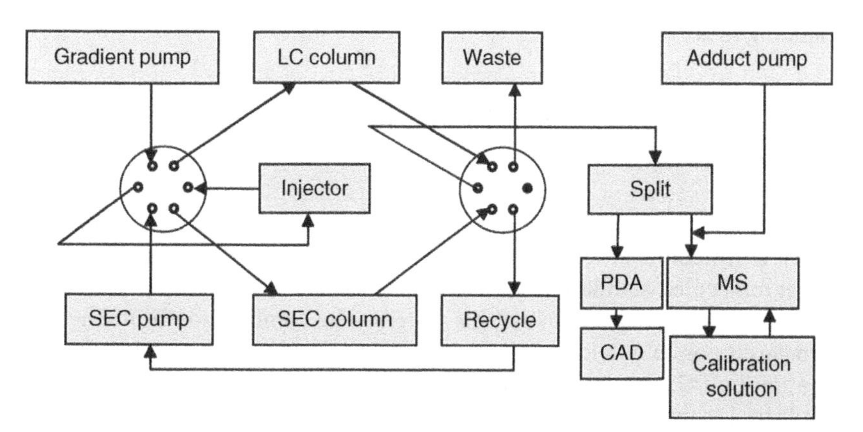

Figure 15.9 Blueprint of LC-MS system. The system consists of an ESI-TOF-MS combined with two gradient pumps, two isocratic pumps, one single column oven, a PDA, and a CAD.

The use of CAD in combination with MS appeared to be crucial. In general, SEC is performed in combination with only a dRI detector, and gradient elution is applied in combination with UV and CAD. The dRI is highly sensitive to fluctuations of the eluent composition, the temperature, and the flow. Consequently, the dRI is not suitable for use in this type of system where isocratic and gradient applications are combined. The LC-MS system had to be versatile in changing solvents/eluents and switching modes. One of the alternatives for dRI detection is CAD, combining the concentration dependency with fast adaptation to solvent changes. As shown in the previous paragraphs, CAD can be used for quantification in SEC systems as well as gradient LC systems.

The selected LC-MS system is in many ways comparable with the system used by Nielen *et al.* [30] but is considerably more advanced due to the continuing developments in MS: more accurate molar mass determination, higher sensitivity, and better quantitative analysis over several concentration decades. However, the limitations of ESI-TOF-MS still arise: molar mass limitation of 10 kDa, multiple charges resulting in complex spectra, and ion selectivity and suppression. Nevertheless, the information found in the low molar mass region can be used to obtain information on the total MMD. For instance, the presence of low molar mass cyclic oligomeric species can have a large influence on the application properties of polyesters [6, 7]. In addition, hydrolysis and side reactions will occur over the complete range of the MMD. Consequently, characterization of the oligomeric region is worthwhile and in most cases the only option to obtain this information. The combination of detectors, UV, CAD, and MS, enables the system to get optimal information from each separation. Together with Fourier transform infrared spectroscopy, gas chromatography mass spectrometry, and nuclear magnetic resonance, LC-MS is used for the elucidation of microstructures. In many ways, the system is flexible without becoming too difficult to use. However, during maintenance the system can be demanding to service.

15.8 Typical Examples of Industrial Applications Using LC-MS-CAD

Within DSM Coating Resins, the LC-MS-CAD system is used in general as primary LC method development for all types of components. Besides the polymer molecules, standard components, such as solvents, raw materials, and impurities, can be analyzed using this system. The combination of the detectors makes it possible to identify and quantify molecules. The combination of separation by LC and detection by UV, CAD, and MS demonstrates to be a versatile tool to unravel complex materials and obtain essential information on batch-to-batch variations, kinetic studies, and impurity profiling. Within resins applications, raw materials, intermediate products, and end products usually consist of oligomer or polymer distributions. Commercial raw materials are often technical grade; the material seldom consists of one single molecule or even the target molecule. Exceptions are the common vinyl monomers like acrylate monomers that seldom contain more than 1 wt% of impurities.

In this paragraph, some applications of the LC-MS-CAD system will be addressed. The examples can be categorized into raw material analysis, intermediate analysis, and end product analysis. Depending on the application and the analytic question, SEC or gradient LC will be applied in combination with UV, CAD, and MS detection. For all systems the chromatograms of the different detectors will be compared. For UV, different wavelengths can be applied and for MS, the total ion current (TIC) chromatogram as well as mass range or single mass extraction is used. By applying LC-MS-CAD system, strengths of both MS and CAD are used and give complementary information.

15.8.1 Raw Material Analysis

In many polymer production processes, raw materials, which are also polymers, are used. The characterization of these and other raw materials is often a vital step in understanding how raw material structure is related to polymer processing. Information supplied on Safety Data Sheets is often incomplete or inaccurate and often does not explain processing effects as seen in production environments (gelation, exothermic reactions, etc.).

In Figure 15.10, a multifunctional acrylate raw material is shown, measured in SEC-MS-CAD mode. The separation is based on differences in molecular size. The large molecules elute first, followed by small molecules. As can be seen from the chromatograms, the distribution is rather broad. Chemically, these types of materials do not consist of one single molecule but are a mixture of various degrees in functionality and polymerization caused by side reactions. Titrations (e.g., acid number or hydroxyl number) on this type of material would give average functionality data, but average numbers are, in most cases, insufficient information to predict the behavior of the material.

In Figure 15.10, an overlay of signals from the MS, CAD, and two different UV wavelengths is shown. The use of UV is not always an option, since not all molecules contain chromophores. The combined use of all detectors gives most information needed. CAD and UV signals show comparable responses, with an underestimation on the UV signal for higher molar mass fraction. The MS signal shows less response in the low molecular weight region, which can be caused by poor ionization of the molecules, possibly due to ion suppression effects caused by the THF. By comparing the information of the MS (identity of the various constituents) with CAD (quantity of the various constituents), an accurate mass analysis of the material is obtained. It is always preferable to use more than one detector, since assumptions on identity and/or quantity are easily made that may lead to inaccurate conclusions. In Figure 15.10, a distribution is visible,

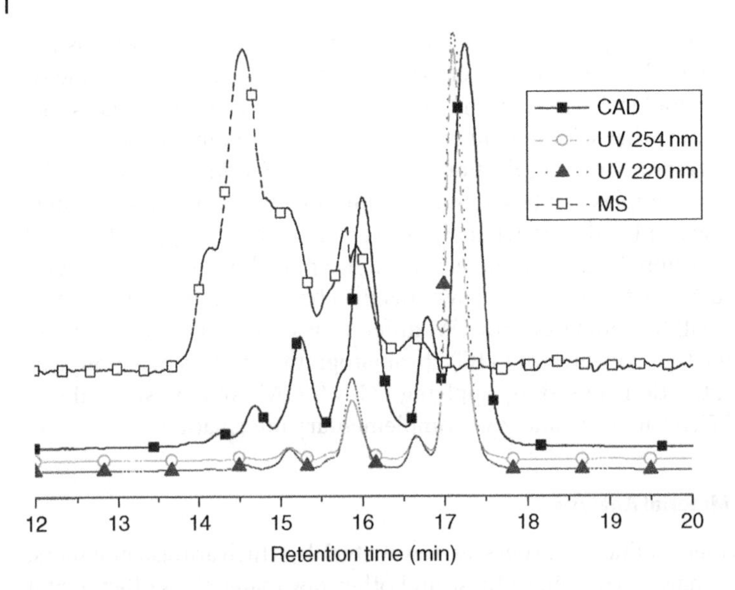

Figure 15.10 Overlay of CAD, UV (220 and 254 nm), and MS signals of a multifunctional acrylate obtained with SEC.

which has a significant fraction of higher molecular weight material. By using the accurate mass data from the MS spectra, the identity of the components can be obtained. The higher molecular weight fraction can cause viscosity increase resulting in off-spec production. For the analysis of raw materials, the SEC-MS-CAD combination gives detailed insight in the molecular buildup, which is essential to perform successful product and process development.

15.8.2 Intermediates

Besides the analysis of raw materials, the monitoring of polymerizations and analysis of intermediate products is also important.

In Figure 15.11, the results of a gradient LC separation of an intermediate are shown. The intermediate is a typical example of the preparation of a urethane polymer, synthesized by urethanization of a polyol with a diisocyanate. Since none of the reactants have chromophoric groups, the UV detector only shows the UV signal due to the change in eluents from water to MeCN to THF (dotted line). Both MS and CAD show a distribution of the urethane oligomers and polyol distribution. The MS shows a more intense signal around 30 min compared with CAD. This is probably caused by selective ionization of the higher molar mass oligomers (containing the well-ionizable urethane groups), resulting in an overestimation of the higher masses. CAD shows two molar mass fractions, which is in close agreement with theoretical expectation of the composition of the oligomer sample.

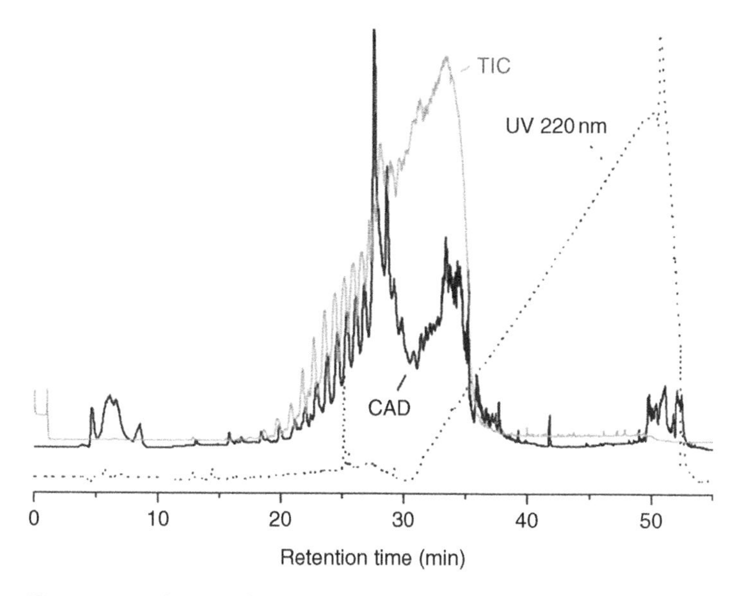

Figure 15.11 Overlay of gradient LC chromatograms of an intermediate product of TIC MS and CAD. Gradient as described in Table 15.4 is used.

15.8.3 End Products

The application of LC-MS-CAD for end products is not so straightforward. As discussed in the SEC and GPEC sections, CAD can be used successfully for poly distributions. However, the MS cannot be used to analyze high-molar mass polymers (>10 kDa). Nevertheless, many resin end products are low-molar mass polymers. In 2007, the European Community introduced the *R*egistration, *E*valuation, *A*uthorization and *R*estriction of *Ch*emical substances (REACh) project [32], which started to regulate and register substances and chemicals. In order to register products in REACh, more detailed information on the chemical composition and the molecular structure needs to be known. The effect of the legislation is to force companies to show that the end products are safe for humans and for the environment. One of the topics within REACh is the polymer definition. If an end product can be registered under the polymer definition, no additional registration is needed. If the end product cannot be registered under the polymer definition, additional tests, especially toxicology tests, need to be performed. Consequently, the end products need to be analyzed in detail to define the product as polymer. A detailed description of the polymer definition can be found on the European Community website [32].

One of the main issues with the polymer definition in REACh is the definition of the repeating units and the quantification of the oligomers with the number of repeating units below $n = 3$. Similar to Toxic Substances Control Act (TSCA) registration (United States substance registration legislation) [33], the preferred method to identify whether a substance falls under the definition of a polymer is SEC, most often combined with a dRI detector. As described in previous paragraphs, a CAD detector can also be used in combination with SEC to quantify distributions. In standard SEC-dRI systems, assumptions have to be made on peak identification, purely based on retention time and reference molar masses. By applying additional detectors such as MS, identification of the different peaks can be performed, resulting in a more accurate definition of the oligomer peaks. This drastically increases the reliability of the REACh analysis, since no more assumptions have to be made on molar mass and number of repeating units. In the following example, an end product that was subject to REACh registration was analyzed using SEC-MS-CAD. The product contains chromophore groups so the UV detector could also be used.

As seen in Figure 15.12, a wide range of components and a distribution is clearly visible in all signals. The TIC MS signal does not reveal the higher-molar mass species in contrast to the results of the raw material in Figure 15.10. In this case the MS detector is used to identify the different structures present in the product. It was possible to identify the components that have $n \geq 3$ by the MS detector and quantify these as a percentage of the total product by integrating the CAD signal. By using CAD, the end product was successfully classified as polymer under the REACh definition.

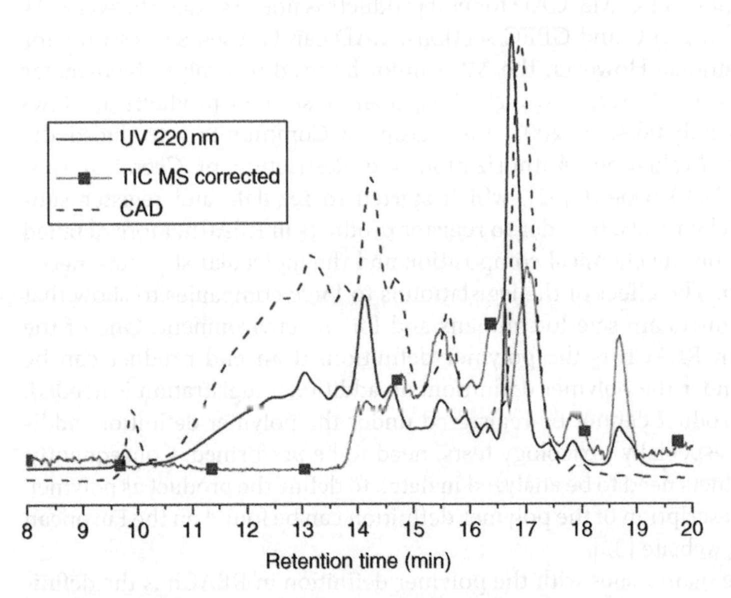

Figure 15.12 SEC-MS-UV-CAD chromatograms of oligomeric resin.

15.9 Epilogue

For the analysis of polymeric distributions, CAD has shown to be extremely valuable. CAD is used routinely in the described LC-MS system for troubleshooting and molecular structure elucidation. The LC-MS-CAD system is a valuable addition to the analytical toolbox for the structural elucidation of oligomers and, in combination with nuclear magnetic resonance, is very powerful. CAD turns out to be useful as a concentration detector for distributions in SEC as well as in gradient applications. In combination with an LC-MS system, CAD is better suited than the dRI detector. On the other hand, the quest for a universal detector for LC applications with linear concentration dependency has not ended, since the concentration dependency of CAD is nonlinear. This can be overcome by using log-log scale conversion, but in most commercial software, this is not included, or difficult to achieve. Overall, for the determination of distributions or for comparison purposes, CAD is an important addition to other detectors, such as dRI and UV.

Acknowledgments

The authors would like to thank Harry Philipsen for the useful discussion and scientific input. Tijs Nabuurs en Frank Vaes from DSM Coating Resins (Waalwijk, The Netherlands) are acknowledged for preparing the polymer samples.

References

1 Striegel, A., Yau, W.W., Kirkland, J.J., and Bly, D.D.; "Modern Size-Exclusion Liquid Chromatography, Practice of Gel Permeation and Gel Filtration Chromatography," Second Edition; John Wiley & Sons, Inc., Hoboken, NJ, 2009; 978-0-471-20172-4.
2 Staal, W.J., Cools, P., van Herk, A.M., and German, A.L.; Journal of Liquid Chromatography, Volume 17, Issues 14 & 15, 1994, pp. 3191–3199.
3 Cools, P.J.C.H., Maesen, F., Klumperman, B., van Herk, A.M., and German, A.L.; Journal of Chromatography A, Volume 736, Issues 1–2, 1996, pp. 125–130.
4 Philipsen, H.J.A., Klumperman, B., and German, A.L.; Journal of Chromatography A, Volume 746, Issue 2, 1996, pp. 211–224.
5 Staal, W.J.; "Gradient Polymer Elution Chromatography"; PhD Thesis, University of Technology Eindhoven, Eindhoven, 1996, 90-386-0126-3.
6 Philipsen, H.J.A.; "Mechanisms of Gradient Polymer Elution Chromatography and Its Application to (Co)polyesters"; PhD Thesis, University of Technology Eindhoven, Eindhoven, 1998, 90-386-0578-1.

7 Cools, P.J.C.H.; "Characterization of Copolymers by Gradient Polymer Elution Chromatography"; PhD Thesis, University of Technology Eindhoven, Eindhoven, 1999, 90-386-0970-1.
8 Schunk, T.C.; Journal of Chromatography A, Volume 656, Issues 1–2, 1993, pp. 591–615.
9 Glöckner, G.; "Gradient HPLC of Copolymers and Chromatographic Cross-Fractionation"; Springer-Verlag, New York, 1991.
10 Gorshkov, A.V., Verenich, S.S., Evreinov, V.V., and Entelis, S.G.; Chromatographia, Volume 26, 1988, pp. 338.
11 Cools, P.J.C.H., van Herk, A.M., German, A.L., and Staal, W.J.; Journal of Liquid Chromatography, Volume 17, Issues 14–15, 1994, pp. 3133–3143.
12 Pasch, H., Brinkmann, C., Much, H., and Just, U.; Journal of Chromatography, Volume 623, 1992, p. 3153.
13 Entelis, S.G., Evreinov, V.V., and Gorshkov, A.V.; Advances in Polymer Science, Volume 76, 129, 1986.
14 Pasch, H. and Trathnigg, B.; "HPLC of Polymers"; Springer, Berlin, 1997.
15 Schultz, R. and Engelhardt, H.; Chromatographia, Volume 29, Issues 11–12, 1990, pp. 517–522.
16 Snyder, L.R., Kirkland, J.J., and Dolan, J.W.; "Introduction to Modern Liquid Chromatography", Third Edition; John Wiley & Sons, Inc., Hoboken, NJ, 2010.
17 Vehoveca, T. and Obrezab, A., Journal of Chromatography A, Volume 1217, Issue 10, 2010, pp. 1549–1556.
18 Hutchinson, J.P., Li, J., Farrell, W., Groeber, E., Szucs, R., Dicinoski, G., and Haddad, P.R.; Journal of Chromatography A, Volume 1217, Issue 47, 2010, pp. 7418–7427.
19 Lee, H.C. and Chang, T.; Polymer, Volume 37, Issue 25, 1996, pp. 5747–5749.
20 Martin, A.J.P.; Biochemical Society Symposium, Volume 3, 1950, pp. 4–20.
21 Flory, P.J.; "Principles of Polymer Chemistry"; Cornell University Press, Ithaca, NY, 1953.
22 Huggins, M.L.; "Physical Chemistry of High Polymers"; John Wiley & Sons, Inc., New York, 1958.
23 Hoang, N.-L., Landolfi, A., Kravchuk, A., Girard, E., Peate, J., Hernandez, J.M., Gaborieau, M., Kravchuk, O., Gilbert, R.G., Guillaneuf, Y., and Castignolles, P.; Journal of Chromatography A, Volume 1205, Issues 1–2, 2008, pp. 60–70.
24 Macko, T. and Pasch, H.; Macromolecules, Volume 42, Issue 16, 2009, pp. 6063–6067.
25 Roy, A., Miller, M.D., Meunier, D.M., Willem Degroot, A., Winniford, W.L., Van Damme, F.A., Pell, R.J., and Lyons, J.W.; Macromolecules, Volume 43, Issue 8, 2010, pp. 3710–3720.
26 Evans, C.E. and McGuffin, V.L.; Journal of Chromatography, Volume 503, Issue 1, 1990, pp. 127–154.
27 Van Doremaele, G.H.J., Kurja, J., Claessens, H.A., and German, A.L.; Chromatographia, Volume 31, Issues 9–10, 1991, pp. 493–499.

28 Allen, L.B., Koropchak, J.A., and Szostek, B.; Analytical Chemistry, Volume 67, 1995, p. 659.

29 Biosolve B.V., http://www.biosolve-chemicals.com/page.php?content=product_line_lcms (accessed January 30, 2015).

30 Nielen, M.W.F., Buijtenhuijs, F.A.; Analytical Chemistry, Volume 71, 1999, p. 1809–1814.

31 Staal, B.B.P.; "Characterization of (Co)polymers by MALDI-TOF-MS"; PhD Thesis, University of Technology Eindhoven, Eindhoven, 2005, 90-386-2826-9.

32 Registration, Evaluation, Authorisation and Restriction of Chemical Substances (REACH), European Community, June 2007, http://ec.europa.eu/environment/chemicals/reach/reach_intro.htm (accessed March 6, 2017).

33 Toxic Substances Control Act (TSCA), United States Environmental Protection Agency, 1976, http://www.epa.gov/regulations/laws/tsca.html (accessed March 6, 2017).

Index

Charged Aerosol Detection for Liquid Chromatography and Related Separation Techniques, First Edition. Edited by Paul H. Gamache.

i

l

m

t

u

Printed and bound by CPI Group (UK) Ltd, Croydon, CR0 4YY
16/04/2025
14658344-0002